DIVIDED WE GOVERN

SANJAY RUPARELIA

Divided We Govern

Coalition Politics in Modern India

HURST & COMPANY, LONDON

First published in the United Kingdom in 2015 by
C. Hurst & Co. (Publishers) Ltd.,
41 Great Russell Street, London, WC1B 3PL
© Sanjay Ruparelia, 2015
All rights reserved.
Printed in India

The right of Sanjay Ruparelia to be identified as the author of this publication is asserted by him in accordance with the Copyright, Designs and Patents Act, 1988.

A Cataloguing-in-Publication data record for this book is available from the British Library.

ISBN: 9781849042123

www.hurstpublishers.com

This book is printed using paper from registered sustainable and managed sources.

"Men make their own history, but they do not make it just as they please; they do not make it under circumstances chosen by themselves, but rather under circumstances found, given and transmitted."

Karl Marx, *The Eighteenth Brumaire of Louis Bonaparte*

"In decisive historical moments, political capacity (which includes organization, will, and ideologies) is necessary to enforce or to change a structural situation. Intellectual evaluation of a given situation and ideas about what is to be done are crucial in politics. The latter is immersed in the shady area between social interests and human creativity. At that level, gambles more than certainty line the paths through which social forces try to maintain or to change social structures. Briefly, in spite of structural 'determination', there is room for alternatives in history. Their actualization will depend not just on the basic contradictions between interests, but also on the perception of new ways of turning a historical corner through 'a passion for the possible.'"

Fernando Henrique Cardoso and Enzo Faletto,
Dependency and Development in Latin America

"[The] very probability of committing mistakes presupposes simultaneously a political project, some choice among strategies, and objective conditions that are independent with regard to a particular movement. If the strategy of a party is uniquely determined, then the notion of 'mistakes' is meaningless: the party can only pursue the inevitable ... [But the] notion of mistakes is also rendered meaningless within the context of a radically voluntaristic understanding of historical possibilities ... if everything is always possible, then only motives explain the course of history ... 'Betrayal' is indeed a proper way of understanding social democratic strategies in a world free of objective constraints. But accusations of betrayal are not particularly illuminating in the real world."

Adam Przeworksi, *Capitalism and Social Democracy*

"What were the arguments used by those opposed to you being Prime Minister?

... Our argument was: this cannot last five years. If we are there, much more than the others we can make them accept some policies, put them before the country, whatever the limits are. You can't remove every obstacle, that is not possible: but we could do something for self-reliance, for the countryside, for panchayats, all that we can push through. Anti-poverty programmes: it is there but it does not reach the people. ... But it is a political blunder. It is a historical blunder ... We do not accept many of their policies, they do not accept many of ours. But the minimum programme was there, and we could have implemented it much better than others. Because we have the experience, nothing more, nothing personal."

Jyoti Basu, former chief minister of West Bengal

CONTENTS

Acknowledgements	ix
List of Tables	xv
Abbreviations	xvii
Glossary	xxiii
Introduction	1
1. The Paradoxes of India's Coalition Politics	15

PART I
THE GENESIS OF THE THIRD FORCE

2. The Roots of the Broader Indian Left (1934–1977)	45
3. The Janata Party (1977–1980)	67
4. The Rise of the Regions (1980–1989)	89
5. The National Front (1989–1991)	103

PART II
THE MATURATION OF THE THIRD FORCE

6. The Crystallization of the Third Force (1991–1996)	125
7. The Formation of the United Front (May 1996)	147
8. Establishing Political Authority (June–September 1996)	181
9. Exercising National Power (September–December 1996)	213
10. Reform amid Crisis (January–April 1997)	235

PART III
THE FALL OF THE THIRD FORCE

11. The Decline of the United Front (May 1997–March 1998)	257

CONTENTS

12. The Dissolution of the Third Force (1998–2012)	287
Conclusion	319
Notes	345
Bibliography	419
Index	451

ACKNOWLEDGEMENTS

This book seeks to explain the rise and fall of the broader parliamentary left, and the dynamics of national coalition governments more widely, in modern Indian democracy since the 1970s. It has taken far longer to finish than I ever imagined. Teaching commitments and family obligations, routine and unexpected, demanded extra attention at numerous junctures. Opportunities to conduct new research and pursue collaborative projects also emerged, shifting my timeline. Yet after finishing a first draft of the manuscript three years ago, which exclusively focused on Indian coalition politics in the 1990s, the focus of my Ph.D. dissertation, I realized the need to analyze the history of the broader Indian left to assess its advances, setbacks and futures properly, which entailed much new work. Given how many years have passed, I have accrued many debts, personal as well as intellectual, which have left an imprint. I apologize for anyone that I may have inadvertently missed.

My interest in modern Indian democracy emerged during my graduate studies at the University of Cambridge. Unconventionally, I began my doctoral research at Yale University, where I spent a very enjoyable year as a Fox Visiting Fellow in 1996–97, thanks to Anthony Badger and Brian Carter. Seminars with Benedict Anderson, Joseph LaPalombara, Frances Rosenbluth, Alexander Wendt and Eric Worby, and numerous conversations with Eric as well as Casiano Hacker-Cordon, John Gould, Jeff Miley and Jonathan Rodden, exposed me to many contending perspectives on the role of ideas, interests and institutions in politics. Yet it was my years of study in Cambridge that naturally had a far greater impact. The greatest influence was my supervisor,

ACKNOWLEDGEMENTS

Geoffrey Hawthorn, whose views on the importance of power, skill and judgment in politics, and what it means for how we explain the politics of any historical context, deeply shaped my intellectual outlook in ways that I only fully realized much later. David Lehmann provided guidance, encouragement and humour over many years in his official role as my faculty advisor. But it was his friendship that I came to value the most. John Dunn and Helen Thompson helped me to make sense of the material I had gathered during my original fieldwork, pushing me to weigh its possibilities and limitations, and suggested the idea of writing a narrative. Jim Whitman counselled me on how to write it, supplying numerous selections of music to inspire the process. Stuart Corbridge provided critical feedback at various stages, and has been a supportive mentor and inspiring colleague ever since. Christophe Jaffrelot and the late Raj Chandavarkar, my Ph.D. examiners, showed tremendous generosity during my viva and helped me see how to turn the dissertation into a book. And I was enormously lucky to have discovered fellow travellers in Cambridge—including Subho Basu, Bela Bhatia, Eric Breton, Caroline Brooke, Lisa Brown, Jude Browne, Rebecca Eldredge, Fredrik Galtung, Lawrence Hamilton, Jeesoon Hong, Jaeho Kang, Andrew Kuper, Christine Minas, Deirdre Collings-Rohozinski, Glen Rangwala, Rafal Rohozinski and Helen Yanacopulos—many of whom inhabited the 'Attic' and made it such a convivial place. Despite leaving Cambridge long ago, I continue to learn from Reuben Brigety, Zeev Emmerich, Thushara Hewage, Eleanor O'Gorman, Adam Smith and Sigal Spigel. John Clarke and Justin Robertson, friends from home, remain steadfast mates. And Vinayak Chaturvedi has patiently engaged almost every half-baked idea I raised in conversations with him over the last fifteen years, even when he clearly disagreed, encouraging me to trust my instincts.

My understanding of Indian politics, of course, has been decisively shaped by repeated visits to India. I first conducted fieldwork in New Delhi between 1998 and 2000, and have returned virtually ever year since 2006. I remain indebted to the many party leaders, senior government officials and seasoned journalists for granting numerous confidential interviews. It is their distinct voices, concerns and perspectives that animate the narrative at the heart of this book. If my analysis has any sense of reality, it is due to their respective insights. I am extremely grateful to a variety of institutions—the Commonwealth Scholarship Commission, Gilchrist Educational Trust and Sir Ernest Cassel

ACKNOWLEDGEMENTS

Educational Trust, Smuts Study Travel Fund and Worts Travelling Fund, and the Faculty of Social and Political Sciences and Sidney Sussex College in Cambridge—whose financial support made these early ventures possible. My extended family, Vicky, Kavita, Kabir, Siddharth and the late Nome Seth, took great care of me during my early years of fieldwork in Delhi. Kuldeep Mathur, C.P. Bhambri and Deepak Nayyar at the Jawaharlal Nehru University steered me away from various research pitfalls early on, while Shiv Viswanathan, Chandrika Parmar and Kaushik Sunder Rajan entertained never-ending questions on Indian political history. E. Sridharan at the University of Pennsylvania Centre for the Advanced Study of India (CASI) generously shared his many insights into the politics of coalition in India with me over many years. Yet my greatest intellectual debt is to the Centre for the Study of Developing Societies (CSDS), a beacon for the study of contemporary Indian democracy and its alternatives. Yogendra Yadav believed in my project when others expressed scepticism, repeatedly sharing his penetrating insights, becoming a cherished friend in the process. Himanshu, Oliver Heath and Alistair McMillan demonstrated the value of electoral statistics and provided me with original data that I needed. And Rajeev Bhargava warmly accepted a request to spend the summer of 2008 at CSDS, inviting me to present the key theoretical arguments of the book to the faculty, which advanced them at a critical stage. A subsequent presentation at the Centre for Policy Research in Delhi, arranged by Pratap Bhanu Mehta in 2011, allowed me to reconsider their scope. Discussions with both of them continue to challenge my understanding of Indian politics. And I greatly benefitted from various exchanges with Madhulika Banerjee, Neera Chandhoke, Sanjay Kumar, Partha Mukhopadhyay, Aditya Nigam, Dhirubhai Sheth and Achin Vanaik on various occasions.

Finally, I have had the immense good fortune to have spent my post-graduate life in New York city, a remarkable crossroads for anyone studying South Asian politics outside the subcontinent. Philip Oldenburg invited me to become a visiting scholar at the South Asian Institute at Columbia University in 2001, where I thereafter began to work and teach. His knowledge, experience and passion for Indian politics, and willingness to read virtually everything I sent him, provided constant stimulation. Partha Chatterjee, Sudipta Kaviraj and Alfred Stepan offered much encouragement, criticism and guidance during my time at Columbia and since. Similarly, I owe considerable

ACKNOWLEDGEMENTS

gratitude to several colleagues who took me under their wings after I joined the New School in 2005. Andrew Arato, Vicky Hattam, Courtney Jung and Jim Miller provided critical intellectual feedback and moral support during especially trying moments. Comments from Arjun Appadurai, Richard Bernstein, José Casanova, Carlos Forment, Nancy Fraser, Mala Htun and David Plotke highlighted ways I could strengthen the manuscript as it evolved. And I have greatly benefitted from many exchanges over the years with colleagues and friends at institutions in New York, or those visiting the city, including Lopa Banerjee, Mukulika Banerjee, Banu Bargu, Sanjib Baruah, the late Carol Breckenridge, Vidya Dehejia, Faisal Devji, Nicholas Dirks, Amitabh Dubey, Ashok Gurung, Julia Foulkes, Thomas Blom Hansen, Ron Herring, Craig Jeffrey, Andreas Kalyvas, Anush Kapadia, Rob Jenkins, Madhav Khosla, Jim Manor, Anjali Mody, Veena Talwar Oldenburg, Gustav Peebles, Arvind Rajagopal, Vyjayanthi Rao, Rathin Roy, Leander Schneider, Anwar Shaikh, Aseema Sinha, Vinay Sitapati, Nidhi Srinivas, Vamsi Vakulabharanam, Ashutosh Varshney, Gina Walker, Deva Woodly and Rafi Youatt. My gratitude to John Harriss and Christophe Jaffrelot and to Tim Pachirat and Sanjay Reddy, for their camaraderie and example, is as much personal as intellectual.

In the end, finishing the manuscript required uninterrupted stints of reading, fieldwork and writing. I was extremely fortunate to receive a visiting fellowship from the Kellogg Institute for International Studies at the University of Notre Dame in the winter and spring of 2009, whose commitment to studying democracy comparatively made it an ideal setting. I thank Ted Beatty, Michael Coppedge, Frances Hagopian, and Robert Fishman and Scott Mainwaring in particular, for their collegiality during my stay and for encouraging me to frame my research in broader terms. Cecelia Van Hollen and Isabelle Clark-Decès kindly invited me to present aspects of the book to the South Asia seminars in Syracuse and Princeton in 2012 and 2013, respectively, compelling me to sharpen the text upon revision. Atul Kohli and Deborah Yashar graciously allowed me to spend a portion of my fellowship with the Project for Democracy and Development at Princeton in the 2012–2013 academic year to expand the scope of analysis, taking advantage of the terrific holdings of the Firestone library. Neeraj Priyadarshi and Rajkumar Srivastava of the Indian Express generously made time in their hectic schedules to help me locate suitable photographs in their archive. Shabnam Hashmi and Mansi Sharma kindly sent me a copy of the public appeal to the Left Front made in 2004. I must thank Marianna Assis, Lisa Bjorkman, Pete Galambos, Samarjit Ghosh, Jennifer Terrell, Douglas Voigt and Amanda Zadorian for their research assistance at various junctures. And I am enormously grateful to my editor, Michael Dwyer, for taking on the manuscript enthusiastically from the start, encouraging me to tell a much larger story than I had originally planned and providing ample time to do justice to the task. I thank Jon de Peyer, Daisy Leitch, Rob Pinney and Prerana Patel for answering my umpteen concerns, hastening the process of producing the book at crucial stages, and Mohini Gupta for highlighting persistent ambiguities. The gracious criticisms of two anonymous reviewers highlighted important limitations of the book, which I have sought to address in the Conclusion, albeit very briefly. Needless to say, despite the contributions of so many individuals, any remaining mistakes of fact and interpretation are mine alone.

It is impossible to acknowledge the debt I owe to my family with any measure of adequacy. I doubt my parents, Indu and Vasant, and my brother, Raju, ever fully understood what I was doing. Yet their love, patience and belief have never wavered. Durgasankar and Tia Mukhopadhyay, my parents-in-law, have always shown me great affection. Yet it is my wife, Tanni, who shared my preoccupations, for better and worse, on a regular basis. Wary of high political ideals that neglect the demands of real social life, in every sense, she has shaped my understanding of politics in India and beyond through years of conversation, argument and debate. I have often wondered whether working on this book for so long was worth the myriad costs involved, especially after the arrival of our children, Siddharth and Tara, who similarly endured my bouts of inattentiveness and periods of absence. I hope its publication redeems these sacrifices in some measure. I would never have seen it through without their love, joy and exuberance for life.

LIST OF TABLES

Table 2.1:	Electoral Performance of Partisan Blocs, 1951–1977	48
Table 3.1:	Sixth General Election, 1977	68
Table 3.2:	National Vote Shares of Parties by Type, 1951–1977	69
Table 3.3:	National Seat Shares of Parties by Type, 1951–1977	70
Table 3.4:	Janata Party Government	75
Table 5.1:	Ninth General Election, 1989	104
Table 5.2:	National Front Government	111
Table 6.1:	Number of Parties Contesting National Elections, 1980–2009	127
Table 6.2:	Effective Number of Parties in Parliament, 1980–2009	128
Table 6.3:	Decline of National Party Vote Share, 1980–2009	128
Table 6.4:	Decline of National Party Seat Share, 1980–2009	129
Table 7.1:	Eleventh General Election, 1996	148
Table 7.2:	Interest in Level of Government, 1996	150
Table 7.3:	Preference for Regional Party State Governments, 1996	151
Table 7.4:	Loyalty to Region before Nation, 1996	152
Table 7.5:	Social Bases of Party Alliances, 1996	155
Table 7.6:	The United Front	171
Table 7.7:	The H. D. Deve Gowda Ministry	173

LIST OF TABLES

Table 8.1:	The Expansion of the Deve Gowda Ministry	185
Table 8.2:	Opposition to Building an Atomic Bomb, 1996	192
Table 8.3:	Knowledge of Economic Policy Changes, 1996	194
Table 8.4:	Opinion of Economic Policy Changes, 1996	195
Table 8.5:	Loyalty to Region before Nation, 1996	199
Table 8.6:	Strategy towards Kashmir, 1996	201
Table 9.1:	Support for Backward Caste Reservations in Government, 1996	216
Table 9.2:	Support for Reservations for Women in Parliament, 1996	217
Table 9.3:	Desirability of Friendship with Pakistan, 1996	220
Table 10.1:	Desirability of Unrestricted Foreign Investment, 1996	239
Table 10.2:	Desirability of Privatization, 1996	240
Table 11.1:	Twelfth General Election, 1998	283
Table 12.1:	Thirteenth General Election, 1999	293
Table 12.2:	Fourteenth General Election, 2004	299
Table 12.3:	Fifteenth General Election, 2009	313

ABBREVIATIONS

AAP	Aam Aadmi Party
AASU	All Assam Students Union
ABVP	Akhil Bharatiya Vidyarthi Parishad
AC	Arunachal Congress
AFSPA	Armed Forces (Special Powers) Act
AGP	Asom Gana Parishad
AIADMK	All India Anna Dravida Munnetra Kazhagam
AICC	All-India Congress Committee
AIIC(T)	All India Indira Congress (Tiwari), or Congress (Tiwari)
AIKS	All-India Kisan Sabha
AITC	All India Trinamool Congress
AITUC	All India Trade Union Congress
AJGAR	a caste alliance involving Ahirs, Jats, Gujjars and Rajputs
APHC	All-Party Hurriyat Conference
BKD	Bharatiya Kranti Dal
BKKP	Bharatiya Kisan Kamgar Party
BJD	Biju Janata Dal
BJP	Bharatiya Janata Party
BLD	Bharatiya Lok Dal
BSP	Bahujan Samaj Party
CBI	Central Bureau of Investigation
CITU	Centre of Indian Trade Unions
Congress(I)	Congress (Indira)
Congress(O)	Congress (Organization)

ABBREVIATIONS

Congress(R)	Congress (Requisition)
Congress(U)	Congress (Urs)
CMP	Common Minimum Programme
CPI	Communist Party of India
CPI(M)	Communist Party of India (Marxist)
CPI(ML)	Communist Party of India (Marxist-Leninist)
CPP	Congress Parliamentary Party
CPSU	Communist Party of the Soviet Union
CSP	Congress Socialist Party
CSS	Chatra Sangharsh Samiti
CTBT	Comprehensive Test Ban Treaty
CWC	Congress Working Committee
DC	Disinvestment Commission
DMK	Dravida Munnetra Kazhagam
EC	Election Commission
FBL	All India Forward Bloc
FEMA	Foreign Exchange Management Act
FERA	Foreign Exchange Regulatory Act
FICCI	Federation of Indian Chambers of Commerce and Industry
FIPB	Foreign Investment Promotion Board
FMCR	Fissile Material Control Regime
FPTP	first-past-the-post, or single-member simple plurality (SMSP), electoral system
GJP	Gujarat Janata Parishad
GNLF	Gorkha National Liberation Front
HJC	Haryana Janhit Congress
HMS	Akhil Bharati Hindu Mahasabha
HSS	Harijan Sangharsh Samiti
HVC	Himachal Vikas Congress
HVP	Haryana Vikas Party
IAEA	International Atomic Energy Agency
IAC	India Against Corruption (a movement)
IAS	Indian Administrative Service
IB	Intelligence Bureau
ICS	Indian Civil Service
IFDP	Indian Federal Democratic Party
IMDT	Illegal Migrants (Determination of Tribunal) Act
INC	Indian National Congress

ABBREVIATIONS

INTUC	Indian National Trade Union Congress
ICS(SCS)	Indian Congress Socialist (Sarat Chandra Sinha)
IPF	Indian Peoples Front
IPFK	Indian Peace Keeping Force
ISC	Inter-State Council
IUML	Indian Union Muslim League
KCM	Kerala Congress (Mani)
KMPP	Kisan Mazdoor Praja Party
JAAC	Jharkhand Area Autonomous Council
JD	Janata Dal
JD(S)	Janata Dal (Secular)
JKLF	Jammu and Kashmir Liberation Front
JKN	Jammu and Kashmir National Conference
JMM	Jharkhand Mukti Morcha
JNP(S)	Janata National Party (Secular)
JP	Janata Party
JP(S)	Janata Party (Secular)
KCP	Karnataka Congress Party
KEC	Kerala Congress
KEC(M)	Kerala Congress (Mani)
LaC	Line of Actual Control
LDF	Left Democratic Front
LF	Left Front
LJP	Lok Janshakti Party
LS	Lok Shakti
LTTE	Liberation Tigers of Tamil Eelam
MADMK	M.G.R. Anna D.M. Kazhagam
MAG	Maharashtravadi Gomantak
MAT	Minimum Alternative Tax
M-COR	Marxist–Coordination
MDMK	Marumalarchi Dravida Munnetra Kazhagam
MGP	Maharashtrawadi Gomantak Party
MGJP	Maha Gujarat Janata Parishad
MIM	All India Majlis-e-Ithehad-ul-Mulimeen
MISA	Maintenance of Internal Security Act
MLA	Member of the Legislative Assembly
MNF	Mizo National Front
MP	Member of Parliament
MPVC	Madhya Pradesh Vikas Congress

ABBREVIATIONS

MSCP	Manipur State Congress Party
MTCR	Missile Technology Control Regime
MUL	Muslim League
NAC	National Advisory Council
NCO	Indian National Congress (Organization), or INC(O)
NCMP	National Common Minimum Programme
NDA	National Democratic Alliance
NDC	National Development Council
NF	National Front
NLP	National Loktantrik Party
NLR	Loktantrik Congress
NPF	Nagaland People's Front
NPT	Non-Proliferation Treaty
NRI	Non-Resident Indian
NSG	Nuclear Suppliers Group
NTRTDP(LP)NTR	Telugu Desam Party (Lakshmi Parvathi), or TDP(NTR)
OBC	Other Backward Class
PCE	Partido Comunista de España
PCF	Parti Communiste Français
PCI	Partito Comunista Italiano
PRBP	Peoples Republican Party
PCC	Pradesh Congress Committee
PDS	Public Distribution System
PR	proportional representation
PMK	Pattali Makkal Katchi
PSE	Public Sector Enterprise
PSP	Praja Socialist Party
PSU	Public Sector Utility
PUF	People's United Front
PULF	People's United Left Front
RAW	Research and Analysis Wing
RJD	Rashtriya Janata Dal
RJP	Rashtriya Janata Party
RPI	Republican Party of India
RPI(A)	Republican Party of India (A)
RPK	Republican Party of India (Khobragade)
RSP	Revolutionary Socialist Party
RSS	Rashtriya Swayamsevak Sangh

ABBREVIATIONS

SAARC	South Asian Association for Regional Cooperation
SAD	Shiromani Akali Dal
SAD(M)	Shiromani Akali Dal (Simranjit Singh Mann)
SAP	Samata Party
SC	Scheduled Caste
SDF	Sikkim Democratic Front
SEZ	Special Economic Zone
SHS	Shiv Sena
SJP	Samajwadi Janata Party
SJP(R)	Samajwadi Janata Party (Rashtriya)
SLA	State Legislative Assembly
SMSP	single-member simple plurality, or first-past-the-post (FPTP), electoral system
SSP	Samyukta Socialist Party
ST	Scheduled Tribe
SVD	Samyukta Vidhayak Dal
TDP	Telugu Desam Party
TDP(NTR)	Telugu Desam Party (N.T. Rama Rao), or NTRTDP (LP)
TMC	Tamil Maanila Congress
TPDS	Targeted Public Distribution System
TRC	Tamilaga Rajiv Congress
TRS	Telangana Rashtra Samithi
UC	Utkal Congress
UCC	Uniform Civil Code
UCPI	United Communist Party of India
UDF	United Democratic Front
UGDP	United Goans Democratic Party
ULF	United Left Front
UMFA	United Minorities Front, Assam
UNPA	United National Progressive Alliance
UPA	United Progressive Alliance
USSR	Union of Soviet Socialist Republics
UT	Union Territory
VCK	Viduthalai Chiruthaigal Katchi
VHP	Vishwa Hindu Parishad
VIDS	Voluntary Income Disclosure Scheme
WBTC	West Bengal Trinamool Congress

GLOSSARY

Adivasis	original people; Scheduled Tribes
bandh	strike
benami	a title given to a third party to evade the law
Bharat	the Hindi translation of India in the Constitution; invoked to represent rural India post-independence (as in "the politics of *Bharat*")
Bharatvarsha	term designating ancient India/Bharat
bhoodan	land-gift movement
crore	ten million
crorepati	an individual worth ten million rupees
Dalits	the "downtrodden"; previously untouchable castes; Scheduled Castes
dalitbahujan	the "oppressed peoples"
dharma	principles of discipline, order and harmony
garibi hatao	abolish poverty
gherao	a protest that encircles its target
goonda raj	rule of thugs
hawala	informal network of money exchange
Hindutva	Hindu cultural nationalism
Jan Morcha	People's Front
jati	local hierarchical endogamous caste groupings
kar sevaks	volunteers in the movement to build a Ram temple in Ayodhya
Kashmiriyat	Kashmiri cultural identity
kisan	farmer
kulak	rich peasants; a Russian term meaning "fist"

GLOSSARY

lakh	hundred thousand
lok shakti	people power
mandir	temple
masjid	mosque
navaratnas	nine jewels
nikamma	a person who is good for nothing
panchayat	a village council
panchayati raj	system of government based on village councils
Ramjanmabhoomi	movement to build a Ram temple in Ayodhya
Ram shila pujan	consecration of the bricks collected to build a Ram temple
rath yatra	chariot journey
saamajik samarasata	social integration
sampoorna kranti	total revolution
Sadr-e-Riyasat	nomenclature for "president" in Kashmir
samaj	society
sangh parivar	the Hindu nationalist family of organizations
saptakranti	a seven-fold revolution
sarvodaya	collective welfare through constructive work
satyagraha	non-violent direct action
shudra	the fourth *varna* caste in the traditional social order, largely comprising laborers, artisans and menial service providers
swadeshi	self-reliance
swaraj	collective self-rule
varna	the four broad categories that traditionally define the caste order
Wazir-e-Azam	nomenclature for "prime minister" in Kashmir
zamindari	feudal landed classes who collected taxes under colonial rule

INTRODUCTION

Since the 1970s, a series of national coalition governments, challenging the dominance of the Indian National Congress (Congress), have ruled the world's largest democracy. Significantly, the first three of these multiparty experiments espoused a social democratic vision, inspired in varying degrees by the politics of the broader Indian left.[1] The Janata Party (1977–1980), an amalgamation of four parties from northern India, was the first. Rallied by the eminent Gandhian socialist Jayaprakash Narayan, the Janata coalesced during the electrifying sixth general election to overturn the Emergency (1975–1977), a disastrous experiment in authoritarian rule instigated by Indira Gandhi. It pledged to restore parliamentary democracy and end mass poverty through constitutional reform, political decentralization and small-scale cooperative development in the countryside.

The second coalition government to rule the Union was the National Front (1989–1991), successor to the Janata Party, which came to power after the highly fractured verdict of the ninth general election.[2] The seven-party alliance formed a minority government, supported by two external formations, the communist Left Front and the Hindu nationalist Bharatiya Janata Party (BJP). Led by the Janata Dal (JD), the National Front advocated the interests of propertied middle castes and poorer subaltern classes, championing social justice. Yet the coalition also encompassed ascendant regionalist forces from non-Hindi speaking states, seeking greater political devolution and economic autonomy in Centre-state relations, as well as cultural recognition in the national imaginary. Indeed, its formation constituted a watershed, heralding the start of the "third electoral system" in India's federal parliamentary democracy.[3] The increasing electoral participation of historically subordinate groups, rise of state-based parties and growing threat of militant Hindu nationalism signaled the beginning of a "post-Congress polity", in which neither the Congress nor any other party could muster a plurality of votes across the Union. Fractured electoral verdicts and hung parliaments became the norm. The National Front suggested that a "third force" might emerge vis-à-vis the Congress and the BJP to fulfill the rising aspirations of its most disadvantaged citizens.

Indeed, the inconclusive verdict of the eleventh general election produced a third center-left coalition government in New Delhi, a minority fifteen-party alliance christened the United Front (1996–1998). The primary aim of its left, secular and democratic forces was to prevent the BJP from seizing national power after militant Hindu nationalists destroyed the *Babri masjid* (mosque of Babur) in Ayodhya in 1992, unleashing the worst communal riots since Partition. The politics of *Hindutva* (Hindu cultural nationalism) threatened to tear the country apart. In many ways, the United Front was the culmination of the protracted attempt to forge a progressive Third Front in modern Indian democracy. Its socialist parties, whose predominantly lower-caste leaders had become chief ministers in the most populous states, undermined the presumption of upper-caste rule. The parliamentary communists, who had alleviated rural poverty and expanded local self-government in their regional bastions and advanced a more egalitarian vision of national economic development, comprised key members inside and outside the governing coalition. And its ascendant regional formations, which contested the idea of India championed by different national parties, had shifted the dynamics of power between New Delhi and its peripheries. In short, the United Front symbolized the possibility of a new politics, expanding the foundations of the most populous, diverse and poverty-ridden democracy in the world.

Yet each of these governing coalitions failed to last a full parliamentary term, let alone survive. The Janata Party, encompassing lower-caste socialists, agrarian capitalists and Hindu conservatives, succumbed to high-level intrigues and deeper partisan conflicts. Clashing leadership ambitions within the JD, and growing tensions between the rising political aspirations of the Other Backward Classes (OBCs) and the Hindu right, tore the National Front apart too. Finally, the minority parliamentary status of the United Front left the coalition vulnerable to the

Congress, a rival to many of its constituent parties in the regions. All three experiments, according to critics, "ended in ignominy".[4]

The fall of the United Front allowed the BJP to capture national power at the head of the National Democratic Alliance (1998–1999 and 1999–2004), whose ranks eventually comprised over 20 parties. Its first spell in office, which saw the BJP weaponize India's nuclear capability, lasted just over a year. Yet the party adroitly exploited tensions amongst the remnants of the third force, while attracting new state-level allies that expanded its regional presence and social base, enabling the National Democratic Alliance to secure a rare parliamentary majority after the thirteenth general election in 1999. Its second tenure witnessed increasing violence against Christians and Muslims, efforts to undermine secular institutions and the transformation of India into a strategic nuclear power, throwing the ranks of the broader Indian left into disarray.

Despite the surprising defeat of the National Democratic Alliance in the fourteenth general election in 2004, the prospects of the third force worsened. The Congress put together the United Progressive Alliance (2004–2009 and 2009–2014), enabling the grand old party to recapture national power with the crucial outside support of the communist Left. Conflict over the Indo-US civil nuclear deal fractured their partnership in 2008. The Congress survived, strengthening its position following India's fifteenth general election in 2009, garnering credit for a range of progressive social legislation that its communist allies had pushed for and reflected many longstanding socialist commitments. The historic downfall of both Left Front administrations in Kerala and West Bengal in state assembly elections in 2011 constituted a moment of reckoning. But the sixteenth general election in 2014 inflicted a far more punishing defeat. Led by the combative Hindu nationalist chief minister of Gujarat, Narendra Modi, whose administration had failed to prevent an anti-Muslim pogrom in 2002, a resurgent BJP attacked the steep economic deceleration, high political corruption and paralysis of leadership that engulfed the second avatar of the United Progressive Alliance. Indeed, it vowed to restore rapid growth and national prestige by implementing a muscular version of neoliberalism, centralizing political authority and unleashing capitalist accumulation. Dividing the opposition, the BJP won a stunning electoral victory, becoming the first party to win a parliamentary majority since 1984. The politics of constructing a Third Front in modern Indian democracy lay in ruins.

What explains the rise of socialist, communist and regional parties since the late 1970s? Why have they faced repeated difficulties in constructing a stable Third Front in New Delhi? What explains the politics, policies and performance of its various incarnations, the Janata Party, National Front and United Front governments? To what extent have socialist, communist and regional parties reshaped the agendas, strategies and prospects of their principal national rivals, the Congress and the BJP? Finally, why did the idea of a third force persist in contemporary Indian democracy, despite its waning political fortunes since the late 1990s?

In general, scholars of modern Indian politics advance two perspectives to explain the chronic difficulties of a third force. The first contends that personal ambition, political expediency and struggles for power drive these inherently fragile coalitions.[5] Led by disgruntled former Congressmen, its customary parties come together to capture office for its own sake, with little to distinguish their policies. Hence their predominant image as "stopgap arrangements".[6] The second perspective emphasizes the multitude of personalities, interests and forces that divide its constituents. Conflicts involving caste, region and class, divergent political ideologies and competing electoral incentives in the states explain the volatility of third force politics.[7] Despite these differences, however, most observers agree: the struggle to build a Third Front in modern Indian democracy resembles the chronicle of a death foretold.[8]

These standard interpretations contain obvious truths. By focusing on moments of high political drama, however, they obscure as much as they reveal. This book seeks to break new ground in three realms. The first is to challenge prevailing empirical accounts of the Janata Party, National Front and United Front, each now consigned to the footnotes of history, reviewing old evidence and marshaling new facts to reevaluate their respective tenures in office. Existing scholarly accounts focus on how each coalition struggled to realize professed social democratic ideals, and rightly so. Despite committing more public resources to poverty alleviation, employment generation and egalitarian rural development, all three administrations witnessed progressively greater success in liberalizing industry, trade and investment, benefiting their more prosperous constituents.[9] Yet observers typically disregard how each of them established new institutions and re-engaged contested regions to improve parliamentary democracy and

INTRODUCTION

Centre-state relations.[10] Their respective efforts to forge better ties between India and its smaller neighbors in the subcontinent, while enhancing strategic flexibility in wider foreign relations, attract even less scholarly attention. And no attempt has been made, to date, to synthesize these broad trends into a synoptic historical account. Suffice to say, many of the policy initiatives and institutional reforms unveiled by the Janata Party, National Front and United Front were rushed, inconsistent, sometimes even contradictory, interventions that rarely developed into a clear overarching programme. Failures and disappointments, moreover, dogged each government. Nonetheless, their cumulative records outlined a distinctive political vision, challenging the view that none of them had any idea about how to run the country.[11] In particular, the politics of the third force gradually advanced a more faithful version of the asymmetric "federal nationalist" principle that has shaped the dynamics of the Indian "state-nation" since Independence.[12] The failure to consolidate its possibilities was a missed historical opportunity to forge a more progressive coalition representing the broader Indian left.

Second, analyzing the history of the third force afresh yields new theoretical insights. Recent scholarship on Indian coalition politics, which increasingly studies its dynamics through the prism of comparative theoretical inquiry, emphasizes competing party interests and formal institutional arrangements to explain the formation and demise of different multiparty experiments. Both factors matter. Yet such investigations overstate the stability of and polarity between competing political motivations. Many also neglect internal party debates over whether to share power, and with whom, to what extent and how, which frequently caused real schisms. Lastly, given these complexities, party leaders confronted difficult choices. To grasp the politics, achievements and shortcomings of different coalition governments in India, in other words, requires an innovative conceptual grammar able to illuminate the dynamic interplay between political institutions, social interests and human agency.

In general, the prospects of competing multiparty administrations in India owe much to the evolving logic of its democratic regime, in particular, the ramifications of contesting for power in an increasingly regionalized federal parliamentary democracy with first-past-the-post (FPTP) elections. The reorganization of states along distinct linguistic-cultural lines in the 1950s and 1960s encouraged many political formations to mobilize electoral support in local idioms of caste, region and language.[13] Over time, plurality-rule elections fostered specific distinct party systems in the states, in which two parties or blocs competed for power in the vernacular, creating a federal party system with "multiple bipolarities" that generated parliamentary fragmentation in New Delhi by the late 1980s.[14] The rise of socialist, regional and communist parties was partly a long-term consequence of these complex path-dependent processes.

However, the same regime dynamics that encouraged the proliferation of state-based parties and national coalition governments after 1989 paradoxically tested their ability to survive. Dissimilar electoral incentives in the regions hindered durable alliances at the Centre. High electoral volatility, factional splits and the disproportional effects of FPTP generated mistrust amongst coalition allies.[15] Lastly, structural economic reforms deepened these centrifugal tendencies through the 1990s. In sum, despite its relative institutional stability, India's macro-democratic regime generated tremendous political uncertainty after 1989. Party leaders confronted an intensely competitive federal party system, where marginal electoral swings, tight electoral races and multiparty blocs determined the balance of power.[16] Sustaining a diverse coalition government in such circumstances, especially an alliance of diverse state-based parties seeking to forge a Third Front, became exceedingly difficult.

That said, these crosscutting pressures did not determine the fate of competing national coalitions. As historical institutionalists point out, electoral regimes, governmental systems and different state designs create incentives, opportunities and constraints that political actors have to appraise.[17] In some cases, institutions and rules may fail to specify how to act. In others, complex interaction effects produce multiple possible outcomes. Lastly, even stable political institutions can fail to mitigate uncertainty during moments of significant historical change. All three scenarios are germane for understanding the politics of the third force. India's parliamentary cabinet government supported executive power sharing. But forging political agreement and enforcing collegial responsibility in unwieldy minority administrations requires power-sharing formulas and conflict resolution mechanisms to accommodate rival interests. Party leaders had to devise a number of practices and techniques—from local pacts, friendly electoral contests and joint state-level campaigns to national electoral alliances, common minimum programmes and high-level steering committees—to facilitate bargaining, negotiation and compromise. The superior cohesiveness of the United Front vis-à-vis the Janata Party and National Front, and the even greater durability of the National Democratic Alliance and United Progressive Alliance in turn, underscored their importance.

Ultimately, the vicissitudes of India's coalition politics heightened political agency. In particular, it forced competing party leaders to exercise sound political judgment. They had to comprehend the possibilities and constraints of specific historical contexts, only partly shaped by their beliefs, desires and practices; to distinguish the foreseen, foreseeable and unforeseeable consequences of different political decisions, whatever their intentions; and to seize the possibilities of a given political moment. In general, to judge well in politics requires actors simultaneously to synthesize a vast array of information, yet grasp what differentiates a particular context, and to demonstrate a pragmatic outlook, strategic orientation and detached spirit, yet possess a passion for a cause.

The quality of judgment was especially salient in determining the trajectory of the third force and India's coalition politics more widely. The demise of single-party majority governments in New Delhi after 1989 represented a new era, when the ideas, identities and interests that had hitherto shaped modern Indian democracy became less stable, creating new opportunities. The organizational weakness of many Indian parties enhanced their leaders' political autonomy. Lastly, the number, diversity and fluidity of parties and factions that comprised successive coalition experiments, and the minority parliamentary status of many of the latter, undermined the possibility of strict rational calculation. In short, these circumstances tested the prudence, creativity and foresight of every party in equal measure, yet simultaneously made it difficult to judge well.

What prevented the leaders of the third force from exercising good political judgment, especially those on the broader Indian left, was their tendency to conceptualize power in fixed, indivisible and zero-sum terms. Its manifestation differed. On the one hand, the refusal of many leftists in the Communist Party of India (Marxist) (CPI(M)) to share formal executive power with non-communist parties betrayed a stridently moralistic yet strangely mechanical conception of power, whose majoritarian underpinnings frustrated its possibilities of expansion. On the other, the readiness of many socialist leaders to join hands with myriad forces yet not share organizational power amongst themselves revealed a politics of resistance, insubordination and defiance necessary for genuine subaltern groups, but self-liquidating for its elites. Put differently, the moralistic intentions of the communists as well as the cynical instrumentality of the socialists undermined good political judgment at decisive moments.

The third aim of the book, vital for understanding the dynamics of national coalition politics in modern Indian democracy, is to provide a fine-grained analytic narrative of the rise and decline of the third force since the 1970s. In general, narratives elucidate the nexus between agency, structure and process in a single coherent account. They are exemplary for investigating temporal issues: the timing, sequence and contingency of events, the crystallization of particular historical conjunctures, the passage of an era.[18] Hence narratives provide an excellent technique for creating a "moving picture" of "politics in time".[19] Constructing a narrative proves very useful for demonstrating how multiparty governments actually function—something that existing scholarly accounts and the dominant theoretical models upon which they rely only partly explain. Moreover, analyzing the judgments of real political actors in proper historical context demands a prospective orientation: examining their incentives, opportunities and constraints, and assessing the foreseen, foreseeable and unforeseeable consequences of their decisions, by reconstructing the moment of action itself. If done well, such an approach can minimize the specter of inevitability that beguiles so many explanations in political science today.

Plan of the book

Writing a narrative involves several customary steps: identifying the main protagonists, tracing the chronology of events and plotting their actions from beginning to end.[20] Our intellectual preoccupations always influence how we depict social reality, however. In narrating events, we pose questions, make assumptions, and select concepts and methods to identify facts we deem significant amongst a range of alternatives, suggesting particular arguments versus others. Put differently, narratives do not merely illustrate arguments, but carry them.[21] This is self-evident with analytic narratives, whether rational choice histories of political action, meso-level studies of social mobilization or macroanalytic comparative history, which present theoretical arguments to explain specific historical accounts.[22] Yet even abstract theoretical models encapsulate particular narratives from the real world: no ideal chronicle or ideal chronicler exists.[23] Hence the need for conceptual precision, theoretical leverage and empirical rigor, to allow readers to compare particular arguments vis-à-vis others, enhancing their persuasiveness and reliability.

Chapter 1 examines the dominant explanations of coalition politics in modern India vis-à-vis the comparative theoretical literature that increasingly shapes the contours of debate. Recent scholarship illuminates how myopic party leaders and the complex interaction effects of formal political institutions encourage unstable multiparty governments in India. Yet many existing studies discount the hard political choices party leaders frequently confront, and the incompleteness and indeterminacy of institutions, especially during moments of change. Moreover, coalition leaders must devise formulas, strategies and tactics of power sharing to facilitate consultation, negotiation and compromise. Lastly, they often have to make decisions under conditions of great political uncertainty. Such circumstances demand astute political judgment, a form of reasoning that militates against strict instrumental rationality as well as ultimate moral values. The chapter ends by identifying and assessing the sources of information and variety of methods—including private letters, political speeches and party manifestoes, media reports, electoral and survey data and government documents, and the rare confidential testimonies of key political actors—that furnish and orient the narrative.

Part I of the book deploys these arguments, and synthesizes previous scholarship and selective archival material, to explain the genesis of the third force. Chapter 2 traces its conceptual roots and political manifestations from the anti-colonial movement in the mid-1930s until the Emergency in the mid-1970s. It analyzes the failure of communists and socialists to unite vis-à-vis the Congress. Ideological conflicts over many issues—caste, class and religion, the nationalist movement and the postcolonial regime, and the prospects of socialism, capitalism and democracy—divided the two pillars of the broader Indian left. Yet internal party disputes over power sharing, including whether to align with the progressive wing of the Congress, caused splits within each camp too. The failure of various opposition forces to forge durable multiparty governments in the states in the late 1960s tarnished the image of coalitions from the start. Yet their failure obscured the early regionalization of the federal party system. The centralization of power by the Congress, culminating in the Emergency, allowed opposition forces to coalesce.

Chapter 3 explains the formation, performance and demise of the Janata Party, the first non-Congress government to rule the Union. The myopic political ambitions of its leadership, and the rising power of propertied intermediate castes in the states, ruined its avowed neo-Gandhian ambitions to uplift the rural poor. Yet the Janata also enhanced parliamentary democracy through constitutional reforms and political innovations, reset Centre-state relations and pushed subcontinental relations in positive new directions. Clashing political ambitions and poor tactical choices in New Delhi triggered its downfall. Yet the collapse of the Janata Party also revealed strains between the socialist left and Hindu right, sowing the seeds of the third force.

Chapter 4 analyzes the rise of new state-based parties and the regions more generally during the 1980s. The Congress' return to power, and its massive electoral victory in 1984, suggested renewed dominance. Indeed, the party declared its ambition to modernize the state, castigating bureaucratic corruption and promoting economic liberalization. Yet the growth of various opposition forces, from a reintegrated communist Left and new regional parties in the middle to Hindu nationalists on the right, intensified electoral competition in the states. The failure of the Congress to grasp the nature of these developments, and its cynical tactical misjudgments, stoked growing communal polarization and deteriorating Centre-state relations. Perceptions of high political corruption, general economic mismanagement and mounting opposition unity sealed its demise.

Lastly, Chapter 5 explains the formation, performance and demise of the National Front, the second non-Congress government to rule the Union. Like its predecessor, rival aspirations over the prime ministership threatened the minority governing coalition from the start. Deeper political tensions over reservations for the OBCs ruptured its ranks towards the end. And its leadership failed, despite early signs of goodwill, to resolve growing militant insurgencies in Punjab and Kashmir. Nonetheless, the National Front advanced the politics of the third force in several ways. The coalition encompassed significant regional parties within its ranks and established stronger relations with the communist Left. It also crafted new political institutions to improve

INTRODUCTION

Centre-state relations while extending a more conciliatory approach to subcontinental affairs. Ultimately, the politics of *Hindutva* precipitated its demise, crystallizing the idea of the third force.

Part II of the book analyzes in greater depth the maturation of third force politics, using extensive primary research to explain the rise, performance and demise of the United Front. Chapter 6 explores the prospects of the Third Front in the early 1990s by investigating the political manifestoes, social ideologies and election strategies of its various constituents and their principal national rivals. Power struggles within the Janata *parivar*, and differences with the communists, weakened the prospects of the broader Indian left. Economic liberalization tested wider cross-party solidarities. And strategic disagreements and tactical disputes in the states frustrated the possibility of building a coherent national alliance. Nonetheless, the regionalization of its federal party system rewarded parties that mobilized distinct social constituencies in key state-level arenas, making it hard for national parties to occupy the center of gravity. The menace of growing communal violence by militant Hindu nationalists sharpened the secular commitment of third force politics.

Chapter 7 explains the formation of the United Front, which came together to prevent the BJP from capturing national power, following the eleventh general election in 1996. The political orientation and social bases of the three principal fronts revealed distinct profiles: parties comprising the Third Front disproportionately represented the middle and lower ranks of every major cleavage in society. Yet the composition of the United Front, and the establishment and formulation of its Steering Committee and Common Minimum Programme, failed simply to reflect the parliamentary seat tallies or prior political preferences. The decision by the CPI and more regionalist parties to join government, and of the CPI(M) to reject the prime ministership amidst intense debate, exposed numerous competing interests, strategies and judgments.

Chapter 8 investigates the ways in which H.D. Deve Gowda sought to establish his prime ministerial authority, manage relations vis-à-vis external parliamentary supporters and address economic liberalization, Centre-state relations and foreign affairs. These early months uncovered the array of views on how to address inter-state water sharing, public sector retrenchment and the insurgency in Jammu & Kashmir. In general, the United Front displayed surprising energy in tackling various challenges. Yet tight political relations between Deve Gowda and Narasimha Rao, the embattled leader of the Congress, failed to protect either of them from a confluence of pressures. Political manipulation, alongside the failure to devise agreed power-sharing formulas and decision-making practices, exacerbated latent conflicts.

Chapter 9 examines various events in Centre-state relations that tested the pro-regional credentials of the United Front amidst growing tensions within its ranks. The coalition exploited national power for partisan ends, imposing President's rule against rivals in Gujarat and Uttar Pradesh and failing to ensure free assembly elections in Kashmir, weakening its claim to difference vis-à-vis the Congress and the BJP. Yet its regional inclinations also led to partial efforts to re-engage longstanding grievances in the Northeast. More imaginatively, the United Front resolved a seemingly intractable dispute with Bangladesh, extending asymmetric concessions which the media christened as the Gujral doctrine. In doing so, the coalition exhibited a more regionalized conception of the national interest

Chapter 10 analyzes growing political tensions over national economic policy in the run-up to the 1997–98 Union budget. By joining government, regional parties had acquired control of several key ministries that pushed further economic liberalization, belying the view that a left-of-center governing coalition would stymie the latter. Yet it also threw into question the strategy of the CPI(M), leading Jyoti Basu, its venerable chief minister of West Bengal, to characterize its decision not to join the government as an "historical blunder". Indeed, the United Front proceeded to unveil a "dream budget", according to proponents of liberalization, and to restart crucial high-level dialogue with Pakistan. Deteriorating political relations between the Congress and the United Front precipitated the downfall of the Deve Gowda ministry, however. The vanity, skullduggery and misjudgments of key political leaders caused an unnecessary political crisis, undermining the conditions necessary for further economic liberalization and sensitive bilateral negotiations to succeed.

The final part of the book, the third, explains the decline of the third force. Chapter 11 analyzes the selection of I.K. Gujral as the second prime minister of the United Front and the events that marked his tenure. The changed political circumstances lessened the prospects of accommodation between rival political interests in economic policy, hampered efforts to make clear progress in Indo-Pakistan relations and frustrated progress in Centre-state affairs. Despite internal tensions, the United Front government fell because of ham-fisted machinations by the Congress. Yet the coalition unraveled during the twelfth general election, unable to resolve the divergent electoral incentives facing its parties across the regions, allowing the BJP-led National Democratic Alliance to capture national power in March 1998.

Chapter 12 explains the gradual dissolution of the third force from 1998 to 2012. Regional political calculations, mounting inter-state economic competition and widening social disparities pushed its erstwhile constituents further apart. Yet their declining coalition prospects also owed much to the strategic maneuvers, tactical acumen and political choices of their national rivals. The ability of the BJP to retain national power at the helm of the National Democratic Alliance for six years demonstrated considerable political skill. On the one hand, the party astutely shelved its most controversial proposals, attracting previous members of the third force. Sharing executive power, establishing new consultative institutions and espousing coalition *dharma*, the BJP expanded its regional presence and social base. On the other, the party mollified its militant Hindu allies by trying to shift the center of national political gravity to the right. The Congress party, unwilling to grasp the logic of the third electoral system for several years, suffered a long political winter. The survival of the National Democratic Alliance after the anti-Muslim pogrom in Gujarat highlighted the paradoxical effects of the federal party system, encouraging ostensibly secular parties to weigh the potential electoral backlash in their respective states, belying their professed ideals.

The failure of the BJP to maintain its national alliance, combined with the belated strategic reorientation of the Congress, allowed the United Progressive Alliance to form a minority Union government with the crucial parliamentary support of the Left. Exploiting their position, the communists blocked economic liberalization in New Delhi, supported a range of progressive social legislation pushed by like-minded political allies in the Congress, demonstrating the relevance of concerns that had been raised originally by the third force. Yet it became impossible for the CPI(M) to oppose further reforms in New Delhi yet advance them in the states, stranding the party between government and opposition. In the end, divisions over the Indo-US civil nuclear deal, exposing substantive disagreements as well as crude political machinations, drove the socialists and the communists apart.

The book concludes by situating the politics, successes and failures of the broader Indian left in comparative historical perspective. The tenures of the Janata Party, National Front and United Front deserve critical reappraisal, I argue, as do the conventional terms of art employed by most comparative studies of coalition politics. Crafting stable national coalition governments in India, given the diverse regional forces and complex institutional compulsions that encourage their emergence yet undermine their survival, poses unparalleled challenges when compared to virtually every other consolidated democracy in the world. Equally, complex ideological differences and tougher material conditions made it harder to forge a broad progressive coalition along the lines of the famous "red-green" alliance of urban workers and rural smallholders that enabled social democracy in twentieth century Europe.[24] Indeed, the predicament of the broader Indian left since the 1970s recalls the fate of Eurocommunists, highlighting the constraints of capitalism and democracy. Yet the vicissitudes of Third Front governments in India belie the assumption that regional parties are inherently destabilizing forces, suggesting theoretical insights and practical lessons for other parliamentary democracies where processes of devolution and demands for federalism may fragment national electoral mandates in the future. Ultimately, a more imaginative grasp of the real possibilities at key historical junctures, including unprecedented opportunities to exercise national power, would have enabled reformist elements amongst the socialists and communists to advance the cause of a progressive third force in modern Indian democracy to a greater degree. The book concludes by reflecting on the prospects of social democratic politics in India in the years ahead.

1

THE PARADOXES OF INDIA'S COALITION POLITICS

Traditionally, scholars of coalition politics in the comparative tradition pose the following questions: What explains the formation of coalitions? What causes them to fall apart? A few ask: How do coalition governments perform in office? A substantial corpus of scholarship addresses these broadly framed questions in a comparative theoretical framework. Conspicuously, the explanations put forward are general, too.[1] Guided by the belief that explanations are nomothetic in principle, leading coalition theorists assert that fine-grained idiographic studies cannot establish valid causal inferences. Ascertaining the latter requires explaining—sometimes even predicting—specific coalition experiments through concepts, theories and methods that are applicable to and account for the largest number of cases. Consequently, most scholars of coalition politics tend to dismiss the value of particular case studies; regard them as useful first steps in the process of forming causal hypotheses and constructing general theories; or employ them only to test the latter.[2] The search for general political explanations, marked by conceptual parsimony, theoretical ambition and empirical range, remains an ideal.

The study of coalitions in modern Indian politics has developed over two broad phases. The first examined its emergence in the states in the 1960s and 1970s. A number of scholars analyzed these experiments in comparative theoretical perspective.[3] However, most offered particular historical analyses.[4] In contrast, the second phase increasingly engaged the comparative tradition to explain the advent, character and ramifications of national multiparty governments since the late 1980s. Broadly speaking, three positions emerged. Some observers emphasized the significance of deep contextual peculiarities, particularly social cleavages such as caste, language and region.[5] Others applied classic theories of coalition politics, electoral regimes and party systems in the west, including models of rational choice, to explain the formation of successive minority governments since 1989.[6] In the middle, a number of commentators embedded the politics of coalition in India into comparative theoretical frames without losing sight of its historical particularities.[7]

This book pursues the last path. General political theories guide our investigation into a particular case. Yet the concepts, theories and methods employed in their service often fail to explain contextual features that bear on the problem under study.[8] Indeed, the claim that only general theories yield defensible explanations commonly leads to tendentious descriptions of specific empirical cases—which turns problems into artifacts of theory rather than independently observed puzzles.[9] Arguably, similar problems characterize much of the existing scholarly literature on coalition politics. Its leading theoretical models furnish relatively wooden accounts of the formation and demise of coalitions as discrete political events; formulate strict causal hypotheses from these accounts; and test the latter in large-N analyses on the ground that explanations must be general in scope. In doing so, these highly aggregated models obscure the complex political dynamics that govern many coalitions in the real world. Testing abstract theoretical models against particular cases may advance the state of knowledge within narrow sub-disciplinary parameters. But the costs, ranging from possibly distorted cases to inadequate causal understanding, can be high.

Ultimately, convincing explanations presume valid descriptions. Acquiring detailed knowledge of particular cases, which requires asking questions afresh, is therefore a necessary first step.

The main protagonists

Who are the main protagonists of India's coalition politics? Comparativists focus on competing party organizations, which dominate the struggle for power in modern representative democracies. Many characterize these parties as unitary political actors. Scholars recognize that parties contain parliamentary and organizational wings, whose leadership, functions and prerogatives vary, not to mention political factions. Indeed, most concede that internal coalition dynamics consequently remain mysterious.[10] Yet few analyze them directly.[11] Some claim that studying the politics of leaders or factions would make wide-ranging analyses of many cases harder. Others point out that mutually binding incentives ensure parties' cohesion, especially in western Europe, which overwhelmingly constitutes their region of focus. In particular, backbench politicians obey the whip at crucial moments of decision to further their personal careers, which party membership normally enhances, while party leaders use the resources and rewards at their disposal to ensure their subordinates' allegiance.[12] The most important decision points, of course, concern the formation and demise of government. Most comparative studies focus on these moments, construing each as independent of prior events or future expectations, in order to test statistically which theories yield greatest leverage. Hence the conceptual assumptions, empirical record and methodological parameters of these inquiries reinforce each other.

To grasp the dynamics of India's coalition politics, in contrast, requires us to investigate the actions of senior party leaders: the high-ranking elected representatives and organizational functionaries of competing electoral formations. There are several reasons for doing so. First, characterizing the main protagonists as unitary party organizations is simply untenable. Many of India's "weakly institutionalized" parties split into rival political factions before, during and after the tenures of the Janata Party, National Front and United Front, "on the basis of feuds or deals of leaders".[13] In many cases they comprised local personal networks, which individual politicians formed and disbanded expediently. The ancient conception of parties, as factions engaged in plots of intrigue for their own personal benefit, depicts their character far more accurately.[14] Others had proper organizational structures. Yet their leaders, seeking to enhance their personal political power, undermined their integrity and functioning. Indeed, institutional changes and political developments encouraged these proclivities. The passage of the Anti-Defection Law in 1985, designed by the Congress to support party unity, ironically encouraged many factions across the spectrum to create their own parties as vehicles for power.[15] The resulting electoral fragmentation and advent of national coalition politics, which lowered the threshold for acquiring parliamentary influence, deepened these incentives.[16] That said, it makes sense to designate the main actors of India's coalition politics as party leaders. Many of them performed expected tasks—recruiting electoral candidates, mobilizing social interests, structuring political issues, collecting funds, and forming and breaking governments.[17] Indeed, the fact that even satraps aspired to office by forming or joining parties and contesting elections highlights the significance of structured party competition.

Second, as comparativists acknowledge, viewing parties as unitary political actors obscures their internal power struggles. Intra-party deliberations and alternative organizational structures affect electoral contestation, government formation and the demise of coalitions.[18] Modern Indian politics displays a diversity of parties. These range from the communist parties and Hindu nationalist Jan Sangh/BJP, independent political organizations with relatively disciplined cadres, institutionalized decision-making procedures and strong links with ancillary social organizations, to the various splinter groups of the Congress (I) and Janata *parivar*, clientelistic networks that sometimes acquired independent status by registering themselves with appropriate state authorities. The majority of parties in India occupy the middle range, however, with the most effective combining elements of both. These organizational differences, as the narrative illustrates, had two ramifications. On the one hand, intra-party disunity heightened the autonomy of various party leaders at crucial political moments.[19] On the other, strong party organizations sometimes constrained their respective leaders' options, most obviously in the case of the communist Left.

Lastly, conceptualizing the main protagonists as party leaders is useful methodologically. It provides an ideal-type for appraising their actions: whether personal self-interest, partisan advantage or larger social purposes inspired their conduct.

Preferences, interests and goals

What motivates party leaders' decisions? Three general theories, which construe specific purposes as the clearly defined, mutually exclusive and exogenously determined goals of rational agents, dominate scholarly debate. Power-maximization theories argue that considerations of power *simpliciter* drive coalition politics: in particular, the desire to acquire the greatest share of cabinet posts in government.[20] Assuming that such posts are "fixed prizes" and that parties are fully aware of the various bargains that others are considering,[21] such theories posit that "minimum winning coalitions" are likely to emerge according to the "size principle": parties will form any coalition able to secure a working parliamentary majority with the fewest possible number of partners (the "strategic principle") in order to maximize their relative share of cabinet power (the "disequilibrium principle").[22] Underlying power-maximization theories is a Schumpeterian conception of politics: parties seek power for its own sake.[23]

In contrast, policy-realization theories maintain that coalition formation cannot be explained solely by the will to dominate.[24] Rather, the contest for power involves substantive concerns, specific policy goals that reflect divergent political ideologies and represent distinct social interests, as pluralists and Marxists would contend. Typically, policy-realization theorists claim that economic policy differences matter the most; that party ideologies differ over how much states should intervene in markets to promote economic stability, growth and redistribution; and that social cleavages reflect the nature and degree of class-based stratification. They argue that parties seek to forge coalitions with other parties that share convergent, or at least indifferent, policy goals through incremental negotiation.[25] Hence policy-realization theorists claim that either "minimum connected winning coalitions" or minority governments are likely to emerge.

Lastly, vote-seeking theories contend that since parties contest the ballot in the first instance, the desire to maximize vote share dictates their coalition strategies. Underlying such theories is a Downsian conception of politics: "parties formulate policies in order to win elections, rather than win elections in order to formulate policies".[26] Parties view elected office as the ultimate reward, a prerequisite for maximizing power or influencing policy, which "in turn implies that each party seeks to receive more votes than any other".[27] Indeed, the value of the latter increases in fractured electoral contexts since "the more votes a party wins, the more chance it has to enter a coalition, the more power it receives if it does enter one, and the more individuals in it hold office in the government coalition".[28]

Strikingly, virtually every observer of India's national coalition politics agrees that constant power struggles define its key dynamics. According to some, the absence of disciplined party organizations with clear ideological differences and relatively stable bases of electoral support generates chronic political instability and disappointing policy records.[29] Parties espouse "catchy slogans", others claim, merely to hide "crass opportunism".[30] Others recognize that personal ambition, principled differences and struggles for power frequently co-existed. Nevertheless, the last motivation often proved decisive, especially amongst India's embattled socialists.[31] The "surplus" multiparty governments forged by the communist Left Front in Kerala, Tripura and West Bengal since the 1970s, whose shared policy goals and ideological commitments enabled them usually to outlast single-party majority administrations, are exceptions that prove the rule.[32] Hence most would concur that "from 1977 onwards ... coalitions have been pulled down [at the Centre] only because they did not suit the personal interests of some contenders for power".[33] It is impossible to refute the significance of these ubiquitous power struggles.

Adopting the terms of standard comparative debates has several drawbacks for analyzing India's coalition politics, however. First, the notion of "power-maximization" requires greater conceptual elucidation. Numerous party leaders sought high office for purely private reasons: personal vanity, a political career or to exploit such posts for corrupt purposes. Yet the desire for office frequently served larger partisan interests: to apportion patronage to important clients and distribute governmental posts and administrative prebends to assorted party members. Indeed, many party leaders sought national office to protect turf in their home states, particularly after 1989. Lastly, several justified their pursuit of office on representational grounds, either by advancing the "pure identity claims" of particular communities or by joining forces with ideological opponents in order to block rival social groups.[34] In short, the struggle for power drives a great deal of India's coalition politics, especially within the third force. But such compulsions may serve a range of interests—personal, organizational, social—that warrant attention.

Second, emphasizing the primacy of power in India's coalition politics risks obscuring parties' substantive differences. Policy considerations, the interests of different social groups and ideological conflicts also influenced party leaders' choices, especially in office. In some instances, a personal "sense of vocation" inspired their decisions, particularly in the realm of economic policy and foreign affairs.[35] At other times, specific policy stances reflected partisan beliefs or the interests of particular social cleavages based on caste, class or regional lines.

Indeed, many Indian voters support particular party programmes out of concern for their socioeconomic well-being, rather than simplistic representational claims of caste, region or religion. Moreover, inferring social interests from pre-existing divisions ignores how political mobilization shapes the issues at stake in electoral competition.[36] Disagreements between the socialists and communists, the two axes of the broader Indian left, exemplify these issues. On the one hand, the socialists historically advocated small-scale rural production, a politics of recognition based on lower-caste identities, and non-alignment in foreign affairs. On the other, the communists traditionally championed rapid state-led industrialization, a politics of redistribution based on classes, and an anti-American foreign policy. Reducing their policy conflicts and ideological debates to the pure mobilizational tools of self-interested political entrepreneurs would be misguided. Finally, several party leaders occasionally took decisions that served a perceived national interest, even at their own expense. Differences over India's nuclear strategy, and terms of bilateral trade and resource sharing in the subcontinent, revealed such motivations at key historical moments. In short, power-based accounts fail to explain why different coalition governments in India have pursued distinct policy agendas with varying success.

That said, grasping substantive inter-party differences requires careful analysis. Policy-realization theories normally make two assumptions: policy goals correspond to specific party ideologies as well as particular social interests. Both are contestable. Ideological differences may not directly orient policy choices. On the one hand, the actual range of choice in an issue-area, such as economic policy, may be limited due to structural constraints in the domestic political economy or international economic order. This was clearly the case in India since the 1990s. On the other hand, a party may appeal to the values and identities of particular social groups, yet fail to advocate policies advancing their economic interests. The politics of dignity pursued by lower-caste parties and militant cultural agenda of Hindu nationalist parties, each of which sometimes made symbolic gains on behalf of their respective constituents yet ignored their wider material interests, illustrate these distinctions. Finally, it is difficult to infer political choices strictly from ideological dispositions because most actors take situational factors into account.[37]

Indeed, the constituents of the Janata Party, National Front and United Front advocated distinctive programmatic objectives, and their respective tenures pushed economic policy, Centre-state relations and foreign affairs in new directions. Many of their respective initiatives were rushed, incomplete and sometimes even at odds, revealing tendencies rather than truly fully developed policies driven by a shared political ideology. Nonetheless, all three governments revealed the outlines of a different political disposition, which require acknowledging the concerns of the socialist, communist and regional parties that composed their ranks in varying degrees.

Third, pure coalition theories overstate the polarity between competing political motivations in actual coalition bargaining. In many cases, a party can only realize its policy preferences by securing enough votes to gain office and ensuring their implementation. Conversely, a party may only be able to capture office by pledging certain policies to the wider electorate that its base dislikes. Several comparativists embrace these realities. Party leaders may prefer to enter heterogeneous coalitions in order to maintain their distinctive profiles. Conversely, they may minimize policy compromises in order to maximize party unity, but largely to protect their own positions.[38] Indeed, particular situational constraints may alter prior political preferences. Yet the desire to construct parsimonious explanations, including the presumption that elections are unconnected political events, prevents many scholars from probing these more complex relations. More generally, however, parties may perceive substantive value in maximizing power, influencing policy and acquiring votes: these objectives are not logically exclusive. But conceptualizing them in such terms encourages the view that coalitions are homogeneous blocs that solely seek to maximize power, influence policy or obtain votes. Indeed, assuming that parties may possess a plurality of goals, other questions arise, such as which motive trumps at specific moments, why and its consequences. Likewise, assuming that coalitions often contain heterogeneous interests, similar questions emerge, such as which parties realize their aims, why and to what effect.

A few comparativists investigate these hard political choices by studying how the leaders of several governing coalitions dealt with critical trade-offs in moments of crisis.[39] Their analyses generate several insights, namely, the importance of formal institutional rules and parties' organizational characteristics in shaping outcomes. Yet most avoid cases that exhibit high "bargaining complexity", circumstances that allow situational factors to dominate but make generalization difficult:

[All else equal], the greater the number of negotiating parties, the higher the level of bargaining complexity. But the latter may also be a function of a lack of unity on the part of the organizations involved or the lack of familiarity among the leaders of the relevant parties ... Numerous, disunited, or unfamiliar parties are likely to [have] given rise to information uncertainties among the partners in bargaining ... The more limited their information, the less likely risk-averse party leaders are to gamble on new coalition partners or on moves whose electoral implications are hard to foresee. Thus, in situations of highly imperfect or incomplete information, we may see fewer policy concessions and fewer unorthodox alliances than we might otherwise expect.[40]

Few established democracies rival the bargaining complexity of India's coalition politics. Indeed, the sheer number, fluidity and diversity of parties that oscillate amongst its national coalition governments engender tremendous uncertainty. Contrary to expectation, however, unorthodox alliances are the norm. Moreover, party leaders had to make policy concessions to form many of India's governing coalitions too.

In short, a diversity of purposes animates national coalition politics in India, even though narrow political instincts wreaked havoc all too often. Rather than embracing a "Manichean dualism of soul and body, high-mindedness and the pork barrel,"[41] we need to grasp how a variety of substantive disputes and power struggles shaped the emergence, tenure and demise of India's coalition governments, and the project of building a progressive third force more broadly.

Institutions as rules and incentives, opportunities and constraints

What shapes the opportunities and constraints party leaders face, and the size, character and durability of multiparty governments? Pure coalition theories presume an unconstrained political world. Most scholars realize that intentions rarely determine outcomes, however. Hence they examine the impact of formal political institutions, which generate incentive structures, influence agents' expectations and shape the probability of outcomes.[42] Three particular institutions draw scrutiny.

First, many analyze the ramifications of different electoral rules on party strategies as well as on the size and degree of polarization within the party system. In general, comparativists argue that electoral rules based on proportional representation (PR), the norm in west European democracies, translate votes into seats in relatively predictable ways. Hence such regimes encourage parties to maximize vote shares. In contrast, single-member simple-plurality (SMSP) systems, in which the candidate first-past-the-post (FPTP) wins, inflate and depress the conversion of votes into seats, creating incentives for parties to stick to their preferred policy positions.[43] Constitutional provisions regarding the installation and termination of governments, and legislative-executive relations in general, are the second institutional focus. In particular, comparativists investigate whether a formal investiture requirement exists, and if so, the specific role played by the head of state; the conditions for tabling parliamentary votes of confidence; and the significance of such votes for the continuation of parliament, government or both. Third, most scholars examine the type of government. Generally, presidential systems create few incentives for cooperation since they concentrate formal executive power into a single indivisible office; parliamentary cabinet systems allow governing coalitions to share executive power and set official policy agendas collectively to a greater degree; and consociational regimes, by dividing the spoils of office and providing constitutional vetoes over key policy issues, facilitate the greatest degree of elite cooperation.[44] Taken together, the preceding institutional arrangements determine the bargaining arena and establish incentive structures that influence parties' decisions.

Additionally, institutional considerations generate several hypotheses regarding why multiparty governments collapse.[45] The "coalition attributes model" tests the durability of "minimum winning coalitions" vis-à-vis "minimum connected winning coalitions" and minority governments. "Regime type approaches" focus on the number of parties in and size of the party system, its degree of polarization and the rules that govern policy formulation as well as the formation and demise of governments. The greater the number of parties in the legislature, and the higher the presence of extreme ideological parties, the greater the difficulty of forming durable coalitions. Strong investiture requirements exacerbate the latter. The "political bargaining environment" approach examines coalition attributes along with the degree of polarity in the party system. Coalitions with extreme ideological parties are more likely to collapse. Lastly, the "event-oriented" perspective investigates the susceptibility of coalitions to sudden critical incidents, precipitating their demise.

Comparative investigations yield two major findings. The first is that policy-realization theories best explain why coalition governments form: "minimum winning coalitions" account for only 30 per cent of the total between 1945 and 1999.[46] In some cases, parties construct "surplus" multiparty governments to insure themselves against potential blackmail. In others, policy affinities matter, leading to "minimum connected winning coalitions". Finally, against theoretical expectation, minority governments are the norm in Scandinavia: the relatively high informational certainty and low political risks characterizing these polities encourage parties to extract policy concessions without joining governments, enabling the latter to endure. In short, for many parties coalition pay-offs depend on the cost of office vis-à-vis their respective votes shares and policy credibility, and whether they believe particular governments will last.[47] In particular, the distribution of strength amongst parties in parliament and the difficulty of the political bargaining environment determine the probability of different arrangements emerging:

> Minority governments are most likely to form when bargaining power is concentrated in the hands of a single party, when the costs of forming free-floating coalitions are low, and when the value that parties place on being in government is not too great. Minimum winning coalitions are most likely to form when the value parties place on being in government is high relative to being in opposition, when uncertainty is low and parties are able to credibly commit to each other, when political decisions are made by simple majority rule, and when bargaining power is neither greatly concentrated nor greatly dispersed. Surplus coalitions are most likely when bargaining is greatly dispersed amongst the various parties in parliament, when political decisions require more than a simple majority in the lower chamber of parliament, and when government membership is neither extremely costly nor extremely valuable.[48]

Thus formal theoretical models predict correctly only 40 per cent of the time.[49] The second major finding is that random exogenous shocks, especially in governing coalitions that include extreme ideological parties and last beyond their first year, are likely to instigate their demise.[50] In sum, substantive policy concerns and contextual political factors matter. The significance of uncertainty and contingency at both ends of the coalition game, although rarely emphasized, is quite striking.

Where does India, an asymmetric federal parliamentary democracy with a FPTP electoral regime, stand in light of these comparative findings? The classic expectation regarding FPTP is Duverger's law: its propensity to produce two-party systems and yield single-party majority governments.[51] Winner-take-all elections generate strong incentives for voters to elect a candidate likely to win and, in turn, centripetal pressures for parties to appeal to the median voter by adopting a moderate electoral platform. If no single party can easily capture a plurality of votes, however, FPTP creates incentives to form electoral coalitions that cross the threshold of votes necessary for government formation.[52] Similarly, Westminster-style governments produce a collegial executive whose tenure depends on gaining and retaining the confidence of parliament, which facilitates power-sharing. This is particularly the case in fragmented legislatures where no single party, even the largest, can easily dictate to its coalition partners. Since prime ministers must retain the confidence of parliament, they often have to concede de facto control over ministries to their allies, which retain the capacity to break the government.[53]

Yet India has never fulfilled these theoretical expectations neatly. New Delhi witnessed a succession of single-party majority governments from 1952 to 1989. The Congress party ran most of them, however, in stark contrast to the two-party systems that FPTP normally engender. Indeed, the average number of parties in the Lok Sabha between 1952 and 1984 and between 1989 and 2004 was 18.6 and 31.6, respectively. To put it in perspective, the second most fragmented parliament in the world over a comparable period was Italy, with an average of 10.5 parties.[54] And plurality-rule elections in India, in contrast to western Europe, have created strong incentives for parties to abandon their professed policy preferences for the sake of political survival.

Furthermore, since 1989 India has witnessed a series of national coalition governments, the largest in the world in terms of the number and diversity of parties they contain. The vast majority are minority governing coalitions, however, rare in Westminster-style democracies. Indeed, 87.5 per cent of the Union cabinet governments formed between 1989 and 2004 lacked a parliamentary majority, with a mean of 5.8 parties in each of them. Their size, character and instability make India an extreme outlier amongst even west European polities with PR systems.[55]

Lastly, national coalition governments in India are far more diverse vis-à-vis other federal parliamentary democracies.[56] On the one hand, the majority of national multiparty governments in Canada, Germany, Australia and Belgium are minimum winning coalitions. On the other, out of 237 pre-electoral coalitions that came together in twenty federal democracies between 1946 and 1998, only two had parties with different regional bases. Yet unwieldy minority coalition governments, comprising numerous state-based parties, have been the norm in India since 1989. Between 1991 and 1999, the number of state-based parties in the Lok Sabha increased from 19 to 35, while their relative vote share jumped from 26 to 46 per cent.[57] In short, India's party leaders face the most difficult political bargaining environment of any modern representative democracy.

Given these anomalies, how useful are standard institutional analytics for explaining the vicissitudes of India's national coalition politics? Specifically, why have national multiparty governments become the norm since 1989, despite the persistence of its macro-democratic regime? Given their ubiquitous power struggles, which should encourage "minimum winning coalitions", what explains the fact that virtually every Union government in the post-1989 era lacked a parliamentary majority? And why did national elections in India continue to produce fractured verdicts, despite the instability of particular coalition experiments, until the surprising parliamentary majority won by the BJP in 2014?

We can resolve these puzzles by synthesizing the deft insights of the leading scholars of India's coalition politics. The demise of single-party majority governments at the Centre since 1989 owes much to the complex interaction effects of plurality-rule elections in a progressively regionalized federal parliamentary democracy. These dynamics shape their performance in office too. Its roots lie in the 1950s and 1960s, when New Delhi acceded to growing popular demands to reorganize the federal system into distinct linguistic-cultural zones. Vernacular public spheres developed in many states, leading new parties to employ local idioms of caste, region and language to mobilize historically subordinate classes vis-à-vis the Congress.[58] The vernacularization of federalism in India gradually encouraged the emergence, under FPTP, of distinct political systems in the states in which two parties or blocs competed for power in the 1960s and 1970s. These were slow burning processes. Yet their ramifications became clearer in the late 1980s. The inability of any single party to maintain a dominant presence in every state created a system of "multiple bipolarities" across the Union and parliamentary fragmentation in New Delhi.[59] Complex state-level rivalries, combined with a system of government that divided executive authority in the Council of Ministers and required opposition parties to demonstrate a parliamentary majority in order to defeat sitting administrations, locked in pre-electoral allies and enabled successive minority coalition governments to form.[60] The rise of lower-caste, regional and communist parties in the 1990s vis-à-vis the more elitist, centralizing and nationalist tendencies of the Congress and BJP, and the crystallization of the idea of the third force in modern Indian politics, was partly a long-term function of these complex processes. Put differently, India's federal parliamentary democracy encouraged the development of "centric-regional" and "polity-wide" parties, incentivizing them to craft diverse governing coalitions in New Delhi.[61]

However, the same macro-institutional design that encouraged particular coalition governments to emerge after 1989 paradoxically tested their ability to survive. First, the emergence of specific party systems in the regions impeded stable alliances over time. From the beginning, India's federal system contained local social cleavages, making it hard for opposition parties to create horizontal political alliances across state boundaries. The Congress party, leveraging the symbolic leadership and organizational resources it had accrued through the national movement, found it easier.[62] The reorganization of states gradually cracked Congress dominance, however, notwithstanding a series of national wave elections in the 1970s and 1980s. The de-synchronization of state and national-level elections in 1971 deepened the regionalization of many parties' electoral horizons and social bases. The tendency for the latter to fragment into increasingly narrow segments in the 1980s, as parties began to court particular lower-caste, regional and communal identities, made it harder to build durable national fronts. Finally, the deepening of liberal economic reforms in the early 1990s compelled state-level governments of every persuasion to court scarce private investment domestically and abroad, creating fissures within many parties and widening inter-state disparities too. In short, the federal underpinnings of India's coalition politics make national political bargains the product of nested state-level negotiations. Rational political decisions in the regions, where parties may offer outside support to avoid becoming "tainted" by national government decisions, frequently encourage ideologically heterogeneous and politically unstable alliances in New Delhi.[63]

Second, a FPTP system may paradoxically generate instability in multiparty governments, despite creating powerful incentives to pool votes prior to elections.[64] Small vote swings in a fragmented electoral field can lead to massive seat changes in parliament, generating potential mistrust within particular coalition governments. Therefore parties have incentives constantly to jockey for advantage. In aggregate terms, the degree of electoral disproportionality in India diminished through the 1990s as

THE PARADOXES OF INDIA'S COALITION POLITICS

the number of parties increased.[65] Yet specific parties still often suffered from massive negative swings between elections. Notwithstanding 1984, the lead party of every Union government between 1977 and 1996 lost between 15 and 50 per cent of its vote share in the next poll.[66] Party leaders had good reason, given the persistently high electoral volatility India witnessed after 1989, to feel insecure.

Third, parliamentary cabinet systems are strategically complex. The timing of elections is endogenous—within a customary five-year limit—since the legislature makes and breaks governments. The prime minister commands greater discretion over such matters in single-party majority governments. Power diffuses in multiparty executives, however, to other leaders. And in minority coalition governments, which predominate in India post-1989, every member of the Opposition with an effective parliamentary veto enjoys such influence. The fact that only two minority coalition governments in New Delhi lasted a full parliamentary term to date—both avatars of the United Progressive Alliance—underscores these vulnerabilities.

A paradox emerges. The tripartite logic of India's macro-democratic regime, despite its relative institutional stability, generated political uncertainty after 1989. Party leaders confronted an intensely competitive federal party system where politicians, seeing the outcome simultaneously as close and open, "configure[d] around alternative parties or party blocs"; where small electoral shifts significantly enhanced their bargaining power; and where the stakes were high.[67] Sustaining a diverse multiparty government in such circumstances, especially a minority governing coalition of diverse state-based parties seeking to create a Third Front, became exceedingly difficult.

Formulas, strategies and tactics of power-sharing

Indeed, these macro-level uncertainties had micro-level dimensions too. The somewhat mechanistic conception of institutions that dominates comparative investigations creates several problems. First, it elides the distinction between rules and incentives. Formal decision rules sometimes produce specific outcomes independently of political agents' decisions. But we cannot assume that political actors will fully respond to the incentives generated by such formulae. India's FPTP regime has constantly encouraged opposition parties to form anti-Congress alliances since independence, producing similar inducements for the

Congress party after 1989. Contrary to what historical institutionalists expect, however, the "strategies induced by [FPTP in India] … [failed to] ossify over time into worldviews, which … ultimately shape[d] even the self-images and basic preferences of the actors involved".[68] Both sides struggled to recognize these incentives and overcome other obstacles over these years.

Second, drawing causal inferences from particular institutional matrices presumes that political agents know, accept and follow patterns of interaction that such arrangements encourage.[69] Rather than "enacting scripts", however, political actors may seek to acquire power or challenge the boundaries of authority by "play[ing] the rules as if they were instruments".[70] Parliamentary cabinet government enjoins collegial responsibility. Yet whether individual ministers respect the latter, and whether parliaments criticize government policy, oversee bureaucratic performance and hold the executive accountable as expected, are empirical questions. Alas, ministers in India's coalition governments regularly disputed formal cabinet decisions or bent, circumvented and subordinated them. Similarly, parliamentarians often showed indifference "to executive abuse …, ignore[d] poor drafting of legislation and provide[d] minimal scrutiny of the budget".[71]

Lastly, even when formal institutions establish accepted routines, many cannot fully determine outcomes. In some instances, the rules are partial or ambiguous, unable to stipulate what an actor should do.[72] In others, the rules may only partly determine the range of options available to them. These indeterminacies may arise because an institution has multiple purposes, boundaries and functions that generate "intercurrence",[73] when complex interaction effects of an institutional matrix create multiple equilibria, or due to major social transformations beyond the formal political arena. The growing political significance of the President of India, who had a largely ceremonial role in the formation and termination of governments and passage of various executive orders prior to 1989, is a dramatic illustration of such developments.

In short, unless we embrace a deep structuralism that negates human decision-making or believe that institutions "come with instruction sheets",[74] political agency matters. Explaining India's coalition politics demands a reflexive form of institutional analysis, simultaneously recognizing how the macro-democratic regime shaped historic possibilities, without forgetting that party leaders had to evaluate the often conflicting incentives, opportunities and constraints it generated.[75]

Indeed, even when party leaders shared convergent goals, their perceptions of how to realize them often diverged. What formulas, strategies and tactics did they employ to realize their diverse aims and wider collective interests?

Few parties rivaled the strategic framework of the communist Left in terms of theoretical articulation.[76] Historically, its leading political formations embraced two classical Leninist strategies. The "rightest" anti-imperialist line required a coalition of workers, peasants and the petty and national bourgeoisie against feudal institutions and monopoly capitalism. The "leftist" anti-capitalist approach entailed the first three strata in battle against bourgeois nationalism. Over time Indian communist forces entertained a possible third strategy, supporting the progressive bourgeoisie along the lines of "people's democracy" in eastern Europe or "new democracy" in the People's Republic of China. Strategy concerned winning the war among classes. Hence choosing a strategic line required settling larger questions: the historical stage of capitalist development in India, the roles of different social classes in particular stages and consequently the aim of the proletariat vis-à-vis other strata. Whether the communists allied with non-communist parties through a united-front-from-above, or infiltrated the latter through a united-front-from-below, was a matter of tactics for its political organizations. Crucially, the movement cast these choices in formulaic terms, articulated well by Mao Tse-Tung: "The task of the science of strategy is to study those *laws* for directing a war that govern a war situation as a whole. The task of … the science of tactics is to study those *laws* for directing a war that govern a particular situation".[77]

Suffice to say, most parties failed to develop such an elaborate theoretical discourse. But questions of strategy and tactics consumed them nonetheless. Debates over whether to share power, with whom and to what extent, and how often led to real schisms. The main protagonists had to address these issues at two related junctures. First, party leaders had to coordinate their electoral strategies. Joint election manifestoes and common programmes allowed them to strike compromises amenable to their respective bases and steer government policy. Collective agreements imparted a measure of coherence to several multiparty alliances, highlighted their distinctive agenda and set red lines vis-à-vis more contested issues. The absence of such pacts and failure to adhere to explicitly stated parameters exacerbated latent conflicts, as comparativists find elsewhere.[78] Of course, such pacts could not guarantee solidarity. Yet joint election manifestoes and common minimum programmes stabilized expectations. The lack of either contributed to the fractiousness of the Janata Party and National Front. Conversely, every subsequent coalition government had common minimum programmes, an innovation of the United Front. Their increasing sophistication under the NDA and UPA shaped government business, demonstrating political learning.[79]

Party leaders also employed various practical techniques at the polls: permitting minimal cooperation by avoiding direct contests or allowing "friendly contests" in particular constituencies; putting up a single candidate in order to pool votes and stumping together on the campaign trail to facilitate greater coordination; and demonstrating maximal solidarity by integrating their parties into a single formation. More imaginatively, rival party leaders sometimes allowed mutual criticism and avoided joint campaigning as long as they opposed a mutual opponent. Suffice to say, India's coalition politics witnessed these stratagems employed with varying degrees of success. Several opposition parties coalesced into the Janata Party, seizing national power in 1977. Yet electoral integration failed to tame the conflicting personal ambitions between their respective leaders or deep political differences amongst their rank-and-file. Similarly, the JD allowed the BJP and CPI(M) to oppose each other stridently in 1989 as long as both sides attacked the Congress (I), allowing the National Front to capture national office. But the arrangement failed to prevent the BJP from withdrawing its external parliamentary support one year later. Still, attempts to craft a common electoral front made a real difference at several critical junctures.

Second, India's party leaders had to devise power-sharing mechanisms in government, an issue most comparativists neglect.[80] How party leaders manage their differences in office matters immensely, however. Coalition leaders need to address several tasks upon seizing office: forge interpersonal relationships, respond to new policy issues and unanticipated political events and, not least, run the government.[81] Put more colorfully, the survival of diverse governing coalitions in India "has depended on the adroitness in planning a balancing act—keeping one party happy, tolerating the whims and tantrums of individual leaders, avoid a conflict with another, and reining in the various regional parties in the coalition pulling in different directions".[82]

In theory, the locus of these functions in a parliamentary cabinet government is the Council of Ministers. But forging political agreement in large multiparty governments is difficult: the problems of enforcing collegial responsibility are multiplied when the loyalties of ministers are diverse. The sheer number of parties in India's national coalition governments intensifies the challenge. Moreover, at different junctures various parties supported a parliamentary coalition but refused to participate in government. With the exception of the Janata Party and second administration of the National Democratic Alliance, every governing coalition in New Delhi was a minority, forced to rely on outside support. Hence they had to concoct power-sharing formulas and conflict resolution mechanisms to facilitate collective decision-making, adjudicate competing demands and make binding claims.

Many relied on ad hoc devices: private bilateral meetings between key party officials. Others were regular yet informal: weekly political dinners at the prime minister's residence. Over time informal political institutions evolved, however, as often happens when pre-existing rules are inadequate.[83] The most important was the coordination committee, which acted as a safety valve as well as an integrative mechanism, providing a forum for parties to engage problems without the glare of media or parliament. At their best, it helped party leaders to "learn to play the game", recognizing "the right [of all constituents] to participate and ... be consulted in the decision-making process".[84] Crucially, coordination committees allowed outside supporters to engage with parties in government. Indeed, the United Front was the first to establish such a forum, with a higher-level steering committee too, in order to accommodate the CPI(M), which helped to construct the coalition and draft its common minimum programme but refused government participation. In several governments, disagreements over the status and functioning of the coordination committee vis-à-vis the Union cabinet hampered political consensus, deepened partisan divisions and blurred the locus of responsibility. Both the United Front and United Progressive Alliance encountered such difficulties vis-à-vis the CPI(M). Nevertheless, the fact every national coalition government in India after 1996 set up such a body testified to its relative utility and political necessity.

The necessity and difficulty of exercising political judgment

In sum, senior party leaders deeply influenced the rise, performance and fall of different coalition experiments, given their relative autonomy, the range, diversity and fluidity of interests that they had to manage and the complex interaction effects generated by India's democratic regime. Yet several conjunctural factors heightened the scope for agency. The Janata Party, National Front and United Front represented major turning points in modern Indian politics. The Janata Party, the first non-Congress government to rule the country since independence, consolidated representative democracy. The National Front, whose tenure witnessed the acceleration of liberal economic reform, popular democratic mobilization of the lower castes and the march of Hindu nationalist forces, ended the era of single-party majorities in New Delhi. Indeed, it constituted a critical juncture, a "major watersheds in political life ... which establish[ed] certain directions of change and foreclose[d] others in a way that shape[d] politics for years to come".[85] The United Front, which crystallized the idea of a third force in modern Indian democracy, deepened the logic of national coalition politics. These "structurally induced unsettled times" expanded the possibilities for "consequential purposive action" and "visions of alternative futures".[86]

To analyze party leaders' choices, I employ the concept of political judgment, drawing on the classical realist tradition of Thucydides, Machiavelli and Weber and their contemporary successors.[87] According to realists, politics demands the exercise of judgment. It is a distinct species of practical reason, which tests the capacity of actors to comprehend the causal relations of the world, which are only partly shaped by their beliefs, desires and practices, with a view to action; to distinguish the foreseen, foreseeable and unforeseeable consequences of different political decisions, whatever their intentions; and to seize the possibilities of a given historical moment:

> There is no great mystery in the formidable set of qualities, personal and political, that good political judgment demands: a clear purpose and a practical view of what has to be done to realize it; an achievable idea of how to command the power and resources to succeed, including a sensitivity to the views and likely strength of those who might support one and those who might not; a sense of how and when to tell the truth, varnish it, lie or be silent; confidence, courage, patience and a good sense of timing; the capacity to imagine the next move but one and the choices that this can present; and what, all along, might go wrong.[88]

Evaluating political choices, given the presence of uncertainty and partialities of judgment, is not easy.[89] The fact that mistakes happen, from small gaffes to costly errors, makes this clear. Hence Machiavelli's claim that fortune governs half of our political life.[90]

That said, five intellectual traits encourage good political judgment. First, it requires political actors to focus on particulars as opposed to generalizations or universals, to possess deep contextual knowledge of the situation they face in order to maximize the chances of success. The relevant context may differ: local power relations, the structure of national politics, an historical epoch. Moreover, judgment always entails relating particulars to universals, which requires the "ability to determine which ... theories abstract from crucial aspects of the situation".[91]

Second, good political judgment is pragmatic. Skilful actors exhibit a grasp of "what will work": to see a political situation "in terms of what you or others can or will do to them, and what they can or will do to others or to you".[92] It requires actors to possess a sense of timing, "to [grasp] opportunities that will not present themselves again",[93] foreseeing not simply what will happen but what will seem good to powerful others that matter.[94]

Third, good political judgment demands a synthetic form of causal understanding, "a capacity for taking in the total pattern of a human situation, of the way in which things hang together".[95] Realists view politics as complex and probabilistic: complex since the forces shaping it may be heterogeneous, reciprocal or contingent over time;[96] probabilistic because what is possible in principle, in a world that might yet exist, may not be at specific moments. Indeed, grasping the precise causal relations that define particular historical contexts frequently involves facing the contradictions of the world squarely.[97]

Fourth, good political judgment involves strategic reasoning. Actors have to consider the intentions, capacities and actions of others with partial knowledge. They may face brute factual uncertainty about states of affairs; higher-order uncertainty about the necessity and cost of resolving such factual uncertainty; indecision over what to do because of asymmetric information or the existence of multiple plausible choices; and inadequate causal understanding of how the political world operates.[98] Thus, strictly speaking, to judge is not to calculate.

Finally, good political judgment requires actors to possess a degree of detachment. On the one hand, they must have a passion for a cause. A purely instrumental politician, devoid of any substantive ends, would be a man without a soul. On the other, they must demonstrate an ethic of responsibility for the consequences of their actions, regardless of their intentions. Hence the ability to "maintain a distance from things and events" is crucial.[99]

Employing a classical realist conception of political judgment requires some justification. Indeed, comparativists that seek to elucidate strategic interactions under conditions of uncertainty normally use rational choice theories, praising their parsimony, elegance and universality. Yet they also highlight their suitability:

> It would be difficult to find a bargaining situation in which the principal actors are more familiar with each other than they are in forming governments. The party leaders involved work almost daily, often for many years, in parliament; they know one another's constituency interests; they read one another's speeches; they are aware of one another's political commitments and divisions. It is surely the case that no one observes the behavior of politicians more closely than politicians.[100]

Several observers of India's coalition politics agree.[101] Yet a realist conception of political judgment offers three relative advantages.

First and foremost, party leaders frequently evaluated decisions in the language of judgment. The CPI(M) Chief Minister of West Bengal, Jyoti Basu, famously declared in 1997 that his party's decision not to participate in the United Front government was a "historical blunder".[102] Indeed, the broader Indian left has debated whether its leading parties correctly read key historical moments since the anti-colonial movement, as the narrative shows. Yet commentators rarely analyze the exercise of judgment in depth or consider its significance, a characteristic lapse of the parliamentary left too. The decision to analyze the exercise of judgment takes its cue from the empirics of the case.

Second, rational choice theories encounter severe methodological limitations in complex settings. Strong models assume utility-maximizing actors that possess rationally formed beliefs, desires and fixed transitive preferences, as well as a complete understanding of the consequences of different courses of action, in a stable political environment.[103] Yet there are scenarios where an actor knows "the set of mutually exclusive and jointly exhaustive outcomes, but find themselves neither able to attach any (cardinal) probabilities to them", nor can say how much more likely a particular (ordinal) result is likely to be.[104] Strict instrumental rationality is very difficult in complex social environments,[105] during rapid political change[106] or over extended time horizons where "[because] the odds of correctly predicting … several moves by the other player requires us to multiply the probabilities of correctly predicting each move, problems can arise even when actors are very good at predicting specific moves".[107] Unsurprisingly, recent attempts to explain the formation and collapse of coalitions using non-cooperative game theory restrict their models to three contesting parties.[108] In contrast, the number and turnover of players and diversity and fluidity of interests in successive multiparty governments in India, not to mention their frequent minority parliamentary status, makes it hard to model their behavior without indeterminate results.

Lastly, and perhaps most significantly, rational choice theories confront inherent difficulties in grasping political mistakes, let alone explaining their causes or ramifications. In principle, models of rationality accentuate the imperative of choice, not least the dilemma facing socialists in west European democracies in the twentieth century:

> [The] very probability of committing mistakes presupposes simultaneously a political project, some choice among strategies, and objective conditions that are independent with regard to a particular movement. If the strategy of a party is uniquely determined, then the notion of "mistakes" is meaningless: the party can only pursue the inevitable.... [the] notion of mistakes is also rendered meaningless within the context of a radically voluntaristic understanding of historical possibilities ... [but] if everything is always possible, then only motives explain the course of history... "Betrayal" is indeed a proper way of understanding social democratic strategies in a world free of objective constraints. But accusations of betrayal are not particularly illuminating in the real world.[109]

Yet rational choice explanations habitually insist that "what the agent did was the best or most effective way of pursuing her purposes. And this entails establishing the embedded normative assertion [of complete rationality]".[110] Two problems arise. As behavioral economists show, actors often make irrational choices, sometimes quite systematically.[111] Moreover, many rationalist accounts paradoxically claim that actors could not have chosen otherwise:

> Was the alternative possible? ... Socialists had no choice: they had to struggle for political power because any other movement for socialism would have been stamped out by force and they had to utilize the opportunities offered by participation to improve the immediate conditions of workers because otherwise they would not have gained support among them. They had to struggle for power and they were lucky enough to be able to do it under democratic conditions. Everything else was pretty much a consequence.[112]

The specter of inevitability robs the notion of choice of meaning; indeed, it disappears. So does a great deal else too—suspense, luck and change—animating political life.[113] All that remains to be explained are rational mistakes.

Models of bounded rationality assume incomplete information, less ordered preferences and limited computational abilities, leading social actors to "satisfice" their choices based on various heuristics.[114] Yet even sober accounts acknowledge that "unintended consequences and unforeseen contingencies are [often] beyond the scope of the enterprise",[115] and continue to ignore the role of "errors ... cumulative but relatively invisible effects ... and environmental reverberations" in larger outcomes.[116] Most fundamentally, rational choice theories overlook how politics sometimes requires actors to act inconsistently, depending on context. Analyzing their actual political judgments, including their capacity to distinguish the foreseen and foreseeable from the unforeseeable consequences of action, provides a richer conceptual paradigm for grasping these issues.[117]

Consequently, synthesizing the focus of rational choice theories with a more historical sensibility may yield greater insight. The former illuminates how political actors' preferences and strategic interactions under conditions of uncertainty shape expectations, decisions and outcomes. The latter suggests that preferences may be eclectic, fluid and endogenous; that actors often lack the power to control events due to lack of foresight or because institutions shape individual behavior and power relations in asymmetric, unanticipated ways; and that consequently their choices may well depend "... on the interpretation of a situation rather than ... purely instrumental calculation."[118]

Suffice to say, analyzing the political judgments of real actors confronts its own difficulties. First, appraisals of good political judgment are context-dependent, of what seemed reasonable in light of what was known or foreseeable at the moment of decision. Yet it is often difficult to establish the true intentions of any party leader. We cannot access mental states. Moreover, they may have failed to grasp adequately their own motivations due to limitations all human agents confront. Lastly, party leaders frequently dissimulate, either to protect their reputations and careers or due to the political sensitivity of the subject matter.

Second, good political judgment is neither necessary nor sufficient for success. In some instances, actors may correctly perceive political possibilities but lack the requisite power, skill or determination to realize their desires. In other instances, they may display poor judgment yet secure their aims, either because more powerful forces abetted their designs, deliberately or otherwise, or due to contingencies they neither had foreseen nor could have. We can only resolve these questions through rigorous process tracing, evaluating plausible counterfactuals in light of general theoretical principles and specific historical knowledge, a task that causal narratives generally face.[119]

Lastly, since good political judgment demands contextual reasoning, wider theoretical inferences are harder to draw. For rational choice theorists, "[what] makes a tale compelling is that the causal mechanisms it identifies are plausible ... [which involves] demonstrating their generalizability to other contexts ...".[120] As a result, "rationalists are almost always willing to sacrifice nuance for generalizability, detail for logic, a forfeiture most other comparativists would decline".[121] The belief that universal causal relations govern the world justifies the claim that valid arguments must be general in scope. Conceptual parsimony and theoretical ambition produce maximum explanatory leverage.

Fortunately, narratives provide a very useful technique for addressing some of these problems. If done well, they supply "diverse forms of internal evidence" that mitigate selection bias.[122] A narrative offers a critical plausibility test to examine the presuppositions underlying competing theories, uncover new facts and develop critical redescriptions of previously studied phenomena.[123] Quantitatively-oriented methodologists agree that valid causal explanations require good descriptive inference.[124] But describing events accurately, whether they suggest larger patterns or not, is necessary too.

In addition, narratives facilitate process-tracing, linking causes, mechanisms and effects and generating insights into possible auxiliary outcomes.[125] Such processes take various forms: linear isolated mechanisms that generate constant effects; the concatenation of actors, decisions and structures that produce complex causal chains; path dependent processes in which early contingent events mold historical outcomes over the *longue durée*, to name a few. "The very act of producing an account [of the past] ... virtually requires an often counterfactual neatness and coherence ... with an air of inevitability being given to an act that may have been highly contingent."[126] Uncovering the complex causal chains that may have produced larger outcomes minimizes such illusions.

Sources and methods

Ultimately, a compelling narrative demands good detective work, resting on the quality of observations and inference.[127] Accordingly, this book relies on a variety of methods and sources. Reportage and commentary provided by newspapers and periodicals, and information from politicians' writings, party manifestoes and public government documents, comprise its foundation. Statistical analyses of official electoral data as well as national election surveys by the Centre for the Study of Developing Societies (CSDS), Delhi, provide further insights into the competing parties' electoral fortunes, policy profiles and social bases. Finally, and most importantly, the narrative contains the rare confidential testimonies of senior party leaders, high-ranking state officials from the bureaucracy, judiciary and police, and seasoned political journalists, elicited through dozens of in-depth semi-structured interviews. These interviews help to elucidate the rise and fall of the third force in modern Indian democracy, specifically its politics in the 1990s, while shedding new light on other coalition experiments.

There are well-known advantages and shortcomings of employing such a methodology. Apart from the credibility of different press accounts, skeptics might allege that "first drafts of history" engender too much speculation and not enough knowledge.[128] "Newspapers are the second hands of history," insisted Schopenhauer, "they always tell the wrong time".[129] These caveats must be kept in mind. Yet waiting for the declassification of secret government documents inspires greater speculation, not less, with observers projecting their own predilections into debates over what really happened. Moreover, by using contemporaneous sources of information, we can try to uncover what motivated the chief protagonists as events transpired, not merely what appears significant to us now. To minimize the vagaries of memory and ambition, the narrative balances the insights gleaned from high-level interviews with many other sources of evidence, not least a lively, diverse and erudite media. Fortunately, considerable time has passed, allowing a more disinterested perspective, especially helpful when judging party leaders' judgments. In the end, writing such a narrative entails such risks and dealing with them as best possible.

Using off-the-record claims by key political informants has obvious disadvantages too, which perhaps explains why few comparativists do.[130] It is hard for others to replicate the analysis or verify its inferences exactly. The difficulty of fully recovering actors' motivations, given the possibilities of dissimulation and incomplete self-understanding, exacerbates the problem. Still, it deserves effort. Treating the intentions of real political actors as revealed preferences or conceptualizing their motives simply via theory, as most rationalist accounts do, skirts these challenges while raising problems of its own.[131] By granting confidentiality to key protagonists, moreover, we arguably increase their likelihood of imparting genuine observations. Lastly, no Archimedean vantage exists. We simply have to assess "diverse, complex and sometimes conflicting" claims, judging their credibility, plausibility and trustworthiness according to the best practices of empirical verification.[132]

Ultimately, recognizing the role of judgment in politics carries a significant implication. It requires us to exercise good political judgment ourselves: to ask what it was possible for the actors we study, in the circumstances in which they found themselves, to reasonably do, analyzing the foreseeable consequences of different political decisions by reconstructing the context of action in time, not as spectators after the fact, as faithfully as we can.[133] If done well, our explanations should resist temptations of abstract moralizing and easy historical judgment as well as tales of necessity and flights of fancy, demonstrating the possibilities and constraints surrounding real political events as they actually happened.

PART I

THE GENESIS OF THE THIRD FORCE

2

THE ROOTS OF THE BROADER INDIAN LEFT (1934–1977)

The desire to construct a progressive third force in modern Indian democracy, vis-à-vis the traditionally dominant Congress and Hindu nationalist BJP, acquired political momentum in the late 1980s. Yet its conceptual origins lay in the anti-colonial struggle during the 1930s. The principal catalyst of the third force was the broader Indian left, which sought to establish democratic socialism after the country gained independence in 1947. Many of the diverse multiparty coalitions put together by forces on the left encompassed political formations from the right, however, especially from the late 1960s onwards. Hence the endeavor to create a progressive third force was convoluted and haphazard from the start, shaped by the structure, character and dynamics of India's nationalist movement and the party system that emerged in its wake. Three distinct phases, which reflected deeper systemic changes, marked the evolution of the latter.

The Congress system

The first, which lasted from the first general election in 1951–52 until the fourth in 1967, was christened the "Congress system".[1] It saw the Indian National Congress, which had led the country to self-rule, define the contours of electoral struggle and political debate. Three factors accounted for its supremacy. First, the Congress was a distinctive centrist party, the successor to an encompassing nationalist movement. Its inclusive ideology, open recruitment strategy and weak organizational structure enabled the party to achieve political sovereignty.[2] Politically, the Congress ran the spectrum of ideologies, from socialists on the left and liberals in the middle to conservatives on the right. Socially, the party encompassed a grand coalition of social groups, whose diverse identities and interests cut across gender, class and caste, as well as language, region, religion and tribe.[3] Indeed, its political centrism and social heterogeneity forced opposition parties to adopt strategies of pressure: to influence sympathetic factions in the Congress to advance their respective agendas.[4] Hence the claim by the latter that it embodied the nation.

Second, the Congress pledged to build a secular federal democracy committed to achieving broad social development.[5] Politically, the party contested for power in a single-member simple-plurality (SMSP), or first past the post (FPTP), electoral system based on universal adult franchise. Despite his stature as India's first prime minister, Jawaharlal Nehru respected the authority of state and district-level elected representatives and party officials over their respective domains. Moreover, the Congress largely recognized the legitimacy of leading Opposition figures. Economically, the party unveiled a series of Five-Year Plans, which provided blueprints for achieving rapid development through import-substitution-industrialization (ISI), public sector dominance and pervasive government control over the private sector. Like many developmentalists in the 1950s, Nehru believed planned economic development would secure national self-reliance and geopolitical security while alleviating mass poverty. Lastly, the Congress introduced various measures to empower historically disadvantaged groups. The 1950 Constitution sanctioned reservations (numerically proportionate quotas) for formerly untouchable castes, or Dalits, and various indigenous peoples, or Adivasis, as Scheduled Castes (SCs) and Scheduled Tribes (STs), respectively, in educational institutions, legislative assemblies and the public sector. The Congress also introduced a model of secularism that sought to contain the menace of potential Hindu supremacy.[6] It reformed a number of traditional Hindu practices that violated the dignity of lower-caste groups—banning untouchability, allowing Dalits the right to enter temples, legalizing inter-caste marriage, divorce and equal inheritance rights for daughters and sons, and prohibiting the practice of polygamy—while granting minority religious communities autonomy over their own customary laws.

Third, the Congress' apparatus comprised a series of "vertical faction chains" that competed for power within the party across the country.[7] Proprietary high-caste notables mobilized vote banks and distributed patronage in the districts. The political bosses that ran Pradesh Congress Committees (PCC) elected its organizational leaders and influenced the workings of legislative assemblies in the states. Their proximity to and knowledge of local affairs allowed the party apex to gather information from and distribute patronage to the peripheries.[8]

Taken together, these three factors enabled the Congress to mediate conflicts of interest within the party, formulate consensus by absorbing rival claims, and occupy the centre of gravity in the wider political arena. Crucially, the party transformed its vast political capital into electoral prowess. The Congress won a majority of seats in every parliamentary election, and virtually every state assembly contest, from 1951 to 1967 (see Table 2.1).

It never exercised political hegemony, however. Indeed, the combined popularity of the opposition always surpassed the Congress' own vote share. Many parties challenged its presumption to rule, demanding a national imaginary and developmental agenda that carried greater possibilities of deeper social transformation.[9] Despite its relatively progressive rhetoric, the Congress' politics of accommodation frustrated its more ambitious goals. The capitulation to big business in the run-up to independence when the party, facing massive strikes organized by the All India Trade Union Congress (AITUC), passed legislation that limited strikes and collective bargaining, and split the labor movement through the creation of the Indian National Trade Union Congress (INTUC).[10] Following independence, the party accommodated propertied interests further through various measures.[11] It failed to reform the Indian Civil Service (ICS), the colonial bureaucracy designed to ensure political stability, simply renaming it the Indian Administrative Service (IAS). The Constitution recognized private property as a fundamental right but relegated socioeconomic entitlements to the unjusticable Directive Principles of State Policy and made agriculture a responsibility of the states within the Union. As a result, the dominant social elites that mobilized the Congress' vote banks and ran its state-level administrations were able to protect their interests against the more radical designs being hatched in New Delhi. Hence most states abolished the *zamindari*, the large feudal landlords that had collected taxes under the Raj, but stymied the redistribution

Table 2.1 Electoral Performance of Partisan Blocs, 1951–1977[12]

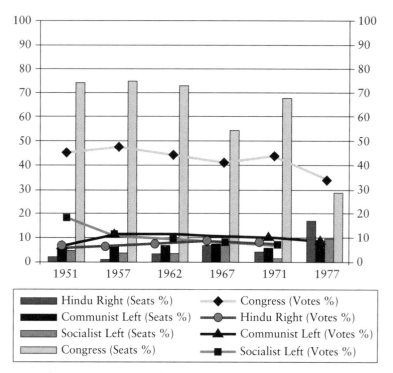

Source: Election Commission of India. Author's calculations.

of land to the tiller. The socialist rhetoric of the party, an organization "enmeshed in society", remained largely unfulfilled in practice.[13]

In addition, the Congress failed to uphold its professed democratic commitments vis-à-vis certain opposition parties and regional movements. The party refused to hold a plebiscite on independence in Jammu & Kashmir, suppressed Naga separatist groups in the Northeast and dismissed the communist government in Kerala in the late 1950s.[14] In addition, the Congress reneged on its longstanding pledge to refashion the provinces of the Union into linguistically organized states upon securing independence. The frenzy of violence unleashed by Partition convinced many in the party high command, especially Nehru himself, to maintain order and stability. It was only in the face of a genuine popular movement demanding the creation of a majority Telugu-speaking state of Andhra Pradesh that New Delhi agreed to the States Reorganization Committee in the 1950s and 1960s. Constitutionally, however, India remained a quasi-federal democracy, concentrating political authority and economic resources at the Centre. Moreover, the Congress' internal power-sharing mechanisms paradoxically allowed various federal institutions to wither from relative neglect.[15] Thus its national vote share never crossed 50 per cent even during its heyday. The party faced serious electoral rivals across the country.

The most important members of the Opposition, comprising the two main flanks of the broader Indian left, were the Praja Socialist Party (PSP) and Communist Party of India (CPI). The origins of the PSP lay in the Congress Socialist Party (CSP), formed in the mid-1930s by Narendra Deva and Jayaprakash Narayan in an effort to radicalize its parent organization during the nationalist movement. In the beginning the CSP, a radical Marxist formation committed to workers' struggles, had revolutionary aims. It objected to the Congress' participation in the 1937 provincial elections and subsequent decision to take office, as well as the 1946 Cabinet Mission Plan, seeing both as a ruse by the Raj to emasculate the anti-colonial struggle.[16] The radicals' stock rose dramatically after they took the lead during the Quit India movement, which had seen many of the Congress' leading figures jailed. But the transformation of the latter from a national movement into a political party, whose new constitution disallowed the persistence of separate party organizations within its ranks, allowed conservatives to oust the CSP in 1948. The coup de grâce was complete when Nehru, a Fabian socialist, abandoned his sympathies for the sake of party and nation.[17] Yet national independence, and Gandhi's assassination by Hindu extremists, also convinced the Socialists to embrace democratic politics and forsake violence. Following its relatively disappointing parliamentary tally in the first general election in 1951, the Socialists merged with the Kisan Mazdoor Praja Party (KMPP), a splinter group of Gandhian dissidents led by former Congress president Jivatram Bhagwandas Kripalani, creating the PSP.[18]

According to Ram Manohar Lohia, one of its leading figures, the Congress' Westernized elite could not help India achieve *saptakranti* (a seven-fold revolution). Complete national independence required genuine social equality based on gender, race and caste, a polity that permitted maximal civil disobedience, democratic rights and national sovereignty, and "maximum achievable economic equality".[19] The PSP advocated meeting the basic needs of the poor through the promotion of labor-intensive agriculture, small-scale technologies and cottage industries, and wholesale political decentralization. The party also sought to banish the use of English and ensure proportional caste-based reservations in public sector institutions.[20] And it championed a "third force" approach in international relations, demanding strict neutrality between the United States and the Union of Soviet Socialist Republics (USSR).[21] The combined popular vote of the larger socialist camp in the 1950s, which reached approximately 10 per cent, was second only to the Congress.

The electoral presence of the CPI, which had formed in the 1920s during the second world congress of the Comintern, was equally significant. Unlike the socialists, the party had suffered far greater repression during British colonial rule, and was banned between 1934 and 1942. The CPI also underwent dramatic strategic reversals during the national movement. The coming of independence led the party general secretary, P.C. Joshi, to advocate an anti-imperialist strategy on grounds that the Congress was a "popular" formation and Nehru a "progressive".[22] Fierce internal disputes led to his downfall two years later, however. According to B.T. Ranadive, his successor, Joshi's tactical line represented a "right reformist deviation": "[the] so-called transfer of power was one of the biggest pieces of political and economic appeasement of bourgeoisie" and "does not mean that the Indian people have won freedom or independence, nor does it insure [sic] they will be moving in the direction of democracy".[23] Indeed, the CPI adopted an anti-capitalist strategy in 1948, denouncing the "slave constitution" of the new republic for codifying "fascist tyranny".[24] On the one hand, the party called for a national rail strike and general insurrection on 9 May 1949, invoking the Russian path to socialism. Mass police arrests and military suppression inflicted an ignominious defeat, however, compelling Ranadive to resign. On the other, his successor, Rajeshwar Rao, saw wider revolutionary potential in the agrarian revolt in Telengana in the southern region of Hyderabad, calling for direct guerrilla action along the Chinese path in June 1950. Yet it too was overwhelmed. The army crushed the revolt within a year, while Congress passed the 1950 Preventative Detention Act, which targeted communists in particular. The "fiasco" reflected a "misreading of the historical conjuncture".[25] Arguably, the cause lay deeper. The "illusionary revolutionary visions" of the CPI "prevented it theoretically from coming to grips with reality [that India had become an independent democratic state]."[26] The tendency of India's communists to misjudge the might of the postcolonial Indian state, as well as the possibilities of its democratic regime, would become a perennial point of contention.

Democratic politics and national independence forced the CPI to adopt a new political thesis at its 3rd party congress in 1951. The party argued that India remained a satellite run by semi-colonial rulers at the behest of a dependent bourgeoisie. Hence the Russian path of urban working class insurrection as well as the Chinese path of rural guerrilla warfare remained tactically relevant.[27] But the CPI also recognized the importance of mobilizing popular classes and fighting for change through formal democratic institutions. Thus it simultaneously contended that a "national democratic front", comprising workers and peasants as well as the petty and national bourgeoisie, could usher in a "national democratic revolution".[28] The party's new strategy paid quick dividends. In 1957, the CPI emerged as the single-largest party in state assembly elections in Kerala, becoming the first democratically elected communist government in the world. The party introduced far-reaching legislation concerning land reform, educational institutions and industrial relations.[29] Its momentous victory in Kerala bolstered rightists in the party who dreamed of seizing national power through the ballot. Indeed, it raised the possibility of mounting a wider popular front. Lohia repeatedly characterized the Congress as a "motley" political coalition, held together by "87 year old tradition, the inherited habit of working together and the shared loyalty to the Nehru-Gandhi family".[30] He believed that non-Congressism—fielding a single opposition candidate against the Congress in each district—would dispel its aura of supremacy. The combined popular vote of the socialists and communists, roughly 20 per cent in every general election from 1951 to 1967, made it plausible.

Yet the Opposition failed to dislodge the Congress in any other state, let alone New Delhi, in these first two decades. The primary reason was the division of the broader Indian left. The differences between the communists and socialists erupted during their very first attempt to forge a united front in the CSP when Narayan invited the CPI to join its ranks in the mid 1930s.[31] A significant portion of the CPI remained wary, reflecting the views of their founder, M.N. Roy, given the Congress' bourgeois elements.[32] However, P.C. Joshi convinced his comrades to join. The evolving strategic line of 7th Comintern instructed communists to find progressive national allies in the anti-imperial struggle. Labor militancy and peasant mobilization were also on the rise. The collaboration bore immediate rewards: the formation of All-India Kisan Sabha (AIKS), the first national organ of the peasantry, whose radical agenda influenced the Congress' manifesto for the 1937 provincial elections.[33]

The radical socialists' rise within the Congress met stiff opposition, however. Conservatives isolated their forces. But significant cleavages divided the socialists and communists too. The most obvious were ideological. The socialists abhorred the single party hegemony of the USSR.[34] Conversely, many leading figures in the CPI viewed Gandhi and his followers as reactionaries, given the Mahatma's prescription of *satyagraha* (non-violent direct action), belief that personal ethical transformation and upper-caste trusteeship could eliminate untouchability and deep social exploitation, and valorization of local self-sufficient development in the countryside.[35] Many socialists were equally skeptical that constructive work could overturn rural power relations. Gandhi's emphasis on indigenous social reform appealed to them, however.

An equally significant fissure was the question of power-sharing. Communists quickly dominated the ranks of the CSP, AIKS and its trade unions, and soon controlled a number of party's state-level units in the south, leading the latter to join the CPI. The socialists accused the communists of conspiring to take over the CSP; the communists castigated their new associates for "organiz[ing] a witch-hunt".[36] Narayan sought to broker peace by offering one-third of the seats on the CSP's national executive to the CPI, but it allegedly wanted half. The discovery of two documents, declaring the CPI as the true custodian of Indian socialism and outlining its plans of capturing the CSP, intensified socialists' fears. The perception that the communists were pursuing a united-front-from-below damaged the possibility of a united-front-from-above.

Finally, sharp differences arose over the character and strategy of the Congress on the eve of World War II, dealing a fatal blow. According to two CPI leaders, Puchalapalli Sundarayya and Makineni Basapunniah, the idea of Congress socialism was a contradiction, presaging fascism.[37] Socialists retorted that the German communists' failure to form a united front with their socialist comrades enabled the rise of Hitler. The non-aggression pact between the USSR and Germany in

1939 led, according to E.M.S. Namboodiripad, to "a hysterically anti-Soviet and anti-Communist campaign" by Lohia, Narayan and others.[38] Yet the CPI's decision not to join the Quit India movement called by Gandhi in 1942, on grounds that Britain had to be supported after Germany invaded the Soviet Union, deepened the socialists' long-held suspicion that the communists were ultimately anti-national forces. Gandhi's assassination, the violence unleashed by Partition and the specter of national disintegration convinced them to abandon militant class struggle.[39]

Relations improved somewhat in the 1950s. The two sides frequently brokered local alliances in state assembly elections. Sporadic attempts at deeper political rapprochement also occurred. Khrushchev's famous denunciation of Stalin at the 20th congress of the Communist Party of the Soviet Union (CPSU) in 1956 created an opening. The CPI published an article by Narayan, asking how the party could have been unaware of his crimes, in the *New Age*, its main journal.[40] Its election in Kerala in 1957 provided greater impetus. Visiting the newly inducted Chief Minister, Namboodiripad, the Gandhian Vinoba Bhave declared that "communism and socialism, like the Jumna and Ganges rivers, could join together in the ocean of *Sarvodaya* [collective welfare through constructive work]". The party responded by publishing Bhave's articles in *New Age* too.[41] The CPI had stitched together a broad coalition of tenants and laborers, comprising distinct class, anti-caste and nationalist mobilizations that had marked the peculiar agrarian history of the region over many decades, successfully translating "popular idioms of moral outrage and social justice ... [into] a class of the agrarian poor".[42] Indeed, caste and community comprised "the geology of the party structure", influencing the CPI's selection of candidates in the 1950s.[43] The fusion of these distinct movements, however, was "more a product of spontaneous oppositions than theoretical insight".[44] Differences over official communist ideology and anti-caste strategies persisted. The invasion of Tibet by the People's Republic of China in 1959, which the radical wing of the CPI supported, led Narayan to denounce "Chinese aggression" as "the first phase of a new imperialism", ridiculing Lenin's thesis.[45] By the late 1960s, with the exception of Kerala, the leadership of the CPI remained the preserve of highly educated upper-caste elites.[46]

The second major reason why the broader Indian left failed to dislodge the Congress in any state except Kerala was the inability of either the socialists or communists to maintain their cohesion as parties. High-level desertions and intra-elite disunity afflicted the former severely. Narayan's decision to join Bhave's *bhoodan* movement shortly after the PSP formed, restricting his involvement in formal politics to writing letters to Nehru periodically and supporting electoral candidates, robbed the party of its leading political figure.[47] But as damaging was the tendency of its other leaders to split and reunite the party with lamentable frequency.[48] Disagreements over whether to cooperate with the Congress in the wake of its Awadi declaration, which called for "establishment of a socialistic pattern of society", split the PSP in three in 1955. Lohia, adamantly opposed to the Congress, created a new Socialist Party. His erstwhile colleague, Ashok Mehta, whose political thesis, "political compulsions of a backward economy", advocated supporting the main bourgeois formation, joined the Congress, taking one-third of the PSP's cadre with him.[49] The two sides reunited in 1964, following their weak electoral performances in the 1962 general election, as the Samyukta Socialist Party (SSP). The latter split merely one year later, however, reviving the PSP. Self-destructive tendencies gradually destroyed India's second largest parliamentary force.[50]

Three factors caused the socialists to implode. Individual skirmishes bedeviled its ranks. Principled differences divided its leadership. Yet personal ambition and the struggle for power ultimately proved decisive.[51] A significant underlying factor, however, had to do with the fissures of caste. On the one hand, rivalries within particular caste segments undermined the SSP. Lohia bemoaned that "ever and ever again, the revolt of the down-graded castes has been misused to upgrade one or another caste rather than destroy the entire edifice of caste".[52] On the other, the PSP's aversion to promoting social justice by expanding caste-based reservations reflected the dominance of highly educated upper-caste urban professionals.[53] Finally, differences over the Congress created much dissension. The SSP maintained a consistent anti-Congress stance, in contrast to the PSP, which the Congress partly absorbed. Indeed, this was perhaps "the basic issue" dividing the two socialist groupings.[54]

The CPI exhibited far greater cohesion.[55] The party enjoyed a powerful organizational apparatus with clear decision-making procedures, and a large and committed full-time cadre. Moreover, it followed the Leninist principle of democratic centralism, enforcing party unity.

Finally, both the leadership and rank-and-file displayed a strong ideological commitment to creating a communist society. These powerful assets enabled the CPI to avoid the ruinous internal splits that befell the socialists through the 1950s.

Nonetheless, the party struggled over the Congress question too. Fundamental strategic and tactical issues, of whether to seek a non-capitalist transition to socialism through electoral struggle and partisan alliances with progressive social forces in the formal parliamentary arena or to support popular militant insurgencies against the newly independent state in India, eventually tore the CPI apart. The Congress' Awadi declaration and professed non-alignment in the mid-1950s, which coincided with Khrushchev's declaration that a peaceful non-capitalist path to socialism was possible, buttressed voices advocating political cooperation. The party endorsed the latter at its 5th congress in Amritsar in 1958.[56] A concatenation of events at home and abroad exacerbated unresolved tensions within the CPI, however, between its rightists, who comprised many of its theoreticians, leftists, who largely controlled the organization, and centrists, who represented the party electorally.[57] Domestically, Nehru's failure to persuade the Congress to support the Nagpur resolution, which advocated land reform and cooperative agricultural production, and the growing food crisis in West Bengal stoked leftist agitation. The notorious dismissal of the CPI government in Kerala in 1959 by New Dehli underscored the hazards of cooperation. Several factors led the Congress to impose President's rule: concerted political opposition from the Church, landlords and its state-level unit to the CPI's legislative reforms, growing social conflict on the ground, and machinations by Nehru's daughter, Indira Gandhi.[58] But the prime minister's growing hostility to the Soviet Union, following the breaking of ties with Yugoslavia in 1948 and invasion of Hungary in 1956, also played a role: the CPI had officially endorsed both moves.

Indeed, the CPI's domestic predicament became entangled with wider international conflicts. The first Sino-Indian border clash in 1959, alongside mounting Sino-Soviet tensions and the Chinese invasion of Tibet, had exacerbated anti-communist sentiment in India. The stronger electoral showing of the CPI's leftists in the 1962 general election, just before the Sino-Indian war, only deepened tensions within the party. Rightists supported Nehru; leftists denounced him. The rightists entrenched themselves within the party after Nehru jailed many leftists under the Defence of India Rules for their allegedly treasonous act. The leading centrists of the CPI, Jyoti Basu and E.M.S. Namboodiripad, attempted to avert a split. Indeed, they even suggested establishing a "new third force" to reconcile their differences.[59]

Yet their efforts were in vain. The party split at its 7th congress in 1964. The pro-Chinese leftist wing rechristened itself the Communist Party of India (Marxist) (CPI(M)). The rightist CPI resolved to obtain power, overcome the domination of the CPI(M) and attract a mass following through a national democratic front with the "progressive section of the Congress".[60] Ultimately, three factors were to blame. Primary was the question of how to relate, strategically and tactically, to the Congress. The growing antagonism between the PRC and USSR, which formally triggered the division, was a secondary matter. Yet these ideological-strategic differences within the party also reflected, to some extent, the underlying material strength of each faction in these years. The rightists hailed from Uttar Pradesh and Maharashtra, regions where the communists had little political support. In contrast, the leftists came from regions where the CPI was relatively strong, such as West Bengal, Kerala, Andhra Pradesh and Punjab.[61] Genuine ideological fissures within the CPI(M), however, would shortly challenge such materialist explanations. Ultimately, the primary division within the broader Indian left was the Congress, underscoring its centrality.

Lastly, the third major reason why the Congress dominated the party system in these first two decades was the country's electoral regime. Plurality-rule elections normally encourage two-party systems.[62] But the fragmentation of the Opposition into "an inchoate front" at the national level, comprising an amalgamation of different state parties based in particular regions with relatively narrow bases of support, lowered the threshold of winning.[63] The logic of first-past-the-post elections in a diverse multiparty system enabled the Congress, facing a disunited parliamentary left, consistently to muster a plurality of votes across the Union in the 1950s and 1960s.

The breakdown of the Congress system

Its fabled political machine encountered difficulty, however, in the fourth general election in 1967. Two general processes of "awakening" and "decay", working dialectically, explained the emergence a "dominant multiparty system".[64] A variety of groups began to contest the

Brahminical upper class structure of authority that upheld the Congress' vast patronage network. Economically, *zamindari* abolition and the beginning of the Green Revolution enhanced the strength of the middle and rich peasantry, reconfiguring class relations in the countryside.[65] Politically, members of *shudra* castes, *Dalits* and *Adivasis* realized the potency of their votes and signaled discontent with their subordination.[66] In addition, successive crises struck the country in the 1960s. Some were natural: drought wreaked havoc across the countryside in 1965 and 1966. Others, such as India's humiliating defeat by China in the 1962 border war, were man-made. But the rise in inflation and corruption that marked these years, forcing New Delhi to rely on external food aid and devalue the rupee, revealed the deeper systemic inadequacies of the Congress system. Nehru's death in 1964 led the party's state-level bosses, the so-called Syndicate, to choose Lal Bahadur Shastri as his replacement. Shastri's unexpected demise two years later saw them install Nehru's daughter, Indira Gandhi, in order to signal dynastic continuity while retaining political control. Yet it failed to stem the country's wider transformations. The rising political consciousness and socioeconomic aspirations of intermediate proprietary castes and historically subordinate classes culminated in the fourth general election in 1967. For the first time, the Congress' proportional tally of seats in the states fell below 50 per cent, revealing the breakdown of its extant patronage network.[67] Moreover, the crises of the 1960s had made issues more salient at the polls.[68] The demise of the Congress system allowed a variety of non-Congress governments to form in eight important states. Three patterns emerged.

In Tamil Nadu, the Dravida Munnetra Kazhagam (DMK) won a legislative majority, becoming only the second ethnic party to form a state-level government since independence.[69] The DMK was a legatee of the egalitarian self-respect movement that had mobilized non-Brahminical groups in southern India from the 1920s, a reaction against the colonial ideology and positive discrimination policies of Raj, and the domination of the north over the south.[70] In particular, the party had successfully used anti-Brahmanism and the Tamil language to forge a broader "ethnicized" identity of Dravidianism.[71] The party frequently entered local electoral alliances with the socialists and communists, starting with the United Democratic Front in the 1951 state assembly election.[72] Its decision to support the growing anti-Hindi agitation by students and lawyers seeking "polity-wide careers" in the mid-1960s, and the party's growing organizational links with the regional film industry and reach through the countryside, catapulted the DMK into office.[73] The decision to rename the state of Madras as Tamil Nadu symbolized its ethno-regional demands.

Every other non-Congress state-level administration, however, required a coalition of parties. The communist parties played a major role in two states. In Kerala, the CPI and CPI(M) temporarily overcame their mutual hostility to form a united front government, mediated by a coordination committee. Despite being deposed in 1959, the leftist faction of the CPI had organized "near-continuous popular mobilization and open class struggles", ushering the demise of "landlordism, the attached labor system and caste domination".[74] Following its split, the CPI(M) and its secondary organizations captured temple festivals and reading rooms, shaping new literary forms in "the trenches of civil society".[75] These political struggles forced the Congress to improve wages, working conditions and social entitlements through incremental legal reform.[76] In West Bengal, the CPI(M)-led United Left Front (ULF) forged a second united front government with People's United Left Front (PULF), comprising the CPI and dissident Congress leader Ajoy Mukherjee. The formation of two left fronts betrayed mutual suspicion. The CPI(M) characterized the PULF as a coalition of "vested interests" and "revisionists". For hardliners like B.T. Ranadive, united front administrations were simply a "weapon of struggle".[77] Nonetheless, criticizing the "dogmatic, sectarian and wrong attitude" it had taken in the past, the party also proposed to table progressive legislation, while mobilizing pressure outside to ensure their passage.[78] Crucially, the CPI(M) enjoyed the single largest block of seats and a majority in the ULF itself, conferring substantial influence.[79]

The majority of these new state-level governments, in contrast, required the formation of motley political coalitions. The People's United Front that emerged in Punjab, encompassing the CPI, SSP and CPI(M) on the left as well as the factions of the Sikh-oriented Akali Dal and the Hindu chauvinist Jan Sangh on the right, was perhaps the most incoherent. More significant for the project of constructing a national third force were the Samyukta Vidhayak Dal (united parliamentarian group, SVD) coalitions that took office in Uttar Pradesh, Madhya Pradesh and Bihar. Socially, the SVD represented aspiring backward castes that sought power commensurate with their growing economic clout: Jats in Haryana and Uttar Pradesh, Kurmis and Koeris in Bihar, and Lodhs in Madhya Pradesh, as well as Marathas and Vokkaligas in Karnataka, Vellalas in Tamil Nadu, Reddys and Kammas in Andhra Pradesh and Ahirs/Yadavs across various states.[80] Politically, the composition of each SVD coalition varied state to state, ranging from PSP, SSP and CPI on the left and the classically liberal Swatantra Party in the middle to the Jana Sangh on the right, with discontented former Congressmen in between. The most important amongst the latter was the Bharatiya Kranti Dal (BKD), formed in 1967 by Charan Singh, the towering Jat leader who represented the rich agricultural proprietors in rural north India. A longstanding Congressman who held senior ministerial posts in Uttar Pradesh, Singh had major policy differences with the party, selectively implementing land reform in the state and subverting Nehru's proposal to form agricultural cooperatives in the 1950s.[81] But his basic reason for leaving the Congress was frustrated political ambition: he was unable to rise through its ranks. According to Lohia, the premise underlying the SVD coalitions was twofold. On the one hand, despite their limited programmatic appeal, power-sharing allowed its parties to secure a plurality of votes and thus take new political risks.[82] On the other, Lohia believed that "working together [might] smoothen out the extreme edges of fanaticism".[83] The emergence of several non-Congress governments furnished a major political opportunity to advance progressive causes.

Alas, few did. In Kerala, the Left Front introduced the 1967 Land Reforms Amendment Act, which radicalized the Congress' earlier version, conferring broad ownership rights and, to a lesser extent, redistributing surplus land.[84] But disagreements between the CPI and CPI(M) over industrial strategy, labor relations and agricultural reform, as well as growing partisan conflicts over the distribution of patronage and charges of corruption, undermined their truce by 1969.[85] Charging the CPI(M) with "super-bossism" and "political gangsterism", the CPI joined hands with the Congress, compelling the Namboodiripad ministry to resign.[86] In West Bengal, persistent animosity between the two allowed a breakaway faction of the Bangla Congress to briefly lead a Progressive Democratic Front ministry.[87] The CPI(M) led a second united front administration in 1969 after the Congress' machinations undermined the first. Controlling the ministries of home, labor and land, the party ordered the police not to interfere in class struggles in the cities—where labor militancy secured better wages, the registration of new trade unions, and the legalization of *gheraos*—as well as the countryside—which saw efforts to register sharecroppers and grab *benami* lands. But escalating political violence against landlords and competitive radical mobilization by its coalition partners, which accused the CPI(M) of exercising "hegemony", forced the latter to unleash the police.[88] The most damaging schism occurred in Naxalbari in northern West Bengal, where leftists from the CPI(M) had organized tribals and peasants over the right to land and the forests through the Krishak Samiti (Peasant's Organization). The so-called Naxalites accused the party high command of "neo-revisionism"; the latter responded by accusing their comrades of "sectarianism" and "adventurism". Attracting the participation and support of many urban middle class youth in the face of mounting state repression, the Naxalites formed the Communist Party of India (Marxist-Leninist) (CPI(ML)).[89] New Delhi responded by imposing President's rule, inaugurating a succession of administrations that unleashed brutal state repression.[90]

The collapse of both united front experiments reflected conventional ideological differences. Yet it also revealed the deeply absolutist conception of power that infused the CPI(M)'s political vision. The leadership of virtually every party that participated in all these non-Congress administrations held that strategic maneuvers and tactical decisions shaped the balance of power within their specific alliances. The CPI(M) was the only party that believed that its vote tally and seat ratio determined its relative coalition power.[91] The party's reluctance to share formal power unless it commanded a majority would merely intensify in the coming decades.

Finally, the SVD governments in northern India proved deeply unstable, viewed as "highly personalized clientelistic parties with extremely weak organizational structures".[92] Ideological differences—over the status of English, agrarian issues such as land revenue and food grains, and Hindu chauvinism in general—divided their constituents.[93] The vacillation of various SVD administrations in the face of rising communal violence engendered growing social turmoil too.[94] Yet most damaging was the desire for money or office by individual legislators causing repeated splits and defections:

[the] irony of the failure of the Indian Socialist movement is that it has not disintegrated because it could not achieve power, but because its leaders could not agree on the appropriate tactics to achieve [maintain] power when it became available ... [and thus] offer no long term prospects for themselves or for their supporters of playing a major role in bringing about political or social change.[95]

Madhu Limaye, a leading socialist, agreed: "They were their only enemy".[96] The power-mongering that bedeviled the SVD experiments inscribed themselves in public imagination and scholarly analyses in equal measure.

The road to Emergency rule

In the short run, the fortunes of the Opposition worsened. Alarmed by the Congress' waning electoral prowess and seeking to buttress her own position, Mrs Gandhi pursued two strategies. Initially, the fledgling prime minister tacked ideologically to the left. In 1969, she supported the CPI's candidate for the presidency, V.V. Giri, against the wishes of the Syndicate, effectively splitting the party into two, with the latter now in charge of the Congress (Organization, O), pitted against her new vehicle, the Congress (Requisitionists, R). Mrs Gandhi also nationalized the country's fourteen largest banks after deposing the Syndicate's putative leader, Moraji Desai, as the Union finance minister. Her maneuvers won the support of the Congress' so-called Young Turks, who had earlier pressed Mrs Gandhi to accept their Ten Point Programme, which sought to nationalize various economic sectors, ensure public distribution of food grains, impose limits on urban property and rural landholdings, and abolish the princes' privy purses. They even advocated "united action" with all "progressive" forces versus monopoly and feudal elements.[97] Crucially, Mrs Gandhi's adroit political moves also attracted the CPI and even temporarily earned the support of the CPI(M), fearing the Syndicate would seek to ban it.[98]

These realignments split the Opposition, attracting the CPI and remaining half of the PSP to the Congress (R), as well as the DMK and Muslim League.[99] To some extent, the emergent alliance reflected converging views. In Tamil Nadu, the DMK had launched its own brand of "assertive populism", including reservations for *shudra* castes as Other Backward Classes (OBCs), food subsidies and cheap housing for the urban poor, as well as the extension of agricultural loans, well irrigation and land ceilings.[100] In Kerala, Mrs Gandhi's leftist turn facilitated a governing coalition with the CPI, which implemented the Land Reforms Amendment Act.[101] Yet it also demonstrated her skill in acquiring personal authority through radical pledges and exploiting internecine rivalries for political advantage.

The prime minister quickly centralized power. She began by cancelling internal elections in the Congress (R) and appointing weak chief ministers. Yet the process soon extended to the wider state apparatus. Mrs Gandhi consolidated the powers of central intelligence organizations and government police forces, and of the Prime Minister's secretariat and Youth Congress, vis-à-vis the central bureaucratic apparatus. Moreover, she demanded that public officials display personal loyalty to her regime. A contentious example was the idea of a "committed judiciary", espoused by Mohan Kumaramangalam, a former ideologue of the CPI.[102] But perhaps her most notorious decision was the passage of the Maintenance of Internal Security Act (MISA), 1971, which authorized law enforcement agencies to perform indefinite preventive detention, search and seizure of property without a warrant, and other civil rights violations. In short, she sought to transform India's federal parliamentary democracy into a far more unitary presidential system,[103] undermining its checks and balances.

Mrs Gandhi's decision to call an early general poll in March 1971 highlighted these ambitions. The Congress (R), trumpeting the slogan *garibi hatao* (abolish poverty) in response to the Opposition demanding *Indira hatao* (get rid of Indira Gandhi),[104] crushed the latter. The electorate, fearing that a hung parliament would exacerbate the politics of opportunism and extremism that had marred the SVD governments, voted for "change and stability".[105] Her newfound allies paid dearly. The poor electoral showing of the PSP effectively sealed its demise.[106] The CPI fared better, maintaining its tally of parliamentary seats. But the party lost votes everywhere, especially in Andhra Pradesh, Assam and Maharashtra.[107] The rest of the Opposition, comprising the Congress (O), Swatantra, Jan Sangh and SSP, "was in a state of utter demoralisation."[108]

Yet ramifications of the 1971 general election, and the way in which Mrs Gandhi sought her victory, proved more complicated. Paradoxically, her decision to centralize political authority and personalize its basis of legitimacy undermined the power of the state.[109] The Congress (R)'s massive electoral triumph generated high expectations. But the systematic dismantling of the party, and by extension its influence over the apparatus of the state, rapidly undermined the capacity of both to govern effectively. Moreover, by dividing the larger Congress party, Mrs Gandhi allowed marginalized politicians to join opposition parties seeking to mobilize social discontent towards her rule.

Indeed, her sweeping personal victory obscured the accelerating regionalization of India's federal party system. The reorganization of states in the 1950s and 1960s, which encouraged flourishing regional cultures and greater administrative efficiency, had strengthened national unity by the early 1970s.[110] But the development of vernacularized electoral domains simultaneously encouraged the growth of new regional parties and distinct party systems across the Union. In eight of eighteen states, the first and second largest parties in the 1971 general election in terms of the national popular vote comprised a regional political formation. The rise of the latter came at the expense of their national counterparts. The collective vote share of the Jan Sangh and Swatantra on the right, and SSP and PSP on the left, had fallen from approximately 26 per cent in 1967 to 14 per cent in 1971.[111] In addition, Mrs Gandhi's decision to de-synchronize the timing of elections between the Centre and the states benefited her immediately. Indeed, the 1970s saw her refashion the old Congress system from a national catch-all organization to a party that sought to impose a macro-level class-based cleavage upon the polity, yet rested upon different state-level social bases, such as lower OBCs in Karnataka, the KHAM social coalition in Gujarat and a Brahmin-Dalit alliance in Uttar Pradesh.[112] None of these electoral tactical changes, however, could arrest rising political aspirations and social demands in an increasingly regionalized federal parliamentary democracy.

The Opposition soon regained its voice. The global oil price shocks in 1973, combined with two successive crop failures, fueled mass inflation. Deteriorating economic conditions encouraged protests and demonstrations by industrial workers, lower middle classes and the urban poor. Student insubordination in the universities grew.[113] Mounting social disaffection with Congress rule crystallized in Gujarat through the Nav Nirman, the Movement for Regeneration, forcing the deeply tainted Chimanbhai Patel ministry to resign in early 1974. But the Centre's imposition of President's rule failed to dissuade other mass demonstrations. A coalition of left-wing groups, The Bihar State Struggle Committee of Workers and Employees Against Price Rise and Professional Tax, demanded "give us work and give us food, or else we will bring life to a standstill".[114] Yet their struggle also attracted Hindu nationalists, such the Akhil Bharatiya Vidyarthi Parishad (ABVP), the student union of the Jan Sangh, and the Rashtriya Swayamsevak Sangh (RSS), its principal organization, leading to the formation of the Chatra Sangharsh Samiti (CSS). Crucially, the CSS asked Jayaprakash Narayan to lead it. The aging *sarvoday*-ite had supported various oppositional movements in recent years, including the Naxalites. The Hindu right shared his critique of the Nehruvian modernist project of state-led development, voicing the need for greater moral reform, social welfare and political decentralization. But Narayan successfully demanded that the emergent movement abjure violence and remain politically open. It was a momentous decision.

In April 1974, Narayan, now simply called JP, held a rally in Patna, calling for the people of India to join continuous mass action leading to *sampoorna kranti* (total revolution): the creation of a peaceful revolutionary society based on *lok shakti* (people power), freed of caste, class, communalism and corruption, with youth as its catalyst. Economically, he called for greater land reform, trade unionization and cooperative agriculture, and the imposition of ceilings on income and wealth. Politically, JP demanded the introduction of party-less electoral candidates, a right of recall and the devolution of power to villages, blocks and wards.[115] In May, the socialist president of the All-India Railwaymen's Federation, George Fernandes, launched a national rail strike. The three-week action met with heavy repression. Narayan's ongoing dialogue with Mrs Gandhi had deteriorated into personal acrimony, with the latter accusing him of living with posh businessmen and receiving support from the Central Intelligence Agency.[116] In early June, Narayan repeated his call for a march on parliament on June 25, invoking Gandhi's Salt March against colonial rule, urging students to boycott classes, citizens not to pay their taxes and administrators to renege their duties. The groundswell of opposition on the streets coincided with high legal wrangles. On 12 June, the Allahabad High Court charged Mrs Gandhi with electoral malpractice, declaring her previous election invalid. Twelve days later, the Supreme Court rejected her appeal. On 26 June, the embattled prime minister declared a state of internal emergency, suspending democratic rule.

The Opposition coalesces

The justification for the Emergency was twofold. Mrs Gandhi emphasized the need to remove mass poverty through radical political action. Her Twenty Point Programme for Economic Progress pledged to implement land reform, abolish bonded labor and liquidate rural debt amongst other measures.[117] Implementing radical transformation required suppressing the "forces of disintegration". Many observers contended that both Jayaprakash Narayan and Indira Gandhi, ignoring the imperatives of representative democracy and the modern state, shared responsibility for the outcome.[118] Yet few justified the depth of reaction. In early July, the prime minister banned the RSS and Jamaat-e-Islami.[119] Indeed, she proceeded to arrest thousands of individuals under MISA, amending the Act to allow their detention for three years without trial, while censoring the media. Her campaign against the Opposition, many of whose leaders she had jailed, included the imposition of President's rule in Tamil Nadu and Gujarat. In short, the radical economic slogans and massive political centralization that had characterized Indira Gandhi's tenure had reached their culmination.

The Emergency attracted support from several quarters. Mrs Gandhi received backing from the All Indian Anna Dravida Munnetra Kazhagam (AIADMK), which had broken away from the DMK in 1972 over growing corruption and the dominance of intermediate backward castes within its ranks.[120] Her pledge to restore public order appealed to the urban middle classes and industrial capital, which initially enjoyed greater profits.[121] But the most important supporter was the CPI, which had stridently denounced Narayan's *sampoorna kranti* as "a typical fascist mass movement designed to destroy elected assemblies, subvert parliamentary democracy and create a constitutional crisis".[122] Socialists refuted the charge vehemently, saying that "J.P. could by no stretch of the imagination be considered as a sympathizer of communal forces ... he honestly felt that total mobilization of vote ... was the highest priority.... During the Second World War Stalin, Churchill and Roosevelt ... forged unity to defeat fascism". But they also conceded that "communal parties, which were marginalised, gained some ascendancy."[123] Indeed, the pact was a "veritable godsend" for the Jana Sangh, which took advantage of well-organized RSS activists.[124] The CPI believed that it could pressure Mrs Gandhi to implement the Twenty Point Programme. Yet the party's stance exposed its dependence on the Congress (R) for electoral growth, perhaps even its political survival.[125] The grand strategic vision of the CPI, which prioritized high socialist rhetoric over democratic values and failed to grasp the prime minister's real intentions, led it to commit a fantastic political mistake.

The draconian character and inherent limitations of authoritarian rule gradually revealed themselves. Economically, Mrs Gandhi backtracked from her grand promises, ruling out further economic nationalization and revoking her pledge to grant a minimum guaranteed bonus for industrial workers.[126] Politically, the prime minister continued to incarcerate critics of her rule, delaying parliamentary elections for another year. Indeed, she consolidated executive power further, ramming through several constitutional amendments. The most controversial included the 38th and 39th, which respectively empowered the executive to pass ordinances and barred the Supreme Court from challenging the election of a prime minister, and the 42nd, which curtailed the fundamental rights and re-described India as a "socialist secular republic". All the while, the coercive apparatus of the state expanded. By 1976, the combined manpower of the Central Reserve Police, Border Security Forces, Central Industrial Security Forces, and Home Guards had reached approximately 600,000 troops. Hence seasoned observers claimed, "it seems likely that the era of multiple centers of power and decision by consensus in India is now over."[127]

But the growing political arbitrariness of Emergency sowed deepening resentment, even amongst its initial beneficiaries. The Congress(R)'s Young Turks, led by Chandra Shekhar and Mohan Dharia, increasingly admired JP.[128] The local implementation bodies of the new regime sidelined the CPI's cadres, engendering frustration amongst their ranks.[129] The most important force, though, was by far the largest. The draconian family planning and slum clearance programmes instigated by Sanjay Gandhi, who enjoyed immense extra-constitutional authority as the prime minister's heir apparent, earned particular opprobrium amongst many sections of the poor.[130] Yet their growing hostility failed to register in the corridors of power. In January 1977, Mrs Gandhi surprised her political opponents, releasing them and calling for elections to be held in March, anticipating victory. It was a colossal blunder that isolated autocrats often make. On 23 March, the masses astounded the country by sweeping the Janata Party into office, electing the first non-Congress Union government since independence.

3

THE JANATA PARTY (1977–1980)

The Janata Party, an amalgamation of the Socialists, Bharatiya Lok Dal (BLD), Jan Sangh and Congress(O), won 43 per cent of the national vote and 298 parliamentary seats. In contrast, the Congress(R) managed to retain only 154 seats in the Lok Sabha, with 34.5 per cent of the vote. Their respective seat tallies underscored the importance, as Lohia repeatedly stressed in the 1950s and 1960s, of oppositional unity. The constituents of the Janata either fielded a single candidate or avoided direct electoral contests with several opposition parties, inflating its overall parliamentary strength due to the disproportional mechanical effects produced by the country's FPTP electoral system (see Table 3.1).

Ideologically, the Janata Party espoused a Gandhian socialist vision. It framed the issue in stark terms: "BOTH BREAD AND FREEDOM: A Gandhian Alternative": "It is a choice between freedom and slavery; between democracy and dictatorship ... Bread cannot be juxtaposed against liberty. The two are inseparable."[1] Vowing to end "the nightmare of fear and humiliation", the party manifesto presented three charters. Politically, it pledged to restore fundamental liberties and the rule of law, reestablish judicial independence, freedom of the press and the right to peaceful dissent, and repeal MISA and the 42nd amendment. These measures would restore India's parliamentary democracy and separation of powers. Yet the new governing coalition also promised to enhance the power of the states by amending Articles 352 and

Table 3.1: Sixth General Election, 1977

Party	Seats	Vote
BLD	295	41.32
CPI	7	2.82
CPI(M)	22	4.29
INC	154	34.52
NCO	3	1.72
ADK	18	2.90
DMK	2	1.76
FBL	3	0.34
JKN	2	0.26
KEC	2	0.26
MAG	1	0.06
MUL	2	0.30
PWP	5	0.55
RSP	4	0.45
SAD	9	1.26
UDF	1	0.07
JKP	1	0.07
RPK	2	0.51
Independents	9	5.50

Source: Election Commission of India.

356, which respectively allowed the Centre to declare a state of emergency and to impose President's rule upon a state, and of ordinary citizens by legislating a right to recall. Indeed, its economic pledges reflected an avowed Gandhian outlook to a greater degree. The Janata promised to end destitution and enhance *swadeshi* (economic self-reliance) by deleting property as a fundamental right, increasing redistributive taxation and affirming the right to work, and to promote rural development through the use of appropriate technology, greater agricultural investment and small-scale and cottage industries that emphasized wage goods production. Finally, the party stated the need to construct a "new society". It vowed to eradicate illiteracy, universalize access to safe drinking water, public housing and social security, and guarantee the rights of the poorest by creating new statutory commissions, establishing public ombudsman bodies such as Lok Pal and Lokayukta, and enhancing greater legal aid. Narayan's final appeal to the electorate, "Make India a free, progressive, Gandhian country",[2]

THE JANATA PARTY (1977–1980)

crested a wave against emergency rule. The election of the Janata represented a critical juncture in modern Indian democracy.

The disaggregated results yielded a complex verdict, however. First, the Congress(R)'s vote share amongst the total eligible electorate remained largely constant, disproving the claim that it was a critical election in a technical sense.[3] Indeed, the share of votes and seats accentuated national parties' dominance (see Tables 3.2 and 3.3).[4] That said, a significant sociological shift had occurred in the national political class. For the first time, upper-caste individuals comprised less than 50 per cent of all MPs from the Hindi belt, while agriculturalists now comprised 36 per cent of the Lok Sabha.[5] The national electoral breakthrough of intermediate proprietary castes symbolized a changing political order in New Delhi. Third, the verdict exposed a profound north-south divide. The Congress(R) suffered a complete rout in the northern states, losing every seat in Uttar Pradesh and Bihar, and winning only a single constituency apiece in Rajasthan and Madhya Pradesh. The party dominated the south, however. It virtually swept Andhra Pradesh and Karnataka, won a legislate majority in Kerala and held its own in Tamil Nadu. Put bluntly, the Janata Party won only six parliamentary seats in the south, despite increasing its relative vote share.[6] Indeed, the Congress(R) won 92 parliamentary seats in these states, improving its tally of 71 in the fifth general election.[7] The worst

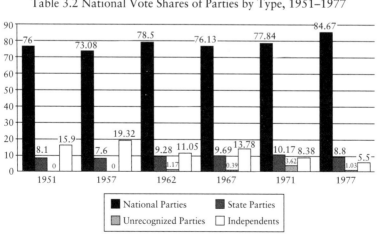

Table 3.2 National Vote Shares of Parties by Type, 1951–1977

Source: Election Commission of India. Author's calculations.

Table 3.3 National Seat Shares of Parties by Type, 1951–1977

Year	National Parties	State Parties	Unrecognized Parties	Independents
1951	418	34	0	37
1957	421	31	0	42
1962	440	28	6	20
1967	440	43	2	35
1971	451	40	13	14
1977	481	49	3	9

Source: Election Commission of India. Author's calculations.

abuses of the Emergency—forced sterilization, slum clearances and civil rights violations—had compelled Muslims and Dalits to support the Janata in the north.[8] The Congress(R) won only 16 of 78 and 12 of 38 parliamentary seats reserved for SCs and STs, respectively, and merely 20 of the 81 constituencies with sizeable Muslim communities.[9] Yet the party had effectively mobilized backward castes in Andhra and Karnataka through competitive populist measures such as government reservations, political tickets and higher social expenditures.[10] Additional local factors, including land reforms by the CPI-Congress(R) coalition in Kerala, the corrupt reputation of the DMK in Tamil Nadu and the general fear of the Jan Sangh across the south, helped stem the popular tide.[11]

The question of uniting the parties into a single political formation had first arisen in 1975. Only the BLD had expressed interest. But the Emergency compelled a "change of heart" among the Socialists, and altered the strategic outlook of the Jan Sangh, which wanted to rescue the RSS.[12] Indeed, according to Biju Patnaik, a former Congressman from Orissa who had joined the BLD, "our experience in gaol made us forget what had divided us."[13] The truth was, arguably, less romantic. JP's political leadership proved decisive. According to Dandavate,

nobody was very keen on giving up the symbols of their parties and to form one party ... [but] J.P.... issued a public statement: If, after such sacrifices, struggles and detentions, there is not total mobilization of votes against the government ... there is no chance of the Emergency going out [sic]. Immediately within five minutes all people changed their views.[14]

The parties' leaders, fearing that Narayan would abandon his campaign, fell into line.[15] The Janata Party, officially launched on 23 January 1977, contested under the symbol of the BLD across most of the country.[16]

To consolidate its prospects, the Janata agreed tactical seat adjustments with several opposition parties. The media paid greatest attention to the Congress for Democracy (CFD), led by Jagjivan Ram, which formed in early February. Ram, a senior Congressman and the country's leading Dalit politician, had served as the Union minister for agriculture and irrigation during the Emergency. But he reproached Indira Gandhi for the civil liberties violations and decline of intra-party democracy during its reign.[17] His support carried great symbolic weight. The party formally joined the Janata during the campaign.

In addition, the Janata entered an alliance with the two most important regional parties, the Shiromani Akali Dal (SAD) in Punjab and the DMK in Tamil Nadu. The 1973 Anandpur Sahib Resolution, which demanded greater political devolution and recognition of cultural minority rights,[18] resonated with the JP movement. Moreover, the SAD was perhaps the only party to mount a sustained agitation against the Emergency.[19] The DMK had supported the Congress(R) in the 1971 general election. However, the party's advocacy of caste-based reservations and greater political devolution resonated with the Janata as well. Mrs Gandhi's imposition of President's rule in 1975 pushed the leading Tamil formation into the Opposition. The electoral pact secured the SAD and DMK, which respectively won 1.26 and 1.76 per cent of the national vote, a total of eleven seats in the Lok Sabha.

Lastly, the Janata brokered an electoral agreement with the CPI(M), which captured 4.29 per cent of the national vote and 22 parliamentary seats. The party stridently opposed the Congress(R). Moreover, several of its leaders admired Narayan personally. The most important was Jyoti Basu. A prominent centrist, the newly elected chief minister of West Bengal had supported a national united-front-from-above and the radical federal decentralization of power and resources in the late 1960s.[20] Moreover, Basu had served as JP's vice-president in the All India Railwaymen's Federation in the 1940s. But the CPI(M) remained steadfastly opposed to the Hindu right. The RSS' participation in the JP movement made formal cooperation difficult. According to Basu,

... JP said: "You are making a mistake. If you join, these fellows [RSS] will run away; our party is not there, nothing is there, and they have an organisation. So they are helping me, but if you people come they won't be there later on." I said: "No, we cannot join, but please keep us informed of your programme, where we can join, what we can do about it to make your movement a success."[21]

The CPI(M) supported the Janata by agreeing not to field candidates against it during the general election. Observers debated its political motivations and strategic judgments. According to some, the party had failed to grasp the significance of the JP movement. Its embrace of the "contiguous area thesis" at the Muzaffarpur session in 1973, which prioritized mobilizing support in peasant belts around key industrial centers, arguably constrained its political imagination.[22] In contrast, others underscored the serious political debate within the CPI(M) over whether to radicalize the JP movement. Although none of its leaders were arrested during the Emergency, many went underground, while cadres belonging to Centre of Indian Trade Unions (CITU), its labor wing, suffered political repression and large-scale arrests.[23] Gradually, however, the party became "rudderless, unable to choose between insurrection and parliamentary forms of government or between joining and opposing the JP movement ... the CPI(M) gradually reduced itself to a state of political indecisiveness and passivity".[24] Ultimately, "[the] principal objective of the party was to safeguard its hard core support base" through mass action and the establishment of people's democracies and united fronts in the states.[25] The question of whether the CPI(M) should have joined the JP movement remained a live political issue. Indeed, the 1978 Review Report at the 10th party congress admitted, "we were unable to evolve a correct tactical line regarding the JP movement—a line which would have helped us to strike better relations with the people who were participating in the movement" because we "grossly underestimat[ed] the conflicts and contradictions between the ruling Congress ... and the bourgeois opposition parties ...."[26] It would not be the last time the party high command regretted not taking the risk of participating in a new oppositional formation. Nonetheless, given that it had not joined the JP movement, the CPI(M)'s decision to limit its cooperation with the Janata Party to avoiding direct electoral contests demonstrated good judgment. The party had only captured 22 parliamentary seats, less than one-tenth of the Janata's tally, making it virtually impossible to sway the balance of

THE JANATA PARTY (1977–1980)

forces to the left. The CPI(M)-led Left Front had just captured power in West Bengal after years of repression by the Centre and proxy Congress(R) forces. The changing of the guard in New Delhi provided an historic opportunity to consolidate the party's forces and test the limits of reform.

Indeed, despite its historic election, the new Janata government faced difficult prospects. "The Janata Party," Narayan optimistically proclaimed, "is no greater hotchpotch than the Congress". He had few illusions, however, confessing that it consisted of "all types of vested interests and ... is seething with internal differences".[27] Its five constituents had distinct ideological positions and social roots.[28] The Socialists, whose electoral base comprised the rural poor and urban labor, championed Gandhian values. The BLD comprised Charan Singh's erstwhile BKD, factions of the Swatantra Party and the SSP (Raj Narain) and several lesser organizations.[29] Formed in 1974, the BLD represented the rising agrarian power of the AJGAR coalition– Ahirs, Jats, Gujars, and Rajputs–which Singh advanced throughout his career.[30] In contrast, the Jan Sangh espoused Hindu chauvinist demands, supported by the high-caste urban middle classes in the Hindi belt. The Congress(O), a conservative formation, was sympathetic to the Jan Sangh. Yet it was essentially a splinter group, led by Mrs Gandhi's longstanding rival, Morarji Desai. Lastly, the CFD was a Congress offshoot too. Yet it represented poor low-caste communities in the countryside, at odds with the Congress(O). Multiple cross-cutting fissures pervaded the Janata.

The first of these involved caste. On the one hand, the CFD, Congress(O) and BLD "broadly shared Congress ideology". But they had "conflicting [political] ambitions", partly reflecting the differential caste status of their leaders and rank-and-file.[31] On the other, the Socialists' desire to expand reservations for OBCs in Bihar and Uttar Pradesh caused unease in the Jan Sangh, which sought to maintain traditional social hierarchies.[32] The second cleavage in the Janata concerned socialist ideology. Both the BLD and the Socialists relied on backward castes' support. The "kisan [farmer] politics" of the former appealed explicitly to the interests of peasantry, however, ranging from small agricultural proprietors to dominant rural classes. The "quota politics" of the latter, in contrast, invoked the longstanding socialist belief that lower caste groups deserved equal self-representation through reservations in the public sector.[33] Charan Singh had complained to JP during the Emergency that "the idea of a total plan or total responsibility of the state for the welfare of the people, of which it [JP's proposal] smacks throughout is likely to end up in totalitarianism".[34] Indeed, Singh countered a resolution moved by JP in 1976 seeking the "establishment of socialist state" by proposing the "establishment of egalitarian society, consistent with individual freedom".[35] Yet the BLD leader's egalitarianism had severe economic limits. Genuine land reform was a non-starter: "if a man is landless he cannot be called a farmer, peasant. Then he is a labourer ... there is no land for giving to the labourer."[36] It was an ominous sign. The third fissure in the Janata was the communal question, in particular, the role and standing of the RSS in the Jan Sangh.[37] According to his followers, JP had enlisted the support of right-wing Hindu organizations for largely expedient reasons. Apparently, he believed that integrating the Jan Sangh into the Janata would slowly de-communalize the RSS. But the ideological fervor and organizational discipline of the latter created anxieties amongst the Socialists and the BLD. Hence they advocated integrating the RSS with other volunteer organizations during the Emergency. The two senior leaders of the Jan Sangh, Lal Krishna Advani and Atal Bihari Vajpayee, had reluctantly agreed to merge their party in order to oust Mrs Gandhi. But the RSS refused to relinquish its organizational autonomy, leaving the loyalties of the Jan Sangh open to question.

To complicate matters, the balance of power within the Janata was both highly uneven and at odds with its popular mandate.[38] The Socialists embodied its ideological hopes. Yet the party had only secured 51 parliamentary seats, making it the smallest grouping. Both the Congress(O) and BLD, far less progressive, enjoyed greater weight, with 55 and 68 seats, respectively. Indeed, the largest parliamentary formation within the Janata was the Jan Sangh, with 90 seats. The CFD and independent Congress dissidents, which respectively secured 28 and 6 parliamentary constituencies, contributed the rest.

In the end, however, the balance of forces within the Janata's parliamentary coalition failed to determine executive power (see Table 3.3).[39] The Congress(O) was the largest recipient with six cabinet posts, followed by the BLD, which had four. Despite its overwhelming seat advantage, the Jan Sangh obtained just three cabinet positions, equivalent to the Socialists' tally. The CFD, which contributed the fewest parliamentary seats, secured two, the SAD one. The under-represen-

THE JANATA PARTY (1977–1980)

Table 3.4: Janata Party Government

Name	Party	Prior Affiliations	Post	Rank
Morarji Desai	BLD	INC	Prime Minister	Cabinet Minister
Charan Singh	BLD	BLD	Home Affairs	Cabinet Minister
Jagjivan Ram	BLD	INC	Defence	Cabinet Minister
A.B. Vajpayee	BLD	BJS	External Affairs	Cabinet Minister
L.K. Advani	BLD	BJS	Information and Broadcasting	Cabinet Minister
H.N. Bahuguna	BLD	BLD	Petroleum, Chemicals, and Fertilizers	Cabinet Minister
Sikandar Bakht	BLD	INC(O)	Works, Housing, Supply and Rehabilitation	Cabinet Minister
Sardar Prakash Singh Badal	SAD	SAD	Agriculture and Irrigation	Cabinet Minister
Sardar Surjit Singh Barnala	SAD	SAD	Agriculture and Irrigation	Cabinet Minister
Shanti Bhushan	BLD	INC(O)	Law & Justice and Company Affairs	Cabinet Minister
Dr. P.C. Chunder	BLD	INC(O)	Education, Social Welfare and Culture	Cabinet Minister
Madhu Dandavate	BLD	PSP	Railways	Cabinet Minister
Mohan Dharia	BLD	PSP	Commerce, Civil Supplies and Cooperation	Cabinet Minister
George Fernandes	BLD	SSP	Communications/Industry	Cabinet Minister
Purushottam Kaushik	BLD	SSP	Tourism and Civil Aviation	Cabinet Minister
Raj Narain	BLD	SSP	Health & Family Welfare	Cabinet Minister
H.M. Patel	BLD	SWA	Finance	Cabinet Minister
Biju Patnaik	BLD	UC	Steel and Mines	Cabinet Minister
P. Ramachandran	NCO	INC	Energy	Cabinet Minister
Ravindra Varma	BLD	INC	Labour and Parliamentary Affairs	Cabinet Minister
Brij Lal Verma	BLD	PSP, BJS	Industry/Communications	Cabinet Minister
Satish Agarwal	BLD	BJS	Finance	Minister of State
Renuka Devi Barkataki	BLD	INC(O)	Education, Social Welfare and Culture	Minister of State

Arif Beg	BLD	SSP	Commerce, Civil Supplies and Cooperation	Minister of State
Chand Ram	BLD	HSS	Shipping and Transport	Minister of State
Krishna Kumar Goyal	BLD	BJS	Commerce, Civil Supplies and Cooperation	Minister of State
Sardar Dhanna Singh Gulshan	SAD	SAD	Education, Social Welfare and Culture	Minister of State
Ram Kinkar	BLD	INC, BKD	Works, Housing, Supply and Rehabilitation	Minister of State
S. Kundu	BLD	PSP	External Affairs	Minister of State
Abha Maiti	BLD	INC	Industry	Minister of State
Dhanik Lal Mandal	BLD	SP, SSP	Home Affairs	Minister of State
Janeshwar Mishra	BLD	SP, SSP, BKD	Petroleum, Chemicals and Fertilizers	Minister of State
Kariya Munda	BLD	BJS	Steel and Mines	Minister of State
S.D. Patil	BLD	INC(O)	Home Affairs	Minister of State
Fazlur Rahman	BLD	FML	Energy	Minister of State
Larang Sai	BLD	BJS	Labour and Parliamentary Affairs	Minister of State
N.P.S. Sai	BLD	RRP, BJS	Communications	Minister of State
Sheo Narain	BLD	INC, INC(O)	Railways	Minister of State
Sher Singh	BLD	INC	Defence	Minister of State
Bhanu Pratap Singh	BLD	BKD	Agriculture and Irrigation	Minister of State
Jagbir Singh	BLD	BLD	Information and Broadcasting	Minister of State
Dr. Ram Kirpal Sinha	BLD	BJS	Labour and Parliamentary Affairs	Minister of State
Jagdambi Prasad Yadav	BLD	BJS	Health & Family Welfare	Minister of State
Narsingh Yadav	BLD	BLD	Law & Justice and Company Affairs	Minister of State
Zulfikar Ullah	BLD	BLD	Finance	Minister of State

THE JANATA PARTY (1977–1980)

tation of the Jan Sangh in the cabinet reflected widespread concern that it would dominate how the new government worked. To some extent, the party received its proper due in state-level coalition arrangements. It struck a tacit bargain, acknowledging the dominance of the BKD in Uttar Pradesh, Bihar, Haryana and Orissa in exchange for greater influence in Delhi, Himachal Pradesh, Madhya Pradesh and Rajasthan.[40] Still, the distribution of power in the Council of Ministers failed to reflect the rather mechanical algorithm of power that inspired the CPI(M). Contextual strategic concerns and tactical maneuvers considerably shaped the internal power dynamics of the new Union government.

The most important bone of contention, the prime ministership, illustrated this vividly. India's parliamentary democracy, which bestowed executive power on the Council of Ministers, facilitated power sharing. Yet prime ministerial office permitted a single occupant at a time. Parliamentary cabinet government also presumed a willingness and capacity to broker political compromise. The constituents of the Janata, alas, failed to display these qualities when it came to the trappings of power. The Congress(O) and Jan Sangh supported Morarji Desai, given his Hindu conservative beliefs.[41] The Socialists and former Young Turks of the Congress(R) wanted Jagjivan Ram.[42] Charan Singh objected, arguing that Desai had suffered during the Emergency, whereas Ram had moved the parliamentary motion approving its promulgation.[43] But Singh betrayed his general contempt for Dalits and landless laborers, which reportedly even surpassed his "hatred" of Brahmins, frequently calling Ram "that Chamar".[44] In the end, the two elder statesmen of the socialist bloc, Narayan and J.B. Kripalani, anointed Desai as the Janata's parliamentary leader, citing his administrative experience and personal integrity.[45] Ram was appointed deputy prime minister and the Union minister of defense. Singh received the Union home ministry after Desai, whom some observers perceived to be temperamentally unsuited to accommodative politics,[46] rejected his bid for the deputy prime ministership.

The outcome highlighted three salient concerns. First, personal ambition and partisan rivalry beset the Janata government from the start. Second, the decision to select a conservative lifelong Congressman as the leader of the first non-Congress government in New Delhi underscored its limitations. The socialist rhetoric of the new governing dispensation had weak political foundations. Finally, the inability of the socialists to present a credible rival candidate for the prime ministership exposed the long-term historical cost that Gandhianism had imposed on their political forces. The decision of Narayan, Kripalani and others to eschew formal politics in the 1950s and 1960s in favor of *Sarvodaya* had weakened their stock of strong electoral leaders.

A succession of events, many involving deeper social conflicts, quickly exposed narrow political rivalries at the top. Severe tensions between Dalit landless laborers and their proprietary overlords, igniting approximately 17,000 incidents, marked the Janata's first year in office.[47] Upper-caste groups attempted to reclaim lands that had been distributed by Mrs Gandhi's regime during the Emergency.[48] Yet many clashes also pitted the rising agrarian bourgeoisie of the new political order against historically subaltern groups. The failure of the Janata government to respond quickly to a horrific attack in Belchi, Bihar, in July 1977 allowed Indira Gandhi to visit the site first. JP responded by meeting the former prime minister, but the Janata government was split on how to respond, enabling her political resurrection to begin. In May, it had appointed Justice J.C. Shah to lead a commission to investigate the Emergency, especially its promulgation, abuses of authority and shadow power structures. Yet Mrs Gandhi disputed its legality. High-level tensions exploded over the next year. Charan Singh chastised his colleagues for not pursuing her arrest.[49] He accused Kanti Desai, the son of the prime minister and one of his key aides, for peddling influence on behalf of big industrial houses. And the former BLD chief reportedly abetted the publication of explicit sexual photographs involving Jagjivan Ram's son, discrediting his rival.[50] In June 1978, Singh publicly lambasted his own government for being "a bunch of incompetent people", forcing Moraji Desai to demand his resignation.[51] The implosion led JP to lament in private:

> In 1977, the unprecedented zeal seen in the public was not just for the change for power at the center, but a symbol of ambition for our transformation into a new society.... To lose this opportunity because of personal ambitions would be serious breach of the public trust. If we continue to stay embroiled in minor issues ... then the country will be in danger of falling to the forces of dictatorship once again.... we must prove ourselves equal to the task.[52]

The competition for high political office unleashed the "single-mindedly self-destructive" tendencies of a "squabbling gerontocratic triumvirate".[53] Worse, the Government's ministers became enamored with privileges and sinecures, reflecting the "death of idealism".[54] Renewed

THE JANATA PARTY (1977–1980)

political demonstrations by students and workers and a growing police revolt against poor working conditions in a number of states ruled by the Janata created an "impression of a dangerous drift, a galloping anarchy, especially to the vocal middle classes".[55] The desire for power for its own sake seemed to be its driving political motivation.

Yet the Janata had a substantive agenda, inspiring partial victories and deeper partisan conflicts, with the latter ultimately contributing to its defeat. Its most important achievement, almost universally credited, was to restore India's federal parliamentary democracy. Upon capturing office, the Government terminated the state of internal emergency declared by Mrs Gandhi in 1975, as well as its external dimension, which had been promulgated during the 1971 Indo-Pakistan war. The Desai ministry also shepherded through parliament two critical bills that reversed the most despotic features of the 42nd amendment to the Constitution passed by the Congress(R). The 43rd and 44th amendments reinstated five-year terms to parliament and the state legislative assemblies and recognized the right of Supreme Court to adjudicate all elections. Re-establishing India's constitutional system, given the excesses of the Emergency, was the most important task.

That said, the Janata also sought to expand the realm of democracy. The Government granted the official leader of the Opposition cabinet rank in both houses, enhancing the power and legitimacy of the office.[56] It appointed the so-called Ashok Mehta Committee to explore ways of reinvigorating *panchayati raj* institutions, which duly submitted its report in August 1978, influencing experiments in Karnataka in the 1980s.[57] And the Janata shepherded through parliament several measures to protect states against unwarranted central intervention. The 43rd and 44th amendments stipulated that only "armed rebellion", rather than an "internal disturbance", permitted New Delhi to declare a state of emergency under Article 352, required a two-third parliamentary majority for its promulgation, and demanded that it had to be authorized every six months. These provisions also imposed an analogous six-month limit on the use of President's rule under Article 356.[58] Perhaps nothing demonstrated the Janata's political commitment to improving Centre-state relations better than its conduct in Jammu & Kashmir. The Government oversaw the first relatively free election in the region in March 1977, allowing the Jammu and Kashmir National Conference (JKN) to sweep 46 of 75 seats in the legislative assembly.[59] Indira Gandhi and Sheikh Abdullah had laid the foundation for the historic breakthrough by orchestrating the Delhi Accord in 1975, which freed the so-called Lion of Kashmir to contest democratic elections as long as he repudiated the right to political self-determination. The Accord began to reverse several decades of autocratic rule by New Delhi, whose numerous constitutional orders and parliamentary laws violated the integrity of Article 370 of the Constitution, which granted the Muslim-majority region special asymmetric rights in the Union.[60] Yet the fact that the Janata held a credible election was not a foregone conclusion. Mrs Gandhi had engineered the downfall of Abdullah's short-lived ministry on the eve of the 1977 general election once it became clear the Janata would triumph. Moreover, the Jan Sangh had opposed Article 370. Jayaprakash Narayan's public embrace of Abdullah upon his release in 1975—he had consistently opposed the militarization of Kashmir—had provoked the ire of the Hindu right.[61] Hence observers rightly credited the Janata, following Prime Minister Desai's pledge to punish attempted vote-rigging, for the breakthrough.[62]

Needless to say, many of these measures served its political interests. Restricting national power and endorsing political devolution appealed to aspiring rich farmers increasingly well placed to capture state governments.[63] Moreover, the Government did not always wield its new-found power wisely. It exploited Article 356 to dismiss nine state governments run by the Congress(R) on grounds that the national electoral verdict undermined their legitimacy. The Janata's victory in elections subsequently held in each of these states aided its claim.[64] But the move exposed political inconsistency, setting a bad precedent that would eventually boomerang.

Notwithstanding these mistakes, however, the general conduct of the Janata suggested two important motivations. First, its relative self-restraint towards the Congress(R) and efforts to strengthen the foundation for loyal opposition arguably reflected a genuine desire to bolster India's parliamentary democracy. Second, the effort to recognize states' authority and explore political devolution revealed a more decentralized conception of the Indian state nation, less afraid of its peripheries. Put differently, the Janata opposed the idea that a centralized political authority should transform India into a Westernized modern society through technocratic economic development and social engineering, a vision that inspired Jawaharlal Nehru from the start and intensified under Indira Gandhi.[65] The distance between these overarching political imaginaries, which each had their respective merits and shortcomings, was nonetheless striking.

THE JANATA PARTY (1977–1980)

The neo-Gandhian orientation of the Janata's political initiatives informed some of its economic proposals too. The government's draft Sixth Five-Year Plan (1978–1983) sought to raise employment opportunities in agriculture and allied services, consumer goods production by small industries, and the income of the poorest through programmes that targeted their minimum needs such as *Antyodaya*, a food for work scheme.[66] Apart from emphasizing local developmental processes, many bureaucrats noted that the so-called "Rolling Plan" took only three months to complete, in stark contrast to the Fifth Five-Year Plan (1974–1979), which Mrs Gandhi's administration submitted to the National Development Council (NDC) after two years. Moreover, the Janata encouraged various states' input during NDC meetings, which had previously "rubber-stamped" Congress diktat.[67] According to some, the episode reflected high administrative proficiency, the product of state-level experience and deliberate co-operation.[68] The 1977 Industrial Policy Resolution introduced by the Union Industry Minister George Fernandes advanced these concerns to some extent. It emphasized the need for decentralized industrialization in the rural sector, spearheaded by small-scale and cottage industries. That said, it also argued for simplifying the licensing regime, raising import quotas in order to enhance the autonomy of industry, while maintaining the public sector's leading role in supplying essential capital goods. Indeed, Prime Minister Desai sought to placate major industrial houses, which had opposed the resolution, by offering tax concessions in exchange for their help in implementing rural development programmes.[69]

Lastly, and perhaps most surprisingly, the Janata Party sought to recalibrate foreign policy. Historically, India's socialist parties favored non-alignment, taking an assertive stance towards Pakistan and China.[70] The Jan Sangh was even more strident. Yet the Government embraced a policy of "genuine non-alignment" between the Soviet Union and the United States while promoting a "good neighbor policy" in the subcontinent on the basis of reciprocity.[71] The Janata consolidated longstanding relationships during its first year in office. On the one hand, Moraji Desai and his foreign minister, A.B. Vajpayee, traveled to the USSR in October 1977, reaffirming the 1971 Indo-Soviet Treaty of Peace, Friendship and Cooperation and bolstering several trade, scientific and technical agreements.[72] On the other, the government appointed Nani Palkhivala, a well-known jurist who favored stronger ties with the west, as ambassador to the United States, encouraging President Carter to visit India in January 1978.[73] Prime Minister Desai reciprocated six months later, facilitating the resumption of American development aid, which had been suspended following the 1971 Indo-Pakistan war, and several joint commissions.[74] In addition, prime ministerial visits to Nepal, Sri Lanka and Bangladesh underscored the desire of the Janata Party to improve subcontinental relations, even if it never jettisoned India's traditional aspirations to regional dominance.[75] In 1974, the Congress(R) had unilaterally diverted the waters of the Ganga river by building the Farraka dam, at the expense of Bangladesh. To make amends, the Janata Party signed a five-year agreement in November 1977, guaranteeing the lower riparian state a greater proportion of the waters in lean seasons.[76] Indeed, the second year in office emboldened it to go further.[77] The Government agreed to Nepal's longstanding demand to separate its bilateral trade accord from their transit agreement.[78] More ambitiously, Foreign Minister Vajpayee traveled to Islamabad in February 1978, signing commercial treaties and offering a non-aggression pact, while maintaining diplomatic silence on the July 1977 military coup. He also visited Peking in October 1978 to discuss the border issue and various other issues. In doing so, Vajpayee became the highest-ranking Indian official to visit either country in almost two decades. Lastly, although it refused to foreclose India's nuclear program, the Desai ministry objected to conducting further tests, despite the presence of several BJS ministers in the cabinet and growing evidence that Pakistan was developing its nuclear capability with Chinese assistance.[79] In short, contra the view that its tenure failed to witness foreign policy changes,[80] the Janata Party sought to shift India's bilateral relations and nuclear strategy.

Several factors helped these initiatives, which also invited critical scrutiny.[81] The US Secretary of State Henry Kissinger recognized India's official non-alignment and regional pre-eminence during his visit in 1974, improving Indo-US relations. Indira Gandhi resumed diplomatic links with Pakistan and China in 1976 after a hiatus of fourteen years. The Carter administration, despite championing universal human rights, pushed bilateral rapprochement as well. Simultaneously, seasoned observers highlighted the Janata Government's various diplomatic setbacks. The Pakistani military rejected its proposed non-aggression pact. China humbled India by invading Vietnam shortly after Foreign Minister Vajpayee's historic visit.[82] And despite warmer political relations, the United States continued arms shipments to Pakistan and insisted that nuclear fuel supplies remained conditional upon India recognizing stringent international protocols, partly reflecting differences between Vajpayee and Desai on whether to retain the nuclear option.[83] Hence foreign policy realists denounced the Janata for allegedly ignoring genuine security threats.[84] Yet other skeptics, while agreeing that "radical change could not come out of goodwill [by the Government] alone", conceded that it partly succeeded in pursuing national interests differently.[85] "India should not only remain non-aligned", Vajpayee proclaimed, "but must also appear to be so".[86] The personal vision and political standing of the Jan Sangh leader, whose party was the most vociferous critic of China and Pakistan, may have given the Government political room to push these initiatives. Yet Vajpayee had pledged to pursue the aims of the Janata as a whole.[87] Its various overtures underscored a more general commitment to practice genuine non-alignment.

Alas, such unity proved exceptional. The basic political cleavages dividing the Government were domestic. In November 1978, Mrs Gandhi won a by-election in Karnataka. Her victory came less than a year after a split in the Congress(R), pitting the Congress (Indira, I) against the Congress (Urs, U), named after the Karnataka chief minister. The Janata had "seriously misjudged" her political resilience.[88] The powerful inner tensions of the Janata soon erupted, aiding her resurrection.

In December 1978, Morarji Desai appointed Bindheshwari Prasad Mandal to head the Second Backward Classes Commission, following pressure from the BLD and the Socialists to extend reservations in central public institutions to *shudra* castes. The first commission, led by Kakasaheb Kalelkar in the 1950s, concluded that OBCs constituted 32 per cent of the population. It identified 2399 such classes using four criteria: low educational achievement, inferior social status and inadequate representation in both government and the organized private sector. But disagreement amongst the commissioners, including a dissent note from the chairman, had allowed Nehru to let the states devise their own criteria for backwardness and implement reservations as they saw fit.[89] Hence Prime Minister Desai charged the so-called Mandal commission, all of whose members were OBCs, to articulate measures for backwardness and make national recommendations. Its remit fuelled antagonism between the Jan Sangh and the BLD in the states, however, whose power struggles and clashes over Hindu-

Muslim violence had already worsened relations.[90] As a result, the former withdrew support from state ministries led by the latter in Bihar, Haryana and Uttar Pradesh, causing their collapse.

The BLD reacted swiftly. In December 1978, Charan Singh mobilized one million farmers in New Delhi, immobilizing the city for two days. The demonstration of brute political force, alongside broader efforts to reunify the Janata, compelled Desai to appoint Singh as Union finance minister and deputy prime minister in February 1979.[91] The powerful Jat leader seized his opportunity. The 1979–1980 Union budget rewarded intermediate proprietary castes, lowering interest rates for loans and indirect taxes on mechanical tillers, diesel and fertilizers, while increasing subsidies for irrigation and electrification.[92] Singh's budget also introduced limited economic liberalization, helping to accelerate industrial development, which reached 8 per cent in 1980.[93] The second global oil price shock, combined with a failed monsoon, had halted growth on both fronts. Nonetheless, the resurgence of production in agriculture and industry demonstrated that aggregate economic growth required neither Congress dominance nor authoritarian rule.[94]

Yet the Budget largely overlooked the question of economic redistribution and caste stratification in the countryside. Its failure to protect the interests of landless labor and poor tenants belied the radical economic rhetoric of the "Rolling Plan". If anything, the "elaborate network of patronage and subsidies" that had disproportionately benefitted "rent-seeking proprietary classes" and "silt[ed] the channels of surplus mobilization and public investment" grew more skewed.[95] Moreover, although real public fixed investment rose 3 per cent in 1979–1980, the failure to find new revenues to cover the expanding subsidy bill accelerated the post-1975 deterioration in India's general fiscal position.[96]

The failure to invest in critical public infrastructure and basic human capabilities reflected a long-term trend. But the asymmetric social impact of the 1979–1980 Union budget reflected the interests of rural propertied classes that Singh represented. Narayan had expressed doubts while in prison whether normal democratic politics could alter the basic structure of society.[97] Yet he acknowledged after the Janata came to power that reducing economic inequalities would require organizing landless Dalits along class lines.[98] Hence observers rightly blamed Narayan for eschewing hard political questions of leadership

THE JANATA PARTY (1977–1980)

and organization, and never translating his radical economic pronouncements into specific policies and programmes, forcing him to rely on the erstwhile BLD.[99] The prime minister and finance minister had no desire to enact redistributive measures.[100] Singh's maiden budget underscored the importance of office for pushing specific agendas. In the end, it symbolized the "march of the kulaks [rich peasants] through the door of politics".[101]

Indeed, their assertion eventually caused the Janata to split. Singh and his lieutenant, Raj Narain, began to denounce the prime minister for pushing a Hindu chauvinist agenda. In December 1978, Desai had introduced the Freedom of Religion Bill, which sought to prohibit forced conversion and censor history textbooks written by notable Marxist scholars. In May, he supported a draft constitutional amendment that put the question of cow slaughter on the concurrent list in order to grant states the authority to impose the law. Neither legislation, longstanding demands of the Hindu right, passed.[102] Nonetheless, the BLD demanded the expulsion of the Jan Sangh from the Council of Ministers for failing to renounce their membership of the RSS, a matter that had genuinely preoccupied the Socialists from the start. The conflict over "dual membership" reflected principle as well as opportunism. Charan Singh had opposed the autonomy of the RSS within the proposed Janata organization as early as 1976.[103] Yet he vacillated. Hence many observers saw his renewed opposition as a "rationalization" in the quest for power.[104] The Socialists' anxiety towards Hindu chauvinism ran deeper. In May 1979, Madhu Limaye organized a conference to explore possibility of constituting a "third force" against the authoritarianism of the Congress(R) and the communalism of the RSS. The meeting attracted secular members of the Janata, Congress(O) and various communist parties, all of whom pledged to "cooperate on issues of common concerns such as national unity, democratic freedoms, communal harmony and social justice".[105] It was perhaps the first such articulation of the idea of a progressive third front incorporating both halves of the broader Indian left. Yet the participants failed to actualize it. The BLD and Socialists forced a vote over the dual membership issue. On 23 June 1979, Raj Narain mobilized 47 MPs to form the Janata Party (Secular) (JP(S)).[106]

On 10 July, Y.B. Chavan of the Congress(U) tabled a no-confidence motion in the Lok Sabha. Many saw the vote as *pro-forma*. Indeed, Chavan himself reportedly confessed to Dandavate, a leading Socialist,

"I thought it would be a ritual; how the hell we can win! [sic] You are having a brute majority in the House, how can it be passed?"[107] But the motion led others from the erstwhile BLD and the Socialists to join the JP(S). Desperate to maintain its majority, the Janata party nominated Jagjivan Ram as its parliamentary leader. But Desai refused and, on 15 July, he resigned. On 28 July, after two weeks of frantic partisan scrambling, President Neelam Sanjiva Reddy swore in Charan Singh as India's sixth prime minister, giving him three weeks to demonstrate his parliamentary majority. Taking the oath of office, Singh reportedly declared, "my life ambition has been fulfilled".[108]

His crowning achievement was short-lived. The JP(S) had cobbled together a disparate supporting coalition: leftists such as the CPI, PWP and a handful of Socialists, the Congress(U) and the Muslim League. But the prime minister-elect also had to rely on the Congress(I). Crucially, Mrs Gandhi demanded that Singh impede various cases against her and her son, Sanjay. But the Congress(U) objected, leading the Congress(I) to renege its support.[109] On 20 August, unwilling to propose that Jagjivan Ram be given the chance to show his parliamentary strength, Singh recommended the dissolution of the Lok Sabha and resigned. Controversially, President Reddy accepted the recommendation, provoking observers to speculate that he did not want to allow his former rival, Ram, who reportedly had enough support on the floor, to become the first Dalit prime minister.[110] Instead, Reddy requested Singh to lead a caretaker administration and called for a mid-term general election. The fiasco highlighted Singh's "relentless drive to exercise power and his contempt for most of his political associates and rivals."[111] "Human conduct is highly random", he later claimed: "we are all quite unpredictable."[112] Yet his readiness to accept Mrs Gandhi's support, despite having castigated her relentlessly, was a remarkably myopic step for a highly seasoned politician.

The Government's collapse unleashed immense rancor. Jagjivan Ram accused Charan Singh of conspiring with "forces of disintegration" and "despotism and dynastic dictatorship".[113] Yet the senior Dalit leader, denied the prime ministership twice, eventually joined the Congress(U).[114] A.B. Vajpayee cast blame more widely:

In retrospect, the responsibility for this state of affairs must be shared by all ... Group loyalties and personal ambitions marked the very first steps of the infant party ... The performance of the Janata governments, both at the Centre as well as in the states, was better than that of the earlier Congress regime. But

... factional quarrels within the party and public airing of grievances ... vitiated the atmosphere and sullied [its] image.[115]

Others were harsher. The president of the Janata Party, Chandra Shekhar, viewed the "sordid drama" as the

culmination of creeping malady that had infected the Janata Party from the beginning.... Almost every group tried to expand their own base. This was bound to create factional rivalries.[116]

JP simply lamented: "The garden is destroyed".[117] He died in October 1979.

High political wrangling continued into the campaign for the seventh general election.[118] Two blocs emerged. On the one hand, the JP(S) of Charan Singh and various socialists forged the Lok Dal, and struck electoral agreements with the Congress(U) and a leftist alliance led by the CPI(M). On the other, the rump Janata Party, consisting of Chandra Shekhar and Jagjivan Ram and the old Jan Sangh, pledged to "complete its unfinished task" of "bread with liberty" and "stability with freedom".[119] Yet both groups proved unstable. In the former, several Congress(U) members had opposed Charan Singh from the start, defecting to the Congress(I). In the latter, rumors that Ram might rejoin Mrs Gandhi and conflicts between him and the Jan Sangh over tickets and funds undermined political credibility. Sadly, neither side highlighted the Janata Party's record in office. Their persistent acrimony and uncertain composition seemed to vindicate the Congress(I), which called the Janata "a coalition of convenience, bound by common desire to cling to power", and promised to "stem the all-round deterioration and drift resulting from [its] misrule" through "a government that works".[120]

The voters agreed. In January 1980, the Congress(I) recaptured national power, winning 42.7 per cent of the total popular vote and 353 seats in the Lok Sabha. In stark contrast, the truncated Janata Party secured only 19 per cent, capturing just 31 parliamentary seats. The Lok Dal, officially renamed the Janata National Party (Secular) (JNP(S)), won 9.4 per cent of the vote. However, its concentration of support in western Uttar Pradesh, Haryana and Punjab allowed the party to win 41 constituencies. The remaining votes and seats went to six other parties.

India's first non-Congress Union government had suffered a resounding political defeat. Many credited "the Indira factor".[121] The former prime minister campaigned skillfully. The Congress(I) came first or second in almost every state across the Union, confirming its self-touted national reach.[122] The party stayed relative popular in the south. The outcome seemed to prove that "the long-term character of political ideology, electoral politics and public policy ... remained centrist".[123] A more important cause of the Congress(I)'s rebound, however, was the exasperation and disgust many voters showed towards the Janata Party.[124] Its unseemly leadership squabbles prevented the Government implementing the programmes it had promised upon capturing power and recognizing the increasingly negative attitude of the public after its collapse. Moreover, neither of the two main groupings that subsequently emerged resolved their leadership or composition, and persisted to fight until the eve of the polls. Finally, the contest for power in the Hindi belt proved decisive, benefitting the Congress(I). On the one hand, political fragmentation lowered the threshold of winning, which factor had catapulted the Janata Party into national office just three years earlier. Yet on this occasion, the Lok Dal challenged the rump Janata Party in 164 of 209 parliamentary constituencies, which simultaneously witnessed a massive increase in independents standing for office. As a result, very few winning candidates secured an absolute majority of votes at the constituency level, regardless of their party affiliation.[125] On the other, the downfall of the Janata in the Gangetic plains exposed growing social polarization. Observers noted a massive electoral swing of 7.5 per cent towards the Congress(I).[126] Many communities feared the rise of the intermediate proprietary castes.[127] The slogan of stability and order appealed to the metropolitan industrial capitalists and predominantly upper-caste urban middle classes that opposed the pro-agrarian rhetoric of the Janata parivar and its desire to expand reservations to the OBCs. Dalit laborers feared the growing class power of the rich peasantry in the countryside. And the failure of the Janata Party to prevent rising communal violence and promote the status of Urdu convinced many Muslim voters to return to the Congress(I). "With the disintegration of the Janata", one seasoned observer surmised, "Indian politics has returned to normal".[128]

1. Ram Manohar Lohia, the eminent socialist, who championed anti-Congress coalitions and *saptakranti* (a seven-fold revolution to attain social justice).

2. Jayaprakash Narayan, who left the Gandhian path of Sarvodaya to call for *sampoorna kranti* (total revolution), addressing a mass rally before the Emergency (1975–1977).

3. Atal Bihari Vajpayee, a key leader of the Hindu chauvinist Bharatiya Jan Sangh (BJS), and Jayaprakash Narayan, who persuaded the BJS to join the Janata Party in January 1977.

4. The Bharatiya Lok Dal (BLD) leader Charan Singh and Jagjivan Ram, who left the Congress after the Emergency, following the momentous 1977 general election. Acrimonious relations between Singh and Ram—respectively home minister and deputy prime minister in the Janata administration (1977–1980)—caused persistent turmoil and hastened its demise.

5. The Congress(O) leader Morarji Desai and A.B. Vajpayee, respectively prime minister and external affairs minister in the Janata government, advocated 'genuine non-alignment' in foreign affairs during their tenure.

6. Charan Singh, seeking to rejoin the Janata ministry, demonstrated his massive political clout in the capital in December 1978.

7. The Telugu Desam Party (TDP) chief, N.T. Rama Rao (right), congratulating the Janata Dal (JD) leader V.P. Singh (center) upon becoming prime minister of the National Front government (1989–1990). Disagreements between Singh and his deputy prime minister, the Lok Dal chief Devi Lal (left), created much tumult towards the end of its term.

8. A meeting of the National Front: sitting from left to right are N.T. Rama Rao, V.P. Singh, S.R. Bommai, Ram Vilas Paswan and Lalu Prasad Yadav.

9. The longstanding Janata Party president Chandra Shekhar became prime minister in November 1990 after instigating the fall of his nemesis, V.P. Singh. Yet he proved unable to answer many questions during his brief tenure.

10. The three principal leaders of the Communist Party of India (Marxist) in the 1990s: its general secretary, Harkishan Singh Surjeet; West Bengal chief minister Jyoti Basu; and E.M.S. Namboodiripad, the chief minister of Kerala.

11. The formation of the United Front in May 1996: standing from left to right are P. Chidambaram, N. Chandrababu Naidu, M. Karunanidhi, Biraj Sarma, G.K. Moopanar and H.D. Deve Gowda.

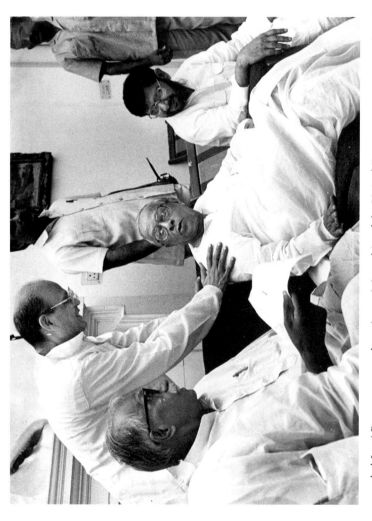

12. V.P. Singh persuaded Jyoti Basu to accept the prime ministership of the United Front government. But the CPI(M), fearing its limited parliamentary strength, rejected ministerial participation.

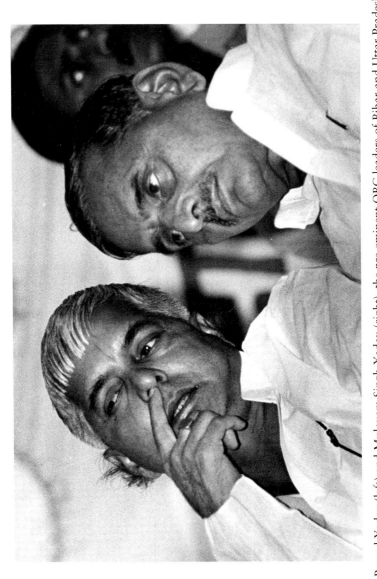

13. Lalu Prasad Yadav (left) and Mulayam Singh Yadav (right), the pre-eminent OBC leaders of Bihar and Uttar Pradesh in the 1990s, allies and rivals in equal measure.

14. Meeting of the United Front Steering Committee: seated from left to right are Mulayam Yadav, N. Chandrababu Naidu, N.D. Tiwari, Indrajit Gupta, H.S. Surjeet, G.K. Moopanar and Lalu Yadav.

15. Renuka Chowdhury of the TDP, inducted into cabinet in May 1997 by I.K. Gujral, the external affairs minister and second prime minister of the United Front. Despite his brief prime ministership, Gujral significantly improved subcontinental relations by extending unilateral concessions to neighboring countries, subsequently christened the 'Gujral doctrine'.

16. The venerable West Bengal chief minister, Jyoti Basu, eventually portrayed the decision of the CPI(M) not to participate in government a 'historical blunder'.

4

THE RISE OF THE REGIONS (1980–1989)

The downfall of the Janata Party threw its forces into disarray. First, the Jan Sangh rechristened itself the Bharatiya Janata Party (BJP). Its members had contested the 1980 general election with their socialist colleagues. The national executive of the Janata had amended Article 5 of its constitution in July 1979 to read, "The concept of a religious State is against the creed of the Janata Party and no member of an organisation having faith in a theocratic State can be a member of the Janata Party", apparently settling the dual membership issue in favor of the socialists.[1] But Jan Sangh candidates performed well in the subsequent polls, accounting for 16 of the 31 parliamentary seats the Janata won, which convinced Hindu conservatives to break away.[2] Second, Jagjivan Ram himself joined the Congress(U) during the campaign. Yet his attempted coup led its president, the former Karnataka chief minister Devraj Urs, to expel Ram, pitting the Congress (Jagjivan) against the renamed Congress (S).[3] Third, the JP(S)-Lok Dal itself split the following year, with Raj Narain and H.N. Bahuguna forming the Democratic Socialist Party.[4] The various elements of the Lok Dal would continue to implode and regroup for several years.

The Congress(I) remained politically vulnerable, however. The twin processes of political awakening and organizational decay that had begun in the late 1960s continued to accelerate in the early 1980s.[5] The party reclaimed critical votes amongst Muslims and Dalits in the Hindi belt. Yet its hold over both strata had weakened.[6] Despite its unruly short-lived tenure, the ideological rhetoric of the Janata Party had stoked political expectations amongst historically subordinate groups. Indira Gandhi unveiled new anti-poverty programmes and bolstered longstanding subsidies upon recapturing office.[7] Yet she simultaneously initiated a more fundamental "pro-business", "growth first" economic policy shift, downplaying her earlier redistributive rhetoric.[8]

Indeed, the Congress(I) rapidly encountered opposition in the regions. Many states returned the party to office after Mrs Gandhi dismissed their respective administrations in February 1980, citing the precedent set by the Janata in 1977, in the wake of her resounding national mandate. Yet these victories failed to prove the persistent national basis of state-level politics. If anything, the early 1980s revealed a groundswell of diverse oppositional formations. In the northeast, longstanding Assamese grievances against both Bengali domination in the urban economy and state administration as well as New Delhi for treating Assam as an "internal colony" coalesced with a new mass campaign against "foreigners".[9] The All Assam Students Union (AASU) demanded the deportation of all illegal migrants, particularly Bangladeshis arriving after 1971, and the revision of electoral rolls. The Congress(I) had imposed President's rule several times between 1979 and 1982 in response to the agitation. But it also promoted anti-Muslim rhetoric to court the Hindu vote. Its decision to hold state assembly elections on the basis of the unrevised 1979 lists, an estimated one-third of which compromised "foreigners", provoked the AASU to call for a boycott. "[The] last struggle for survival" triggered the deaths of thousands in the Nellie massacre in February 1983, forcing the Centre to re-impose Article 356.[10]

In addition, the emergence of a united communist bloc in West Bengal and Kerala, following the decision of the CPI and CPI(M) to pursue "left and democratic unity" in 1978, exposed the Congress(I)'s earlier developmental pretensions.[11] In West Bengal, the Left Front government pursued a strategy of "class conciliation" through "electoralist, gradualist and constitutitionalist" means.[12] On the one hand, the administration introduced Operation Barga, extending legal recognition to sharecroppers and offering greater protection to landless laborers. It also reinvigorated the *panchayats*, helping to break the influence of local power holders, replacing them with disciplined party workers. On the other, the Left Front state government provided agricultural subsidies for middle and large farmers, reversing its earlier opposition to higher remunerative prices,[13] and restrained labor militancy for industry and business, appointing a representative from each chamber of commerce on the West Bengal Industrial Development Corporation. Several factors explained its moderate redistributive posture. The repression of the 1970s highlighted the dangers of radicalism. Governing required political stability, moreover, and a longer time horizon. Lastly, the structure of powers in the Union made radical structural change impossible. According to the CPI(M) chief minister, Jyoti Basu:

> We said ... we can give better governance. We can assure democracy; we can give much more relief and even within this context of the capitalist feudal system [and Centre-state relations], some alternatives can be given, but no change of fundamentals.[14]

In short, Left rule in West Bengal was "communist in name ... [but] social-democratic in its ideology, social program and policies".[15] Its experiment in Kerala, which saw the Left Democratic Front (LDF) alternate in office with the Congress(I)-led United Democratic Front (UDF), was more radical, however. The LDF formented labor militancy after 1975, inducing capital flight and a crisis of accumulation by the early 1980s. However, it also strengthened primary education and public health and promoted greater workers' rights and gradually improved industrial relations, spurring labor productivity.[16] Stiff electoral competition, and the relatively greater independence of powerful trade unions and social movements in civil society, enabled remarkable social development outcomes.[17] Notwithstanding these differences, both Left experiments comprised genuine attempts to lessen absolute poverty and economic inequality in the countryside. The convergent political line of the CPI and CPI(M) towards national popular fronts bolstered anti-Congress forces. Indeed, both parties now sought to devolve the powers of the Union in economic, fiscal and administrative matters through radical constitutional amendments.[18]

Two longstanding strongholds of the Congress(I) in the south also witnessed the emergence of significant opposition parties. In Andhra Pradesh, a powerful backlash against high political manipulation led to the formation of the Telugu Desam Party (TDP) in 1982, formed by the regional film star Nandamuri Taraka Rama Rao (widely called NTR). The imposition of a succession of hapless chief ministers by Mrs Gandhi was the immediate catalyst. T. Anjiah, widely decried as the least competent, simply confessed: "I came in because of Madam, and

I am going because of her. I do not even know how I came here."[19] Her "abuse of power" was a decisive factor:

> The TDP is basically an anti-Congress(I) party. The Congress(I) party represents a self-seeking politics and promotes corruption ... They can't stand being out of power, and will use any means to illegally throw out anybody who wins the support of the people ... Mrs Gandhi misused the provision of Article 356.[20]

Yet the TDP also tapped growing regional pride:

> There was a strong and long-standing perception in Andhra that Telugus were not getting their due ethnic recognition ... that their self-respect was being assaulted ... This was an emotive issue and provided a foundation for our party.[21]

Finally, the TDP harbored an overriding desire to rework the distribution of powers in Centre-state relations towards the states:

> We wanted greater state autonomy and financial resources. For instance, we want[ed] to end the need to have concurrence from the Centre for many our programmes and initiatives ... We also want a greater devolution of [national] revenue ... It takes too much time to get central approval. And why should we need it? ... We know the needs of our state better than they do. They just interfere.[22]

The TDP, declaring that Andhra Pradesh was not the "branch office" of the Congress(I), rode to victory in the 1983 state assembly polls. Upon capturing office, the party introduced massive food subsidies, in contrast to the increasingly rightward shift of its opponent in New Delhi.[23] The Congress(I) suffered an analogous setback the same year in neighboring Karnataka, another southern bastion, where assembly polls catapulted the new Janata Party's Ramakrishna Hegde into the chief ministership. Finally, the incumbent JKN won the relatively free election held in Jammu & Kashmir in 1983, which saw Farooq Abdullah inherit the mantle of his recently deceased father. Hence by the early 1980s most states in the Union had experienced a spell of non-Congress rule.[24]

More importantly, the election of various opposition parties heralded a substantive agenda, raising demands for greater political devolution, economic decentralization and cultural recognition. Constitutionally, the Union divided political authority along federal lines.[25] The Centre enjoyed overriding authority, however. Politically, New Delhi had the power to appoint state governors, who could reserve state bills for Presidential consideration and veto, and to dissolve legislative assemblies. Moreover, although several bodies existed to address Centre-state relations—the Chief Ministers' Conference, National Integration Council and several zonal councils—successive Congress(I) governments had largely convened them on a makeshift basis. Economically, the Centre controlled the most buoyant revenue sources—income and excise taxes, custom duties and foreign aid—which amounted to approximately two-thirds of the total. In addition, several institutions distributed resources amongst the states, yet warranted reform. Many states viewed the Planning Commission, which allocated macro-level resources, as unresponsive to their concerns. The findings of Finance Commission, which issued recommendations on how to distribute central revenues, lacked statutory power. Lastly, the National Development Council, which prescribed guidelines and reviewed the functioning of the Planning Commission, had met sporadically at best. Formally, Mrs Gandhi had revived the National Integration Council in November 1980. Yet the organization remained inert. Far more telling was the 1980 President's speech to parliament, which highlighted the dangers posed by "antinational forces of regionalism, linguism and communalism".[26]

Hence in 1983 the chief ministers of leading opposition parties proclaimed a "constitutional revolt", organizing a series of conclaves in Bangalore (March), Vijayawada (May) and Srinagar (October).[27] Their deliberations included proposals to reform the political economy of Centre-state relations through greater cooperative planning and a new fiscal commission, to enhance state autonomy by devolving various powers and establishing an Inter-State Council and, most provocatively, to restrict or abolish Article 356. Mrs Gandhi, recalling her battles with the Syndicate, responded in two ways. On the one hand, she appointed Justice Ranjit Singh Sarkaria to lead a commission to examine possible reforms in Centre-state relations. It progressed slowly, however, leading many to see it as a containment strategy. On the other, she cynically stoked communal anxieties of various minority groups in different regions for electoral gains. In doing so, the increasingly embattled prime minister sought to demonstrate that only the Congress(I) could defend national unity.

These short-run tactical maneuvers had disastrous strategic consequences. Although the party had frequently mobilized vote banks by encouraging group identities in state and local elections in the 1950s and 1960s, Nehru rarely made religious appeals in national political discourse. By questioning the legitimacy of regional political formations, bolstering rivals with more sectarian agendas and giving explicit recognition to various communal identities, however, Mrs Gandhi roused extremist forces that gradually escaped her control. In the predominantly Hindu region of Jammu, the Congress(I) fanned communal anxieties, engineering a coup within the JKN to punish Farooq Abdullah for attending the conclave in Vijayawada and hosting its follow-up in Srinagar.[28] But it was in Punjab that her actions led to personal tragedy.[29] The prime minister undermined the moderate leadership of the SAD, which had been demanding greater political devolution, by supporting the Sikh extremist Sant Jarnail Singh Bhindranwale. The increasingly violent anti-Hindu campaign pursued by the latter, orchestrated from the Golden Temple in Amritsar, generated demands for central intervention. In June 1984, Mrs Gandhi ordered the Indian army to end Bhindranwale's operations by launching Operation Blue Star, causing his death. The desecration of the revered temple instigated two of her Sikh bodyguards to assassinate the prime minister on 31 October 1984, provoking an anti-Sikh pogrom in New Delhi, and leading the eighth general election to be held in December.

The Congress(I) won a landslide, capturing 49 per cent of the vote and 404 parliamentary seats, its most dominant performance ever. Its sources were threefold.[30] First, as the slain prime minister's younger son whom the party appointed as its parliamentary leader, Rajiv Gandhi enjoyed an unusual personal advantage. Mr Gandhi was a relative latecomer to national politics, projecting a fresh political image associated with youth and integrity. Yet he was simultaneously the heir to the Nehruvian political dynasty, thus representing continuity. Second, many commentators claimed the national electorate desired political stability and national unity, fearing the various regional conflicts that had grown since 1980. The Congress(I) chastised the opposition parties' ostensibly "backward looking, communal and reactionary policies and programmes", claiming that it alone stood "between unity and disintegration … stability and chaos … and self-reliance and economic dependence".[31] The fact that Mrs Gandhi's rule exacerbated multiple conflicts, and that politicians belonging to the Congress(I) directed, abetted and exculpated the anti-Sikh pogrom that followed her assassination, failed to break the wave of support that surged towards the party during the polls. Indeed, Rajiv Gandhi vilified regional opposition parties as anti-national. Lastly, the Opposition suffered from disunity, despite various efforts to build multi-party coalitions. On the one hand, the BJP and the Lok Dal overcame their earlier rift, forming the National Democratic Alliance. On the other, the Janata Party formed a United Front with the Democratic Socialist Party, Congress(S) and several minor outfits.[32] Yet the two party alliances, each harboring divergent social interests, failed to cooperate. The Opposition as a whole, moreover, had no prime minister designate. The electorate had good reason to distrust the capacity of its disparate constituents to govern collectively. Their internal divisions, disproportionately magnified by India's plurality-rule electoral system, bolstered the Congress(I).

The 1984 general election arrested neither the regionalization of the federal party system, however, nor the increasing political assertiveness of various backward castes and lower classes. A majority of national parties' supporters, especially of the Congress(I) and the BJP, cited national unity as their most pressing concern. But the voters of leading regional formations gave more importance to inflation, corruption and improving federal relations.[33] Moreover, various state-based parties either withstood the Congress(I) wave in their home regions, such as the AIADMK in Tamil Nadu, TDP in Andhra Pradesh and Left Front in West Bengal, or quickly rebounded in assembly polls the following year, such as the Janata Party in Karnataka.[34] Indeed, the TDP emerged as the single largest party in the Opposition, unprecedented for a genuinely regionalist formation. Hence Mr Gandhi faced persistent regional forces and escalating social demands for better political governance.

The new Congress(I) administration began well, expressing a zeal for modernization. Upon taking office, the prime minister declared that internal elections would be held to reinvigorate the party. His ministry also shepherded through parliament the 52nd Constitutional Amendment Act. The so-called Anti-Defection Law, stating that defectors would have to stand for a by-election unless they carried at least one-third of their erstwhile party, sought to encourage party cohesion. Simultaneously, Gandhi championed economic liberalization, surrounding himself with several like-minded technocratic advisors.[35] The 1985–86 Union budget, introduced by Finance Minister Vishwanath Pratap Singh, liberalized trade and lowered taxes, and simplified licensing and deregulated various sectors of the economy. Significantly, it failed to mention the mantra of socialism. Further concessions to big business followed. In addition, the Congress(I) sought to resolve intense regional conflicts in the summer of 1985. The prime minister signed the Rajiv Gandhi-Harcharan Singh Longowal Accord, recognizing several longstanding demands of the Akalis and agreeing to pursue the perpetrators of the anti-Sikh pogrom in Delhi.[36] The SAD secured a massive legislative majority in the October 1985 assembly polls. Similarly, the Government negotiated the so-called Assam accord, which partly accommodated the electoral demands of the AASU while pledging to establish higher educational institutions in science and technology.[37] The Asom Gana Parishad (AGP), a successor to the AASU, swept to power in the ensuing 1985 polls. Finally, the Rajiv ministry courted liberal opinion by backing the landmark Supreme Court judgment in the Shah Bano case, which recognized the right of a Muslim widow to alimony under civil law. Collectively, these various overtures seemed to herald a new beginning.

But the failure of the Congress(I) to implement its various proposals, and key electoral defeats in several regions, exposed its wayward political commitments and weak organizational roots. First, the Government encountered mounting resistance within the party against economic liberalization and internal reform, allowing each initiative to wither.[38] Various quarters expressed anxiety that its "pro-rich" urban image would alienate the middle peasantry and the rural poor. Sporadic agrarian mobilizations against the new reform agenda in various states justified such concerns politically. Poor monsoons in 1986 and 1987, stoking food inflation, lent them credibility. Hence the prime minister gradually reverted to using socialist rhetoric and "muddling through", creating a discordant policy framework after 1986. The drive to reform the party, moreover, petered out. The old clientelistic office-bearers of the Congress(I) sabotaged the electoral process by creating bogus membership rolls. Mr Gandhi also protected many unpleasant characters that had entered the party under his mother and brother. The prime minister, realizing that his position partly depended upon loyal subordinates that lacked an independent political base, eventually abandoned his proposed reforms.

Second, although the Congress(I) secured legislative majorities in many state-level polls in 1985, several regional formations consolidated their strength. The party lost heavily to the Janata Party in Karnataka in 1985, and to the communist Left Fronts in Kerala and West Bengal and to the Lok Dal in Haryana in 1987. Critically, all three fronts attacked the Government's technocratic liberalizing reforms. In addition, the Congress(I) revised the terms of the Ninth

Finance Commission by holding states responsible for fiscal indiscipline but not the Centre, generating widespread opposition. Its poor handling of the NDC, which impeded deliberation, produced further acrimony.[39] Lastly, the Anti-Defection Law backfired. Rather than improving party cohesion, the amendment gave political factions an incentive to form smaller cohesive groupings prior to elections. Small factional parties, with relatively tiny bases of support in a single state, began to form.[40]

Third, the Congress(I)'s old political reflexes regarding communal questions and Centre-state relations gradually reemerged. On the one hand, ham-fisted short-term attempts to placate religious conservatives undermined its professed secular credentials. The Government reversed its earlier support for the Shah Bano case by passing the Muslim Women (Protection of Rights on Divorce) Act, 1986, which nullified the Supreme Court judgment. Expedient calculations drove the Congress(I)'s volte face. The party, facing mounting opposition from conservative Muslim organizations, feared greater losses in upcoming state elections.[41] The BJP president, Lal Krishna Advani, denounced Congress(I) for practicing "minority appeasement" and "pseudo-secularism".[42] Rather than defending its concession to orthodox Muslim opinion, however, the Congress(I) sought to placate the Hindu right. In 1984, the Vishwa Hindu Parishad (VHP) and its new militant wing, the Bajrang Dal, had resumed an old campaign to build a *Ram mandir* (temple for the Hindu god Ram) on the grounds of the Babri masjid in the town of Ayodhya in Uttar Pradesh. The ethno-religious mobilization threatened to polarize communal tensions and provoke violent conflict. Yet the state-level Congress(I) administration failed to challenge a controversial verdict by a district court judge to unlock the gates to the disputed site. Indeed, in due course the party authorized the VHP to perform its *shilanyas* (foundation stone laying ceremony), which Rajiv Gandhi eventually approved, energizing the *Ramjanmabhoomi* movement.[43]

On the other, the Government's vaunted regional breakthroughs never materialized. The prime minister failed to implement the Rajiv-Longowal Accord in Punjab, due to administrative incompetence, narrow electoral considerations and the assassination of the Akali leader. Terrorist activities resumed, which the Centre exacerbated through coercive repression.[44] Similarly, the goodwill created by the Congress(I) in Assam quickly dissipated. The absence of a modern identification regime, the contested status of "plains tribal" groups such as the Bodo in the greater Assamese population and the ramifications of the Illegal Migrants (Determination of Tribunal) Act, 1983, made it immensely difficult to implement the accord.[45] The AGP, perceived as incompetent and corrupt, was a factor. But claims that the Research and Analysis Wing (RAW) was training Bodo militants in opposition to the AGP recalled the Congress(I)'s earlier scheming. Yet most damaging was the party's decision, in alliance with the JKN, to rig the 1987 Jammu & Kashmir assembly polls. Farooq Abdullah, the JKN chief, shared much blame. Given its control over the central state apparatus and recent political machinations in the region, however, the Congress(I) shouldered the burden of responsibility. The fraudulent polls ignited a massive popular uprising in the Valley, which the Centre's subsequent heavy-handed response transformed into a violent insurgency. The massive electoral victory won by Rajiv Gandhi in 1984, in short, failed to mask the exclusion of various regional voices and the closure of the Congress(I) to their concerns.[46] Thus many opposition parties participated in a second round of conclaves, organized by the communists and the socialists, in Bangalore (January 1988) and Calcutta (December 1987 and January 1989).[47]

Ironically, the Government's modernizing zeal ultimately proved its undoing. In his bid to liberalize the economy, the Union finance minister, V.P. Singh, had sought to capture tax evasion. His highly publicized raids on well-known industrial firms, however, netted key donors to the Congress(I). Unwilling to lose their support, the prime minister shifted the crusading minister to Defense in January 1987, forcing his resignation in April after allegations arose that Germany had paid commissions for the sale of HDW submarines to India.[48] But concurrent media reports that a Swedish arms company, Bofors, had offered kickbacks worth Rs. 64 crore to the Government the previous year for the supply of Howitzer field guns entangled the prime minister. According to reports, an Italian businessman with close ties to the Gandhi family, Ottavio Quattrocchi, was the middleman for the deal.[49] The scandal resurrected the Opposition.

The formation of the National Front

On 2 October 1987, V.P. Singh launched the Jan Morcha (People's Front) with Arun Nehru, Arif Mohammad Khan and several other Congressmen that had been expelled for dissidence. On 16 June 1988,

Singh routed the Congress(I) candidate and son of former Prime Minister Lal Bahadur Shastri in the Allahabad by-election. His victory carried symbolic weight: although his previous home constituency, Allahabad was an old Congress bastion, originally held by Nehru himself. On 6 August 1988, the Jan Morcha became the fulcrum of the National Front (NF), a diverse multiparty alliance of largely state-based parties, encompassing the AGP, TDP and DMK as well as the Congress(S). The regional parties shared a peculiar affinity with their new allies. On the one hand, both groupings opposed the high-handed centralizing proclivities of the Congress(I). On the other, they generally supported the politics of non-Brahmanism and the rights of minority religious communities, with the exception of the AGP. On 11 October 1988, Jayaprakash Narayan's birth anniversary, Singh transformed the Jan Morcha into the Janata Dal (JD), incorporating the rump Janata Party and the two main factions of the Lok Dal.

In order to defeat the Congress(I), however, the National Front had to maximize opposition unity. Crucially, Singh secured electoral agreements with the communist Left and Hindu right. The primary aim of both formations was twofold: to expand their respective electoral bases and defeat the Congress(I). The CPI(M) organized a major opposition rally in Calcutta following Singh's momentous victory in Allahabad. Significantly, the West Bengal Chief Minister Jyoti Basu shared the stage with him as well as A.B. Vajpayee of the BJP.[50] Yet many communists had severe misgivings about such "unprincipled unity".[51] The growing influence of the RSS in Kerala, their southern fortress, caused much concern.[52] The decision by the VHP to conduct *Ram shila pujans* (the consecration of bricks collected for building a Ram temple) in late 1988, which the BJP endorsed, facilitated mass communal mobilization. The cycle of rioting that ensued in Uttar Pradesh propelled the Left towards open confrontation. At its 13th party congress in December, the CPI(M) general secretary E.M.S. Namboodiripad declared that the party would seek "to bring about the unity of Left and secular Opposition parties ... to oust the Congress(I) and isolate the communal and divisive forces".[53] The electoral pact in September 1989 between the BJP and the Shiv Sena, a nativist regional party in Maharashtra that propagated Hindu extremism, intensified the communist parties' discomfiture: "[it is] regrettable that some elements in the Janata Dal are pulling wool over their eyes ... [which gives] the Congress(I) the opportunity to parade as the only secular force ... and will only push the minorities in [their] lap".[54] But the Left was unwilling to oppose the National Front. Its ambivalent position, according to some, reflected confusion.[55] Yet it exposed a genuine political dilemma:

> For both [the CPI and CPICM] to merge and join the Centre would represent greater co-optation, yet would also give a more meaningful social-democratic and populist angle to the rudder of government as it steered through various policies ... a more principled and radical mainstream left would resist ... yet allow the weight of its redoubled force to exercise pressure on government policies ... [yet both parties exhibit signs of] disturbing complacency, especially the CPI(M) ... it cannot afford to stand still.[56]

The BJP was divided too. A section of the party sought full membership in the National Front. The Left's fierce opposition to its participation provoked L.K. Advani, the BJP president, to declare:

> EMS [Namboodiripad] has an obsession about the BJP. He is worried about [our] growing strength and is conscious of his own party's limitations... In more than 425 Lok Sabha seats ... the Communist Parties just do not matter... nevertheless they keep trying to dictate to other parties what they should do or not do. It is high time the Communist bluff is called... [During the Emergency] they were either active supporters of dictatorship or passive spectators thereof. I would like to caution other Opposition parties to beware of this potential Trojan Horse".[57]

Despite his taunt, Advani opposed joining a prospective government. A majority of the party cadre, whose "memory of [their] stepmotherly treatment in the 1977–79 government still rankled", supported his position.[58] Moreover, the BJP, which had courted political respectability during the Janata experiment by suppressing its Hindu nationalist proclivities, was now unwilling to do so. The party sought to ban cow slaughter, impose a uniform civil code, which would nullify the special judicial arrangements regarding personal laws governing minority religious communities, and abrogate Article 370 of the Constitution, which gave the disputed Muslim-majority state of Jammu & Kashmir special asymmetric rights in the Union. V.P. Singh had excluded the BJP during the formation of the National Front. But its strong electoral showing in local elections in Uttar Pradesh in late 1988, where the party focused attention on cancelling peasants' debts, establishing agricultural cooperatives and other Gandhian socialist programs, forced him to acknowledge its growing power.[59]

Ultimately Singh pursued a delicate tactical approach, displaying a capacity for ruthlessness, to retain the support of the communist Left and Hindu right.[60] He consistently referred to the former as his "natural ally" at public rallies, maintaining extensive communication with the CPI(M) West Bengal Chief Minister Jyoti Basu. The JD leader highlighted the communist parties' calls for greater labor participation in industrial relations, an anti-corruption programme and the necessity of electoral reforms. Conversely, Singh attacked the RSS, and refused to share the stage at electoral rallies with the BJP. Yet he maintained regular contact with Vajpayee, the former BJP president, whom the media portrayed as a moderate. The JD chief stressed that their electoral understanding was not a power-sharing agreement, while defending political secularism and minority rights, "statements that should have driven the BJP away, had they been possessed of a thinner hide and a weak or vacillating leadership".[61] Singh's tactical ploys and the shared desire of the communist Left and Hindu right to defeat the Congress(I) allowed them to suspend, for the moment, their intense mutual antipathy.

The coming together of major opposition parties spelled the end of the Congress(I). In June 1989, their respective MPs collectively resigned from parliament, launching mass political mobilizations through August. In November 1989, the National Front defeated Rajiv Gandhi in India's ninth general election, becoming only the second non-Congress formation to capture national power since independence.

5

THE NATIONAL FRONT (1989–1991)

The ninth general election, which produced a hung parliament, unveiled a new political landscape.[1] The Congress(I) remained the single largest presence in the Lok Sabha, winning 39.5 per cent of the vote and 197 seats. It contested more constituencies, 510, than any other party. Few of its candidates lost their deposits. Hence President Ramaswamy Venkatraman invited Rajiv Gandhi to try forming a Union government. But the party's electoral performance paled in comparison to the stunning parliamentary majority it had lost. The coalescence of the Opposition was a major factor. Yet the theme and approach of the Congress(I) campaign revealed deep political limitations.[2] The party managed its campaign from New Delhi, focusing attention on Rajiv Gandhi and highlighting the importance of political stability and national unity. As a result, it often failed to address regional concerns or employ local idioms during campaign speeches. Moreover, the Congress(I) had re-nominated approximately 80 per cent of its sitting MPs, symbolizing its satisfaction with the status quo. The massive swing against the party represented a severe electoral repudiation. Thus its high command eschewed government formation, shifting attention to the National Front.

The most important force in the incipient governing coalition was the JD (see Table 5.1). The party won 17.8 per cent of the national vote and 143 parliamentary seats. Significantly, the majority of the latter came from the Hindi belt, with 54 in Uttar Pradesh and 32 in Bihar.

Table 5.1: Ninth General Election, 1989

Party	Seats	Vote
BJP	85	11.36
CPI	12	2.57
CPI(M)	33	6.55
ISC(SCS)	1	0.33
INC	197	39.53
JD	143	17.79
ADK	11	1.50
FBL	3	0.42
JKN	3	0.02
MAG	1	0.04
MUL	2	0.32
RSP	4	0.62
SSP	1	0.03
TDP	2	3.29
GNLF	1	0.14
HMS	1	0.07
IPF	1	0.25
JMM	1	0.34
KCM	1	0.12
M-COR	1	0.08
MIM	1	0.21
SAD(M)	6	0.77
SHS	1	0.11
Independents	12	5.25

Source: Election Commission of India.

Indeed, the ratio of upper-caste MPs fell below 40 per cent for the first time.[3] Crucially, the party also won Muslim support in the wake of the rising communal violence, expanding its traditional AJGAR social base.[4] Of the remainder, the party secured sixteen constituencies in Orissa, eleven apiece in Rajasthan and Gujarat, and an equivalent number between Haryana and Maharashtra. Significantly, it captured only one seat in the south, in Karnataka. The second largest force supporting the National Front was the BJP. The party captured 11.4 per cent of the vote and 85 parliamentary constituencies, far above its performance in 1984, when it secured 7.7 per cent and merely two seats. Its militant Hindu posture paid rich dividends, especially in Madhya Pradesh, where it captured 27 seats. Of equal importance, though, was

THE NATIONAL FRONT (1989-1991)

its electoral pact with the JD. The parties agreed upon which candidates to field in 231 of 252 seats in the Hindi belt, leading to thirteen BJP victories in Rajasthan, twelve in Gujarat and eight apiece in Uttar Pradesh and Bihar.[5] The third most important force was the communist Left. The CPI and CPI(M) won 2.6 and 6.6 per cent of the vote and 12 and 33 parliamentary seats, respectively. Several issues separated the two. The CPI(M)'s unwillingness to distribute more tickets to the CPI and their smaller communist allies during assembly elections, and accusations of growing administrative corruption, generated political conflicts in West Bengal.[6] Moreover, the CPI displayed more democratic instincts regarding the potential for reform. The party supported *perestroika* in the USSR, and was "saddened" by the Tiananmen Square massacre in the PRC. The CPI(M) supported the status quo in both regimes.[7] Yet both parties maintained a convergent political line. Their combined electoral showing underscored the benefits of cooperation. Finally, despite their recent upsurge in the states, many regional parties suffered disappointing results. Growing political violence in Assam, which revolved around the demand for a Bodo homeland, delayed polling in the state, thereby shutting out the AGP. A growing internal feud within the TDP, which saw N.T. Rama Rao dismiss his cabinet, cost the incumbent party dearly. Many dissidents, given tickets by the Congress(I), won their constituencies.[8] The TDP, despite capturing 43.6 per cent of the vote in the 33 seats it contested, managed to survive in only two. Similarly, despite capturing office in the January 1989 state assembly elections in Tamil Nadu on a platform of cultural nationalism and social welfare, the DMK suffered a terrible rout at the hands of the Congress(I)-AIADMK.[9] Nevertheless, the National Front stuck together politically.

Indeed, its formation signaled the arrival of the "third electoral system", or pattern of national party competition, in modern Indian democracy.[10] The ninth general election was the first normal contest in decades. In contrast to the single overriding issues and sense of crisis that had dominated national polls since 1971, the contestants highlighted everyday concerns, such as corruption, inflation and general political mismanagement.[11] Yet the verdict suggested deeper political transformations. First, unlike the national waves of the past, states increasingly comprised the effective units of electoral choice. Second, distinct two-party/multi-party systems had arisen in most states, yielding a complex parliamentary outcome in New Delhi. Third, the electoral participation of historically subaltern groups had begun to surpass the voting propensities of more privileged classes. In sum, the increasing regionalization of the federal party system and popular democratic mobilization of previously marginalized groups had produced a critical juncture in modern Indian democracy, despite the absence of formal institutional change.[12] As a result, no single party could dominate political competition in every state or set the terms of rule in New Delhi, not even the Congress(I). Put differently, the prospect of a diverse multiparty government at the Centre heralded the possibility of a radical new politics led by a national third force.

Ideologically, the new governing coalition was a successor to the Janata Party. "The main object of the National Front", according to Article 2 of its constitution, "is to build up a Democratic, Secular and Socialist State in India and provide a clean and efficient administration for the country with an emphasis on equality and social justice."[13] Its manifesto dwelt upon corruption, inequality and violence. Politically, the alliance emphasized federal relations, reflecting its pronounced regional orientation. The National Front promised to create an Inter-State Council, bolster the NDC, prevent the abuse of Article 356 and legally empower *panchayats*. It also pledged to improve governance by passing a Lok Pal Bill, ensuring judicial independence and protecting the right to information in the media. Economically, the National Front vowed to pursue labor-intensive small-scale industrialization and curb multinational corporations, bolster agricultural development through greater public investment, remunerative prices and crop insurance, and tackle the growing debt trap. Socially, the alliance promised to promote equity by implementing central reservations for OBCs as well as women, granting statutory authority to the Scheduled Castes and Scheduled Tribes Commission, and introducing a right to education, housing and work.[14] Finally, it pledged to maintain an independent foreign policy, while establishing a national security council to coordinate strategic activities.[15] In short, while the National Front's manifesto lacked the fire of the original Janata Party, their expressed ideological continuities were still clear.

Several challenges confronted the new governing coalition, however. First, as the Janata had demonstrated, the "inability to act in unity" was a chronic problem of anti-Congress formations. The "ambitions of leaders for senior posts", whose "mercurial politics" seemed devoted to "power-first" principles, required astute political management.[16] This particularly concerned the JD. Seething personal rivalries endangered its cohesion. V.P. Singh had earned the mantle of leadership by rousing public support in the Gangetic plains and bringing the complex multiparty alliance together. Yet he was a somewhat odd choice to lead the socialists' latest avatar. Despite being a former sarvodayite Singh was the scion of a Rajput princely family, who had risen to prominence in the Congress(I) under Indira Gandhi. Significantly, he viewed the Emergency as harsh, but never censured it. Indeed, Singh valiantly defended Mrs Gandhi during the Shah Commission proceedings, earning him the chief ministership of Uttar Pradesh in 1980.[17] Consequently, the newly minted president of the JD had many competitors. Chief amongst them were Devi Lal, the chair of its central parliamentary board and erstwhile leader of the Lok Dal(A), and Chandra Shekhar, the original Janata Party president. The establishment of the JD had involved "long days assuaging hurt feelings, imagined slights, political differences".[18] Following its formation, the prime minister asked several trusted colleagues in the party, Biju Patnaik, George Fernandes and Ramakrishna Hegde, to handle interpersonal relations. Arguably, Singh's willingness to delegate political responsibility showed personal humility as well as organizational acumen. But his main political adversaries, according to one close observer, saw it as a sign of his weakness.[19]

Second, the National Front was a minority coalition government, whose survival depended upon the Left and the BJP. Singh had managed to balance their interests during the campaign. The former declared unconditional backing of the Government after the verdict, while the latter expressed "general but critical" support.[20] Nevertheless, powerful intersecting tensions pitted these mortal political enemies against each other, threatening rifts across the wider alliance. The most important fissure was the increasingly militant posture of the BJP. The Left steadfastly opposed the *Ramjanmabhoomi* movement and other Hindu nationalist demands. The latter also unnerved the socialist members of the JD. Apart from matters of principle, the promotion of backward caste aspirations clashed with the conservative Sanskritic order championed by the Hindu right. Lastly, the unitary national vision of the BJP collided with the regional assertiveness of the TDP, AGP and DMK. Tellingly, the BJP had not participated in the regional conclaves in the mid-1980s.[21] To some extent, the strong federal orientation generated concern amongst some communists too. Jyoti Basu had played a leading role in driving the conclaves' agenda, and hosted its forums in Calcutta. Still, several CPI(M) leaders feared that demands for "regionalism" could lead Western "imperialists" to threaten national sovereignty by encouraging separatism.[22]

National economic policy was the second potential fissure. Rhetorically, the National Front adopted a more traditional stance regarding industrial policy and external protection.[23] Nonetheless, V.P. Singh had introduced several liberalizing measures during his stint in the Union finance ministry, worrying several parties. Socialists in the JD continued to promote egalitarian measures in agriculture, industry and the public sector. A more powerful critic was the Left. The policies of the latter belied its anti-capitalist rhetoric to some extent. In 1985, the West Bengal government had allowed joint industrial projects with monopoly houses and foreign capital, hurting its support amongst workers. "Committed to the pledge to give 'immediate relief' to the people...", noted one commentator, "the CPI(M) finds itself ... [with] the very forces which, according to their party programmes, they should resist in order to prevent them from making inroads into the Indian economy".[24] The mid 1980s had witnessed declining labor militancy in Kerala too. The deradicalization of the Left had several causes. Proponents rightly highlighted external constraints. Successive Congress(I) governments at the Centre had discriminated against West Bengal in terms of industrial licensing, regulation, and finance, compelling the state to rejuvenate its faltering industrial economy through foreign investment. Hence many claimed the CPI(M)-led administration was in office but not in power.[25] A similar pattern of underinvestment and interference with transfers of finance and food mounted in Kerala too.[26] Yet critics attacked the Left in West Bengal given its suppression of labor in industry, overrepresentation of landowning cultivators and middle class groups in *panchayats*, and failure to pursue cooperative experiments in industry, tax higher agricultural incomes or utilize Central funds.[27] Both sides of the argument had a point. Nonetheless, further economic liberalization at the Centre risked provoking the Left. In Kerala, negotiated industrial agreements recognized the importance of productivity and growth for redistribution. The balancing of wages and profits reflected constructive class compromise, laying the foundation for shared economic rejuvenation after the LDF returned to office in 1987.[28] Moreover, the leftist wings of the communist parties often controlled their respective politburos, based in New Delhi. Further economic liberalization would attract their denunciation.

THE NATIONAL FRONT (1989–1991)

The third major disagreement facing the National Front was its pledge to implement the so-called Mandal Commission Report. The demand to extend public reservations for OBCs posed greatest concern for the BJP. The ascent of the backward castes threatened its project of building a united Hindu *rashtra*, based on a hierarchical Sanskritic ideology, as well as the material interests of the high-caste urban middle classes that supported the party. The politics of caste-based reservations also clashed against the class-based politics championed by the Left. Yet the Report also created fissures within the JD.[29] On the one hand, Devi Lal and Ajit Singh of the Lok Dal(A) championed the material interests of intermediate proprietary castes in western Uttar Pradesh. A devotee of *kisan* politics, Lal had prevented the party from being named the Samajwadi Janata Dal (Socialist People's Party). On the other, socialists in the Janata Party advocated quota politics, led by Chandra Shekhar in eastern Uttar Pradesh. They emphasized the need to expand caste-based reservations to address inequalities of status and power. In between lay the Lok Dal(B), led by Mulayam Singh Yadav, who appealed to the backward caste identities of his followers while protecting their material interests in the countryside.[30] As a result, the politics of Mandal portended serious conflicts.

The various tensions gripping the National Front were not "insurmountable".[31] Yet they required astute political skills and coalition management. V.P. Singh had reportedly considered building a sustained grassroots movement over several years. But the temptation to topple the Congress(I) proved too great.[32] Parliamentary cabinet government encouraged power sharing. Yet the desire for particular executive posts posed obstacles. Electing the parliamentary leader of the new ruling coalition, and thus the prime minister, was the first political test. It ended poorly.[33] Many quarters assumed the post belonged to V.P. Singh. Crucially, he was also the first choice of the Left Front and BJP, which provided external parliamentary support. Chandra Shekhar opposed the move, however. Associates of the Janata Party chief, who had longed for the prime ministership, contended that he had earned the chance given his presidency of the original Janata formation, persistent leadership of the socialists during the 1980s and seniority. Thus he agreed to support Devi Lal instead. But the latter nominated Singh, in a jointly planned ruse, receiving the Union agricultural ministry and deputy prime ministership in return. On 2 December, Singh was sworn in as India's seventh prime minister since independence. His political elevation was inevitable. But the trick humiliated Chandra Shekhar, who simply remained a member of the central parliamentary board of the JD.

The newly elected prime minister handled the distribution of portfolios in the Council of Ministers better. Several criteria influenced its composition.[34] First, the Singh ministry reflected the primacy of the Janata Dal and its bastions in the north. More than half of its ministers represented Bihar and Uttar Pradesh, the two biggest states in the country.[35] Second, Singh appointed several able hands to key economic ministries, such as Madhu Dandavate to Finance, Ajit Singh to Industry and Arun Nehru to Commerce. He also appointed the Kashmiri politician Mufti Mohammed Saeed as home minister, making him the first Muslim to occupy the post. Finally, despite their disastrous electoral performance, Singh appointed members of the TDP, AGP, Congress(S) and DMK to his ministry. Several motivations played a role. The JD had only one outpost in the south, Karnataka. Every regional party had participated in the regional conclaves in the mid-1980s, moreover, becoming full members of the National Front. Lastly, and perhaps most importantly, it underscored the diverse regional orientation of the latter, symbolized by its launch in Madras (see Table 5.2).

Finally, the prospective governing coalition had to coordinate its relations with the Left Front and the BJP, its two crucial parliamentary allies. The prime minister instituted an informal political institution, namely, a weekly dinner at his residence.[36] He also proposed formal cabinet-level panels to address critical issues, such as containing inflation, implementing the right to work, empowering local government, reforming electoral finance, and resolving the conflicts in Jammu & Kashmir and the Punjab, amongst other concerns.[37] Yet neither Singh nor any of his colleagues tried to formulate a common minimum programme or high-level coordination committee that could provide, respectively, a working policy agenda or a dispute resolution mechanism to unify the diverse parliamentary coalition upon which the Government's tenure relied. Singh's failure to introduce formal decision-making mechanisms and durable inter-party agreements revealed, according to one seasoned observer, a "distaste for organisation building ... that made him vulnerable [to partisan conflicts]".[38] The tenure of the National Front was likely to be short-lived and unimpressive, according to another, unable to institutionalize the growing political mobilization of interests and groups that had characterized the 1980s.[39]

THE NATIONAL FRONT (1989–1991)

Table 5.2: National Front Government

Name	Party	Post(s)	Rank
Vishwanath Pratap Singh	JD	Prime Minister	Cabinet Minister
Devi Lal	JD	Agriculture	Cabinet Minister
Ajeet Singh	JD	Industry	Cabinet Minister
Madhu Dandavate	JD	Finance	Cabinet Minister
George Fernandes	JD	Railways (also Kashmir Affairs)	Cabinet Minister
Dinesh Goswami	AGP	Steel & Mines. (also Law & Justice)	Cabinet Minister
I.K. Gujral	JD	External Affairs	Cabinet Minister
M.S. Gurupadaswamy	JD	Petroleum & Chemicals	Cabinet Minister
Arif Mohammad Khan	JD	Energy (also Civil Aviation)	Cabinet Minister
Murasoli Maran	DMK	Urban Development	Cabinet Minister
Mufti Mohammad Syed	JD	Home Affairs	Cabinet Minister
Arun Nehru	JD	Commerce (also Tourism)	Cabinet Minister
Ram Vilas Paswan	JD	Labour (also Welfare)	Cabinet Minister
K.P. Unnikrishnan	ICS(SCS)	Surface Transport (also Communications)	Cabinet Minister
P. Upendra	TDP	Parlimentary Affairs (also Information and Broadcasting)	Cabinet Minister
Sharad Yadav	JD	Textiles (also Food Processing Industries)	Cabinet Minister
Nilamani Routray	JD	Health & Family Welfare, Environment & Forests	Cabinet Minister
Nathu Ram Mirdha	JD	Food & Civil Supplies	Cabinet Minister
Manubhai Kotadia	JD	Water Resources	Minister of State
Maneka Gandhi	JD	Environment & Forests	Minister of State
M.G.K. Menon	JD	Science & Technology (also Atomic Energy, Electronics, Ocean Development and Space)	Minister of State
Dr. Raja Ramanna	CPI(M)	Defence	Minister of State

Rashid Masood	JD	Health & Family Welfare	Minister of State
Chimanbhai Mehta	JD	Human Resource Development	Minister of State
Bhajaman Behra	JD	Petroleum & Chemicals	Minister of State
Hari Kishore Singh	JD	External Affairs	Minister of State
Upendra Nath Verma	JD	Rural Development	Minister of State
Subodh Kant Sahay	JD	Home Affairs	Minister of State
Satya Pal Malik	JD	Parliamentary Affairs (also Tourism)	Minister of State
Bhagey Goverdhan	JD	Planning and Programme Implementation	Minister of State
Nitish Kumar	JD	Agriculture and Cooperation	Minister of State
Srikant Jena	JD	Small Scale, Agro and Rural Industries	Minister of State
Arangil Sreedharan	JD	Commerce	Minister of State
Ram Poojan Patel	JD	Food & Civil Supplies	Minister of State
Ajay Singh	JD	Railways	Deputy Minister
Usha Singh	JD	Women & Child Development	Deputy Minister
Anil Shastri	JD	Finance	Deputy Minister
Bhakt Charan Das	JD	Youth Affairs & Sports	Deputy Minister
Jagdeep Jhankhar	JD	Parliamentary Affairs	Deputy Minister

THE NATIONAL FRONT (1989–1991)

The politics of Centre-state relations, and its foreign policy entanglements, occupied the Government at the start. In early December, the Jammu and Kashmir Liberation Front (JKLF), a secular pro-independence movement that championed *Kashmiriyat* (Kashmiri cultural identity), kidnapped Mufti Mohammad Saeed's daughter. The Government responded in a contradictory manner. On the one hand, it acceded to the demands of the JKLF, releasing some of its associates from prison. The National Front also convened a special all-party meeting on Kashmir, affirming Article 370, and established an advisory committee led by George Fernandes.[40] And it proposed high-level dialogue with Pakistan. On the other, the Singh ministry reappointed Jagmohan as governor, despite his past record of repression. It also continued to blame Pakistan for sponsoring "terrorism" in Kashmir. Bellicose statements by Benazir Bhutto in March, alongside evidence that Pakistan had assembled a nuclear device and established training camps in Pakistani-administered Kashmir, escalated tensions drastically and drew American intervention.[41]

Indeed, the February state assembly elections had witnessed much anti-Pakistan rhetoric, escalating tensions between the JD and BJP.[42] The two parties struck electoral adjustments in Madhya Pradesh, Himachal Pradesh, Rajasthan and Gujarat. In the former two states, the BJP won legislative majorities; in the latter two, it formed coalition governments with the JD as a junior ruling partner. The parties contested each other in a majority of seats in Bihar and Orissa, however, allowing the JD to form their respective administrations. Moreover, only in Orissa was the latter able to form a government on its own. The outcome of these state-level contests highlighted the rising political fortunes of the BJP and its growing electoral competitiveness vis-à-vis the JD.

The Left pressured Prime Minister Singh to recall Jagmohan in May 1990 after he had unleashed "a regime of terror" in the Valley, converting an underground insurgency into a mass political movement for self-rule, which provoked most parties to call for restraint by the armed forces.[43] The move riled hardliners in the BJP, leading to the dissolution of the Fernandes committee.[44] The Government refrained from attacking training camps on the Pakistani side and, setting a precedent, agreed to initiate confidence-building measures.[45] But in September the Lok Sabha passed the draconian Armed Forces (Jammu and Kashmir) Special Powers Act, 1990, granting soldiers legal immunity to conduct search operations, seize property and detain without warrant. Violence steadily escalated.

Early moves by the prime minister promoted some goodwill in Punjab.[46] Singh visited Amritsar, held an all-party rally in Ludhiana and appointed a new governor. He also promised to establish special courts to prosecute the perpetrators of the 1984 anti-Sikh pogrom. But growing violence between state armed forces and various insurgent groups undermined the small opening that had been created. In May 1990, the Lok Sabha reimposed President's rule for another six months, extending it again in November.

The National Front managed to rescue some of its federal ambitions. In late May, the Government established the Inter-State Council (ISC) under Article 263 of the Constitution, a longstanding demand of the states that successive Congress administrations had denied. It empowered the ISC, an advisory body, to investigate subjects, make recommendations and initiate general deliberations on all matters regarding Centre-state relations.[47] Observers expressed concern that it lacked statutory independence.[48] Yet the wide membership of the ISC—which included the prime minister, all chief ministers and six cabinet members by nomination—and its *modus vivendi*—all meetings and decisions would be held and made, respectively, in camera and by consensus—underscored two advances. First, it signaled the arrival of state-based parties into the seat of national power. Second, creating a formal political institution devoted to Centre-state relations indicated the decentered political vision of the new ruling formation. Its decision to constitute the Cauvery Waters Dispute Tribunal in early June, to adjudicate water-sharing disputes among Karnataka, Tamil Nadu, Kerala and Pondicherry, manifested a similar political spirit.[49]

Indeed, the federalist vision that animated the National Front influenced its foreign policymaking to some extent too. The National Front and the Congress(I) expressed few substantive disagreements over foreign policy during the 1989 campaign. Some observers noted their distinctive rhetoric, however, leading others to discern more substantive differences.[50] Indo-Nepal relations provided the first test. In March 1989, the Congress(I) had unilaterally decided not to renew the trade and transit treaties between the two countries, attempting to force Nepal to reconsider its growing economic ties and security relations with China. The resulting blockade caused major hardship to the smaller landlocked state. Sino-Indian relations divided the National

THE NATIONAL FRONT (1989–1991)

Front. The External Affairs Minister, Inder Kumar Gujral, a former Congressman whom Sanjay Gandhi had dismissed from his post in the ministry of Information & Broadcasting during the Emergency for being too soft,[51] had previously expressed support for turning the *de facto* line of control into the *de jure* border. However, many socialists remained opposed to Chinese sovereignty over Tibet. Thus V.P. Singh largely sought to maintain the status quo.[52] The Government initially proposed that India retain the sole right to train Nepal military officers and provide oversight of other countries' projects in exchange for new transit rights. Neither ministerial-level talks between India and Nepal in February 1990, nor between their respective foreign secretaries in April, made any breakthrough.[53] But Gujral eventually renegotiated the Indo-Nepal trade and transit treaties, allowing bilateral relations to return to their status quo ante of 1 April 1987.[54] Some attributed these overtures to the external affairs minister, and his unusual political autonomy, given Singh's relative lack of interest.[55] Yet they reflected the conciliatory stance of the National Front more widely.

Similarly, the Government oversaw the withdrawal of the Indian Peace Keeping Force (IPKF) from Sri Lanka.[56] The 1987 Indo-Sri Lankan Accord, signed by Rajiv Gandhi and Junius Richard Jayawardene, agreed to recognize Tamil minority rights and devolve various powers within a united Sri Lanka. Controversially, it also acknowledged India's right to be the sole bilateral force and defend its perceived national interests in Sri Lanka. But deadly clashes between the LTTE and the IPKF, and growing hostility towards the latter by the Premadasa administration in Colombo, convinced the Singh ministry to withdraw Indian forces in March 1990.[57]

Ultimately, however, the defining interventions of the National Front involved economic policy and social affairs. On 19 March, the finance minister had introduced the 1990–91 Union budget.[58] Rhetorically, it "reject[ed] the trickle-down theory of development" in favor of employment-oriented growth and political decentralization. Practically, however, Dandavate balanced traditional spending priorities with tentative economic liberalization. The Budget allotted 50 per cent of all investible resources towards the agricultural sector and rural development, increasing their combined share from 44 to 49 per cent. The finance minister also promised to initiate employment guarantee schemes in drought-prone areas, raise fertilizer subsidies and waive farmers' loans up to Rs. 10,000, as well as instruct the Commission on

Agricultural Costs and Prices (CACP) to revise its formula for input charges by incorporating minimum wage rates and costs of entrepreneurship.[59] That said, the Budget sought to bolster the manufacturing sector by reducing corporate taxation and furthering industrial deregulation. It also rewarded urban middle classes by raising the income tax threshold. Tellingly, the finance minister stressed the need for pragmatism and "non-doctrinaire socialism". The *Approach Paper to the Eighth Five-Year Plan (1990–95)*, titled *Towards Social Transformation*, reiterated a commitment to targeted employment growth, expanding welfare services and greater political decentralization.[60] In late May, however, the Industry Minister, Ajit Singh, unveiled a new industrial policy statement, doubling asset thresholds in the small-scale sector, raising the ceiling for license-free investments and automatically permitting foreign equity investments up to 40 per cent.[61]

The tensions between the 1990–91 Union budget, its related economic policy statements and the *Approach Paper to the Eighth Plan* provoked divergent reactions. Many observers agreed the National Front had inherited a difficult economic situation.[62] The previous Congress(I) administration had bolstered aggregate economic growth via partial liberalization and higher public spending. Yet high net imports and greater commercial borrowing from abroad created growing pressures on the fiscal deficit and the balance of payments. Between 1984–85 and 1989–90, the share of total debt-to-GNP expanded from 17.7 to 25.5 per cent, while the ratios for debt servicing and debt-to-export grew from 18 to 27 per cent and 210 to 265 per cent, respectively. Foreign exchange reserves had fallen to two months' worth of imports when the National Front took office. Moreover, the average level of current expenditure as a percentage of GDP had increased from 18.6 per cent in the first half of the 1980s to 23 per cent in the second, driven by the rising current expenditures on subsidies, defense and debt interest payments. Real fixed capital formation in the public sector suffered as a result.

The Left criticized the pro-liberalization measures introduced by the 1990–91 Union budget. Chandra Shekhar reproached the Government's new industrial policy statement, saying it would please the World Bank and multinationals, exacerbate pressures on the rupee and boost wasteful investment in luxury goods.[63] Yet these ideological-policy differences, though real, posed no danger. The communist parties had promised to support the National Front on the understanding that it would implement its own manifesto.⁶⁴ Nonetheless, the Government failed to tackle the structural pressures squeezing its finances and hence the scope for productive capital formation.⁶⁵ The ratio of current expenditure to GDP mounted in relative and absolute terms. Agricultural subsidies and loan waivers far outstripped spending on primary education and basic healthcare.⁶⁶ The former largely rewarded intermediate proprietary castes; the latter benefitted the rural poor. Finally, despite Dandavate's proposal to eliminate investment allowances, reduce tax evasion and raise custom duties in order to lower the fiscal deficit by 1 per cent of GDP, resource mobilization remained severely inadequate. The country's revenue deficit, and its increasing reliance on external commercial borrowing, simply worsened.

In the end, the National Front sought to protect the interests or recognize the claims of historically marginalized groups through various administrative reforms and symbolic measures. In April 1990, Prime Minister Singh posthumously conferred the Bharat Ratna, the country's highest civilian award, upon Bhim Rao Ambedkar. His ministry marked the occasion by placing a portrait of the eminent Dalit leader in the main hall of parliament, alongside other nationalist heroes, and by declaring his birthday a national holiday.⁶⁷ In May, the Government introduced the National Commission for Women Act, 1990, which granted the body statutory powers to review constitutional safeguards, advocate remedial legislation and facilitate the redressal of grievances and acts of injustice. The measure gained Presidential assent in early August, which then saw the National Front extend analogous powers to the National Commission for Scheduled Castes and Scheduled Tribes, by passing the Constitution (Sixty-Fifth Amendment) Act, 1990.

The centerpiece of its social policy agenda, however, was the Mandal Commission Report. On 7 August, V.P. Singh announced his decision to implement the Report by ordinance, which included reserving 27 per cent of the seats in public sector employment and higher educational institutions for members of the OBCs. Critically, the Report identified 3743 specific castes as backward on the basis of state-level surveys. The rationale, according to B.P. Mandal, was twofold:

> To treat unequals [in terms of social advantages] as equals [to make them compete on basis of merit] is to perpetuate inequality ... and to hold a mock competition in which the weaker partner is destined to failure right from the start.⁶⁸
>
> By increasing the representation of OBCs in government services, we have given them an immediate feeling of participation in the governance of the

country. When a backward caste candidate becomes a Collector or Superintendent of Police, the material benefits accruing from his position are limited to the members of his family only. But the psychological spin off of this phenomenon is tremendous; the entire community of that backward caste candidate feels elevated. Even when no tangible benefits flow to the community at large, the feeling that now it has its "own man" in the "corridors of power" acts as a morale booster.[69]

Historically, the policy of reservations rested on the principle of "compensatory discrimination" for groups that had suffered grave injustice, which originally comprised SCs and STs.[70] But over time its proponents advanced a more general claim: individuals serving in positions of authority had a bias towards the interests of communities to which they belonged. Thus effective political representation required institutions to mirror the composition of society. The condition of marginalized social groups would only improve if their own parties, led by members from their own communities, captured public office and manned the state apparatus. The presumption of impartiality was either an illusion or a guise for domination.

The Report generated fierce intellectual debate. Opponents argued that strict representational parity undermined the norms of impartiality that governed public institutions.[71] Some contended that reservations for OBCs primarily served backward caste politicians, offering them an effective mobilization strategy and source of patronage. This was especially true in the relatively under-industrialized north, where Brahmins comprised a greater share of the population and enjoyed overrepresentation in state administration, making "government jobs the main vehicle of upward social mobility".[72] Others claimed that caste-based reservations, while addressing status inequalities regarding dignity and self-worth, could not challenge the underlying material causes of exploitation and poverty.[73] Each of these criticisms had merit. Indeed, the desire of many OBC leaders to expand the ambit of reservations reflected a belief that formal state power was the most effective means of redressing social inequality, which stemmed from their experience as the beneficiaries of reservations in the states and public sector dominance of the economy.[74] Nevertheless, the ascendance of the OBCs challenged the belief that high public office was the preserve of upper caste groups, as well as the stigma of inferiority and subordination that marred wider social relations.

In the end, the mainsprings of the Mandal Commission Report were twofold. Most observers viewed Singh's decision as a personal power ploy vis-à-vis Devi Lal. The deputy prime minister had opposed the Report, which excluded several dominant proprietary castes from its ambit, including Jats.[75] Relations between the two leaders had soured in February 1990 after Singh forced Lal's son, Om Prakash Chautala, to resign his chief ministership of Haryana following massive poll-related violence and alleged vote-rigging in the Meham by-election.[76] Attempts to accommodate Chautala failed. In late July, the former Lok Dal leader attacked Singh publicly, compelling the latter to dismiss him from the Council of Ministers on 1 August. Chandra Shekhar accused Singh of implementing Mandal to upstage Devi Lal, who had planned to rally mass support in New Delhi on 9 August.[77] The misdeeds of political leaders' sons; the mobilization of mass agrarian support by a famous Jat leader; the invocation of grand social ideologies to camouflage everyday power struggles: the tussle between Singh and Lal seemed to replay the antics of Moraji Desai and Charan Singh that beset the Janata Party in the late 1970s.

Yet Singh's move, which undoubtedly sprang from his rivalry with Devi Lal, had larger compulsions. The demand for extending public-sector reservations to OBCs had been a longstanding socialist demand. The Mandal Commission had submitted its recommendations to President Reddy on 31 December 1980. Mrs Gandhi only tabled it in parliament in 1982, however, ignoring it thereafter.[78] Moreover, the 1989 JD election manifesto had pledged to implement the Report within one year of capturing national power.[79] The party's state administration in Uttar Pradesh, led by Mulayam Singh Yadav, had promulgated an ordinance in July 1989 allowing OBC reservations. "Far more important than economic well being", claimed the rising Yadav leader, "is status and honour in society".[80] The Mandal report had a genuine political following.

Its proposed implementation ignited turbulent public demonstrations and heated political debate. Singh consulted neither his cabinet nor his external parliamentary supporters before making his decision.[81] Prime ministerial office granted him discretionary power. Moreover, he justified his action by saying "history has shown that any real change has come through an element of ruthlessness or it has not come at all. We must realise that we cannot argue away interests."[82] Unlike a directly elected presidential system, however, Singh's minority government was beholden to parliament for its survival.

The CPI(M) West Bengal chief minister, Jyoti Basu, argued that material poverty was a better indicator of social backwardness.[83] Some

JD leaders, most prominently Biju Patnaik and Ramakrishna Hegde, agreed.[84] They were only partly right. The communists had historically ignored caste and the politics of dignity and self-representation as a fundamental social cleavage. Indeed, leading Dalit figures had long criticized the communist parties for using the term *harijan* in the 1950s and 1960s, the failure of the West Bengal government to create an OBC list before 1980, and the absence of prominent backward caste leaders in their hierarchy. According to the BSP chief Kanshi Ram, communists were "green snakes in green grass".[85] Nevertheless, many scholars rightly feared that privileged members of the OBCs would corner the benefits.[86] Significantly, Singh had failed to link the case for reservations to land reform or public investments in basic education, as the Mandal report had done.[87]

The prime minister sought to placate aggrieved constituencies through various measures. In early September, the Government unveiled the Agricultural and Rural Debt Scheme, 1990, waiving all farm loans under central jurisdiction worth Rs. 10,000 or less.[88] It also introduced an important democratic reform. The Prasar Bharati (Broadcasting Corporation of India) Act, 1990, sought to expand the right to information and freedom of the press by granting autonomy to All India Radio and Doordarshan. The Act came in the wake of the Constitution 67th Amendment Bill, introduced in May, which sought to depoliticize the appointment and transfer of higher judges by establishing a National Judicial Commission.[89] Lastly, in early October the National Front proposed an Employment Guarantee Bill, which sought to make the right to work a fundamental constitutional right, as well as agreeing to allocate greater resources to the states based on their per capita income and special developmental problems.[90]

But none of these initiatives placated the BJP. The politics of Mandal directly challenged its political agenda and social base. The Report unleashed a fierce backlash amongst many upper castes in northern India that increasingly supported *Hindutva*. "We are supporting a government," remonstrated the RSS leader K.R. Malkani, "which is writing our epitaph".[91] On 15 September, the BJP president L.K. Advani launched a modernized *rath yatra* (chariot journey) to rally mass support for a Ram *mandir*, proposing to reach Ayodhya by the end of October. Its deeply charged iconography and rhetoric provoked riots in many states. On 22 October, the JD Chief Minister of Bihar, Lalu Prasad Yadav, arrested Advani for inciting communal disorder while

THE NATIONAL FRONT (1989–1991)

police forces under the control of Mulayam Singh Yadav, his counterpart in Uttar Pradesh, clashed with *kar sevaks* in Ayodhya. The BJP responded by withdrawing its external parliamentary support. On 7 November, the prime minister put his ministry on the line. His willingness to face a vote of confidence contrasted sharply with the actions of Moraji Desai and Charan Singh during the Janata Party ministry, underscoring the stakes. After an intense eleven-hour debate, the Government collapsed, allowing Singh's bête noire to secure his overriding ambition. On 16 November, with the support of 58 JD parliamentarians and 195 MPs from the Congress(I), Chandra Shekhar became India's eighth prime minister.[92]

Yet his tenure, replaying the endgame of the Janata Party, was extremely short-lived. The Iraqi invasion of Kuwait in August had precipitated a growing financial crisis. The rising price of oil and petroleum, combined with falling exports to and remittances from the Gulf, put severe pressure on the balance of payments. The Chandra Shekhar ministry responded by raising taxes and implementing sweeping cuts. But the credit rating agencies downgraded India's sovereign debt in November, diminishing its access to further international credit, and generating massive capital flight. Overwhelmed, the new Union administration could only manage to present a vote-on-account, postponing its inaugural Union budget past the customary deadline. On 6 March 1991, following reports that his administration had put Rajiv Gandhi's residence under surveillance, Chandra Shekhar resigned his prime ministership. Shortly thereafter, President Venkataraman accepted his recommendation to dissolve parliament, setting the state for India's tenth general election.

The collapse of the National Front government exposed the faultline that had separated the socialist left and Hindu right during the original Janata experiment. The *Ramjanmabhoomi* movement crystallized the natural ideological gravity of the third force. Its leaders declared: "The National Front, instead of bowing before communal fundamentalism to save its Government, took a principled stand and allowed itself to be voted out of office", reiterating a pledge to fulfill its mandate "in close cooperation with the Left Parties".[93] Indeed, V.P. Singh believed that history would repeat itself, with the BJP suffering the same fate as the erstwhile Jan Sangh if it had not been for the politics of Ayodhya.[94] It was a moot counterfactual, however, at least in the short term. The implementation of the Mandal Commission Report energized the *Ramjanmabhoomi* movement. The BJP, a far stronger political organization, recruited additional resources. The JD, whose relative paucity of posters, banners and graffiti exposed its poor finances, paled in comparison.[95]

The Congress(I) revived its slogan of stability, a cynical maneuver given its machinations vis-à-vis the Chandra Shekhar ministry. Yet few in the party could have anticipated the political tragedy that personified its meaning with terrible irony. In late May, a suicide bomber from the LTTE assassinated Rajiv Gandhi, rupturing the field of play. The chief election commissioner, T.N. Seshan, postponed further polling for three weeks. The electorate, half of whom had not yet cast their votes, swung to the Congress(I). Even the stalwart general secretary of the CPI(M), E.M.S. Namboodiripad, publicly intimated the possibility of switching political sides.[96] In the end, the Congress(I) won 36.3 per cent of the national vote and 232 parliamentary seats. The BJP emerged as the second largest formation, capturing 120 seats. The party contested twice as many seats as it had in 1989. Nonetheless, it posted gains across many states, massively increasing its vote share to 20.1 per cent.[97] The combined tally of the seven-party National Front-Left Front alliance, whose ranks now included the Jharkhand Mukti Morcha (JMM), secured just 135 parliamentary seats, eight less than the JD had captured by itself less than two years earlier. On 21 June, President Venkataraman invited Narasimha Rao to form a minority Union government. The second major attempt to form a centre-left alternative to the Congress(I) in New Delhi had suffered a clear defeat.

PART II

THE MATURATION OF THE THIRD FORCE

6

THE CRYSTALLIZATION OF THE THIRD FORCE (1991–1996)

The return to power of the Congress(I), and its ability to complete a full parliamentary term, appeared to signal another return to normality. Yet major shifts occurred during its tenure. The first concerned the fate of modern Indian secularism. The BJP openly advocated *Hindutva* as central to the national political imaginary:

> Our nationalist vision is not merely bound by the geographical or political identity of India, but defined by our ancient cultural heritage. From this belief flows our faith in "Cultural Nationalism" which is the core of *Hindutva*. That, we believe, is the identity of our ancient nation—*Bharatvarsha*. Hindutva is a unifying principle which alone can preserve the integrity and unity of our nation. [It] is also the antidote to the shameful efforts of any section [of society] to benefit at the expense of others. We are resolved to put an end to the politics of competitive communalism, of appeasement and of casteism.[1]

On the one hand, the BJP castigated the Congress(I) for violating the principle of equality amongst religions in the public domain and for capitulating to various minority communities. Far from separating religion from politics, however, the party espoused the maxim of "one nation, one people, one culture," endangering India's cultural heterogeneity.[2] It was in the name of restoring Hindu honor that zealots belonging to the VHP demolished the Babri masjid in Ayodhya on 6 December 1992. Senior BJP leaders sought to distance themselves to some extent. The party's Chief Minister of Uttar Pradesh, Kalyan

Singh, relinquished his post. L.K. Advani, its national president, resigned as leader of the Opposition in parliament. However, he also publicly "sympathized" with the "frustration" of Hindus that supported the demolition.[3] Despite muting its ethno-religious campaign, the party suffered heavy losses in several states during the 1993–1995 assembly elections, convincing its leadership to pursue social engineering—via "direct Mandalization" (appointing lower castes within the party organization) and "indirect Mandalization" (seeking electoral alliances with lower-caste dominated parties)—with the backing of the RSS.[4] Nonetheless, the violence in Ayodhya convinced many secularists that allowing the party to seize national office would tear the country apart.

The second major shift concerned the economy. The severe balance of payments crisis that confronted the Congress(I) when it took office in June 1991 forced the Government to stabilize the economy. Yet the minority Rao ministry exploited the crisis to introduce far-reaching structural changes, dismantling many pillars of the license-permit raj. Many opposition parties denounced the reforms. Given the severity of the crisis, however, it was doubtful whether any government would have entirely forsaken similar measures.[5] In addition, avowedly secular parties feared toppling the minority Congress(I) administration. The threat of militant Hindu nationalism paradoxically assisted the process of reform. Lastly, economic liberalization generated its own political momentum. Declining public investment and central economic assistance forced states to compete for scarce private investment with each other, in contrast to the more vertical competition for permits and resources from New Delhi that shaped earlier dispensations, compelling parties of every stripe to play a two-level game of national opposition, state-level accommodation.[6] Greatest progress occurred in relatively technical policy-areas: the liberalization of trade, exchange rate and investment regimes, as well as capital markets.[7] Most of these reforms primarily concerned "elite politics" rather than "mass politics": they failed to impinge directly upon a large number of people or organized interests in a visible manner. However, despite the resumption of aggregate economic growth by the mid-1990s to its pre-crisis level, the reform process lost considerable momentum. The Congress(I) suffered heavy defeats in a number of state assembly elections between 1993 and 1995. Many within the party claimed the results constituted a backlash against the reforms, weakening their impetus.[8] The Congress(I)'s reluctance to push liberalization was also related to the difficulty of "second-stage" reforms involving agriculture, privatization, labor and the fiscal deficit. Many state governments had managed to implement policy changes in these domains through stealth and obfuscation.[9] Nonetheless, advancing such measures at the Centre proved far more difficult, given that potential benefits would arise, if at all, in the medium- to long-run. Lastly, many observers pointed to the "partiality" and "instability" of liberalization, which had exacerbated social inequalities amongst classes, sectors and regions.[10] Thus, even though many parties agreed on its irreversibility, the future pace, scope and management of further economic reform ignited genuine debate.

The third major development was the continuing regionalization of the federal party system. The Congress(I) remained the only party capable of garnering electoral support across the country. The deterioration of its electoral performance in various state assembly elections between 1993 and 1995, however, highlighted a new political landscape. The number of parties contesting national elections had dramatically increased since 1984. Relatively few performed successfully enough to enter the Lok Sabha at the start. Nonetheless, many progressively did, increasing their effective share of votes and seats (see Tables 6.1 and 6.2).

Table 6.1 Number of Parties Contesting National Elections, 1980–2009

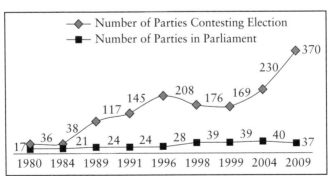

Source: Sanjay Kumar, "Regional parties, coalition government, and functioning of Indian parliament: the changing patterns," *Journal of Parliamentary Studies*, 1, 1 (2010): 75–91.

Table 6.2 Effective Number of Parties in Parliament, 1980–2009

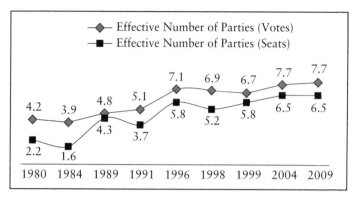

Source: Sanjay Kumar, "Regional parties, coalition government, and functioning of Indian parliament: the changing patterns," *Journal of Parliamentary Studies*, 1, 1 (2010): 75–91.

Moreover, the declining proportion of votes and seats won by national parties in general elections accelerated after 1991 (see Tables 6.3 and 6.4). A pattern of multiple bipolarities emerged in the states, in which no single party exerted similar influence across the Union, increasing the prospects of fragmentation in New Delhi.[11] The Congress(I) could no longer presume to determine the terms of competition, debate and rule. Hence there was a strong expectation that no

Table 6.3 Decline of National Party Vote Share, 1980–2009

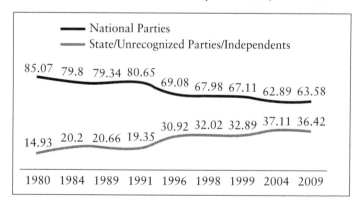

Source: Election Commission of India. Author's calculations.

THE CRYSTALLIZATION OF THE THIRD FORCE (1991–1996)

Table 6.4 Decline of National Party Seat Share, 1980–2009

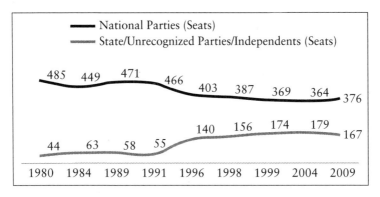

Source: Election Commission of India. Author's calculations.

single party would emerge from eleventh general election in April–May 1996 with a parliamentary majority.

Manifestoes, strategies, prospects

The Congress(I)'s fortunes were in serious decline. The party appealed to the anti-colonial struggle when its leadership reconciled "differing and divergent interests" into "a single harmonious national mosaic".[12] It claimed that it was the only party capable both of providing "stable government"—compared to the "shifting" and "unstable" coalitions of "one-issue parties" in the opposition—and defending secularism.[13] Yet the rhetoric sounded worn on both counts. The party's declining electoral popularity exposed its diminished capacity to manage a diversity of interests. The absence of state police forces in Ayodhya in December 1992, the failure of the Congress(I) to act swiftly after the destruction of the mosque and its prevarication about the future of the disputed site severely damaged its credibility.[14] The party's decision to induct H.K.L. Bhagat into a senior organizational post following his threat to run against it independently, despite his role in the anti-Sikh pogrom following Mrs Gandhi's assassination,[15] further undermined its secular pronouncements.

The Congress(I)'s stance towards economic liberalization enjoyed greater credibility. The party recalled the precarious conditions under which it came to office in 1991, defended its record in averting a protracted economic crisis and stressed its skill in implementing structural reforms. However, sensitive to the charge that liberalization had harmed the most disadvantaged, the Congress(I) defended its various social programmes, such as the Integrated Rural Development Programme, Indira Awas Yojana (a scheme subsidizing housing for the rural poor) and Development of Women and Children in the Rural Areas Scheme, asserting that poverty-alleviation remained its primary objective.[16]

Apart from its embattled secular credentials and reluctance to trumpet economic liberalization, the main threat to the Congress(I) was the weakness of and disunity within its organization. Public skirmishes between leading party figures and deeper internal fissures marred its image preceding the campaign. Notably, many of these involved the Prime Minister, P.V. Narasimha Rao, leader of both the All-India Congress(I) Committee (AICC) and Congress(I) Parliamentary Party (CPP), the organizational and legislative wings of the party, respectively. Rao had assumed both posts following Rajiv Gandhi's assassination. The party bosses pressed Sonia Gandhi, Rajiv's politically inexperienced widow to accept the mantle of leadership. Her refusal enabled Rao—an elderly man in poor health on the verge of retirement, who was only the second person outside the dynasty, and the first from the south—to take over the helm. Many Congressmen believed he would be a transitional figure.[17] Rao surprised his skeptics. The Jharkhand Mukti Morcha (JMM), a regional party from Bihar, supported the Congress(I) during a critical motion of confidence in July 1993, earning the latter a parliamentary majority and enabling it to complete a five-year term.

However, a series of scandals endangered Rao's exploits. Of the various criminal investigations pursued by the Central Bureau of Investigation (CBI), the most sensational was the *hawala* scandal, involving allegations of bribes and kickbacks from illegal foreign exchange transactions between April 1988 and March 1991, incriminating many leading figures of the political class.[18] According to reports, a businessman named S.K. Jain claimed to have funneled Rs. 3.5 crore to the prime minister in March 1995, allegedly to bribe the JMM during the 1993 confidence vote. The most astonishing feature of the scandal was that many suspected Rao himself of instigating it against not only opposition politicians but also vis-à-vis his own colleagues. According to one senior Congress(I) official,

Many of those suspected of being guilty in the *hawala* charges blamed Rao personally ... They believed that he was out to get them and actually pushed

THE CRYSTALLIZATION OF THE THIRD FORCE (1991–1996)

the CBI to investigate them ... And then by simply not intervening with the inquiries, he allowed the perception to persist that they were in fact guilty. You could say that judicial activism filled a vacuum created by executive inertia.[19]

The "single directive" principle enjoined the CBI to gain the concurrence of the prime minister's office before pursuing investigations.[20] Hence many surmised that Rao encouraged the investigations to enhance his personal control over the Congress(I).[21] It was difficult to corroborate. On the one hand, Rao had attempted to revive intra-party elections in 1992.[22] On the other, the CBI only began to issue charge sheets (formal criminal accusations) in January 1996, three years after the Supreme Court directed it to pursue the case. The early sluggishness of the investigations, compared to the subsequent bout of charge-sheeting, prompted several observers to discern the prime minister's involvement. The suspicion that Rao was "out to get them" appeared to match a wider perception:

> Rao wasn't really concerned about the party, but about his own skin ... He didn't want to have to face any rivals and so always tried to undermine their positions ... In that sense, his strategy was paradoxical: Rao could only be politically secure individually to the extent that Congress(I) as a party was divided and weak.[23]

The prime minister's alleged ploy encouraged key regulatory institutions to exercise their formal powers, however.[24] Greater scrutiny by the Supreme Court, as well as efforts by the Election Commission (EC) to lessen malfeasance during the 1996 campaign,[25] ensnared Rao himself in a range of investigations: the St. Kitts' forgery affair,[26] the Lakhubhai Pathak cheating case[27] and the JMM pay-off scam.[28] Rao denied involvement. Indeed, despite growing calls within the Congress(I) for his resignation, he refused to step down, allowing family members of tainted party figures to contest the election as well.[29] The specter of rot engulfing the party corroded public trust of the political class more widely.[30]

Inept tactical decisions, which caused important factions to break away, worsened its prospects. In 1994, Narayan Datt Tiwari, a senior Congressman who had held several ministerial portfolios in New Delhi and the chief ministership of Uttar Pradesh thrice, formed the All India Indira Congress (Tiwari) (AIIC(T)):

> We came together because we were opposed to Rao ... He was constantly trying to sideline Sonia [Gandhi]—he felt threatened by her. He simply wanted power to himself ... we were against the fact that he was both leader of the party and the organization. The Congress(I) is a vast organization. You cannot

concentrate power in a single person's hands ... We were also upset with the manner in which he was implementing liberalization, which wasn't paying any attention to the plight of the poor. So we raised our voices ... The result was that in March'93 [several of us] were suspended ... But he still tried to destroy us, by implicating us in *hawala* for almost two years ... In the end, of course, we were all rightly exonerated. But he tried to end our political careers.[31]

Another senior figure implicated in the *hawala* scandal was Madhavrao Scindia. Unable to absolve himself, Scindia formed the Madhya Pradesh Vikas Congress (MPVC) in 1996, contesting for office in his home state. Essentially, both parties were temporary political vehicles, established in opposition to Rao.

A far more damaging split, however, occurred in Tamil Nadu. In January 1996, the Congress(I) leadership rejected the possibility of renewing its electoral pact with its former ally, the AIADMK. G.K. Moopanar, who led the Congress(I) in the state, vociferously opposed Jayalalitha, the leader of the AIADMK. But the party high command soon reversed its decision:

The options were very clear for us. We needed a tie-up with somebody in Tamil Nadu ... We had to choose between the AIADMK and the DMK [Dravida Munnetra Kazhagam] ... The DMK was out of the question ... it was suspected of collusion with the LTTE in Rajiv's death. On top of that, the Jain Commission inquiry [set up to investigate Rajiv Gandhi's assassination] was going on.[32]

Given Mr Gandhi's stature as a "figure of veneration", Rao allegedly perceived the AIADMK's corrupt image to pose a lesser risk than the "stigma attached to the DMK".[33] But he misjudged political sentiment on the ground. In late March, Moopanar formed the Tamil Maanila Congress (TMC), forging a seat-sharing pact with the AIADMK's arch political rival, the DMK:

We had our misgivings about the DMK ... But we knew she [Jayalalitha] would lose in 1996. The stench of corruption around the AIADMK was terrible ... So our party workers demanded that we set up our own party ... something that was well nigh impossible given the time constraints. But we did it ... and discretion dictated that we join hands with the DMK.[34]

The electorate would eventually reward the acumen of the TMC. The Congress(I)'s blunder reflected the folly of ignoring the wishes of local party leaders—a feature of its over-centralized apparatus since the early 1970s—who correctly read the electoral mood. Ultimately, these various political mistakes exposed the extent to which the party high command was out of touch with both its own ranks and the wider electorate.

The electoral prospects of the BJP, in contrast, were on the rise. The party vowed to implement several controversial proposals: to adopt a uniform civil code, abrogate Article 370 of the Constitution, and construct a Ram *mandir* on the site of the razed mosque in Ayodhya. It pledged to disband the Minorities Commission and amend Article 30, which gave minority religious communities special rights in educational matters.[35] The realization of *Hindutva*, claimed the BJP, would "restore to our state its authority ... its honor and its prestige".[36] Domestically, the party supported increasing defense expenditure and providing a "free hand" to security forces to deal with "malcontents" in the Northeast and Jammu & Kashmir, in addition to introducing identity cards, deporting illegal immigrants and spurring economic development. Internationally, it opposed the Comprehensive Test Ban Treaty (CTBT), Fissile Material Control Regime (FMCR) and the Missile Technology Control Regime (MTCR), vowing to conduct a nuclear test.[37] Implementing these changes would radically alter the character of state-society relations as well as India's posture in international affairs.

The BJP claimed that *Hindutva* distinguished its economic prescriptions too. Invoking the principle of *swadeshi* (self-reliance), the party pledged to create

[a] self-reliant India, asserting its national economic interests, not as autarchy, but as a pre-requisite to meeting the challenge of globalization; of preserving our identity without compromising our sovereignty and self-respect. We reject unbridled consumerism and believe in adherance [sic] to sustainable consumption and growth.[38]

The BJP accused the Congress(I) of precipitating the crisis in 1991, worsening the bias against agriculture and exposing the economy to the vagaries of globalization.[39] Evoking its earlier Gandhian tropes, the party vowed to abolish "preferential treatment" for multinational corporations, ban foreign entry into the consumer goods sector and maintain reservations for small-scale enterprise.[40]

The slogan of *swadeshi* failed to suppress tensions within the BJP, and between it and the *sangh parivar*, however. Sections of the party advocated various pro-liberalization measures which served the interests of the small traders and petty industrialists that traditionally supported it, attacking the management of liberalization. It criticized the

Congress(I) for pruning government spending through lower capital expenditure as opposed to generating public revenue through tax reform. Similarly, the BJP reproached the incumbent administration for relying on foreign portfolio capital to balance the external deficit.[41] The party argued for a stronger capital market and regulatory framework to entice foreign direct investment, welcoming external capital in infrastructural development and domestic partnerships in strategic, export-oriented and high technology sectors of the economy.[42] Lastly, while it sought to protect workers' interests by introducing worker participation in both public sector utilities (PSUs) and foreign companies operating in India and disallowing firms from "exiting" without fulfilling their obligations to labor, the BJP also proposed a Disinvestment Commission to PSUs that require private capital.[43] Proponents of liberalization harbored misgivings over the *swadeshi* lobby's influence over the party. Yet the record of BJP state governments in the regions also demonstrated a readiness to push further economic reform.[44]

Institutionally, the BJP boasted a tightly disciplined organization. Divisions existed within the party between pragmatists and ideologues, however. The pressures and opportunities of electoral politics and democratic rule, moreover, exposed tensions vis-à-vis its associates in the *sangh parivar*. Hence the BJP pursued a two-pronged strategy.[45] Certain figures pushed a militant *Hindutva* agenda, flanking hardliners in the *sangh parivar* and targeting minorities in local electoral campaigning. But the party high command adopted a more urbane tone in the national media. Crucially, it chose A.B. Vajpayee, given his image of moderation, as its parliamentary leader. Vajpayee's demand for probity and stability in government, and aspiration for greatness in international affairs, appealed to many sections of the urban middle class, weary of corruption and apprehensive of the rising political clout of the lower castes. In addition, the BJP forged electoral agreements with several state-based parties: the Shiv Sena in Maharashtra, Haryana Vikas Party (HVP) in Haryana, and Samata Party (SAP) in Uttar Pradesh and Bihar. Tactical provincial considerations, rather than genuine ideological affinities, compelled the latter two into the alliance— to defeat their erstwhile colleagues in the Congress(I) and JD, respectively. In sum, the strategic outlook of the BJP positioned it to fare well in the eleventh general election.

The final major formation was the Third Front, which now revolved around two blocs, the National Front and the Left Front. The National

Front consisted of the JD and the Samajwadi Party (SP). Charging the Congress(I) with "equivocation" over its defense of secularism and attacking the BJP for its "naked communal appeal", the JD vowed to protect the rights of religious minorities by bringing *Wakf* properties (donated for purely religious purposes) under the Public Premises (Eviction) Act, granting the Minority Commission full legal status, and referring all disputes regarding religious shrines to the Supreme Court.[46] Both parties also defended the interests of intermediate- and low-caste groups, by capturing political office, influencing policy decisions and extending the policy of caste-based reservations:

In promoting social justice, *democratisation of the power structure is crucial*, for that will empower the sections [of the caste order] which have been historically denied all privileges and opportunities.[47]

The National Front justified its defense of secularism and pursuit of social justice under the rubric of legitimate minority rights and thwarted democratic aspirations. "The challenge is nothing less than accomplishing a political transition to the post-Congress India", proclaimed the JD, "while steering clear of communal and reactionary forces."[48]

To some extent, similar arguments informed its stance towards economic liberalization. Yet they failed to mask contradictions and disagreements within the coalition. Advocates of liberalization in the JD, particularly from Karnataka, whose state unit represented powerful agricultural interests, highlighted the Congress(I)'s failure to restrain the fiscal deficit, encourage greater levels of FDI or push further deregulation.[49] Yet its socialists attacked the Congress(I)'s record on inflation, agricultural development and public sector employment, espousing a "definitive role for the State" in order to tackle the country's "vast inequalities, economic backwardness and lopsided regional development".[50] Hence the JD promised to extend the principle of reservations to cover Dalit Christians and incorporate the judiciary and ensure that disinvestment would not undermine previous gains.[51] It also vowed to implement a minimum support price for agricultural produce and an insurance scheme for crops and livestock, allocate 20 per cent of financial sector outlays and 60 per cent of total plan expenditure to the agricultural sector, and restore subsidies for fertilizers and other agricultural inputs to pre-liberalization levels;[52] to defend the interests of labor by establishing independent commissions to oversee industrial disputes, enforce labor laws, and index wages and benefits to the cost of living;[53] and to maintain the small-scale sector, insulate the financial sector from foreign capital and grant more operational autonomy to public sector units.[54] It was uncertain which of these rival political visions would prevail if the party came to power.

The other half of the Third Front comprised the Left Front, consisting of the CPI, CPI(M), Revolutionary Socialist Party (RSP) and All-India Forward Bloc (AIFB). Claiming that anti-communist sentiment within the "bourgeois opposition forces" "was on the wane" due to "conflicts and contradictions in the ruling classes", the CPI(M) justified forging a "people's democratic front" on grounds of social justice and national unity after the fall of the NF.[55] The Left remained the most powerful opponent of militant Hindu nationalism. The bloc condemned the Congress(I) for allowing Hindu nationalist brigades to destroy the mosque, which it depicted as "an act of medieval frenzy", driven by "feudal and unscientific ideas, obscurantism, mysticism, obscenity, glorification of crimes and violence".[56] Indeed, it argued that

every single element of the BJP's cultural nationalism is aimed against the minorities, and is a denial of the pluralism and multifacetedness of Indian society and culture, which is the basis of our secularism.[57]

Accordingly, the CPI(M) demanded a series of measures to safeguard political secularism, similar to those advocated by the JD.[58] Of equal concern to the Left, however, was the question of economic liberalization. Indeed, of all the parties contesting the polls, the communists expressed the most vociferous opposition to further reform. The CPI(M) argued that

... the Congress(I) government under the IMF/World Bank dictates has led the country on a path that severely mortgage [sic] our economic sovereignty. During the last five years, the entire self-reliant basis of independent India is being [sic] systematically dismantled. The country is neck deep in foreign debt. The plight of the working people has worsened with relentless price rise, growing unemployment; poverty is increasing, the disparities between the rich and the poor further widened alarmingly [sic]. And in the process, the ruling politician-businessman-bureaucrat nexus has been looting the country through unprecedented institutionalized corruption.[59]

Hence the bloc vowed to reverse many extant reforms. It pledged to impede privatization by ensuring the commercial viability of PSUs through internal corporate reforms, rescinding preferential status for private foreign capital into critical infrastructure sectors and reviewing external investment proposals on a case-by-case basis.[60] In addition, the CPI(M) presented various initiatives to protect workers' interests and improve social opportunities. The party pledged to introduce a needs-based minimum wage in the organized sector, index wages and emoluments with inflation and implement the Fifth Central Pay Commission Report for government employees. It promised to push legislative reform and improve tenure records regarding landholdings, increase Plan allocations for irrigation for small farmers and introduce national crop and cattle insurance schemes, and expand credit facilities for poor peasants and landless labor. Lastly, the CPI(M) vowed to make employment, primary education and housing fundamental constitutional rights, and increase public spending on basic health, sanitation and literacy.[61]

The Left's national economic programme warranted scrutiny. The Left Democratic Front (LDF) alliance had actively defended its egalitarian social achievements in Kerala. The 1991 industrial policy statement, offering tax breaks, power subsidies and a streamlined process for approving new projects, envisioned the state as a midwife to capital. Significant investments in tourism, food processing and textiles as well as electronics, minerals and biotechnology, ensued.[62] Nonetheless, despite lower public investment, declining central transfers and volatile commodity prices, the state still secured rising per capita incomes and lower poverty ratios. Historic investments in basic human capabilities clearly helped. Yet these achievements also reflected active political struggles to maintain social expenditures despite a growing fiscal crisis. Indeed, the LDF was on the verge of unveiling a programme of genuine democratic decentralization, deepening the so-called Kerala model of development.[63]

The pursuit of various market reforms in West Bengal, where the Left Front had ruled for almost two decades, undercut its strident anti-liberalization posture, however. State industrial output, despite very low strike activity, had reached its nadir. Leading observers blamed overregulation, poor infrastructure, and insufficient credit for small industrial enterprises.[64] An opening emerged in the early 1990s, though, when the Centre abolished general licensing as well as its freight equalization policy for iron and steel, which had penalized West Bengal in particular. The Left responded by providing tax concessions and empowering the West Bengal Industrial Development Corporation in order to lure private investment.[65] Indeed, its 1994 industrial policy statement, approved by 15th party congress of the CPI(M), "welcomed" private capital, foreign investment and technology.[66] Proponents defended the move on pragmatic grounds. Critics perceived further political deradicalization and greater regional self-interest at play, reflecting a shift from a "united front of left and democratic forces" to "secular and democratic forces" in national coalition strategy.[67] Potential regional allies, liberal in outlook, mocked its ideological beliefs: "The Left was living in another world, preaching one thing, practicing another. US power companies are providing power to West Bengal, yet they still call the Americans imperialists."[68] The longstanding CPI(M) Chief Minister, Jyoti Basu, simply claimed: "I never thought it was necessary to send it [to the Politburo] ... because I did not know that [it] knew nothing about what we were doing in West Bengal for twenty years."[69] Ultimately, structural factors compelled the reforms. Indeed, many claimed the ideological battle over economic liberalization had already been settled amongst the national political establishment, bureaucracy and intelligentsia, removing it from electoral contestation.[70] Nonetheless, the pace, direction and scope of further reform were open questions. Ardent economic liberalizers feared the prospects of the Left gaining political leverage in New Delhi.[71]

Several observers perceived an "historic opportunity" to construct a broader center-left coalition following the rise of Mandal, *Hindutva* and liberalization.[72] The respective stands of the National Front and Left Front vis-à-vis *Hindutva* had converged. Groupings within each alliance differed—albeit to a lesser extent amongst the Left—over the pace and scope of economic liberalization. And India's FPTP electoral system generated strong incentives for the constituents of the Third Front to form pre-poll alliances or make seat adjustments in the run-up to the 1996 general election. The increasing number of contestants for office lowered the threshold of winning in the constituencies. The logic of plurality-rule elections in such conditions disproportionately rewarded strategic electoral coalitions. But the capacity of the broader Indian left to exploit these new opportunities remained ambiguous. Indeed, some observers commented, the "elements of agreement within this "third force" are so narrow and weak, apart from the slogan of "social justice", there seems to be little that will convince anyone that this assortment of parties could run a government on the basis of a credible minimum programme."[73] Discord existed within and between its constituents, caused by internal differences and structural factors, as well as their former political allies.

THE CRYSTALLIZATION OF THE THIRD FORCE (1991–1996)

The Left confronted a peculiar situation. The bloc enjoyed steady popular support in Kerala and Tripura, and uninterrupted political dominance in West Bengal since 1977, its citadel of power. Yet these "fortresses" were, paradoxically, also its "prisons".[74] The formal rhetoric of class-based struggle failed to galvanize support in the larger states, where slogans and issues regarding caste and religion influenced public discourse as well as strategies of mobilization to a greater extent. Hence the Left was compelled to broker a strategic alliance with the National Front.

However, many communist leaders had reservations:

We did not fight the election as a full-fledged national front … Having a national front implies a minimum programme of action. We didn't have this … The third force prior to the elections was better understood as an alternative arrangement.[75]

Indeed, they articulated a Gramscian view of electoral politics, where "the counting of 'votes' is the final ceremony of a long process".[76]

Political struggle isn't just about contesting elections … It is also about pursuing specific struggles, leading common movements and making popular protests against policies and decisions … that are inimical to the interests of the people and the country … If you do these things together, then contesting elections as a unified force is a natural process.[77]

Key political decision-makers explained these failures on various grounds. According to one, an "absence of objective conditions" bedeviled the two main axes of the Indian left.[78] Another, however, claimed:

There was no conception amongst the parties to really contest together … Theoretically there was nothing to prevent it, but the programmatic basis was never worked out.[79]

The reason was partly theoretical, however, especially within the CPI(M). It concerned the persistent ambivalence towards caste as a concept for legitimate social imagination and as a tool for concerted political mobilization. Many communist theoreticians viewed conflicts between lower and higher castes, the working class and the urban bourgeoisie and agricultural laborers and rural landlords in analogous terms: the "most exploited classes" largely constituted "the most socially oppressed castes".[80] They also acknowledged that lower-caste demands for sharing high public office had a "democratic aspect".[81] Nonetheless, the CPI(M) maintained that "feudal and semi-feudal land relations" produced "caste consciousness", rejecting the idea that a politics of recognition based on lower caste reservations could transform the underlying economic relations of production generating their exploitation.[82] Moreover, many communist leaders accused their socialist colleagues of "perpetuating caste divisions" for cynical reasons:

> The appeal is to elect their brethren to power, thus spreading the illusion that coming to power within the same system that protects this existing socio-economic order is a solution to their problems. This may serve the *lust for power* of the leaders but the living conditions of the mass remain as backward as ever.[83]

The Left rightly contended that capturing office and distributing its spoils would not fully redress caste-based inequalities. But its reduction of demands for recognition to class-based struggles was deeply contestable. The distribution of distinction, status and prestige within a society failed to strictly match inequalities of income and wealth. Humiliation and exploitation comprised distinct manifestations, with distinct yet related causes, of social injustice.[84] These longstanding ideological tensions within the broader Indian left persisted.

The other major difference between the two blocs was the disunity of the National Front. This was for two principal reasons. First, the slogan of "social justice" confronted structural barriers.[85] Vertical conflicts divided the lower castes. In part, these concerned relations of class, as Marxists claimed. The desire for agricultural input subsidies and higher food prices amongst proprietary OBCs clashed with demands by Dalit and Adivasi laborers and sharecroppers for higher agricultural wages and greater land redistribution.[86] In part, these conflicts involved questions of status. Many SCs and STs came to resent the Mandal Commission Report for extending reservations to OBCs because it discredited the argument that such measures were only for the most disadvantaged groups.[87] The implementation of the Mandal Commission Report had briefly galvanized a broader horizontal identity. But its eventual political acceptance ironically undermined its capacity to unite lower castes across the country.

These vertical conflicts partly explained the fall-out between the SP, which primarily represented the OBCs, and the Bahujan Samaj Party (BSP), which explicitly stood for *dalits*.[88] Indeed, the respective core base of each party, respectively Yadavs and Chamars, dominated their castes. The two parties joined forces in the 1993 assembly election in Uttar Pradesh, winning a majority. Despite the fact that Dalits had not received important cabinet portfolios in the first JD administration in

THE CRYSTALLIZATION OF THE THIRD FORCE (1991–1996)

1989,[89] the BSP leader Kanshi Ram had supported Mandal, claiming that "Bahujan Samaj is comprised of Scheduled Caste, Scheduled Tribe, Other Backward Classes, and converted minorities ... 85 per cent of the people of this country belong to [it]."[90] The BSP won 11 of 27 cabinet posts in the SP-led 1993 state ministry, led by Mulayam Singh Yadav, who also increased the ratio of *dalits* in the higher bureaucracy.[91] But the SP favored its predominantly Yadav base in district administration, rewarding the community through favorable agricultural policies, while seeking to corral the Muslim vote.[92] Growing material conflicts and social violence between dominant Yadav proprietors and landless *dalit* laborers, and efforts by the SP to split the BSP, fractured the alliance in 1995.[93]

Horizontal divisions further exacerbated their disunity. Despite the rough similarity of formal *varna* classifications, actual caste groups comprised various endogamous *jatis*, whose roles, relations and meanings remained local. The federal party system, which generated distinct political incentives in each state, made aggregating these groups more difficult. This was evident in the JD, which represented the distinct interests of particular social coalitions in different states. In Orissa, the party drew support from a diverse cross-section of groups with little lower-caste mobilization. In Karnataka, it catered to powerful middle-caste farmers. In Bihar, the JD represented an alliance between OBCs and Muslims, which favored Yadavs in particular.[94] A superimposition of social and spatial cleavages, in short, tested its cohesion.

Second, internecine struggles continued to afflict with the wider Janata *parivar*. In early 1992, Ajit Singh had revived his erstwhile Lok Dal, throwing his support to the Congress(I), while Chimanbhai Patel, fearing the surging popularity of the BJP and backlash against Mandal, merged his Janata Dal (Gujarat) with the latter.[95] Soon thereafter Mulayam Singh Yadav broke away from the Samajwadi Janata Party (SJP), following its disastrous performance in the 1991 general election, to form the SP. The SJP, which had only five MPs by late 1993, eventually rechristened itself the Haryana Lok Dal (Rashtriya) under Om Prakash Chautala.[96] Ideologically, the SP espoused a commitment to Gandhian socialism, attacking Hindu communalism and economic liberalization in equal measure. Socially, higher sections of the OBCs and Muslims populated the party's ranks. Yet its formation also reflected, according to one knowledgeable observer, "the familiar tale of men with big ambitions of self-advancement ... drawing sustenance from caste allegiances, intrigues and manipulations".[97] Finally, in 1995 Nitish Kumar and George Fernandes left the JD to form the SAP, contesting for power in Uttar Pradesh and Bihar. A nexus of factors drove the split. Personal factors, according to one JD leader, were the first:

> In the art of splitting ... George Fernandes had acquired mastery and finesse.... The cause of all these splits is temperamental incompatibility ... and, of course, with some it was crude power-motivation.[98]

Ideological differences comprised the second. Equating "social justice" with "democratic socialism", the SAP espoused "a constant struggle against caste-based privileges and discriminations.... the implementation of the Mandal Commission Report cannot be a piecemeal affair without focus on expanded educational opportunities, employment and access to land and water resources".[99] Finally, the politics of Mandal had disproportionately benefitted the Yadavs, who dominated the share of OBCs amongst MPs and MLAs, party officers and cabinet berths, and educational seats and government posts, at the expense of Lodhis in Bihar and Kurmis and Koeris in Uttar Pradesh, despite their roughly equal shares in the general population.[100] The bias was partly the result of successful political lobbying. Lalu Prasad Yadav and Sharad Yadav secured an exemption for richer OBCs from the Supreme Court's decision in *Indra Sawney v. India* to exclude relatively well-to-do intermediate castes.[101] Yet it also reflected a deeper prejudice. Historically the Yadavs had traced their ancestry to Aryan Kshatriyas, placing them above Kurmis and Koeris. As a result, its politicians sought greater positional mobility via the logic of Sanskritization, rather than enunciating radical social values that might usher in complete structural change.[102] A superimposition of personality clashes, social disparities and ideological conflicts, in short, drove the SAP into the arms of the BJP. The ranks of the original National Front, badly depleted, now merely comprised the SP and a vastly truncated JD.

To make matters worse, their relations were strained. According to a senior JD politician, Mulayam Yadav

> left the JD for entirely personal reasons ... He is not a democrat, and doesn't want to share power at all. His ambition to rule Uttar Pradesh is the main thing ... He likes to sit in his chair.[103]

The allegation referred to the upper-caste practice of making lower castes sit on the ground. It highlighted the aversion of many leaders in the Janata *parivar* to share political spoils and foster social esteem on an egalitarian basis.[104] Yet a leading SP official countered that

THE CRYSTALLIZATION OF THE THIRD FORCE (1991–1996)

[the] JD [had] lost its roots in UP. It was never too fruitful for us to stay so close to them ... But they wanted to maintain our unity for the sake of secularism ... we succumbed to their pressure.[105]

Responding to these strident exchanges, a senior JD figure lamented that

the bane of forces outside the Congress(I), if you exclude the BJP and the Left, is that personal ambitions always wreck the experiment. There is no spirit of give and take ... All parties have real power struggles within them. But it is important not to have them too visible. In our party, they are naked.[106]

In short, the presence of genuine social fissures could not fully explain the lack of solidarity within the National Front. Equally important was the failure of its leaders to restrain their personal ambitions for collective gain, or at least accommodate such differences with greater finesse. Alas, matters of contingency exacerbated these inherent difficulties. V.P. Singh remained the most obvious candidate for such a task. Yet his incapacity due to cancer, diagnosed in 1991, prevented him from doing so. The sudden death of the N.T. Rama Rao in January 1996, the former TDP chief minister of Andhra Pradesh, eliminated a worthy potential successor.

Lastly the AGP, TDP and DMK, the former regional members of the National Front, decided not to rejoin its ranks formally. All three parties remained opposed to the Congress(I), which continued to portray most regional formations as antithetical to the nation. Moreover, they continued to support the principles of non-Brahmanism and advocate greater regional autonomy. Historically, the license-permit raj had favored national capitalists with links to the Congress(I), at the expense of regional business houses, whose interests influenced the TDP and DMK.[107] Economic liberalization expanded the field of play. Yet it also convinced regional parties to engage central policy-making: "the limitations of the reform process post-91 are because the states are not adequately involved. We are the final executors: it is we who must package and sell the reforms."[108] The question of redrawing the balance of powers within the Union extended to other issues as well. A senior DMK politician claimed:

Some people say that we have geographically limited aims. This is true, of course. We represent the interests of Tamil Nadu ... But our programme is also more. It is about improving federalism and governing better ... We wanted to implement many of the recommendations of [the] Sarkaria [Commission

143

Report, which recommended various measures to reform Centre-state relations]. In fact, we wanted to go much further.¹⁰⁹

The Rao ministry, like previous Congress(I) administrations, had ignored Justice Sarkaria's recommendations. Reforming Centre-state relations continued to be a significant issue for many regional parties.

The federalized model of India espoused by the TDP, AGP and DMK contrasted sharply vis-à-vis the conceptions of nationhood and modalities of rule championed by the Congress(I) and the BJP. Rhetorically, both parties promised to improve the workings of federalism. Practically, they recognized the advantages of devolving administrative processes. Still, both the Congress(I) and the BJP equated strong regions and multiparty coalitions with national political instability. The regionalists' federal vision also constituted a potential, albeit smaller, fissure within the Third Front. The forerunner of the JD and the CPI(M) had participated in, and even organized, several regional conclaves in the 1980s. They also promoted greater political devolution to the states, and the concomitant dispersal of resources. Yet if social justice on the basis of caste drove the politics of the JD and SP, and if class was the main conceptual prism of the Left, Centre-state relations signified the "question of questions" for the TDP, AGP and DMK.¹¹⁰

Given these various cross-cutting differences, the Third Front failed to release a common manifesto, coordinate a national electoral strategy or pursue collective political campaigns.¹¹¹ Its putative constituents struck agreements in many of the strongholds of the National Front—in Bihar, Karnataka and Orissa—and of the Left—in West Bengal and Kerala. Divisions within the JD over whether to support the SP in Uttar Pradesh split party ranks, however.¹¹² Similarly, in Maharashtra, the JD allied with the Peasants and Workers Party and factions of the RPI, while the Left supported the SP.¹¹³ Lastly, parties within all three blocs contested the polls against each other in the home states of the AGP, DMK and TDP.¹¹⁴ In Assam, the Left partnered with the AGP, only to see the JD compete against the alliance. In Tamil Nadu, the DMK tacitly allied with the CPI and AIFB. But the CPI(M) opposed the "anti-people" economic policy of the DMK,¹¹⁵ joining the JD in backing the Marumalarchi Dravida Munnetra Kazhagam (MDMK) due to the "higher standing of its leader [Vaiko], who was younger, in New Delhi" and his "stance against the [Congress(I)] dynasty".¹¹⁶ Lastly, in Andhra Pradesh, following the split in the TDP after Rama Rao's death,

THE CRYSTALLIZATION OF THE THIRD FORCE (1991–1996)

[the] CPI and CPI(M) sided with Chandrababu Naidu [his son-in-law] while the JD stayed with the [TDP] faction of Lakshmi Parvati [his widow] ... There was a perception that she would gain a sympathy vote after NTR's death. Not everyone agreed ... [Some] were against it because [they] realized that the rank-and-file of the TDP was with Chandrababu Naidu [the other leader] ... But many in the party felt that this is what NTR would have wanted. We wanted to respect his sentiments.[117]

Ideological unease, personal obligations to the dead, divergent perceptions of local political strength: all these factors played a role in weakening the cohesion of the putative third force. Yet the centrifugal logic of the federal party system that bedeviled its ranks mired the aspirations of the BJP and Congress(I) too. Indeed, the country had entered a state of unprecedented political flux, a "decisive historical conjuncture".[118] Hence many observers anticipated that the eleventh general election would produce the first genuinely national coalition government in an evolving "post-Congress polity".[119]

7

THE FORMATION OF THE UNITED FRONT (MAY 1996)

The eleventh general election produced a deeply fractured verdict. Twenty-nine parties and nine independents managed to secure a presence in the Lok Sabha. Seven of eight national parties accounted for approximately three quarters of the total: 403 of 543 parliamentary seats. 18 of 30 state parties captured 127 seats, a fifth of the overall prize. Notably, only four of 170 registered (unrecognized) parties, each winning one constituency apiece, managed to survive. Indeed, merely nine of the 10,635 independent candidates standing for office, virtually all of whom lost their deposits, prevailed.[1] Yet new parliamentarians comprised more than half of the eleventh Lok Sabha.[2] The final verdict highlighted the inability of putatively national parties to achieve a mandate to rule the Union on their own and the challenges of building electoral coalitions in a regionalized federal party system.

The electoral verdict

The incumbent Congress(I) suffered its worst electoral defeat to date. The party won 140 seats, a third less than the 232 it had secured (despite contesting fewer constituencies) in 1991. Moreover, its limited electoral pacts completely failed. None of its partners—the Indian Union Muslim League (IUML), Kerala Congress (KEC) and AIADMK—managed to win a seat (see Table 7.1).

Table 7.1: Eleventh General Election, 1996

Alliance	1991 Votes	1991 Seats	1996 Votes	1996 Seats
INC+				
INC	35.50	232	28.80	140
AIADMK	1.62	11	0.60	0
IUML	0.01	0	0.04	0
KEC	0.12	0	0.10	0
BJP+				
BJP	20.10	120	20.30	161
SHS	0.80	4	1.40	15
SAP	N/A	N/A	2.20	8
HVP	0.12	1	0.40	3
Third Front				
JD	11.84	59	8.08	46
SP	N/A	N/A	2.86	17
CPI(M)	6.16	35	6.12	32
CPI	2.49	14	1.97	12
FBL	0.42	3	0.37	3
RSP	0.64	4	0.52	5
TDP	3.00	13	2.97	16
DMK	2.09	0	2.15	17
TMC	N/A	N/A	2.19	20
AGP	0.54	1	0.76	5

Source: Election Commission of India.

Further, the Congress(I)'s vote share fell dramatically since 1991 from 36 to 28 per cent. The electoral backlash was even stronger in large states—such as Bihar, Karnataka, Madhya Pradesh, Maharashtra, Tamil Nadu and Uttar Pradesh—which saw negative swings ranging from 10 per cent in Uttar Pradesh to 24 per cent in Tamil Nadu.[3] The massive decline of its national vote share underscored the scale of defeat.

The traditional refrain of the Congress(I), of being the only genuine national formation, was accurate.[4] The party won seats in 26 of the 31 territories (comprising both states and Union Territories), out of a possible 32, it had contested. With the exception of Orissa, however, the Congress(I) tended to dominate either very small states or single-constituency UTs. The party was one of the two main parties/blocs in several larger states—Andhra Pradesh, Assam, Gujarat, Kerala and Rajasthan—where it possessed a chance of winning—as well as in others—Delhi, Madhya Pradesh, Maharashtra and West Bengal—where it did not. In other regions, such as Karnataka and Punjab, it was the minor player in a triangular contest. Lastly, in several regions its presence was negligible, including Arunachal Pradesh, Goa, Sikkim, Tripura and Tamil Nadu. Crucially, it was weak in Bihar and Uttar Pradesh, the two largest arenas in the country.

Ironically the Congress(I)'s sweeping presence damaged its electoral prospects. This paradoxical outcome was partly due to the effects of the FPTP electoral system. The collapse of its general vote share, which fell below the critical threshold of 30 per cent for the first time, hurt the party.[5] The dispersion of electoral support, and its concentration in smaller electoral arenas, magnified the negative conversion of votes into seats.

But the Congress(I)'s electoral debacle also reflected deeper systemic changes. A representative national survey, conducted by the Centre for the Study of Developing Societies (CSDS) shortly after the polls, captured these trends.[6] In general, respondents declared relatively greater concern with their respective state governments than the national government (see Table 7.2). Similarly, although a plurality expressed ambivalence, a higher proportion of respondents claimed that regional parties ran state governments better than national parties (see Table 7.3). Indeed, respondents expressed greater loyalty to their regions before India as a whole. Supporters of the third force overwhelmingly took this view. But it held for the Congress(I) and BJP too (see Table 7.4).

The upsurge of various sectional demands in the 1990s and continuing regionalization of the federal party system had ruptured the center of politics. The Congress(I)'s previous advantage as an encompassing catch-all party had paradoxically become a liability. Its cross-sectional profile betrayed a residual character: the party won the support of groups that its partisan rivals had not mobilized.[7] The national orientation of the Congress(I) made it difficult to capture the multiple "fragments"—the ideas, interests and identities of diverse social groups—that now animated Indian democracy.[8] The changing political landscape provided strong incentives to seek coalition partners. The Congress(I)'s status as the traditional ruling party, and the fact of its wide national presence, limited such possibilities. Yet inept tactical judgments by the party high command reflected a deep-seated belief

Table 7.2: Interest in Level of Government, 1996

Category	Third Front		Congress+		BJP+		Other		Total
	N	I	N	I	N	I	N	I	
None	28.5	38.0	32.7	39.1	27.2	36.2	11.7	43.6	38.4
State Gov	32.3	26.9	30.6	22.5	26.0	21.3	10.6	24.4	23.6
Both	30.1	22.2	29.9	19.8	31.8	23.5	8.2	16.9	21.3
Central Gov	19.2	7.5	35.4	12.5	35.7	14.0	9.7	10.7	11.3
Other	27.3	5.2	36.5	6.2	26.9	5.1	8.7	4.6	5.4
Total	28.3	100.0	32.1	100.0	28.8	100.0	10.3	100.0	N=8295

Note: N represents the national vote share given to each alliance. I reflects the internal vote share within each alliance.
Source: CSDS National Election Survey, 1996.

Table 7.3: Preference for Regional Party State Governments, 1996

Category	Third Front		Congress+		BJP+		Other		Total	
	N	I	N	I	N	I	N	I	N	I
Disagree	23.4	16.4	59.0	37.2	48.7	34.2	15.9	31.1		20.2
Don't Know	27.5	43.2	28.6	40.2	24.7	38.7	11.8	51.6		45.2
Agree	33.6	40.4	20.9	22.6	22.6	27.2	5.1	17.2		34.6
Total	28.8	100.0	32.1	100.0	28.8	100.0	10.3	100.0		N=8295

Source: CSDS National Election Survey, 1996.

Table 7.4: Loyalty to Region before Nation, 1996

Category	Third Front		Congress+		BJP+		Other		Total	
	N	I	N	I	N	I	N	I	N	I
Disagree	21.5	15.8	31.6	20.8	39.6	29.1	7.2	14.9	21.2	
Don't Know	27.6	23.7	31.8	24.5	27.3	23.4	13.4	32.2	24.8	
Agree	32.2	60.5	32.4	54.6	25.3	47.5	10.1	52.9	54.1	
Total	28.8	100.0	32.1	100.0	28.8	100.0	10.3	100.0	N=8295	

Source: CSDS National Election Survey, 1996.

THE FORMATION OF THE UNITED FRONT (MAY 1996)

that it could still govern the country independently. The verdict rebuffed its pretensions to rule.

The BJP, in sharp contrast, scaled new heights. The party was now the single largest formation with 161 parliamentary seats, 41 more than in 1991, despite contesting almost the same number of constituencies. Furthermore, its limited coalition strategy paid dividends. Collectively, the SAP, Shiv Sena and HVP captured 26 seats, raising the overall parliamentary strength of the BJP coalition (BJP+) to 187 seats (see Table 7.1). However, despite being the single largest bloc of votes, the BJP+ fell 85 seats short of a parliamentary majority.

It was instructive to disaggregate the vote and seat shares of the alliance. The national vote share of the BJP hardly changed, rising from 20.1 per cent to 20.3 per cent since 1991. In contrast, the combined vote share of its allies equaled roughly 3 per cent, increasing the overall measure of the BJP+ to 24.3 per cent. Thus its positive national swing of approximately 4 per cent was almost entirely due to its coalition partners. The appeal of *Hindutva* amongst India's national electorate had stabilized.

The BJP's regional presence, nonetheless, deserved careful appraisal. The party contested the vote in 28 regions but only survived in twelve.[9] On the one hand, it was strong in the north and west, the main protagonist in Delhi, Gujarat, Haryana, Madhya Pradesh, Maharashtra and Uttar Pradesh, and one of two evenly weighted blocs or parties in Bihar and Rajasthan, respectively. The BJP won only six seats in the south, all from Karnataka, and one in the east, in Assam. Indeed, the party suffered a negative electoral swing of between 5 and 10 per cent in several major states, including Andhra Pradesh, Punjab and West Bengal.[10] In other words, unlike the Congress(I), it enjoyed limited appeal beyond its Hindi-dominated strongholds.

On the other, the BJP dominated the competition in the latter, where its electoral support—excluding Haryana and Delhi—approached 36 per cent. The popularity of the party in these regions was much greater than its national vote share, and approximately 5 per cent greater than its relative strength in 1991. The corresponding figure for the Congress(I), roughly 23 per cent, underscored the difference. Moreover, the vote share of the BJP enjoyed a tremendous positive swing in several arenas, ranging from 9 per cent in Maharashtra to 12 per cent in Bihar and 25 per cent in Haryana. In short, the disproportionality induced by plurality-rule enabled the party to win a greater number of seats than the Congress(I), despite capturing a substantially lower share of the national vote.

Ultimately, the BJP had two strategic advantages. First, its formidable organizational capacity allowed the party to harness its relatively intense appeal in particular regions through street-level mobilizations and the power of crowds.[11] Second, the six electoral strongholds of the BJP constituted major political arenas. Two were crucial for national power: Bihar and Uttar Pradesh. Significantly, large voting sections in both states—especially the latter—backed the politics of *Hindutva*. The presence of the National Front in each region challenged its chauvinistic agenda. Yet the self-destructive tendencies of the JD enabled the BJP to exploit its internal rivalries.

Finally, the parties loosely associated with the Third Front together won 173 seats, emerging as the second largest grouping in parliament. Perhaps more importantly, they collectively secured 28 per cent of the national vote, surpassing and approximating the shares of the BJP- and Congress(I)-led fronts, respectively (see Table 7.1). Indeed, the constituents of the Third Front exhibited a distinctive social profile (see Table 7.5).[12]

First, non-high-caste groups dominated their respective electoral bases. The alliance over-represented OBCs in terms of their national vote shares, as well as Muslims, who swung away from the Congress(I). The Third Front was the second choice amongst Dalits, whose massive support for "Other" parties most likely benefitted the BSP, illustrating the vertical conflicts amongst the *dalitbahujan*. Lastly, the Third Front came third amongst Adivasis and high-caste groups. The limited support of upper-caste voters was unsurprising. However, the fact that *Adivasis* voted in greater numbers for the BJP—reflecting the active recruiting strategy of the RSS and its provision of educational activities and basic health services in predominantly tribal areas—highlighted the failure of the Third Front to mobilize a natural social constituency.[13] Voters belonging to lower occupational strata disproportionately supported the Third Front too. Its parties were the first choice amongst the two lowest ranks, in stark contrast to the BJP. Conversely, the Third Front came third amongst voters from the two highest occupational categories by a substantial margin. The Congress(I) received roughly equal support from lower occupational groups. But the relative unpopularity of the Third Front amongst the higher groups underscored its distinctive class profile of its base. Third,

Table 7.5: Social Bases of Party Alliances, 1996

	Third Front		Congress+		BJP+		Other		Total
	N	I	N	I	N	I	N	I	
Community									
Upper	15.3	13.1	30.6	24.3	50.9	45.1	3.2	8.7	25.5
OBC	37.3	41.1	25.4	25.9	30.4	34.5	6.9	23.8	32.6
Dalit	28.3	18.2	36	21.3	16.2	10.7	19.4	38.9	19
Adivasi	17.8	5.4	45.4	12.9	22.6	7.2	14.1	13.5	9.1
Muslim	51.1	18.2	38.4	12.7	3.7	1.3	6.9	7.7	10.6
Other	37.4	4	29.8	2.9	10.7	1.2	22.1	7.4	3.2
Total	29.6	100.0	32.1	100.0	28.8	100.0	9.5	100.0	N=8258
Occupation									
Highest	15.7	3	30.9	5.5	47	9.3	6.4	3.8	5.7
High	24.8	16.7	30.9	19.3	38.8	26.9	5.5	11.8	20
Middle	29	19.9	33.3	21	29.5	20.7	8.2	17.5	20.3
Low	32.1	35.3	32	32.5	25.1	28.3	10.9	37.7	32.6
Lowest	34.7	25.1	32.5	21.7	19.9	14.8	12.9	29.2	21.4
Total	29.6	100.0	32.1	100.0	28.8	100.0	9.5	100.0	N=8295
Education									
Illiterate	30.1	42.8	33.6	44.1	24.4	35.6	11.9	53	42.1
Middle	31.1	33.7	32.7	32.6	27.4	30.5	8.9	30.2	32.1
College	28.3	19.4	29.7	18.8	35.9	25.3	6.1	13.1	20.3
Graduate	22	4.1	26.4	4.5	45.2	8.6	6.4	3.7	5.5
Total	29.6	100.0	32.1	100.0	28.8	100.0	9.5	100.0	N=8283

Locality									
Village	31.9	82.7	32.2	77.2	25.9	69.1	10	81.7	77
Town	22.2	17.3	31.7	22.8	38.6	30.9	7.5	18.3	23
Total	29.6	100.0	32.1	100.0	28.8	100.0	9.5	100.0	N=8295

Source: CSDS National Election Survey, 1996.

THE FORMATION OF THE UNITED FRONT (MAY 1996)

the popularity of its parties amongst different educational strata was noticeable as well, albeit less so. Unlettered voters and primary school leavers preferred the Congress(I) by a small margin. But voters with secondary-level schooling and college graduates disproportionately voted against the Third Front, underscoring their aversion. Finally, the rural orientation and self-image of many Third Front parties mirrored their social base. Rural voters preferred the Congress(I) by the narrowest of margins at the national level. But the Third Front was the only political formation to over-represent this cleavage within its social base. Conversely, its parties came third amongst the urban electorate by a significant margin. In sum, the parties comprising the Third Front tended disproportionately to represent the middle and lower ranks of every major cleavage in society, except *Adivasis*.

The increasing sociological distinctiveness of the third force as a whole straddled growing electoral differences, however. The National Front secured 63 parliamentary constituencies in six of the 21 regions it contested. The JD suffered conspicuous losses. Its national vote share dropped from 11.8 per cent to 8.1 since 1991. The party retained only 46 of its 59 seats. Two states accounted for most of them. In Bihar, the JD lost ten seats but survived as the single largest formation, locked in a bipolar rivalry against the BJP-front. In Karnataka, where the party had failed to win a single constituency in 1991, the JD posted a stunning turnaround, emerging as the dominant party in a triangular fight. Lastly, the party bagged only a handful of seats in Jammu & Kashmir, Kerala, Orissa and Uttar Pradesh. Its inferior electoral performance partly reflected diminished presence: the JD challenged 196 constituencies in 1996, in comparison to 1991, when it contested 307. Yet the party's electoral decline, having secured 143 parliamentary seats and 17.8 per cent of the national vote at its peak in 1989, owed far more to its political implosion and the rise of the BJP. The party lost 20 constituencies in Uttar Pradesh, where the SP captured 17, coming second to the BJP. Factional strife further cost the JD eleven seats in Bihar and Haryana at the hands of the SAP and HVP, which now supported the BJP. In short, the failure of the contending political leaders of the JD to share organizational power amongst themselves and fuse the aspirations of particular caste groupings into larger political solidarities weakened their prospects of capturing national power. Nevertheless, the party remained the single largest force in the Third Front.

The electoral fortunes of the Left displayed far greater continuity. The bloc won 51 seats—5 less than in 1991—under the leadership of the

CPI(M) and CPI. The former won 32 seats, the latter 12. Their electoral stability was remarkable given the extreme volatility that marked other parties and regions. Yet it also signified an inability to expand beyond West Bengal, Kerala and Tripura. The CPI(M), which contested 75 seats in 19 territories, survived in 5. It was the principal force in West Bengal, where it won 23 seats, and in Tripura, a small bastion that had only two parliamentary constituencies. The party was one of two main blocs in Kerala, the other mainstay of the LF. And it won a single parliamentary constituency apiece in Andhra Pradesh and in Assam, losing its foothold in Bihar, Maharashtra and Orissa. In contrast, the CPI won seats in several regions: Andhra Pradesh, Bihar, Kerala, Tamil Nadu and West Bengal. But it never captured more than a handful in any of them. Lastly, the AIFB and RSP retained a few seats each between them in West Bengal and Kerala. In sum, although the Left failed to expand beyond its traditional political strongholds, it provided crucial ideological ballast to any prospective Third Front dispensation.

Finally, the loosely defined regional bloc posted significant gains. The AGP, which fought approximately one-third of the constituencies in Assam, increased its tally from one parliamentary seat in 1991 to five. The TDP, which had recovered its strength in the 1994 state assembly elections in Andhra Pradesh,[14] increased its tally from 13 to 16. The DMK witnessed a spectacular recovery in Tamil Nadu. It seized 17 parliamentary seats, of the 18 it contested, after winning none in 1991. Its new ally, the TMC, performed even better, becoming the single largest force in the state with 20 constituencies. The combined tally of these four regional parties, an impressive 58 parliamentary seats, enhanced their potential coalitional power. Their ascendance challenged the role historically played by the socialists and, to a lesser extent, the communists in determining the shape of previous anti-Congress formations in New Delhi.

The scramble for office

The publication of the official electoral results on 13 May created deep uncertainty. Constitutionally, the power to appoint a prime minister belonged to the President, Shankar Dayal Sharma. The rules concerning the investiture and defeat of government also created incentives for parties in the eleventh Lok Sabha to form legislative coalitions that could produce a collegial executive capable of gaining its confidence. The inconclusive electoral verdict enhanced President Sharma's role beyond its typically ceremonial nature, however. The rules of investiture failed to stipulate precise criteria in the event of a hung parliament. His predecessor, Ramaswamy Venkataraman, had followed a fairly simple norm. The leader of the largest party should be invited to form a government.[15] Yet the precedent was limited. Moreover, forming a Council of Ministers was the first task. Gaining the confidence of parliament was the second crucial step. Given the opposition it had faced during the campaign, it was unclear whether the BJP could secure the 272 seats necessary to gain a majority in the Lok Sabha. It was equally ambiguous whether the party's rivals could unite themselves. Hence many feared the contending formations would resort to bribery or threats to induce defections. President Sharma had to weigh rival arguments and anticipate the foreseeable ramifications of his decision.

The contending parties had several choices to make as well. They faced three particular imperatives: to clarify their parliamentary leadership, articulate a programme of action and forge a coalition of parties capable of winning a parliamentary majority. Considerations of power were ubiquitous. But substantive policy differences and vote-seeking calculations also loomed. The importance of these motives varied amongst the parties too. Simply put, the circumstances undermined the notion that clearly defined, mutually exclusive and pre-determined interests would settle the final outcome. Senior party leaders had to identify which objectives to pursue, determine the strategies and tactics that would allow their realization, and judge the foreseeable consequences of their decisions.

The BJP was on the threshold of seizing national power. The party was well-aware of the stigma associated with it, however. Accordingly, its high command re-confirmed Vajpayee as its parliamentary leader. His relatively moderate image, in contrast to the more hardline BJP president L.K. Advani, enhanced the prospects of winning new allies. On 13 May, its three pre-poll allies reaffirmed their support. The SAP declared it would be issue-based, though, presumably to convey its opposition to militant Hindu nationalism. The only party to join their ranks after the polls was the Badal faction of the SAD, a longstanding rival of the Congress(I), whose support was primarily driven by considerations of power. The slightly enlarged BJP-led coalition, now commanding 195 parliamentary seats, still faced a substantial deficit.

The state of affairs in the Congress(I) was more difficult. The party could not ignore the scale of defeat: "We knew that the vote was anti-incumbent. This was a virtual consensus within the party."[16] Yet the reasons for, and implications of, its rout stirred intense debate. Several figures insinuated that Narasimha Rao bore personal responsibility. On 4 May, Kannoth Karunakaran, a senior member of the Congress Working Committee (CWC), its highest decision-making body, admonished his colleagues for allowing a single person to lead both the organizational and parliamentary wings of the party. The high-profile criminal investigations against Rao lent additional weight. Yet the defeated prime minister stood firm, receiving the backing of many former cabinet members and standing chief ministers, who cast the challenge as Karunakaran's "personal view".[17] On 12 May, the Congress Parliamentary Party re-elected Rao as its leader.

> We re-elected him because we had no other choice at the time ... Not re-electing him would simply have implied to the public an admission of his guilt, which would have been far more damaging ... It was better to let the courts pursue their investigations.[18]

Defending Rao until proven guilty may have been misguided; propriety suggested he step down. Yet the norms of conduct amongst India's political class had gravely deteriorated. Moreover, from a narrow political vantage, closing ranks avoided further immediate dissension. In any event, Rao consolidated his position. He pitted opponents against each other, asking the Maharashtra party boss, Sharad Pawar, if the latter wanted to become national president of the Congress(I), sowing resentment amongst the rebels. Rao also postponed internal party elections until 31 December 1996.[19] He would remain in charge for some time.

Next, the CWC vowed to support a "secular front", claiming that containing Hindu nationalism was its "paramount consideration".[20] Far more likely, strategic electoral calculations dictated its stance: the BJP was the only party with the potential to supplant it nationally.

> We had only two options after the verdict was declared: one, a government must be formed, and two, we did not want the BJP. Therefore the only thing to do was to provide outside support [to a secular front] ... There was no option but to.[21]

In short, a short-term calculus of survival pushed the Congress(I) to back a prospective Union government led by a third force.

The Third Front confronted several problems, however, beginning with its composition. Questions over its parliamentary leadership, modes of collective decision-making and programme of action also demanded attention. Resolving these issues would be difficult given the range of personalities, interests and parties in the coalition, and the fact that its viability in office would depend on Congress' external parliamentary support.

Determining the make-up of the Third Front was the first priority. Dissident Congress factions—the MPVC, AIIC(T) and Karnataka Congress Party (KCP)—joined its ranks before the final electoral verdict; so did the Maharashtravadi Gomantak (MAG) and United Goans Democratic Party (UGDP). Their pooled support was relatively minor. Most of these parties were small political outfits with just one seat apiece.[22] Far more important were the political calculations of erstwhile regional members of the National Front: the AGP, TDP and DMK. The prospect of the TDP backing a Third Front bid was

> a purely contextual possibility ... [the] main enemy, the Congress, was out of the game. And the BJP was finding it hard to find friends.[23]

The emphasis on contingencies recalled the TDP's underlying political strategy: to decide its stance by assessing particular circumstances. Still, the party shared various affinities with other members of the third force. In addition, the TDP sought

> [to] give credibility to the government, [to gain] credibility in the comity of nations. We can't afford frequent elections ... [given] the economic cost, the policy uncertainty, the investor confidence [sic].[24]

Claiming that national political responsibility required the party to consider its choices cut against the view that regional formations had little interest in the politics of the Centre beyond their state-level interests. Economic liberalization had effectively devolved many powers to the states. But the standing of the next Union government in the eyes of the world elicited genuine concern.

The TMC and DMK, equally concerned about the viability of the Centre, weighed their strategic options differently. According to a senior member of the TMC:

> We had a three-point plan. First, we would accept a Congress without Rao at the helm. Failing that, we would try to promote a Third Front, and explore its feasibility ... which prior to the elections was a moot point. Third, we would stay in opposition.[25]

The DMK's reported willingness to support the Congress(I) was debatable given the alliance between the latter and its archrival, the

AIADMK. The TMC's strategy was also contrasted sharply with the TDP, which refused to prop up a Congress-led administration. The AGP was reluctant too.[26] After all, the Congress(I) was the chief adversary of all three parties.

Crucially, senior figures amongst the socialists and communists cajoled the regional parties into backing the Third Front.[27] According to many, V.P. Singh intervened decisively:

> The question was [sic] taking up the threads and pulling them together ... Both Chandrababu Naidu and Karunanidhi made statements, saying "our choices are open" ... The AGP was also willing to play along. They said, "We want good relations with the Centre"... Meanwhile, the National Front and Left Front already had a number of arrangements with the regionals ... [They were persuaded that] allowing the BJP at the Centre would allow them a doorway into the states. And second ... that their other interests were better served [by the Third Front] ... They said, "We will come together, but without participation in government".[28]

Government participation remained a sticking point. But the regional party chiefs agreed that the Congress(I), given its poor electoral showing and subsequent internal struggles, would not threaten a Third Front administration in the short-run. They also acknowledged their corresponding obligations at the Centre, and that backing a national coalition government provided an opportunity to define its agenda as well as protect their respective turfs in the states.[29] Subsequent events would throw into doubt the first argument. Nevertheless, the negotiations exhibited the range and fluidity of interests at play. They also demonstrated how able political leadership could shape the ways in which others perceived their self-interest.

On 12 May, the various constituents of the Third Front met to determine its parliamentary leader. Many pushed for V.P. Singh. His success in earning the regional parties' trust testified to his skills of persuasion. Moreover, notwithstanding the enmity of many upper-caste groups after Mandal, his recognition amongst the national electorate and experience in negotiating competing interests made him a powerful candidate. The former prime minister had remained a central figure behind the scenes despite his illness, coordinating political strategies and defusing personal antagonisms in the Janata *parivar*. Leading CPI(M) figures—notably the West Bengal Chief Minister Jyoti Basu and the party general secretary Harkishan Singh Surjeet—implored Singh to "manage this crowd" he "had been handling".[30] Containing rampant factional-

THE FORMATION OF THE UNITED FRONT (MAY 1996)

ism within the Janata and managing its relations vis-à-vis other coalition partners could easily overwhelm less experienced hands. But Singh's illness prevented him from taking such responsibility. Instead, he nominated Basu, who won general approval based on his "experience, reputation and respect".[31] The venerable chief minister expressed his readiness. He was a terrific candidate. A lifelong centrist within the CPI(M), Basu had explored the possibilities of building a broader Indian left, from his ties with Jayaprakash Narayan and personal opposition to the "left sectarian line" of B.T. Ranadive in the 1940s and efforts to forge a national democratic front in the late 1960s, to his role in the regional conclaves of the mid 1980s and negotiations with V.P. Singh and A.B. Vajpayee during the National Front.[32] Moreover, the Left was committed from the start to supporting a Third Front government. But the CPI(M) Central Committee, while agreeing to establish a steering committee and provide external parliamentary support, rejected government participation. According to Basu,

> ... I think by 8 or 10 votes, we lost—our general secretary and I were in the minority. We thought ... though it may be for a few months, it would be politically advantageous.... the majority thought otherwise[,] that it would be a great risk ... but we said: "Already we had worked out Common Minimum Programme for West Bengal. Now we will have [one] at the Centre". We said: "As people saw in West Bengal United Front Government, similarly, on an all India's scale [sic] it will help our Party, it will help the Left forces, democratic forces.... In the Centre the Prime Minister wields a lot of influence and we can for the time being influence them [the other parties]. Other partners you see the World Bank is there, the IMF is there; they are blindly accepting all that advice given to them, which we shall not do. The people will have a new experience. Within these limitations so many things could be done. Then if we are thrown out and we shall leave a new experience for the people ... people will judge who is to blame ... this is how people will understand with whom lies the responsibility." But this argument was not accepted by the majority.[33]

The decision elicited many reactions within the party. Proponents of formal government participation argued it would enable the CPI(M) to guide legislative change and broaden its appeal.[34] Basu had put the case cogently. Opponents cited mitigating factors, however. Some feared that it smacked of "parliamentary opportunism", "a disease created in the atmosphere of working in bourgeois parliamentary institutions", according to the late E.M.S. Namboodiripad, breeding individualism, factionalism and corruption.[35] Yet the CPI(M) had taken the parliamentary road long ago. A more pressing issue was its control over policy:

We knew that a [Third Front] government, with the inclusion of certain parties within it ... would pursue certain policies and actions representing landlords and kulaks ... This we couldn't support because we would then be responsible.[36]

The numerical strength of the regional parties, now controlling 58 parliamentary seats, caused particular concern:

Our decision to give outside support was based on [a judgment]: to participate only if we can influence policy ... We were not in a position to do so because we didn't have sufficient numbers, we didn't have enough strength ... Rapid reforms hurt the common man and we would take the blame for policies we actually opposed.[37]

Afraid that "we could not push our views" without a parliamentary veto, a majority of the Central Committee sought to "demarcate [its] position within the coalition" by providing outside support.[38] The failure to pursue an independent political line through grassroots mobilization, it later surmised, had forced the CPI(M) to "tail behind the bourgeois parties" at its own expense.[39] Renouncing high office and performing the role of an "honest broker" represented the party's "accumulated moral hegemony".[40]

The resolution evoked various reactions amongst the parties. Senior JD figures recognized genuine ideological considerations:

Their reasoning was solid on orthodox Marxian grounds ... Basu and Vajpayee have the same problem ... They realize the philosophical limitations of their parties, but lack the courage to break away from them.[41]

Figures in the Congress(I) perceived ulterior motives:

the underlying rationale was to keep West Bengal secure. Without Basu, they perceived a threat to their dominance over the state.[42]

Personal charisma could not explain the Left's hegemony in West Bengal. Central Committee members conceded that staking the "prestige" of its leader for elusive and partial gains at the Centre would be a strategic mistake, however.[43] Other rivals accused the CPI(M) of hypocrisy:

[They] wanted to have their cake and eat it as well. They wanted the credit without the responsibility.[44]

Yet a CPI(M) politburo member rejoined, "if we have strength, like we do in West Bengal, then we display responsibility".[45]

THE FORMATION OF THE UNITED FRONT (MAY 1996)

Independent observers presented divergent views. The readiness of Basu and Surjeet to join a Third Front government was pragmatic, according to some, seeking to advance the secular values of social democracy within the conditions of possibility. But "the CPI(M) ... is not willing to take the risks of aggressively pursuing such a line... Its longer term electoral prospects are, therefore, much more intimately tied to the consistency and depth of such extra-electoral activities and influence."[46] Others criticized its ruling apparatchiks for evading the "messy practical realities" of multi-party democracy. By rejecting the prime ministership, the CPI(M) lost the opportunity to set the agenda, earn political credibility with the "new forces of radicalism in the northern heartland" and "educate itself about the intricacies of Dalit/OBC politics." It represented a victory of "high theory" over "the murky business of running a government".[47] A "sense of being besieged, together with rearguard actions," dominate[d] the CPI(M).[48]

Ultimately, the Central Committee made an instrumentally rational decision that simultaneously betrayed a curiously apolitical vision. It was risky for the CPI(M) to assume formal political responsibility in a diverse governing coalition. Protecting its vote base and tenure in the states, while seeking to exercise policy influence at the Centre, was a shrewd political stance. United front tactics did not require "ideological non-aggression pacts".[49] Radicals could denounce reformists where necessary. The party could "have its cake and eat it" too. Yet the medium-term prospects of the CPI(M) expanding beyond its regional bastions were relatively poor. Party membership had grown to 600,000, yet extremely slowly and mostly in West Bengal. A similar fate bedeviled the base and trajectory of its ancillary organizations, with the All-India Kisan Sabha and All-India Democratic Women's Association confined to West Bengal and the All-India Agricultural Workers Union in Kerala.[50] Moreover, the Left's success in these states had different bases, offering no obvious national blueprint. Indeed, voters saw each front in distinct regional terms.[51] Finally, the compulsion to liberalize, a new reality that increasingly pressed every state in India's federal political economy, would not diminish on its own. New Delhi set the terms of liberalization.

Thus much depended on whether external parliamentary support would enable the CPI(M) to steer the national economic agenda. The party enjoyed a veto over any Third Front government. But it was unlikely to exercise such power with the BJP in the wings. In short, the

CPI(M) was stranded between government and opposition. The majority of the party high command opposed government participation because they could not fully control its policy agenda. Such a situation was inevitable in a diverse multiparty coalition. But their unwillingness to share formal political responsibility in a prospective governing coalition suggested a fixed, indivisible and zero-sum conception of power. Only when the party led a majority ruling alliance, as in West Bengal and Kerala, would it be prepared to participate in office. By emphasizing its "accumulated moral hegemony", moreover, the CPI(M) asserted that ethical principle guided political action. It was an admirable stance given the sheer political opportunism of many parties. Nonetheless, the party's self-proclaimed authority indicated a didactic tendency as well as a moralistic conception of political responsibility, where the purity of intentions eclipsed the consequences of action. These distinct conceptions, of governmental power and political responsibility, skewed how the CPI(M) judged its actual possibilities.

V.P. Singh asked the Central Committee to reconsider, convening a second meeting the following day. But he absconded before the party leaders arrived in order to compel them to select another candidate.[52] His ruse worked. Desperate to meet President Sharma's midnight deadline, the assembled chiefs approved Haradanahalli Doddegowda Deve Gowda, the JD chief minister of Karnataka, as their parliamentary leader.

His selection was "a terrible surprise".[53] Barring his election to parliament in 1991, Deve Gowda, a *shudra* from a dominant agricultural caste, had spent his entire political career in his home state.[54] Members of his political clique, extolling "his performance as chief minister", asserted that "nobody [else] was prepared to accept responsibility".[55] In truth, a variety of crosscutting factors eliminated other contenders. Prafulla Kumar Mohanta, the AGP chief minister of Assam, lacked the requisite stature. His party, with five seats, "was simply too small".[56] Larger considerations thwarted G.K. Moopanar of the newly formed TMC: "a polite inquiry was made on [his] behalf and ... was politely rejected, especially by the Left".[57] Others pushed forward the candidacy of the TDP Chief Minister Nara Chandrababu Naidu. But he reportedly quashed the possibility himself, "saying that 'I just came to Andhra [Pradesh]. I need time to establish myself politically'".[58] Naidu had spent his entire political career in his state. But his section of the recently split TDP had just emerged victorious in the polls. Consolidating his political base was a clear imperative.

THE FORMATION OF THE UNITED FRONT (MAY 1996)

The difficulty of agreeing upon a regional party chief threw attention onto the National Front. The bloc housed several aspirants, such as the Bihar JD Chief Minister, Lalu Yadav, and Karnataka JD leader S.R. Bommai. But their associations with ongoing criminal investigations scuttled further discussion:

> Lalu was not in a good position since the fodder scam had quietly surfaced. S.R. Bommai also was not in a position due to his implication in [the] *hawala* [scandal].[59]

The final contender was the SP chief, Mulayam Singh Yadav, the former chief minister of Uttar Pradesh. However, several regional leaders found his "style" of politics "unacceptable", while Lalu Yadav opposed his candidature.[60] The interpersonal rivalry between the two Yadav leaders, who both sought to lead the OBCs in northern India, underscored the one-upmanship that bedeviled the Janata *parivar*. In the end, Mulayam Yadav proposed Deve Gowda, a colleague from the Chandra Shekhar faction in the JD in the late 1980s and early 1990s.[61] The SP chief "sold the idea to the Left, basically to Surjeet ... who was for Basu taking over but ... never thought that the party would overrule them".[62] The CPI(M) Central Committee refused to change its stance.

Deve Gowda's ascendance was a mixed personal triumph. Many party leaders saw him as "an instrument ... the weakest of possible choices".[63] Indeed, the Karnataka chief minister's inclinations generated suspicion. Rumors circulated that a section of the JD, which included Deve Gowda himself, had considered supporting a Congress-led administration:

> Deve Gowda, Lalu [Yadav] and Mulayam [Yadav] were actually willing to let Congress lead ... The regionals had an inkling of it ... But someone sabotaged it, saying "some leaders within us are willing to support the Congress" ... Later they were sheepish and on the defensive.[64]

Sources within the Congress(I) confirmed the allegation:

> [We were] reluctant to lead a secular coalition ... though we were asked to ... the main opposition was against Rao ... others wanted him changed. Basically, most of the parties would have accepted a Congress without Rao, except for the TDP, AGP and the Communists ... The Congress is their main opponent. If they compromised, they would lose face.[65]

The incident highlighted an important fault-line within the Third Front. The Congress(I) had lost its presence in Uttar Pradesh and Bihar in the wake of mobilizations around Mandal and *mandir*. Similarly, the party was a subordinate player in Tamil Nadu, a deeply regionalized state party system dominated by the DMK and AIADMK. But it remained the key electoral rival of the Left in West Bengal, Kerala and Tripura, of the TDP in Andhra Pradesh and of the AGP in Assam. Gaining the trust of his colleagues would test Deve Gowda.

The Third Front sought to convince President Sharma that it could muster the confidence of parliament. Deve Gowda, who reportedly enjoyed a personal rapport with Rao, asked the Congress(I) to send a letter of support. But it failed to reach in time.

> The Congress made a statement of support ... They didn't send a formal delegation to us. Margaret Alva [a Congress official] came on her own and said that the letter had been sent to the President ... But later we discovered they didn't send it ... The excuse they gave later was that Rao signed the letter but fell asleep and therefore did not send it.[66]

It was hard to credit such comic incompetence. Reportedly, Rao feared that many of his detractors would defect if President Sharma invited the Third Front to form a government.[67] The incident foreshadowed Rao's readiness to engage in machinations to ensure his political survival, while the underlining the Congress(I)'s parliamentary veto over the fledgling coalition.

On 15 May, President Sharma invited the BJP to form a Union government and prove its majority on the floor within a fortnight. Leaders of the Third Front contended that haste, imprudence and latent communal sentiments informed his decision:

> [It] surprised us since he asked us before asking Vajpayee to form the government. We lacked time ... But he was also inconsistent in his positions. And he was under the influence of [former president] Venkatraman, who was a [Hindu] revivalist ... The President is supposed to have so many eyes to increase his ambit of discretion [sic].[68]

These accusations ran against the President's secular credentials in the eyes of the media.[69] Furthermore, Sharma had restrained the Congress(I) from exploiting its incumbency during the campaign on several occasions, strengthening his reputation for impartiality.[70] That said, he reportedly claimed that "the Third Front was based on an anti-BJP plank ... it was simply negative".[71] Observers questioned his political judgment. The motivations driving the Third Front were less relevant than its viability. But its leaders protested in vain.

THE FORMATION OF THE UNITED FRONT (MAY 1996)

The isolation of the BJP

On 16 May, President Sharma inducted a twelve-member BJP ministry. It faced two critical issues. The first was to win a parliamentary majority by 31 May—a formidable task. The second issue was whether the party would temper any of its controversial measures to lure possible allies. The most important included its pledge to construct a Ram *mandir* in Ayodhya, implement a Uniform Civil Code (UCC) and abrogate Article 370 of the Constitution. India's parliamentary form of cabinet government generated strong incentives for the BJP to moderate its stance, while providing many power sharing opportunities in the Council of Ministers. Yet the party faced a dilemma. Political moderation risked losing the extremists in the *sangh parivar*.

The BJP responded by pursuing a two-fold strategy. On the one hand, it attacked the motives and stability of its various opponents. On 19 May, Vajpayee assailed the "unprincipled" agreement between the Third Front and the Congress(I), whose single-point agenda of keeping the BJP out of office ostensibly "defeat[ed] the mandate given by the people".[72] The Third Front released a statement, "Platform for secular democratic alternative", which gave Vajpayee some ammunition. It reiterated that the "economic policies [of the Congress(I)] ... have eroded economic sovereignty, provided a bonanza for big capitalists and forcing [sic] multinational corporations and heightened the sufferings of the people", echoing Left rhetoric. Yet the declaration also indicted the BJP, which "poses a threat to the secular fabric of our society and national unity", and outlined a program committed to the "principles of democracy, secularism, federalism, socialism and social justice".[73] The BJP responded by highlighting Rao's personal quandaries. On 18 May, Ram Jethmalani, the newly inducted Minister for Law, Justice and Company Affairs, declared that nobody, including the former prime minister, would be spared if found guilty in the St. Kitts' affair.[74] His remark underscored the risks of aligning with such a discredited leadership.

On the other, the BJP sought to mollify fears that its rule would unleash communal violence. The President's address to both houses of parliament on 24 May, which customarily outlined the policies of the incoming government, pledged to introduce a ban on cow slaughter, family planning measures (to counter the allegedly higher birth rate in Muslim communities) and a re-evaluation of the country's nuclear policy. Yet the speech failed to mention Ayodhya, Article 370 or the

UCC. Furthermore, on 27 May, Vajpayee appealed to Manohar Joshi, the Shiv Sena chief minister of Maharashtra, to reinstate the Srikrishna Commission inquiry into the causes of the December 1992–January 1993 Bombay riots that followed the destruction of the Babri masjid. The Shiv Sena had failed to extend its investigations in early February 1996 after the Commission accused its leader, Bal Thackeray, of instigating anti-Muslim violence.[75] Reinstating the inquiry would signal responsible political leadership.

Ultimately, the fate of the BJP ministry rested with the Opposition. The reliability and extent of the Congress(I)'s support to the Third Front, rechristened the United Front (UF), was doubtful over the medium-term. Rao commented that its "unconditional" external parliamentary support was not given "with closed eyes".[76] Its seat tally provided a crucial veto. Still, the Congress(I), which had its own P.A. Sangma elected parliamentary speaker, had strong incentives in the short run to secure the defeat of the BJP. Hence the United Front simply had to stick together:

> We were disappointed after Sharma invited the BJP ... But we knew that it would fall if we held our unity [sic] ... Therefore our strategy was simple: to show that our unity had no chinks in it, the projection [sic] that we are totally solid.[77]

On 27 May, Vajpayee moved a motion of confidence in the Lok Sabha, offering conciliatory measures. The Maharashtra chief minister agreed to restore the Srikrishna Commission. The prime minister-designate also agreed to "freeze" the most controversial pledges of the BJP during its prospective tenure.[78] Observers expected these last-minute gestures, given the weak ideological moorings of most parties, would enable him to lure requisite support. Yet not a single party, or faction thereof, defected from the United Front during a stormy two-day debate. Thwarting the march of Hindu nationalist forces, rather than acquiring the spoils of office, was their principal collective aim. Realizing the futility of his endeavor, Vajpayee resigned without tabling a vote. The BJP's first grasp of national power lasted less than a fortnight.

The formation of the United Front government

Accordingly, President Sharma invited H.D. Deve Gowda to form a government and demonstrate its parliamentary majority by 12 June.

THE FORMATION OF THE UNITED FRONT (MAY 1996)

The various leaders of the United Front proceeded to determine their power relations and general policy orientation. On 31 May, three of the first four members of the newly christened Federal Front—the TDP, TMC and DMK—reversed their opposition to participating in government. The AGP would follow suit in due course. V.P. Singh once again assuaged their misgivings of being at the mercy of the Congress(I), demonstrating how skilled leadership, processes of negotiation and shifting political circumstances could rapidly alter perceptions of interest. Direct government participation at the Centre provided the regional parties, now the single largest bloc in the third force, with opportunities to further their respective states' interests and collectively improve Centre-state relations. Their decision signaled the reintegration of the original National Front while expanding its regional dimension (see Table 7.6).

Table 7.6: The United Front

Party	Seats	Principal State Arenas	Main Opponents
JD	46	Bihar, Karnataka, Orissa	Congress(I), BJP
SP	17	Uttar Pradesh	BJP, BSP, Congress(I)
CPI(M)	32	West Bengal, Kerala, Tripura	Congress(I)
CPI	12	West Bengal, Kerala	Congress(I)
RSP	5	West Bengal, Kerala	Congress(I)
FBL	3	West Bengal	Congress(I)
TDP	16	Andhra Pradesh	Congress(I)
DMK	17	Tamil Nadu	AIADMK, Congress(I)
TMC	20	Tamil Nadu	AIADMK, Congress(I)
AGP	5	Assam	Congress(I)
AIIC(T)	4	Madhya Pradesh, Rajasthan, Uttar Pradesh	Congress(I)
MPVC	1	Madhya Pradesh	Congress(I)
KCP	1	Karnataka	Congress(I)
MAG	1	Goa	Congress(I)
UGDP	1	Goa	Congress(I)

Constituting formal executive power was the next step. On 1 June, President Sharma invested a 20-member ministry, including twelve ministers with full cabinet rank and eight ministers of state. Conventionally, three factors influenced cabinet formation: parties' relative strengths, individual ability in specific policy matters and their respective political clout.[79] Representing various social groups also traditionally figured into the calculus. However, the Deve Gowda ministry drew particular attention given the pronounced regional character of the alliance and its diversity of personalities, groups and parties (see Table 7.7).

Politically, it rewarded the JD. Despite capturing less than a quarter of the United Front's 181 parliamentary seats, the party secured roughly half of the Council of Ministers, and managed to get Ram Vilas Paswan, its *Dalit* leader in Bihar, appointed as the parliamentary floor leader. Several factors accounted for its overrepresentation. Prime ministerial privilege gave Deve Gowda formal political control. Moreover, the JD remained the single largest party in the coalition, underlining the historic role of its predecessors in forging previous anti-Congress administrations. V.P. Singh manufactured agreement behind the scenes, while the Bihar Chief Minister Lalu Yadav used his clout among important backward castes to secure cabinet berths for his nominees.[80] Lastly, the disproportionate weight of the JD reflected the refusal of the CPI(M) to join the ministry and the regional parties' belated decision to do so.

Sociologically, the Deve Gowda ministry symbolized the historic transfer of power at the Centre by not including a single person of Brahmin origin. Regionally, the Council of Ministers represented six states: Andhra Pradesh, Bihar, Karnataka, Punjab, Tamil Nadu and Uttar Pradesh. Parties that opted for direct government participation demonstrated a strong presence in all of them. More significantly perhaps, the United Front boasted seven chief ministers amongst its ranks: Jyoti Basu from West Bengal, Lalu Yadav from Bihar, M. Karunanidhi from Tamil Nadu, P.K. Mahanta from Assam, N. Chandrababu Naidu from Andhra Pradesh, E.K. Nayanar from Kerala and J.H. Patel from Karnataka. Further rounds of cabinet expansion would redress other regional imbalances.

Policy-wise, the Deve Gowda ministry elicited both surprise and praise. The prime minister designate appointed Balwant Singh Ramoowalia, a member of the National Minorities Commission last elected in 1989, as minister of welfare largely to bolster Sikh representation.[81] Deve Gowda also selected the recent JD president S.R. Bommai, his old patron from Karnataka, as minister of human resource development despite his implication in the *hawala* scandal, which had thwarted the aspirations of Biju Patnaik, his peer from Orissa.[82] Senior JD officials justified the decision by saying that Bommai had not been charged

THE FORMATION OF THE UNITED FRONT (MAY 1996)

Table 7.7: The H. D. Deve Gowda Ministry

Name	Party	Post(s)	Rank
Deve Gowda	JD	Prime Minister	Cabinet Minister
C.M. Ibrahim	JD	Civil Aviation, Food, Civil Supplies,	Cabinet Minister
D.P. Yadav	JD	Consumer Affairs	Cabinet Minister
Inder Kumar Gujral	JD	Foreign Minister	Cabinet Minister
Ram Vilas Paswan	JD	Railways	Cabinet Minister
S.R. Bommai	JD	Human Resource Development	Cabinet Minister
Mulayam Singh Yadav	SP	Defence	Cabinet Minister
Beni Prasad Verma	SP	Communications	Cabinet Minister
P. Chidambaram	TMC	Finance	Cabinet Minister
M. Arunachalam	TMC	Labour	Cabinet Minister
Murasoli Maran	DMK	Industry	Cabinet Minister
T.G. Venkatraman	DMK	Surface Transport	Cabinet Minister
Balwant Singh Ramoowalia	Independent	Welfare	Cabinet Minister
C.P. Verma	JD	Rural Areas and Employment	Minister of State
J.N.P. Nishad	JD	Environment and Forests	Minister of State
Kanti Singh	JD	Human Resource Development	Minister of State
M. Taslimuddin	JD	Home Affairs	Minister of State
S.I. Shervani	SP	Health and Family Welfare	Minister of State
S. Venugopalachari	TDP	Power	Minister of State
U. Venkateswarlu	TDP	Urban Affairs, Employment, Parliamentary Affairs	Minister of State

by the CBI, whereas Patnaik and others had.[83] But seasoned journalists also noted that installing Bommai into the Union ministry pre-empted a possible alliance between him and Ramakrishna Hegde, Deve Gowda's nemesis, in their home state.[84] The maneuver was the first of many occasions in the course of this ministry when personal battles and provincial intrigues affected political decisions in New Delhi.

That said, Deve Gowda made astute policy-related choices too. He reappointed his JD colleague I.K. Gujral as minister of external affairs, earning widespread approval.[85] According to one senior figure, "we felt that we've got to sail together ... so let's delegate authority to those with expertise".[86] Others contended that

> very few ministers took interest [in foreign policy]. Nobody demurred or objected [to Gujral] ... Deve Gowda had no hang-ups about Gujral taking the lead.[87]

The lack of competition facilitated a smart appointment. Nonetheless, it also indicated a desire to push a progressive line on foreign relations in the subcontinent, as events would soon illustrate.

Perhaps more importantly, the prime minister-designate installed Palaniyappan Chidambaram of the TMC as finance minister. The former Union commerce minister was a staunch advocate of economic liberalization:

> The Left said ... give Chidambaram anything but finance. But [V.P. Singh] convinced them of the need for continuity in the reforms process. There was a lack of confidence abroad.[88]

It was unlikely that Singh won over his counterparts. Simply put, having rejected government participation, the Left could not veto the decision. Several factors advanced Chidambaram's elevation, nonetheless. Deve Gowda had campaigned against economic liberalization—particularly against recommendations by the General Agreement on Tariffs and Trade (GATT) to deregulate the agricultural sector—during Rao's prime ministership. His stance changed upon becoming the Karnataka chief minister in 1994, however, leading him to court foreign private investment and weaken land ceiling laws in order to win greater private investment, reforms which benefited the dominant rural classes Deve Gowda represented.[89] The TDP chief Chandrababu Naidu, confronting fiscal pressures caused by generous food subsidies and foregone alcohol taxes in Andhra Pradesh, proposed greater economic liberalization too.[90] Chidambaram enjoyed their respective support. Still, many doubted whether a national coalition government of primarily state-based parties could display fiscal restraint towards particular interests and formulate a coherent national strategy beyond the short-term.[91]

The potential battle lines over policy grew sharper the following day. On 2 June, the CPI agreed to join the Deve Gowda ministry, granting it formal executive power. It was a historic milestone. Compared to its Left allies, the party had always favored united-front-from-above-tactics, leading to the original 1964 split. Its decision to support Indira Gandhi during the Emergency, suffice to say, was a disastrous political misjudgment. Joining a diverse multiparty government, inspired by the project of building a national third force in the wake of the second democratic upsurge, was a different prospect however.

The CPI National Council, its general decision-making body, overwhelmingly favored the decision:

[t]here was a heated debate ... The arguments for joining were twofold: it would help the government function and strengthen the position of the party. The argument against was that we would be in a minority and relatively uninfluential ... But we had a limited capacity to influence alone in any case.[92]

Another CPI politician elaborated, "the days of coalition government have started ... revolution is not on the agenda".[93] Thus

it was logical for us to share power [at the Centre]. We were not in a position to come to power on our own ... and we wanted to change the equations of power—for instance the composition and conduct of cabinet—to represent the weaker classes.[94]

Historically, the party had always believed coalitional power could enhance the struggle for greater socioeconomic equality. It also had several experienced parliamentarians to field as potential cabinet ministers. Yet its specific political judgments contrasted sharply with those of the CPI(M), whose prospects of capturing national power on its own or through a larger Left coalition were equally slim. Whether the CPI could secure important ministerial posts, steer the policy agenda and devise an effective strategy with its larger communist partner vis-à-vis other parties remained to be seen. Nonetheless, senior CPI functionaries expressed cautious optimism. Moreover, they stressed a larger political imperative:

of course we could foresee the difficulties of working with bourgeois regional forces ... so we were cautioning our comrades: "mass movements are the real source of strength" ... The parliamentary wing is only a part of it.[95]

The CPI's dispersed electoral presence compelled the party to accept a subordinate role in the Left in Kerala and West Bengal, and work at the margins in a handful of other states. Government participation gave the CPI an opportunity to demonstrate its commitment to the United Front and advance progressive legislation, beyond any electoral advantage it might derive. Whether the party could do so, or whether the CPI(M)'s political line would yield greater dividends, was the question.

Having put its house in order, the UF delivered a series of key policy statements. On 3 June, Chidambaram stressed the need to achieve economic growth and poverty alleviation through a package of fiscal austerity, public-sector reform and agricultural liberalization.[96] Apart from bearing his personal imprint, the approach bore the stamp of the most senior mandarins in the ministry of finance who had formulated economic policy since 1991, such as Montek Singh Ahluwalia and Shankar Acharya. By retaining them, Chidambaram attempted to signal general policy continuity, despite the presence of the Left.

In addition, the government-in-waiting maintained India's traditional stance on major international questions whilst seeking to improve sub-continent relations. I.K. Gujral reiterated that India would retain its nuclear option, a stance that enjoyed cross-party support, against concerted attempts by the United States and others to pressure India into signing the "discriminatory" CTBT.[97] The JD had pledged to oppose the latter and the Non-Proliferation Treaty in pursuit of total disarmament.[98] Yet the external affairs minister also sought "to end the hostile rhetoric" that marred Indo-Pakistan relations.[99] He announced several conciliatory measures, such as increasing the number of family visas and trade permits, recalling his gestures during the National Front government. The Pakistani prime minister Benazir Bhutto had declared in April 1996 that substantial progress would be unlikely in the event of a "weak coalition government" in New Delhi, lowering expectations.[100] Still, positive incremental change was conceivable—no small thing given the state of Indo-Pakistan relations and the conflict in Kashmir. The decision to soften cross-border rhetoric and implement small confidence-building measures was a good first step.

Most importantly, the United Front produced an agenda of governance, "A Common Approach to Major Policy Matters and a Minimum Programme", which quickly became known as the Common Minimum Programme (CMP). Claiming the general election had mandated "the formation of a secular, liberal and democratic coalition

THE FORMATION OF THE UNITED FRONT (MAY 1996)

government at the Centre",[101] the coalition highlighted three broad issue-areas. First, it vowed to practice greater federalism in political, administrative and economic matters. The United Front promised to implement the main recommendations of the Sarkaria Commission Report and appoint a high-level commission to review other suggestions given the impact of liberalization. In particular, it sought to give states greater autonomy in determining their development priorities, transfer most centrally sponsored schemes to the states and grant assent promptly to legislative bills concerning their regional affairs, as well as explore the possibility of greater financial devolution.[102] The coalition also pledged to reactivate moribund institutions, such as the NDC and the ISC, which had languished under the Congress(I). And it undertook to amend Article 356 of the Constitution in light of the Supreme Court's 1994 Bommai judgment, which declared that a state legislative assembly could not be dissolved without parliamentary assent, while asserting that secularism was a part of the basic structure of the Constitution. In short, the CMP underscored a determination to improve Centre-state relations.

Second, the United Front addressed the question of political secularism and social justice. The CMP contained few specific measures regarding the latter, only extending the principle of reservations to Dalit Christians and to women (on a non-caste basis) in parliamentary and state legislatures.[103] Moreover, it failed to increase the ambit of existing reservations for *dalits*, *adivasis* and OBCs, which the JD had pledged. But the United Front vowed to follow the Supreme Court over the Babri masjid, implement the Protection of Places and Worship Act and safeguard personal religious codes.[104] The coalition also promised to conduct state assembly elections in the strife-torn state of Jammu & Kashmir, maintain Article 370 in the Constitution and allow the region to realize a "maximum degree of autonomy"—a tantalizing if elusive phrase.[105] In short, it attacked militant Hindu nationalism, defended special minority rights and maintained the principle of reservations in public institutions.

Lastly, the United Front presented a bold strategy of further economic liberalization:

Growth with social justice will be the motto of the United Front government. There is no substitute for growth. It is growth which creates jobs and generates incomes ... The United Front is committed to faster economic growth ... at over 7% per year in the next ten years in order to abolish endemic poverty and unemployment.[106]

In order to do so, the coalition promised to lower the fiscal deficit below 4 per cent of GDP—by reducing current spending, cutting the public-sector-borrowing ratio and targeting subsidies towards the most destitute—and pursue rapid, labor-intensive, industrialization.[107] It also sought to achieve an annual industrial growth rate of 12 per cent through further deregulation and higher levels of public and private investment. Given that socialists within the JD and many quarters of the Left opposed foreign private capital, the CMP stipulated that external investment would be channeled into core and infrastructural sectors, with the "bulk of the [sic] industry's requirements [coming] from within the country".[108] Moreover, the United Front pledged to support profitable state-owned enterprises while exploring how to rescue their loss-making counterparts. Nonetheless, the coalition proposed to establish a Disinvestment Commission to provide strategic advice, and redirect revenue generated from privatization towards social expenditure.[109] Indeed, it claimed that

> at the margin, the country cannot do without foreign investment particularly if that foreign investment will also bring modern technology and management practices and create new markets for products manufactured in the country. The nation needs and has the capacity to absorb at least $10 billion a year as foreign direct investment.[110]

The latter amounted to roughly 3 per cent of India's gross domestic product in the mid 1990s.[111] Indeed, despite stressing fiscal discipline, revenue mobilization received little attention.[112]

The significance of the CMP was threefold. First, its twin commitment to political secularism and social justice marked a sharp break with the Hindu nationalism of the BJP, sharpening the idea of a third force. Second, the manifesto reflected its deepening regional orientation. Finally, it registered the sharp decline of socialist policy ideas in an era of liberal economic reform. Indeed, by declaring the primacy of high economic growth, the CMP inverted the traditional Gandhian commitment to generating mass employment through small-scale rural development that earlier Janata dispensations had proclaimed. Several factors shaped its content. First, a troika of senior party figures wrote the document—Chidambaram from the TMC, Sitaram Yechury of the CPI(M) and S. Jaipal Reddy from the JD, who was also the official spokesperson for the United Front. Reportedly, Chidambaram usurped the task of drafting it himself, citing the need to make haste, compelling Yechuri to amend the text. Jaipal Reddy reconciled their compet-

THE FORMATION OF THE UNITED FRONT (MAY 1996)

ing views.[113] As a result, the CMP contained several inconsistencies and suppressed a number of economic policy disagreements. A senior government official close to the process attributed subsequent political clashes to these formative imperfections.[114] Second, however, finding common ground on several issues would have been extremely difficult in ideal circumstances. The pressure of time, given the President's deadline, rendered it moot. Finally, even though members of the CPI(M) expressed frustration, Jyoti Basu endorsed the text.[115] Many of its economic policies resembled his efforts in West Bengal.

Reactions varied. One senior government advisor was critical:

> It was supposed to represent the agreed programme ... [but] it wasn't coherent at all ideologically. It was full of inconsistencies and contradictions ... Yet it was very specific about many things. That was its merit.[116]

A senior regional politician conceded as much, saying that "it was a compromise of ideologies".[117] Yet it was the first time that a multiparty Union government in India had ever produced such a document. Neither the Janata Party nor the National Front had. The CMP represented political learning. Additionally, the manifesto revealed a substantive agenda. S. Jaipal Reddy declared that it sought to restore official Congress(I) ideology.[118] Yet the CMP emphasized many proposals that bore little resemblance to the politics of the latter in practice. Finally, the negotiations challenged the view that such programmatic statements had few political ramifications. Indeed, the writing of the CMP set a precedent that subsequent coalition governments would have to emulate.

Having accomplished these various tasks, the Deve Gowda ministry addressed one final issue: its reliance on the Congress(I). On 9 June, the CBI summoned Rao's son following allegations of kickbacks in the urea scam, emboldening his detractors.[119] Introducing a motion of confidence, Deve Gowda pledged not to interfere with any investigation concerning the former prime minister. The question of whether the two had struck a deal generated serious concern. The potential for blackmail was obvious. But it was unable to derail the momentum of events. On 12 June, the Congress(I) leader declared that his party "would not allow the government to fall under any circumstances".[120] For only the third time since Independence, a diverse coalition government, representing the aspirations to create a progressive third force, would attempt to govern the Union.

8

ESTABLISHING POLITICAL AUTHORITY (JUNE–SEPTEMBER 1996)

The chief priority of the United Front was to establish its credibility to govern. The new prime minister faced a complex strategic calculus: to lengthen the political time horizon of his ministry and shift the balance of power vis-à-vis the Congress(I) without jeopardizing its external parliamentary support. According to one longstanding observer, Deve Gowda was a workaholic who intuitively grasped problems, could listen well and had a penchant for detail. However, he allegedly had a tendency "to focus too much on minutiae, on seeking short-term advantages by forging ties with people in face-to-face dealings ... at the expense of larger strategic thinking". Close associates feared that, bereft of his network of informants in Karnataka, the former chief minister "may be 'flying blind' in New Delhi".[1] It would prove to be an astonishingly prescient judgment.

Relations within the United Front also demanded his attention. First, Deve Gowda had to transform his formal institutional authority into real political clout, as well as accommodate parties that had recently joined his ministry. Second, the Government had to tackle economic policy, Centre-state relations and foreign affairs in a manner that bolstered its standing. Finally, the governing coalition had to devise mechanisms for making collective decisions and resolving disputes to avert potential disagreements and minimize their impact. The formal locus of these functions was the Council of Ministers. In principle, its decisions reflected collegial responsibility. Disciplined single-party cabinets, capable of exercising the whip to enforce loyalty, had an advantage in realizing such an expectation. Forging political agreement and ensuring collective accountability in a diverse multiparty executive was a far more difficult task. It required a willingness to participate in deliberation, strike political compromises and stand by collective decisions—a real challenge given the deterioration of parliamentary cabinet norms since the 1970s.[2] To complicate matters, the United Front had to determine relations between its Steering Committee and the Council of Ministers. Constitutionally, the latter monopolized executive power. Yet the CPI(M)'s decision not to participate in government made it imperative to determine the status and procedures of the Steering Committee. Given the threat of the BJP, few expected the CPI(M) to withdraw external parliamentary support. But disagreements between the party and its partners in the Government could slow policy decisions, deplete political trust and sow governmental instability.

In short, handling these distinct yet interrelated demands required astute strategic judgment, tactical acumen and political skill. Many commentators anticipated that a "power-hungry, unprincipled and opportunistic" Congress(I) and "formidably determined" BJP would hasten the demise of the United Front.[3] Yet some felt that if the governing coalition could last two years, and introduce policy changes that served India's "naturally decentralized, plural and federalized society", it might survive a full parliamentary term.[4] Although unknown at the time, the first three months of the Government involved a wider spectrum of issues than at any other point during its tenure in office. Much depended on the willingness and abilities of its party leaders to address multiple policy challenges, satisfy competing interests and improvise when things did not go according to plan.

The first crisis in the Janata Dal

Unfortunately, a political crisis flared only a day after the Government's investiture. On 13 June, Lalu Prasad Yadav, the chief minister of Bihar and newly appointed president of the JD, expelled Ramakrishna Hegde, its leader in Karnataka, for "anti-party activities, especially his sustained criticism of the United Front", namely its "unholy alliance" with the Congress(I).[5] It was a strange charge: Hegde had previously advocated closer ties between the two parties. Rumors later spread that he begrudged his growing isolation.[6] The JD Political Affairs Committee, its highest decision-making body, had appointed Lalu Yadav as its president in February 1996 after S.R. Bommai was implicated in the *hawala* scandal.[7] Hegde had called the appointment "undemocratic" since the party had failed to obtain the consent of its MPs and MLAs that comprised its electoral college.[8] His unresolved hostility towards Deve Gowda triggered the outburst.

Hegde did not like it [Deve Gowda's prime ministerial appointment], though he did not say it then [the day he was chosen], or the next day. He could not stomach the decision ... It was a personal pique involving a clash over leadership ... and a question of caste difference.[9]

A former member of the Congress(O), Deve Gowda had joined the JD in 1983, significantly contributing to its victory in Karnataka. But he fell out with Hegde by the late 1980s, disparaging "the Brahmin intellectual" for promoting social welfare programmes at the expense of dominant cultivating interests.[10] Their animosity undermined the party in the 1989 assembly elections. Deve Gowda finally realized his ambition to become chief minister after successfully merging the JD and SJP, which he had latterly joined, to contest the Karnataka assembly elections in 1994.[11] Observers speculated that Hegde had sponsored a petition to the Karnataka High Court to investigate possible nepotistic practices in Deve Gowda's administration. Reportedly, Hegde also opposed his other socialist colleagues' orientation. Deve Gowda's surprising political ascent proved too much to bear. The new JD Chief Minister of Karnataka, Jayadevappa Halappa Patel, attempted to reconcile the two. On 15 June, the Political Affairs Committee appointed C.M. Ibrahim, a well-known confidant of Deve Gowda, to replace Hegde. Ibrahim warned the party's MLAs to acquiesce. Less than ten days later, Hegde formed a political vehicle, the Rashtriya Nav Nirman Vedike, supported by 28 loyalists kicked out of the JD.

The incident highlighted three issues. First, it showed how quarrels within a state unit of a party could demand national political attention. Many provincial disputes would pressure the United Front. Yet these seldom arose due to internal party matters. In contrast, conflicts within the JD influenced shifting configurations of power in New Delhi throughout the Government's tenure. The prime minister subsequently appointed Sharad Yadav as "working president" of the JD to appease him—and check Lalu Yadav's influence in Bihar.[12] Second, Hegde's swift removal revealed a penchant for eliminating perceived enemies.

Deve Gowda displayed it again in the first week of July, expelling Maneka Gandhi from the Karnataka state unit for criticizing his approval of a controversial fast-track project by the Congentrix power company, despite environmental concerns.[13] Third, given his longstanding rivalry with Deve Gowda, accommodating Hegde would have been difficult; not to counter him would have damaged the fledgling prime minister. The timing of his outburst made it necessary to act. The fact that few legislators joined him made it easier.[14] Nevertheless, Hedge was a senior political figure in Karnataka. His expulsion generated ambivalence because it was

not according to the rules of the party ... We could have suspended him first, and asked him to explain [himself] before a disciplinary committee.[15]

Following stipulated procedures would have conveyed greater fairness and created an opportunity to compromise. Hegde's brusque dismissal enhanced Deve Gowda's position in the short term. The JD would eventually confront its long term ramifications.

Expansion of the Deve Gowda ministry

Next, the prime minister altered the size and composition of the Council of Ministers. The first round of cabinet formation established the formal relations of power within the Government and its broad orientation. The second aimed to satisfy party quotas, consider parties that had recently joined its ranks and assign outstanding portfolios. On 28 June, the Deve Gowda ministry grew from 21 to 35 members, adding 5 cabinet ministers and 9 ministers of state, with several of the latter enjoying independent charge over their portfolios. Two of the former, Srikant Jena of the JD and Janeshwar Mishra of the SP, had served in the National Front. The other three included Indrajit Gupta and Chaturanan Mishra from the CPI, and Birendra Prasad Baishya of the AGP, all of whom received major portfolios. The expanded number of ministers of state enhanced the dominance of the JD, granted additional posts to most of the regional parties and delegated special portfolios to factions that had left the Congress(I) (see Table 8.1).

Observers noted the absence of members from Kerala, Madhya Pradesh and Maharashtra. Deve Gowda remarked that Rao had influenced the decision, to allow the Congress(I) to remain the dominant rival to the BJP in these states, provoking much speculation.[16] Yet the

Table 8.1: The Expansion of the Deve Gowda Ministry

Name	Party	Rank	Ministry
First Round			
Srikant Jena	JD	Cabinet Minister	Parliamentary Affairs and Department of Tourism
Janeshwar Mishra	SP	Cabinet Minister	Water Resources
Indrajit Gupta	CPI	Cabinet Minister	Home
Chaturanan Mishra	CPI	Cabinet Minister	Agriculture (excluding the Department of Animal Husbandry and Dairying)
Birendra Prasad Baishya	AGP	Cabinet Minister	Steel and Mines
Dilip Kumar Ray	JD	Minister of State	Animal Husbandry and Dairying with independent charge
R.L. Jalappa	JD	Minister of State	Textiles with independent charge
S.R. Balusubramaniam	TMC	Minister of State	Personnel, Public Grievances and Pensions and Parliamentary Affairs
R. Dhanushkodi Athithan	TMC	Minister of State	Human Resource Development—Department of Youth Affairs and Sports
Bolla Buli Ramaiah	TDP	Minister of State	Commerce with independent charge
Muhi Ram Saikia	AGP	Minister of State	Human Resource Development—Department of Education
Ramakant D. Khalap	MGP	Minister of State	Law, Justice and Company Affairs (excluding Department of Company Affairs) with independent charge
Sis Ram Ola	AIIC(T)	Minister of State	Chemicals and Fertilizers with independent charge
Yogendra K. Alagh	Independent	Minister of State	Planning and Programme Implementation with additional charge of Science and Technology
Second Round			
Raghuvansh Prasad Singh	JD	Minister of State	Animal Husbandry and Dairy
N.V.N Somu	DMK	Minister of State	Defence

T.N. Baalu	DMK	Minister of State	Petroleum and Natural Gas
Satpal Maharaj	AIIC(T)	Minister of State	Railways
Changes			
M. Arunachalam	TMC	Cabinet Minister	Urban Affairs and Employment
Chidambaram	TMC	Cabinet Minister	Department of Company Affairs
Dilip Kumar Ray	JD	Minister of State	Minister of Food Processing Industries
Kanti Singh	JD	Minister of State	Minister of State for Coal with independent charge

induction of several prominent AIIC(T) politicians as ministers of state demonstrated the limits of this interpretation too. It was shrewd to accede to such demands if they placated the embattled Congress(I) leader. They cost the prime minister little.

The allocation of important cabinet posts drew scrutiny. The appointment of Indrajit Gupta, general secretary of the CPI, as home minister was particularly noteworthy. Reportedly, Gupta demanded the portfolio as a precondition for joining the Government despite intense opposition from the Congress(I) as well as sections of the JD on the grounds that it was too "sensitive" to be entrusted to a "communist".[17] Given his distinguished parliamentary record and reputation for probity, it was a ludicrous charge. Simply put, the Union home minister had the authority to appoint key state officials regarding law and order and Centre-state relations, prerogatives that many parties desired. Equally noteworthy, Deve Gowda made Gupta's colleague, Chaturanan Mishra, minister of agriculture. The Left Front had successfully reduced material poverty in West Bengal and Kerala through effective land reform, better agricultural wages and improved local self-governance since the 1970s. Whether Mishra could stem the deceleration of agricultural wage growth and general rural stagnation that characterized India's post-liberalization economy by introducing progressive measures at the Centre was a vital policy question.[18] Curiously, the Department of Animal Husbandry and Dairying, normally the responsibility of the Union agriculture minister, went to Dilip Kumar Ray, a JD member from Bihar. The decision warranted critical scrutiny: CBI investigations into the fodder scam, which potentially implicated the JD Chief Minister Lalu Yadav, came within its official purview. Hence who controlled the department was a significant political matter.

Expressions of discontent compelled a second round of cabinet expansion in the following weeks. R.L. Jalappa resigned as minister of state for textiles on grounds that it failed to recognize his seniority. Deve Gowda mollified his Karnataka colleague by elevating his position, and that of Beni Prasad Verma of the SP, to full ministerial rank. The prime minister also inducted four new ministers of state, accommodating the DMK, and reshuffled other portfolios. The most significant change was the appointment of Raghuvansh Prasad Singh, a political confidant of Lalu Yadav, as minister of state for animal husbandry and dairy, leaving Dilip Kumar Ray in charge of food processing industries. Lastly, on 9 July, Maqbool Dar, a JD politician from Jammu & Kashmir, became minister of state for home affairs, following criminal allegations against M. Taslimuddin and pressure from the new home minister.[19]

The last round of cabinet formation enlarged the Council of Ministers to 39 members. By and large, the number and allocation of portfolios reflected the parties' relative parliamentary strength. The CPI, which secured two important ministries despite its small tally, and the CPI(M), which decided against government participation, were the two major exceptions. For the latter and its smaller Left allies, the RSP and AIFB, the Steering Committee, in which they participated, was the highest decision-making body of the United Front. The failure to agree explicitly on its relationship to the Council of Ministers would soon test the new governing coalition.

The Congress(I)-BSP electoral pact in Uttar Pradesh

Ongoing criminal investigations continue to beset the Congress(I). In mid-June, the CBI questioned Rao's son over whether he received financial kickbacks in the urea (fertilizer) scam, enlivening a CWC meeting on 16 June. The former prime minister responded by saying the Congress(I) would withdraw its support to the Government if its activities jeopardized "the interests of the country", inviting party workers to prepare for mid-term polls that could occur "at any time".[20] Yet Rao enjoyed little respite. On 21 June, Surendra Mahato, a former MP of the JMM, testified to the CBI that Rao had bribed him and three party colleagues to vote for the Congress(I) during the crucial 1993 no-confidence motion. The report prompted Dijvijay Singh, the party's chief minister of Madhya Pradesh, to call for Rao's resignation as president of the AICC.

A shrewdly timed announcement momentarily shifted attention. On 24 June, Rao announced a seat-sharing pact with the BSP for the state assembly elections in Uttar Pradesh in September–October 1996, with the Congress(I) contesting only 125 out of 425 legislative seats. According to various party figures, basic electoral calculations and the anticipated strategy of the United Front influenced the decision:

We needed to recover the Dalit vote in the short-term, which the Congress had lost ... We didn't have competing interests with the BSP, in contrast to a Congress-Mulayam [Singh Yadav] pact.[21]

It was sheer compulsion. [AIIC(T) leader N.D.] Tiwari was with the SP. Also, we were ideologically sympathetic.[22]

Assessments differed. Some thought it could help the Congress(I) regain Dalit support in a crucial electoral arena. Others felt that it would destroy the party by further driving upper castes towards the BJP.[23] The detractors had a point. The flagging centrist politics of the catch-all Congress(I), which employed codes such as poverty and inequality to appeal to the SCs, clashed with the explicit *Dalitbahujan* ideology of the cleavage-based BSP.[24] The electoral pact might help to some extent. BSP supporters tended to follow their leadership.[25] But the capacity of the latter to generate political loyalty, and commitment to staffing high political offices with its own figures, highlighted the risk for the Congress(I).

Nevertheless, the Congress(I)-BSP pact complicated the electoral strategy of the United Front. It diminished the prospects of compromise between the SP leader Mulayam Singh Yadav and Mayawati, his counterpart in the BSP. It also threatened to divide the governing coalition. The BSP had withdrawn support to their state-level coalition government with the SP in June 1995.[26] Ground-level material conflicts between *Dalits* and OBCs in Uttar Pradesh, combined with the conflicting personal ambitions of Mulayam Yadav and Mayawati to become chief minister, exposed the limits of lower-caste unity in the state. Sections of the JD had favored backing the BSP prior to the eleventh general election to deepen their social base in the state and revive their electoral prospects vis-à-vis the SP.[27] Discord within the United Front benefited others as well. In late July, the BJP parliamentary leader A.B. Vajpayee demanded the resignation of both Mulayam Yadav and Beni Prasad Yadav from the Council of Ministers on the basis of the Ramesh Chandra committee report, which indicted the SP leadership for authorizing the abduction of BSP MLAs in June 1995 in a bid to save its ministry.[28] The Government rejected the demand, saying the report was a state-level administrative inquiry. It also asked why the subsequent BSP administration had taken no action after the report had been submitted in July 1995. Yet the incident underscored the stakes. The Congress(I)-BSP electoral pact created political space for Rao. Whether the former could successfully gain a presence amongst Muslim groups at the expense of the SP, and retrieve *Dalit* support without alienating upper castes, was the question. The pronounced electoral fragmentation in Uttar Pradesh would likely produce a hung assembly, giving Rao leverage. Cross-cutting pressures in the United Front exacerbated the general difficulty of maintaining national power within a federal party system with de-synchronized electoral battles.[29]

Maintaining India's nuclear policy independence

The United Front addressed several policy issues during its first months in office. The first concerned national security: whether to sign the CTBT. In late June, the United States demanded that India become a signatory to the treaty, an accord that the latter had originally proposed. The Government rejected the draft under negotiation in Geneva, disallowing its progress to the UN General Assembly. It was a momentous strategic decision. The failure to corral India in Geneva forced the five permanent Security Council members to pursue quiet diplomacy in the following weeks.[30]

The rejection of the CTBT reflected an emerging national consensus: to challenge the supremacy and legitimacy of the five declared nuclear weapon states. The draft treaty failed to incorporate a ban on testing within a schedule for gradual nuclear disarmament, stop the development of existing nuclear arsenals through computer-modeling technologies or repudiate their future use.[31] Many US officials had also expressed their desire to "cap, reduce and then eliminate" the nuclear capabilities of non-signatories to the NPT, which included India and Pakistan.[32] Thus the majority of parties in India saw the existing form of the CTBT as a threat to its strategic nuclear doctrine.[33]

The United Front resisted pressure, however, to exercise India's nuclear option. The indefinite extension of the NPT in 1995 and evolving terms of the CTBT had united proponents of the bomb on the militant Hindu right with centrist advocates of Nehruvian non-alignment and critics of Western imperialism on the left in India.[34] Yet the question of testing divided parties. Concerted American pressure had stopped Rao in December 1995.[35] Vajpayee had also seriously considered the option in May 1996: "the BJP was keen", according to one knowledgeable figure, "to demonstrate its will and India's strength".[36] Exercising the nuclear option was integral to its militaristic conception of India and search for great power status. Moreover, the United States' failure to impose sanctions on China after it sold nuclear ring magnets to Pakistan, violating the CTBT, galvanized further pressure to test.[37] But Vajpayee hesitated:

[He] said yes when the high priests of science came. He was ready to test ... But his senior officials convinced him to wait until after the confidence motion.[38]

Veteran analysts contended that Vajpayee was prudent enough not to execute such a radical policy reversal without a parliamentary majority.[39]

ESTABLISHING POLITICAL AUTHORITY (JUNE–SEPTEMBER 1996)

Several factors apparently persuaded Deve Gowda not to exercise the nuclear option. It was partly a question of timing. The entry-into-force provisions of the CTBT would only come into effect three years after its adoption. More pressing concerns demanded his attention, namely the economy. Yet the decision also partly reflected the general posture of the United Front. A greater ratio of its supporters opposed developing an atomic bomb, with a plurality taking no position (see Table 8.2). In contrast, more than half of those supporting the BJP-led coalition backed nuclear weaponization, underscoring the gap.

Indeed, simultaneous moves to improve bilateral relations with Pakistan exemplified the difference. Both countries had articulated their longstanding positions regarding Jammu & Kashmir during the 1996 electoral campaign. Pakistan demanded international mediation. India suggested a wider gamut of direct bilateral relations and highlighted Pakistan's sponsorship of "cross-border terrorism".[40] Nevertheless, in early June Benazir Bhutto had proposed resuming high-level bilateral negotiations, which had been suspended for two years following their initiation by the Congress(I) in 1991.[41] Various factors influenced her offer: growing US pressure, greater domestic support to normalize trade relations with India and New Delhi's determination to hold general elections in Jammu & Kashmir in May 1996.[42] In late June, the Government unveiled several confidence-building measures.[43] It raised the quota of Pakistani journalists from two to five; promoted greater "track-two" contacts by promising to grant visas automatically to intellectuals, lawyers and artists; offered limited visa services for Pakistani citizens in Mumbai in exchange for a similar facility for Indians in Karachi; and invited a delegation of Pakistani parliamentarians to visit New Delhi. Although relatively minor first steps, these initiatives revealed the willingness of the United Front not to view national security concerns through a strictly militaristic paradigm. Adopting a more conciliatory approach towards Pakistan, yet vigorously defending the country's nuclear option, were elements of a coherent foreign policy. Suffice to say, resolving the conflict in Kashmir would require immense diplomatic skill, political courage and substantive compromises on both sides. The opposition of the BJP constrained the Government's room for maneuver. Ideologues in the *sangh parivar* would denounce any measures that appeared to weaken India's presence in the region as they had during the National Front government. Nevertheless, having the United Front at the helm provided a window

Table 8.2: Opposition to Building an Atomic Bomb, 1996

All Alliances

Category	Third Front		Congress(I)+		BJP+		Other		Total
	N	I	N	I	N	I	N	I	
Disagree	25.5	31.8	29.1	32.5	36.0	44.8	9.4	32.9	35.9
Don't Know	31.1	41.1	31.6	37.5	25.5	33.6	11.8	43.7	38.1
Agree	29.9	27.1	36.9	30	23.9	21.6	9.3	23.4	26
Total	28.8	100.0	32.1	100.0	28.8	100.0	10.3	100.0	N=8295

Third Front

Category	National Front		Left Front		Federal Front	
	N	I	N	I	N	I
Disagree	12.5	38.3	6.4	26.5	6.6	28.3
Don't Know	11.9	38.9	10.0	44.2	9.1	41.1
Agree	10.2	22.8	9.8	29.4	9.9	30.6
Total	11.7	100.0	8.7	100.0	8.4	100.0

Note: N represents the national vote share given to each alliance. I reflects the internal vote share within each alliance. Source: CSDS National Election Survey, 1996.

of opportunity to de-escalate tensions and initiate progressive measures in Indo-Pakistan relations.

Signaling the continuity of economic liberalization

Unsurprisingly, economic policy questions generated greater disagreement from the start. The presence of the CPI(M) in the Steering Committee and participation of the CPI in the Council of Ministers threatened many disputes over the direction, pace and scope of reform. Significantly, these disagreements concerned the ends of policy as well as the means, suggesting rival conceptions of where ultimate political authority lay. By and large, the debate over liberalization reflected elite differences. The Left claimed that Congress(I) lost the 1996 general election because of widespread opposition to economic liberalization. Yet less than 20 per cent of voters indicated any knowledge of it, a stunning ratio. Indeed, the relatively poorer supporters of the third force were the least aware (see Tables 8.3). Equally significant, between 46 and 69 per cent of the onefifth professing awareness endorsed it, regardless of their partisan attachments. The ratio of support amongst the constituents of the Left Front and National Front was approximately 36 per cent. Yet it roughly matched the level of opposition in both alliances. More strikingly, more than 70 per cent of the regional parties' supporters favored greater liberalization (see Table 8.4). These findings underscored an important fissure in the United Front. Proponents of further economic liberalization sought to de-reserve the public sector, reorder labor practices and shrink the fiscal deficit. Protecting these domains was the principal task of their opponents.

On 17 June, the finance minister issued guidelines to central government ministries to reduce their expenditure by Rs. 3,000 crore. Chidambaram put forward standard neoliberal arguments. The fiscal deficit and total public expenditure on pay and allowances for central government employees (excluding defense) had nearly doubled in absolute terms since 1991. Central equity investments in PSEs returned meager dividends and compelled the Government to borrow heavily. Accordingly, the new guidelines instructed central government ministries to curtail manpower, ordered PSEs to achieve a minimum 20 per cent dividend on share-holding assets and raised the price of government-related services while phasing out redundant projects and schemes.[44] Public sector unions and the Left accused the finance min-

Table 8.3: Knowledge of Economic Policy Changes, 1996

All Alliances

Category	Third Front		Congress(I)+		BJP+		Other		Total	
	N	I	N	I	N	I	N	I	N	I
Unaware	29.6	83.5	32.4	82.1	27.1	76.3	10.9	86.3		81.2
Aware	25.3	16.5	30.7	17.9	36.5	23.7	7.5	13.7		18.8
Total	28.8	100.0	32.1	100.0	28.8	100.0	10.3	100.0		N=8295

Third Front

Category	National Front		Left Front		Federal Front	
	N	I	N	I	N	I
Unaware	12.3	85.7	16.0	78.7	8.9	85.4
Aware	8.9	14.3	9.8	21.3	6.6	14.6
Total	11.7	100.0	8.7	100.0	8.4	100.0

Source: CSDS National Election Survey, 1996.

Table 8.4: Opinion of Economic Policy Changes, 1996

All Alliances

Category	Third Front		Congress(I)+		BJP+		Other		Total
	N	I	N	I	N	I	N	I	
Disapprove	26.9	31.7	19.8	19.3	44.3	36.3	9	35.9	29.9
Approve	22.2	46.2	39.8	68.6	32.5	47	5.5	38.5	52.8
Can't Say	32.2	22.1	21.5	12.2	35.2	16.7	11.1	25.6	17.4
Total	25.3	100.0	30.7	100.0	36.5	100.0	7.5	100.0	N=1556

Third Front

Category	National Front		Left Front		Federal Front	
	N	I	N	I	N	I
Disapprove	10.8	36	12.3	37.3	3.9	17.6
Approve	6.1	36	7.2	38.6	8.9	71.6
Can't Say	14.4	28.1	13.7	24.2	4.1	10.8
Total	8.9	100.0	9.8	100.0	6.6	100.0

Source: CSDS National Election Survey, 1996.

ister the following day for putting the burden of adjustment on lower-end workers and PSEs deprived of public investment, however.[45] According to Chidambaram, the NDC had made similar recommendations in the past, which many chief ministers in the United Front had endorsed, including Biju Patnaik of the JD.[46] The finance minister also claimed that the cuts protected serving employees' positions, anticipated wage increases and precluded rising costs for basic public utilities. Finally, he emphasized that Rs. 3,000 crore was only a fraction of total revenue expenditure. In short, Chidambaram portrayed the austerity guidelines as representing continuity in policy, a tactic used by many politicians to implement economic liberalization after 1991.[47]

The incident exposed a basic tension in the CMP, which pledged to protect workers' interests while reducing the fiscal deficit.[48] Yet the disagreement also exposed differences over what constituted the political:

Chidambaram suggested it [was] a routine matter, but it wasn't. It was a political issue.[49]

The finance minister allegedly retorted that he

was not obliged to discuss the issue in the Steering Committee, but only required the broad approval of the prime minister ... the Left thought that the steering committee was a super-cabinet.[50]

The dispute also revealed divergent perceptions regarding the locus of power and proper dispute resolution. Chidambaram had exercised his formal ministerial authority. But the CPI(M) believed the appropriate domain to set policies was the Steering Committee. A senior CPI politician defended his Left colleagues, saying they had a right to "have a democratic discussion".[51] The CPI(M)'s readiness to rebuke Chidambaram publicly was a major tactical weapon in its national political strategy. Criticizing economic liberalization in New Delhi allowed the party to distance itself from decisions that potentially threatened its electoral fortunes or might demoralize its rank-and-file in the states.

Disrupting rival announcements enabled the CPI(M) to wrest partial concessions as well. On 2 July, the finance minister raised the price of several petroleum products between 10 and 30 per cent, sparing kerosene given its disproportional use by the poor.[52] Both Chidambaram and Deve Gowda cited immediate compulsions and long-term fiscal stability, including suppliers' threats to stop delivery unless the Government reined in the oil pool deficit, which had surpassed Rs. 11,000 crore because of inaction by the previous Rao administration.[53] The move provoked immediate criticism from the Left. Substantively, the CPI(M) argued that it "would have a cascading effect on inflation", which most harmed the poor, and ignored how "the [oil pool] deficit was caused by increasing [world] prices and the decline of the rupee".[54] The West Bengal Chief Minister, Jyoti Basu, expressed dissent on political grounds, saying "there would be riots in Calcutta".[55] Sympathetic commentators argued that reducing the price given to oil producers, which had secured profit margins in excess of their mandated 12 per cent post-tax return, would have addressed the deficit.[56] Procedurally, the CPI(M) reproved Chidambaram for not consulting the Steering Committee first. The finance minister responded that it lay entirely within his mandate. But other leaders, including some from the JD, forced a partial retreat.[57] On 6 July, the Government halved the rise in the price of diesel.[58] Reportedly, Chidambaram had deliberately set its price high, allowing him to scale back if necessary.[59] It was hard to verify. In any event, the showdown enabled both sides to score limited victories.

The finance minister pushed his agenda in subsequent weeks. In late July, he introduced the 1996–97 Union budget, an interim statement that reflected a delicate balancing act.[60] On the one hand, Chidambaram announced several initiatives to promote rural development and social opportunities: creating a National Bank for Agricultural and Rural Development (NABARD), Infrastructure Development Finance Company (IDFC) and Rural Infrastructure Development Fund (RIDF) with a capital base of Rs. 5,000 crore to mobilize investment in infrastructure; providing higher central outlays for education, health and social services and the decision to direct subsidized food grains to the poorest households, which had been endorsed by the chief ministers' conference in early July;[61] and an increase in subsidies for fertilizer, agricultural machinery and irrigation. However, Chidambaram also cut the capital gains tax from 30 to 20 per cent, lowered tariffs on various intermediate goods and provided a five-year tax holiday for private investment in critical infrastructural activities. To cover these revenue losses he imposed a 2 per cent across-the-board customs duty, introduced a minimum alternative tax (MAT) to plug loopholes on zero-tax companies and announced the constitution of a Disinvestment Commission with a target to raise Rs. 5,000 crore through government restructuring.

Proponents of liberalization noted the budget's increased reliance on capital markets. Critics highlighted the relative decline of both government capital expenditure and total budgetary support for central plan outlays relative to GDP: the former declined from 3.6 to 3.4 per cent, the latter from 2.7 to 2.6 per cent, as a percentage of GDP. They also noted the failure of the budget to tax the rich. In sum, the finance minister attempted to serve various agricultural interests in the governing coalition as well as its stated commitment to redressing social inequalities, while continuing the process of liberal economic reform.

Addressing Centre-state relations

Early policy tussles largely pitted the finance minister against the Left. The agenda of Centre-state relations, in contrast, affected a wider range of parties and interests. Given its composition, the United Front provided an opportunity for greater political devolution and economic decentralization. No less importantly, it raised the prospect of more cooperative relations within the Union in contrast to earlier Congress(I) administrations, which tended to view federal relations in zero-sum terms. Nevertheless, a national coalition government composed of many state-based parties generated problems of its own. Centre-state relations encompassed two distinct issues: conflicts and disagreements between New Delhi and a state or set of states; and clashes between two or more states. Success in tackling the former depended on the United Front practicing an alternative political vision. Dealing effectively with the latter, however, enjoined the governing coalition to act impartially. Its leaders had to exercise sufficient political flair, restraint and far-sightedness in managing inter-state conflicts and devising sustainable compromises. But they faced strong political incentives and constraints in their respective home states. The fact that their supporters prioritized regional loyalties over the nation, an exceptionally powerful sentiment amongst the regional parties' bases, underscored the challenge (see Table 8.5).

Hence Centre-state relations inevitably tested the United Front. Deve Gowda appointed his JD colleague, Biju Patnaik, to lead a high-level review of the Sarkaria Commission recommendations.[62] The prime minister assumed the burden of statesmanship. His office granted national political standing. The Government's minority parliamentary status constrained his maneuverability, however. State-level pressures in Karnataka continued to inform his calculus of survival.

Table 8.5: Loyalty to Region before Nation, 1996

	National Front		Left Front		Federal Front		Third Front	
	N	I	N	I	N	I	N	I
Disagree	11.6	21.1	7.4	18.1	2.5	6.3	21.5	15.8
Don't Know	12.4	26.3	8.6	24.7	6.5	19.1	27.6	23.7
Agree	11.4	52.6	9.2	57.2	11.6	74.6	32.2	60.5
Total	11.7	100.0	8.7	100.0	8.4	100.0	28.8	100.0

N=8295

Source: CSDS National Election Survey, 1996.

Preparing for state assembly elections in Jammu & Kashmir

A series of initiatives aimed at restoring the democratic process in Jammu & Kashmir encouraged cautious optimism. On 12 July, following a trip to Srinagar, Deve Gowda announced that assembly elections would be held in October. It was the first prime ministerial visit to the state capital since the insurgency began in 1989. Resuming electoral politics heralded a significant opportunity to return a degree of normalcy to the region. Throughout the 1990s, the Centre had equated demands for greater political autonomy with violent secessionist movements and sought to quash the insurgency through extreme armed force and direct political rule. The military strategy divided the militants but weakened the political legitimacy of the Union.[63]

Circumstances were favorable. A third of the national electorate supported negotiations to resolve the conflict, regardless of their party affiliations. Approximately 10 per cent favored repression (see Table 8.6). That said, more than half had either not heard of the insurgency or expressed no view. Hence the political class had considerable latitude in determining how to respond.

The Jammu and Kashmir National Conference expressed its position clearly:

> Government of India must take the primary responsibility for what has gone awry in Kashmir. Having done that, it has to take bold initiative and the first step in this direction would be to treat current [sic] Kashmir situation as a political crisis to be resolved through political instruments. Kashmiris cannot be left to be tackled by the army and para-military forces for it is [sic] essentially no law and order problem.[64]

The Centre held two rounds of voting in the region during the eleventh general election. But the JKN had boycotted the polls, demanding political devolution first along the lines of the 1952 Delhi Agreement, which envisioned state-level jurisdiction over all government matters except communications, defense and foreign affairs. Evidence of significant electoral manipulation in the Valley—military personnel "escorted" voters to the polls and "supervised" the casting of ballots—confirmed its reservations.[65] Nevertheless, commentators noted the relatively high level of electoral participation, approximately 49 per cent, which exceeded that of Rajasthan, Uttar Pradesh and Gujarat.[66] They also noted the peaceful conduct of the polls in Jammu and Ladakh, which rewarded the Congress(I) and JD. Aware that rising secessionist forces had damaged his party's standing, and persuaded by the CPI(M)

Table 8.6: Strategy Towards Kashmir, 1996

All Alliances

Category	Third Front N	Third Front I	Congress(I)+ N	Congress(I)+ I	BJP+ N	BJP+ I	Other N	Other I	Total N	Total I
Negotiation	29.1	33.9	31.9	33.3	29.9	34.7	9.0	29.3	33.4	
Can't Say	32.9	37.2	33.0	33.4	23.5	26.4	10.6	33.4	32.5	
Suppression	18.4	7.2	27.8	9.7	45.0	17.5	8.8	9.6	11.2	
Other	28.1	1.8	43.8	2.5	25.5	1.6	2.6	0.5	1.8	
Not Heard	27.4	20	32.1	21	27.1	19.7	13.4	27.3	21	
Total	28.8	100.0	32.1	100.0	28.8	100.0	10.3	100.0	N=8295	

Third Front

Category	National Front N	National Front I	Left Front N	Left Front I	Federal Front N	Federal Front I
Negotiation	11.4	32.6	8.5	32.9	9.2	36.6
Can't Say	11.1	30.8	7.6	28.7	14.2	54.7
Suppression	11.5	11	3.8	4.9	3.1	4.1
Other	15	2.4	6.5	1.4	6.5	1.4
Not Heard	12.9	23.2	13.3	32.2	1.3	3.1
Total	11.7	100.0	8.7	100.0	8.4	100.0

Source: CSDS National Election Survey, 1996.

general secretary H.S. Surjeet to reconsider its strategy, the JKN leader Farooq Abdullah relinquished his demand that power had to be devolved in advance of state assembly elections.[67]

The United Front pursued various measures to prepare for the polls. On 23 July, the prime minister announced several large-scale infrastructural projects.[68] The Government publicized a second package, containing measures to improve daily economic prospects, on 2 August.[69] Both announcements addressed longstanding demands: the need for greater national investment to diversify the economic base of the state and to improve its transportation and commercial links with the rest of the country.[70] According to a senior government official,

> [o]ne of the immense problems of Jammu and Kashmir is that it is isolated geographically ... [therefore] you must ensure that they develop a greater stake in the country. So we suggested to him [Deve Gowda] to build a rail link ... He liked the idea. In fact, he demanded that it become a national railway line after private contractors backed out.[71]

The disbursal of funds represented an important confidence-building measure.

High-level political visits were of equal importance, if not greater. The Home Minister Indrajit Gupta traveled to the region in the first week of August to inspect electoral preparations. Deve Gowda also returned a second time to visit Jammu and the frontier of Ladakh, engaging various local actors and reiterating the United Front's pledge to restore "maximum autonomy" to the region.[72] According to one senior government official accompanying him,

> I don't think he knew precisely [what maximum autonomy meant]. But we told him that it is short of independence ... Part of the internal discussions we had was about what might be required ... So we discussed things like symbols of autonomy, calling its chief minister *Wazir-e-Azam* [prime minister] and its governor *Sadr-e-Riyasat* [president] ... as well as granting the state greater jurisdiction over assembly polls ... Deve Gowda agreed with me. He was keen to sort things out.[73]

The prime minister's lack of knowledge of Kashmir was unsurprising. His political horizons had hitherto been limited to Karnataka. Indeed, the social base of the United Front was relatively unaware of the problems afflicting the region, with more than 50 per cent of the regional parties' supporters expressing no opinion on the matter (see Table 8.6). Nonetheless, Deve Gowda's political instincts and desire to improve the situation won admiration:

ESTABLISHING POLITICAL AUTHORITY (JUNE–SEPTEMBER 1996)

Deve Gowda handled the issue extremely well ... This gentleman visited the area several times ... It was a fantastic effort on his part.[74]

Indeed, these early high-level visits resonated with the broader political orientation of the United Front, whose supporters were the most averse to using repression. Their counterparts in the BJP were the least. The effort was far from perfect. Inadequate political coordination was partly to blame. On 24 June, prior to Gupta's ministerial appointment, the Defense Minister, Mulayam Singh Yadav of the SP, had declared that a bill on "maximum autonomy" was forthcoming, compelling the Home Ministry to issue a clarification.[75] But a more serious rift emerged. On 29 July, Mufti Mohammad Sayeed, who had been the Union home minister during the National Front and still led the JD parliamentary board, quit the party over its alleged "mishandling" of the upcoming assembly polls.[76] Several factors compelled his exit. It was partly due to his well-known rivalry with Farooq Abdullah. In mid-July, the Minister of State for Home Affairs and an ally of Sayeed, Maqbool Dar, had attacked the complicity of the JKN in the rigged assembly elections of 1987, which had instigated the popular insurgency.[77] But Deve Gowda and Surjeet had decided to court Abdullah.[78] Sharad Yadav, the working president of the JD, and Waseem Ahmed, its general secretary, had precipitated Sayeed's revolt by excluding him from the decision.[79] It was a classic failure to follow stipulated organizational procedures.

Sayeed's opposition also involved a wider political concern, however: the prudence in holding state assembly elections before the Centre had reached an understanding with various Kashmiri forces. In particular, he claimed that only by engaging the All-Party Hurriyat Conference (APHC), an alliance of secessionist parties, could the elections install a credible state administration.[80] Sayeed had considerable stature in the region and enjoyed good personal relations with various APHC leaders. The Hurriyat's indecision over whether to contest the polls made sections of the JD nervous, weakening his position.[81] Political wavering by other separatist groups such as the Awami League, which declared a boycott of the polls on 30 July but reversed its stand on 11 August, provided further reasons.[82] Nevertheless, Sayeed's exit robbed the Government of a valuable interlocutor.[83]

In the end, the United Front backed the decision to hold state assembly elections without further delay in order to shore up the perceived legitimacy of the Union.[84] Abdullah was a known political entity, having participated in the regional conclaves of the 1980s. The JKN was a safer partner to engage than secessionist forces. Lastly, the Government's minority parliamentary status gave it little political leeway. Its commitment encouraged the JKN, which perceived the new governing coalition to be a more reliable partner than the Congress(I), to re-engage politically.[85] The proposal to hold state-level elections began to rehabilitate India's position internationally too. Prime Minister Bhutto, while expressing her readiness to resolve "outstanding disputes" with India, maintained that Pakistan would continue to support Kashmiri self-determination.[86] American diplomats praised the resumption of elections, however, encouraging Pakistan to resume bilateral negotiations.[87] The more federal vision of the United Front, combined with Deve Gowda's desire to "make a mark", presented an important political opening. Everything depended on the conduct of the polls and the political forces that emerged in their wake.

Disputes over inter-state water sharing

Inter-state conflicts over common resources comprised a more contentious dimension of federal relations. The first involved Tamil Nadu and Karnataka. On 8 July, the DMK petitioned the Supreme Court to enforce the June 1991 interim award of the Cauvery Waters Disputes Tribunal.[88] The award stipulated that Karnataka, as the upper riparian state, should release 205 tmcft (thousand meter cubic feet) of water to Tamil Nadu per annum. It prescribed monthly and weekly tranches as well. The Tribunal could only issue a directive, however, if Karnataka declared undue hardship in a particular year. Moreover, the court had no powers of enforcing compliance. In general, Karnataka had met the terms of the provisional order, sometimes releasing more than mandated. But wrangles erupted during poor monsoons.[89] In 1995, Karnataka had sought to modify the settlement, citing distress. The Tribunal dismissed its claim. But the state released a smaller than mandated tranche. By July 1996, the situation in Tamil Nadu had become precarious: inadequate waters from the Cauvery, exacerbated by a poor monsoon, had already damaged its Kuruvai paddy crop and endangered its main Samba paddy crop.[90] The 6 July resignation of Justice Chittatosh Mookerjee, the chairman of the Tribunal, compelled the DMK Chief Minister, M. Karunanidhi, to approach the Supreme Court.

His move put the prime minister in a bind. Deve Gowda had frequently disregarded the interim award during his chief ministership.[91] Moreover, an all-party resolution to boycott the Tribunal had passed in the Karnataka legislature in January 1996. On 8 July, Deve Gowda instructed the state irrigation minister to release 5 tmcft of water, below the 15 tmcft requested by Tamil Nadu. But his goodwill measure failed to appease either side. Opposition parties stalled the proceedings of the Karnataka state assembly.[92] The Chief Minister, J.H. Patel, subsequently announced that his delegation would withdraw from negotiations.[93] Facing stiff opposition in Tamil Nadu, Karunanidhi urged the Supreme Court to appoint a new Tribunal chairman, arguing that negotiations would have to address the amount of water in arrears.

The apex judiciary advised both state delegations to negotiate a mutually agreeable interim arrangement until it reached a verdict on the dispute. Following in-camera discussions in Chennai on 5 August, the DMK chief told his legislative assembly that "interim relief" would be provided. Moreover, both delegations arranged to discuss a water-sharing formula of "mutual benefit without prejudice to the interests of farmers of both the States".[94] Substantive progress failed to materialize: the second encounter between the chief ministers in Bangalore on 5 September concluded with an agreement to continue negotiations. Unable to defuse their respective state-level oppositions, the third meeting in Delhi on 25 September ended in deadlock.

Many factors accounted for the impasse. The absence of a scheme to enforce the interim award was the central problem. Contingency also played a role: the unexpected resignation of the Tribunal chair. The tremendous sensitivity of the conflict, moreover, obstructed a negotiated settlement. According to senior government officials,

> no chief minister [could] afford to set up an authority [to implement the award], since it [would] throw them out of office ... Deve Gowda and Karnataka had no interest ... no incentive in resolving the dispute.[95]

The minority parliamentary status of the United Front weighed on both chief ministers and militated against bold initiatives. Others, however, took a more charitable and nuanced view of the prime minister's tactics:

> Deve Gowda took the position that the Government of India would not intervene. He said, "let us leave it to the legal process".[96]

These two interpretations, taken together, illuminated the difficulty. The longstanding dispute over the Cauvery involved complex technical issues and intense regional passions. Neither the Karnataka nor Tamil Nadu governments could make open concessions because it would generate intense hostility from their respective state-level opponents. The episode underscored the growing importance of the Supreme Court in adjudicating jurisdictional disputes and encouraging political negotiation. Still, the two chief ministers demonstrated some deftness. Their readiness to hold in-camera discussions contained the conflict. Karunanidhi retreated from his original position, considering options beyond the interim award, which created some room for maneuver. Indeed, the willingness of both state delegations to resume bilateral negotiations after a hiatus of six years was partly due to the participation of the DMK and JD in the United Front, even if their presence made it harder for the latter to project an image of impartiality.[97]

In contrast, underhanded tactics exacerbated a second longstanding water-sharing conflict. On 8 August, the TDP Chief Minister of Andhra Pradesh, Chandrababu Naidu, threatened "serious repercussions" unless the Centre countermanded its alleged disbursement of Rs. 200 crore to Karnataka from the newly created Rapid Irrigation Benefit Programme.[98] According to Naidu, Deve Gowda had surreptitiously granted funds to his home state to raise the Almatti dam to 524 meters on the upper Krishna River. The TDP chief contended that doing so would enable Karnataka to exceed the 700 tmcft of water allotted to it by the 1976 Bachawat Tribunal Award. His state minister for irrigation warned that

[w]e are not bothered about enjoying power at the Centre. We will take whatever action we deem fit for the occasion but our agitation will be result-oriented.[99]

The situation threatened a serious row.

As with the Cauvery dispute, the longstanding conflict over the Almatti dam involved many technical issues and state-level interests, exacerbated by the failure to create river-basin authorities.[100] Party affiliations mattered far less. In 1994, the Congress(I) Chief Minister of Karnataka, Veerappa Moily, had decided to raise the dam to 524 meters, leading his counterpart in Andhra Pradesh, K. Vijaya Bhaskara Reddy, also a Congressman, to lodge a complaint with Narasimha Rao in New Delhi. The Karnataka delegation now vowed that the extra water capacity would only be used for generating hydroelectric power.

However, apart from doubts over its end use and problems of verification, a two-month delay in receiving the waters could impair the transplantation of the paddy crop in Andhra Pradesh. Moreover, the Bachawat Tribunal Award granted Andhra Pradesh the right to surplus waters, to which Karnataka objected. The conflict pushed both state governments, regardless of the party in office, to defend their perceived regional interests.

That said, allegations that Deve Gowda had funneled central funds to Karnataka aggravated the dispute. On 27 July, the United Front appointed Naidu as convener of its Steering Committee which, combined with an all-party agreement struck in Hyderabad to pressure Karnataka to suspend construction on the dam, provided him with a platform in New Delhi.[101] On 9 August, Deve Gowda convened a meeting to defuse the crisis, but in vain. Denying that he had disbursed the alleged funds, he pleaded impotence, saying

I am the prime minister of the country, not of any state. I can take action only within the ambit of the powers I enjoy. I cannot direct anybody [in Karnataka to stop work on the Almatti].[102]

He proposed to convene another meeting if concurrent efforts by the Planning Commission and the Central Water Commission to resolve the dispute failed. Deve Gowda also directed both states to impart all relevant technical data to both agencies after Naidu accused the chairman of the latter of withholding details regarding Karnataka's irrigation projects. But evidence mounted that the prime minister, who had assiduously directed water to his Vokkaliga constituency for years in Karnataka,[103] had indeed released funds covertly.

The Steering Committee proposed several initiatives to break the impasse. On 11 August, it formed a panel comprising the chief ministers of Assam, Bihar, Tamil Nadu and West Bengal to adjudicate the matter, who recommended establishing a technical experts committee.[104] Questions over its legal status impeded further progress, however. Chief Minister Patel, already mired in ongoing Cauvery negotiations, faced a no-confidence motion in the Karnataka assembly for agreeing to establish the committee.[105] Naidu insisted on granting it legal powers to obtain all necessary documentation. But the Steering Committee declared that neither party had to accept the technical committee's findings. In the end, the prime minister shelved the proposal after the Union attorney general claimed that it would violate the existing provisions of the Bachawat award.[106]

The deadlock over the Almatti highlighted the difficulty of resolving longstanding inter-state conflicts over water sharing, and the tendency of the main disputants to frame the disputes in legal terms, bringing the judiciary into play. The contending delegations encompassed opposition parties from both states. Nevertheless, myopic political decisions exacerbated the conflict. Deve Gowda's refusal to press Karnataka towards a compromise, while agreeing to the expert committee, may have been understandable in terms of his own political roots. But his covert sanctioning of funds to the state, combined with the perceived reluctance of the Central Water Commission to impart details of Karnataka's projects to Andhra Pradesh, had instigated the row.[107] The incident damaged Deve Gowda's credibility with Naidu and undermined his prime ministerial authority in general. Many chief ministers had used stealth and obfuscation to push economic liberalization in the states in the 1990s. However, it was reckless of Deve Gowda, as the fledgling prime minister of a minority coalition government, to deploy such means in inter-state disputes over water. The folly of his actions underscored why context was vital to good political judgment.

The rise of Sitaram Kesri

The United Front pursued various policy objectives, handling the disputes that arose in their wake, with relative autonomy in these early months. The Congress(I) had to contend with its own myriad problems. Nonetheless, the party remained capable of exploiting divisions within the governing coalition as its pact with the BSP displayed. Thus circumstances enjoined the Government to be prudent. Candid political comments and underhanded machinations, which simultaneously antagonized the Congress(I) and the Left Front, made this clear. On 13 July, Indrajit Gupta told the Lok Sabha that Narasimha Rao risked imprisonment. He also forecast political disaster for the Congress(I) if it decided to topple the Government.[108] His remarks were understandable. On 9 July, the chief metropolitan magistrate of Delhi had summoned Rao on charges of complicity in the Lakhubhai Pathak cheating case, leading to further calls for his resignation at the CWC meeting on 12 July.[109] Moreover, the Left was the only political bloc untainted by major corruption scandals in the 1996 general election. Its parties rightly challenged the deteriorating ethical standards and growing criminalization of the national political class. Gupta unnecessarily baited the Congress(I) leadership, however, who portrayed his comments as prejudicial to the investigative process and threatened to review their support to the Government. Chastened, the home minister issued a clarification in both houses of parliament: "I have no desire nor [sic] intention to interfere in the internal affairs of the Congress(I), which is supporting the United Front Government from the outside".[110]

Subsequent events rightly earned the ire of the Left, however. On 15 July, President Sharma appointed Romesh Bhandari as governor of Uttar Pradesh at the prime minister's behest. The former foreign secretary ran unsuccessfully for the Congress(I) in the 1991 general election. He also reportedly attempted to frame some of Rao's opponents during his governorship of Goa, and had connections with the controversial godman Chandraswami.[111] Hence many in the United Front felt that Bhandari was "close to Rao and [Sharad] Pawar".[112] Others contended that Mulayam Singh Yadav of the SP pushed Deve Gowda to pick the new governor.[113] Both sides stood to benefit. Governors had undelineated discretionary powers to decide who to invite to form a government, dismiss chief ministers and dissolve state assemblies.[114] Expectations of an inconclusive electoral verdict in the upcoming assembly elections in Uttar Pradesh made having a pliable governor an asset. Indeed, Deve Gowda selected Bhandari in the middle of the night without consulting Gupta as home minister, fuelling suspicion on the Left.[115] It showed how the CPI could be outflanked on controversial decisions, despite its participation in government and formal political authority over such matters, foreshadowing later tussles.

Accusations of further political malfeasance followed. On 4 August, a senior Congress(I) delegation accused the prime minister of putting several members of their party under surveillance, evoking memories of its pretext for toppling the Chandra Shekhar ministry in 1991. Deve Gowda refuted the charges, ordering a probe by the CBI the following day.[116] Observers suggested that his main concern was to safeguard the Government by protecting Rao.[117] Subsequent events intimated as much. In mid-August, the Supreme Court reprimanded the CBI director, Joginder Singh, for "hobnobbing" with senior politicians associated with the *hawala* scandal.[118] The apex court specially admonished him for visiting the Congress(I) leader. Singh belonged to the Karnataka cadre of the Indian Police Service (IPS). His prior association with Deve Gowda, who appointed him to the post on 26 July, suggested political intrigue. The JD had pledged to grant autonomy to the central investigative agency.[119] But the simultaneous transfer of CBI officials working on the urea scam and other cases intimated further machinations. Indeed, Deve Gowda's unilateral decision to grant all former prime ministers the status of full cabinet ministers without informing Gupta, purportedly to shield Rao from criminal prosecution, intensified such fears.

A CPI(M) Politburo member vouched that "Deve Gowda did not try to relax or impede CBI cases".[120] A senior JD politician contradicted this, however, asserting that he "wanted to protect Rao ... who would support him".[121] A seasoned journalist claimed that

> Deve Gowda brought Karnataka cadre officers to the CBI and the IB [Intelligence Bureau] ... to do his dirty work ... he saw the state machinery as an extension of his personal household.[122]

Neopatrimonialism afflicted the central police services.[123] According to some, Deve Gowda introduced unsavory tactics into national politics:

> Deve Gowda had the political mentality of a *moffusil* ["small-towner"] ... always trying to fix his opponents. His political calculus was completely inappropriate for national politics ... People may have their political differences ... but the political class treats each other with courtesy.[124]

> He had a district-level political outlook ... it was about local machinations, friends, enemies, who you can count on for support ... his main thing was to never allow anything to displace his constituents.[125]

Having a diverse coalition administration at the helm stoked such sentiments. It was debatable whether the "fixing of opponents" only took place in district-level politics, however. After all, Rao was under investigation on similar charges vis-à-vis V.P. Singh in the St. Kitts' forgery case and many rivals within his own party. Moreover, local political struggles neither precluded the pursuit of larger political initiatives in New Delhi nor determined their outcome. The politics of the United Front reflected a complex balancing of multi-level games. Nevertheless, Deve Gowda's attempts to entrap his political opponents failed to obey a crucial Machiavellian precept: to disguise such actions well.

In any event, the judicial process moved ahead independently. On 30 July, the Delhi High Court rejected Rao's petition against the chief metropolitan magistrate, which had named the Congress(I) leader co-accused in the Pathak fraud case. On 21 August, the Supreme Court rejected his subsequent petition against the High Court's summons. In early September, the CBI questioned Rao over his alleged role in bribing the JMM in a crucial 1993 parliamentary no-confidence motion, arresting the four MPs implicated in the scam: Shibu Soren, Suraj Mandal, Simon Marandi and Shailendra Mahato. The Congress(I) leader professed his innocence. Nevertheless, on 13 September, various party leaders assailed the former prime minister at a meeting of the CPP. On 21 September, the Delhi High Court named Rao co-accused in the Pathak cheating case, forcing him to resign as president of the AICC. Rao's growing liability to the party increasingly alarmed many senior Congress(I) figures. Courtly political intrigues also supposedly exerted some influence. According to one JD politician, Sonia Gandhi "was desperately trying to undercut him [Rao]", who was striving "to marginalize her".[126] A confluence of pressures forced the outcome.

On 23 September, the CWC elected Sitaram Kesri as "provisional president" of the AICC. There were few serious contenders. According to one of its members, the septuagenarian treasurer of the party was a safe interim figure:

He was a keeper of the party secrets ... the least unacceptable ... backward [in terms of caste], north Indian ... an uncle to every Congress(I)man ... [and] a family retainer to the Gandhis.[127]

It was hard to verify the last claim. Some claimed that Sonia Gandhi anointed Kesri, on grounds that he would be a compliant "servant".[128] Others contended that his elevation "stunned" her.[129] Reportedly, the CWC sought to appoint a figure from the north, capable of recovering the Muslim vote and galvanizing the party's fortunes in the crucial Gangetic plains. Yet Kesri lacked the requisite political stature, skill or experience in managing the cacophony of personalities, demands and interests that plagued the Congress(I). Observers portrayed him as a trusted insider with "a talent for subtle political manipulation", "a capacity to ingratiate himself with the powers-that-be" and "an ability to keep a low profile while carrying out important tasks".[130] Partisan rivals were far harsher, describing him as "an old servile man, a weakling".[131] Subsequent events would confirm and defy both expectations.

9

EXERCISING NATIONAL POWER (SEPTEMBER–DECEMBER 1996)

The resignation of Narasimha Rao as president of the AICC altered relations with the United Front in three ways. First, it enhanced the party's leverage vis-à-vis the Government. Second, Rao's partial exit provided incentives to various Congress(I) dissidents to return. Their reintegration would not topple the Government. It could deepen political uncertainty, however. Whether such a defection would transpire, and its ramifications, depended partly on a third issue: the nature and degree of communication between the prime minister and the new interim president of the AICC. According to a senior government official, despite their periodic rhetorical threats, Deve Gowda and Rao

> had very smooth relations ... Deve Gowda had great respect for Rao, they were extremely close to each other ... They met twenty-five to thirty times before Kesri came in.[1]

Such constant interaction had stoked fears of collusion. Nevertheless, it ensured communication. In contrast, Kesri was an unknown political quantity. Deve Gowda viewed him as many Congressmen did—as a cipher. Not everyone agreed. As one CPI(M) politician remarked,

> The Congress is a personality-based party ... so we had our doubts once Rao left the helm. But even then we didn't expect such childish behavior.[2]

Callow political antics in subsequent months justified his skepticism. However, Deve Gowda would also bear responsibility for their poor interpersonal relations. Indeed, his failure to take Kesri's measure would cost him the prime ministership.

The majority of activity between September and December of 1996 addressed Centre-state relations. A series of events displayed the range of purposes driving the United Front. Its conduct invited dismay in Jammu & Kashmir and Uttar Pradesh, generated controversy in Gujarat and the Northeast, and led to a surprising breakthrough in India's bilateral relations with Bangladesh that actualized a different federal vision.

Advances and setbacks of the monsoon parliamentary session

The Government pursued other policy initiatives first, however, introducing several bills during the closing stages of the monsoon parliamentary session.[3] P. Chidambaram released the 1996 Union finance bill, which lowered excise and customs duties on various consumer durables and industrial goods, and exempted vulnerable companies and industrial undertakings from the recently introduced minimum alternative tax.[4] The Minister of Industry, Murasoli Maran of the DMK, whose party had introduced a new industrial policy "to speed up the pace of industrialization" and transition to "a free market-oriented competitive and globalized environment" in Tamil Nadu, unveiled a Foreign Investment Promotion Council at a fair co-sponsored by the Federation of Indian Chambers of Commerce and Industry (FICCI).[5] The aim of the Council was to clear infrastructural bottlenecks, create an on-line database and implement authorized projects. The finance bill and trade exhibition concerned first-generation reforms, arousing little opposition. Nonetheless, both illustrated the relative autonomy that Chidambaram and Maran enjoyed to push liberal economic reforms within their remit.

The Government also proposed two constitutional amendments. The first, to enable the delimitation of parliamentary electoral constituencies according to the most recent census, progressed immediately. However, the introduction of the Constitution (81st Amendment) Bill, which proposed to reserve 33 per cent of the seats in the Lok Sabha and state legislative assemblies for women, encountered a setback. Various social organizations and women parliamentarians widely supported the so-called Women's Bill, which sought to extend the principle of gender-based reservations that had been introduced at the *pan-*

chayat and municipality level in the 73rd and 74th amendments to the Constitution in 1993. Every major party professed support. And the United Front had pledged to introduce such a measure in the Common Minimum Programme. Hence the prime minister planned to waive legislative deliberation. But various OBC leaders on both sides of the floor stalled proceedings, saying the proposed bill failed to include explicit caste-based provisions, forcing the Speaker to refer it to a joint select committee.

Various factors caused its deferment. The proximate reason was a last-minute revision. On 10 September, the Government circulated a draft that excluded the Rajya Sabha and state councils from its scope, lowering its chances. However, divisions within the Council of Ministers over the criteria for and percentage of seats to be reserved played a much larger role. According to various OBC leaders, the draft Women's Bill conferred an unfair advantage to parties representing high-caste groups as well as women belonging to SCs and STs, since the latter two groups enjoyed reserved constituencies.

It was a contestable argument. The objection conflated the depth of deprivation suffered by the majority of *Dalits* and *Adivasis* with the far more diverse socio-economic opportunities of OBCs as a whole. Moreover, by saying that female OBC candidates could not win representation through open electoral competition, the predominantly male leadership of the National Front ironically legitimized the widespread social prejudice against intermediate castes. It was a strange argument to make too, given the rising parliamentary representation of the OBCs since the early 1990s.[6] Lastly, their opposition illustrated how the struggle for equality based on lower-caste identities had evolved by the mid-1990s to eclipse other forms and bases of injustice—a far cry from the mutually constitutive links between gender, caste and class that Ram Manohar Lohia and other Indian socialists had perceived.

Indeed, most voters backed reservations for women in national political institutions, comparable to their support for the Mandal Commission Report (see Tables 9.1 and 9.2). Supporters of the National Front expressed less enthusiasm. Nonetheless, more than two-thirds approved. To be fair, parliamentary opposition to the Women's Bill was widespread. Its most vocal opponents comprised only 10 per cent of the Lok Sabha, yet managed to stall debate, indicating tacit support from their male peers.[7] Simply put, the proposed legislation threatened to diminish male privilege in the highest representative bodies of the polity. The decision to defer its passage satisfied cross-party interests.

Table 9.1: Support for Backward Caste Reservations in Government, 1996

All Alliances

Category	Third Front		Congress(I)+		BJP+		Other		Total
	N	I	N	I	N	I	N	I	
Disagree	19.2	12.3	28.8	16.5	45.4	28.4	6.6	12	18.3
Agree	30.8	87.7	32.6	83.5	25.7	71.6	10.8	88	81.7
Total	28.7	100.0	31.9	100.0	29.4	100.0	10.0	100.0	N=7276

Third Front

Category	National Front		Left Front		Federal Front	
	N	I	N	I	N	I
Disagree	9.0	14.2	6.8	14.9	3.4	7.1
Agree	12.2	85.8	8.8	85.1	9.8	92.9
Total	11.6	100.0	8.4	100.0	8.7	100.0

Note: N represents the national vote share given to each alliance. I reflects the internal vote share within each alliance.
Source: CSDS National Election Survey, 1996.

EXERCISING NATIONAL POWER (SEPTEMBER–DECEMBER 1996)

Table 9.2: Support for Reservations for Women in Parliament, 1996

Category	All Alliances									
	Third Front		Congress(I)+		BJP+		Other		Total	
	N	I	N	I	N	I	N	I		
Disagree	29.4	10.3	32.9	10.3	29.6	10.4	8.1	8		10.1
Don't Know	31.3	16.5	27.8	13.1	25.2	13.3	15.8	23.3		15.2
Agree	28.2	73.2	32.9	76.5	29.5	76.4	9.5	68.7		74.7
Total	28.8	100.0	32.1	100.0	28.8	100.0	10.3	100.0		N=8295

Category	Third Front					
	National Front		Left Front		Federal Front	
	N	I	N	I	N	I
Disagree	12.3	10.6	9.8	11.4	7.3	8.7
Don't Know	15.3	19.9	7	12.3	9	16.1
Agree	10.9	69.5	8.8	76.3	8.5	75.1
Total	11.7	100.0	8.7	100.0	8.4	100.0

Source: CSDS National Election Survey, 1996.

The final major initiative concerned high political corruption. On 13 September, the Government presented the Lok Pal Bill, a perennial goal of the socialists. It aimed to create an ombudsman to investigate allegations of corruption against elected representatives and public officials. The bill envisioned a three-member panel of senior jurists with the authority to direct the central investigative apparatus in search-and-seizure operations, censure false complaints and complete inquiries within six months of their commencement. It was a significant initiative, whose purview extended to the prime minister's office, albeit with a five-year statute of limitations. Yet it still had to be passed. The credibility of the United Front in tackling high corruption, hence, rested on its responses to ongoing criminal investigations. Developments in the animal husbandry scam revealed a possible criminal nexus between senior bureaucrats and politicians in Bihar.[8] The affair tested the governing coalition badly.

Lastly, in late September, the external affairs minister unveiled several principles to guide foreign policy, quickly labeled the "Gujral doctrine":

The United Front government's neighborhood policy now stands on five basic principles: firstly, with neighbours like Nepal, Bangladesh, Bhutan, Maldives and Sri Lanka, India does not ask for reciprocity but gives all that it can in good faith and trust. Secondly, no South Asian country will allow its territory to be used against the interest of another country of the region. Thirdly, none will interfere in the internal affairs of another. Fourthly, all South Asian countries must respect each other's territorial integrity and sovereignty. And finally, they will settle all their disputes through peaceful bilateral negotiations.[9]

In short, Gujral declared that India could resolve many long-standing conflicts that had marred bilateral relations by adopting a positive asymmetric stance towards its neighbors. Significantly, the principle of non-reciprocity excluded Pakistan.

The speech provoked diverse reactions within the diplomatic community. Cynics derided it as cant,

[a] complete misnomer ... it was just a media creation, a public relations exercise ... There was little substance to it ... If you ask me, it was just a pursuit of individual glory.[10]

Self-professed realists viewed it with suspicion, if not alarm, calling its underlying principle a "weakness".[11] Other mandarins were less caustic, however:

There is a lot of continuity in foreign policy, regardless of party ideologies and programmes. In that sense, the Gujral doctrine was merely a crystallization of existing doctrine: to create a climate of cooperation through various bilateral initiatives and to maintain our security interests ... Therefore the actual options [were] very limited ... the parameters [were] laid down. It [concerned] a matter of style and language.[12]

Commentators welcomed it to a greater extent. Some viewed the emphasis on accommodation as Nehruvian.[13] Others argued that New Delhi had to resolve disputes with its neighbors given the growing role of China and the United States in the region, the need to seize the opportunities of globalization and the desire for greater international recognition. Gujral essentially articulated a more conciliatory posture that emerged under Vajpayee in the late 1970s, had been advanced by V.P. Singh, and acquired tacit consensus under Rao.[14]

Yet the Gujral doctrine also represented the distinctive outlook of the United Front. Its emphasis on non-reciprocity evoked the professed Gandhian ideal of unconditional Hindu-Muslim solidarity during the nationalist movement, which figures such as Chakravarti Rajagopalachari, a leading Congressman who later formed the Swatantra Party, had embraced after independence to resolve the dispute over Kashmir.[15] Several coalition leaders contended that "we [India] were the larger partner in all of these relations and should be more magnanimous".[16] Indeed, although more than one-third of the electorate failed to express any opinion, third force voters, especially those supporting the JD and SP, backed friendlier relations with Pakistan to a greater extent than their counterparts in the BJP and Congress(I) (see Table 9.3).

India's relatively autonomous strategic enclave, rather than electoral pressures, dominated its foreign policy choices.[17] Still, the Gujral doctrine reflected the particular sentiments of the third force. The Government's range of maneuver and ability to exploit opportunities would soon be revealed.

The temptations of national power in Gujarat, Kashmir and Uttar Pradesh

The drama of Centre-state relations occupied most of its attention in these months, however. In early August, the BJP had expelled Shankersingh Vaghela, a principal architect of its formidable state-level organization in Gujarat.[18] Vaghela had threatened to split the party in September 1995, forcing his rival, Keshubhai Patel, to step down as

Table 9.3: Desirability of Friendship with Pakistan, 1996

All Alliances

Category	Third Front		Congress(I)+		BJP+		Other		Total
	N	I	N	I	N	I	N	I	
Disagree	22	13.2	32.3	17.4	39.1	23.4	6.6	11	17.3
Don't Know	30.3	39.9	32	37.7	26.2	34.4	11.5	42.4	37.9
Agree	30.1	46.9	32.1	44.9	27.1	42.1	10.7	46.6	44.9
Total	28.8	100.0	32.1	100.0	28.8	100.0	10.3	100.0	N=8295

Third Front

Category	National Front		Left Front		Federal Front	
	N	I	N	I	N	I
Disagree	7.9	11.7	8.7	17.4	5.4	11
Don't Know	11.3	36.5	8.5	37.3	10.5	47.1
Agree	13.5	51.8	8.7	45.3	7.9	41.9
Total	11.7	100.0	8.7	100.0	8.4	100.0

Source: CSDS National Election Survey, 1996.

chief minister. But the RSS humiliated him and his associates in the following months, deepening a rift based on personal rivalry and caste differences.[19] The new Gujarat Chief Minister, Suresh Mehta, sought to consolidate his authority, expelling ministers and issuing "show-cause" notices to MLAs that displayed sympathy with the dissident leader. On 18 August, the BJP state unit split. Two days later, the breakaway Gujarat Janata Parishad forged ties with Vaghela's newly founded Rashtriya Janata Party to form the Maha Gujarat Janata Parishad (MGJP), unleashing another round of defections and expulsions.[20] The tumult forced the Governor, Krishna Pal Singh, to confirm the viability of the MGJP in late August. Seeking to end the crisis, Mehta requested an opportunity to prove his majority.

Partisan maneuvers, coercive tactics and unforeseen political developments inflamed the situation. On 3 September, the deputy speaker of the assembly, Chandubhai Dhabi of the Congress (I), circumvented standard procedure by recognizing the MGJP, adjourning the assembly before allowing the chief minister to test his legislative support. Dhabi presided over the session due to the absence of the speaker, H.L. Patel, who was suffering from terminal illness. The ruling faction of the BJP claimed that a communiqué sent by Patel declared the deputy speaker's actions invalid under Article 180(2) of the Constitution.[21] Yet the speaker died one week later. On 18 September, Mehta appointed a political ally as the acting speaker, who proceeded to suspend all opposition parties and forcibly remove their MLAs before calling a vote. The truncated Mehta administration won the measure amidst physical mayhem.[22] The next day, Governor Singh issued a report, asserting constitutional breakdown. The United Front imposed Article 356.

The decision invited scrutiny given the opposition of several United Front constituents towards President's rule and its history of abuse by the Congress(I). The Sarkaria Commission had recommended its use in extreme cases, preceded by the Centre warning the state in question and allowing the latter to respond, and ensuring that governors justified their recommendations in public reports with material facts.[23] Many commentators invoked the so-called Bommai principle of the Supreme Court, which stated that an official ruling party should be allowed first to test its majority on the floor and excluded poor administrative governance, as conditions for imposing President's rule. Some observers claimed that Governor Singh could have recommended the prorogation of the assembly until circumstances permitted a fair vote.[24] A senior government official privy to the decision concurred.[25] Other bureaucrats, however, argued that several factors undermined the constitutional integrity of the assembly.

(T)he [state assembly] speaker was ill, the pro-tem was beaten up, members were bullied ... it was a difficult situation.[26]

The governor and the chief secretary recommended it. The BJP was suffering from a terrible rift ... it was done in good faith.[27]

The situation furnished good non-partisan reasons for implementing Article 356.

The Government justified itself in several ways. High-ranking figures from the Left claimed that "difficult circumstances" obtained, that "we didn't dissolve the assembly" and that "[Article] 356 [was] used only as a last resort" in contrast to "earlier [instances of] summary dismissal".[28] A senior regional politician asserted that his party "objected" to the decision but acquiesced on grounds that "you cannot withdraw [support to the Government] on a single issue".[29] A comprehensive assessment, of all relevant factors, informed its judgment. Some government officials claimed otherwise, however. One contended that "no regional party argued against it [Article 356] in the cabinet meeting" while another vouched that discussion took only a few minutes.[30] It was unclear whether such disagreements occurred, or if so, in what venue.

Ultimately, *bona fide* arguments and convergent political interests led the United Front to impose Article 356 on Gujarat:

[The] situation was very bad ... the governor's report was sent voluntarily ... it suited our political interests.[31]

There was violence in the state legislature for some time ... the governor's position was unclear, the speaker was a sick man ... the Government wanted to cash in on the situation.[32]

The subsequent investiture of Shankersingh Vaghela as the chief minister of Gujarat on 23 October, with the Congress(I)'s outside support, arguably exculpated the governing coalition of serious wrongdoing.

The next test lay in Jammu & Kashmir where, in early October, the JKN swept the state assembly elections. Farooq Abdullah, crediting the United Front for restoring elections, offered his support. The Government received praise from several quarters for seizing the initiative, disbursing resources and making high-level visits to the region prior to the polls, and conducting them under relatively peaceful conditions. The outcome seemed to vindicate its decision to support the JKN. Indeed, observers later credited the United Front for reopening a "small democratic space" in Kashmiri politics.[33]

Allegations of rigging and coercion marred the election, however. The Government's decision to arrest several Hurriyat leaders—including Syed Shah Geelani, Abdul Gani Lone, Yasin Malik and Shabir Shah—for demanding a boycott during the final phase of polling revealed the deep authoritarian reflexes that had characterized India's approach to the region for decades. The newly inducted chief minister, who subsequently put the Hurriyat leaders under house arrest, possessed similar tendencies. Charges of corruption and neglect would slowly tar his administration.[34] The United Front displayed shortsightedness by marginalizing Mufti Mohammad Sayeed and ignoring his advice to engage local separatist groups. The sweeping victory of his People's Democratic Party (PDP) in the 2002 state assembly elections, the first relatively free polls since 1983, eventually demonstrated his superior political judgment.

Ultimately, the Centre could only bolster democratic rule in the region by revitalizing the state bureaucracy, bringing the security apparatus under single unified command and restoring economic activity after years of repressive direct rule.[35] Most importantly, it had to open political dialogue with all parties and end the massive human rights abuses committed by Indian troops. The situation in Kashmir was never solely a domestic affair, however. During the campaign, Prime Minister Bhutto had called for third-party mediation to resolve the disputed status of Kashmir, without curtailing Pakistani sponsorship of external insurgent forces. The mutual expulsion of both countries' envoys in September and October, based on charges of espionage, highlighted the level of mistrust.[36] In addition, ongoing Chinese nuclear assistance to Pakistan substantiated India's longstanding security concerns.[37] Perhaps most importantly, a power struggle within Pakistan's high politics made it impossible to push the agenda. In early November, President Farooq Leghari dismissed Bhutto, his former ally, on charges of graft, misuse of governmental power and her alleged involvement in extra-judicial murders.[38] The resumption of high-level bilateral talks would have to await the outcome of Pakistan's general election in February 1997.

The handling of the state assembly elections in Uttar Pradesh, in contrast, caused greater debate within the United Front. The governing coalition had agreed a seat-sharing pact that acknowledged the predominance of the SP.³⁹ The latter also benefited from the failure of negotiations within the JD over a possible tie-up with the BSP. Reportedly, the SP chief, Mulayam Singh Yadav, agreed to ally with the BSP as long as neither he nor Mayawati would occupy the chief ministership.⁴⁰ A section of the JD—led by Sharad Yadav and C.M. Ibrahim and backed by Deve Gowda—pursued the proposal given their own desire to re-establish a presence in the state. Ram Vilas Paswan, sensing a threat to his Dalit base in Bihar, backed Mulayam Yadav's conditionality, however. In the event, Mayawati rejected the offer. The SP dominated the coalition's electoral pact through deft negotiations. According to a senior JD leader, Mulayam Yadav's stratagem was always

to limit the JD, to give them the least number of tickets ... [H]e then keeps tickets pending until the last moment so that others cannot break off. At the last moment, he pressures the numbers down, and says take it or leave it ... and during the election, ensures your defeat. Initially Mulayam is your ally, a friend ... [but] who in his heart is a rival.⁴¹

Indeed, the SP chief allied with the Bharatiya Kisan Kamgar Party (BKKP), led by Ajit Singh and Mahendra Singh Tikait. Doing so increased the presence of the United Front in the Uttarakhand region of western Uttar Pradesh. However, it was the JD, CPI and AIIC(T) that had to accommodate the BKKP.⁴²

In the end, no single bloc captured the 213 constituencies necessary to form a legislative majority.⁴³ The BJP and the SAP, its coalition partner, formed the single largest formation with 177 seats. The United Front, which captured 134 with the majority going to the SP, emerged as the main contender.⁴⁴ Lastly, the BSP and Congress(I) together secured 100 seats. Significantly, the former won twice as many seats as the latter. Simple electoral arithmetic granted the balance of power to the BSP. The state legislature, which had been under Article 356, remained suspended.

A flurry of political activity ensued.⁴⁵ On 10 October, the United Front repeated its offer to the BSP if Mayawati abandoned her claim to the chief ministership and abandoned the BJP. She rejected both conditions. The upper-caste faction of the BJP sought to ally with the BSP. But Kalyan Singh, its OBC leader, balked. Finally, on 12 October, the BSP leader Kanshi Ram urged the Congress(I) to withdraw its support to the Government unless the latter endorsed Mayawati. The

CWC authorized Sitaram Kesri to apply pressure. But the Steering Committee rebuked the demand.

On 17 October, Governor Romesh Bhandari recommended the annulment and re-imposition of Article 356 after convening the new assembly.[46] The Government accepted his recommendation, reinstating President's rule. Officially, its justification was twofold. The first was that a state government could not be formed. The second was that inviting the BJP would allow it to lure defections with the spoils of office. A CPI(M) Politburo member said:

[D]o you ask the single largest party to form the government, despite the fact that it lacked a majority? ... It gives an incentive for horse-trading and defections ... the Congress, BSP and SP wrote to the governor saying that they were against the BJP.[47]

This was not entirely true. The BSP continued exploring possibilities. Its unwillingness to send a letter of support partly vindicated the Government's stand.[48] Yet the position of the CPI(M) demonstrated, as its general secretary H.S. Surjeet subsequently confessed, its willingness to use Union power until a secular ministry could be formed.[49] "The CPI(M)," charged a senior JD figure, "believes in secularism without democracy".[50]

The CPI saw matters differently. According to its various leaders, Mulayam Yadav pressured his ministerial colleagues into suspending the assembly:

Regional chieftains want to use the Constitution to [suit] their whims ... Mulayam constantly pressed it.[51]

Mulayam, despite his virtues, was not keen to share with the Left or the BSP ... he want[ed] to be seen as the sole champion against the BJP in UP.[52]

They also averred that Governor Bhandari coaxed C.M. Ibrahim to persuade Deve Gowda of its merits:

The Governor recommended it [Article 356] ... he was an errand boy ... [There is] a constant temptation to use the weapon by those in power to occupy and safeguard the chair.[53]

Evidently, the home minister objected.[54] But he could not dictate terms since "[the] non-SP parties were pretty weak [in Uttar Pradesh] ... thinly spread out ... hardly able to mobilize".[55] Put bluntly, despite his formal ministerial authority, Gupta lacked sufficient real power. His continuing marginalization stoked further resentment.

Crucially, the main regional parties backed the imposition of President's rule, despite their generally stated aversion. Several chief ministers in the United Front made repeated overtures to modify, if not abolish, Article 356 at an Inter-State Council meeting on 15 October, where Deve Gowda stressed the need for "cooperative federalism".[56] The DMK, TMC and TDP assented to the decision, however, after learning that the BSP was considering a tie-up with the BJP.[57] In sum, constituents in the United Front that had a direct political interest in Uttar Pradesh exploited the situation for overtly partisan ends, while colleagues with an indirect stake granted their consent.

The move invited a storm of protest. The Congress(I) accused the Government of conspiring to stop "a Dalit woman from becoming chief minister".[58] Several members of the CWC advocated withdrawing parliamentary support.[59] But the majority of the party high command decided otherwise. Rao was still the leader of the CPP. His political associates occupied various positions in the organization. Perhaps most importantly, the party's relative failure in Uttar Pradesh undermined its leverage. The BJP confronted fewer constraints. It denounced the move as a "fraud on the Constitution" and pressed for Governor Bhandari's dismissal, disrupting proceedings in the Lok Sabha and pursuing judicial redress.[60] The United Front, Congress(I) and BSP agreed not to contest each other in the Rajya Sabha elections in the state in mid-November, hinting at a possible resolution. But the BSP again demanded a Mayawati chief ministership.[61] The impasse would persist until March 1997.

Engaging the Northeast

The imbroglio in Uttar Pradesh overshadowed more constructive initiatives. On 22 October, the prime minister began a six-day tour of all seven northeastern states. Unlike in Kashmir, Deve Gowda initiated dialogue with various insurgency groups, including the All Bodo Students Union, People's Democratic Front and Bodoland State Movement Council.[62] He also proposed to establish several high-level bodies to improve infrastructure facilities, generate employment opportunities and implement all existing accords. Finally, the prime minister unveiled an economic package worth Rs. 6,100 crore to assist with these objectives and connect the region to Southeast Asia, and proclaimed the Centre's willingness to repeal the controversial 1983

IMDT Act.[63] In doing so, Deve Gowda sought to placate the AGP, which had criticized the Congress(I)'s militarist approach under Hiteswar Saikia, authorizing paramilitary forces to battle the United Liberation Front of Assam (ULFA), the radical insurgency movement that had established rival power structures across the state since the late 1980s.

The duration and scope of the tour drew considerable attention. Few prime ministers had spent much time in the region. A senior government official portrayed the enterprise as

a very significant contribution ... It sent a message that India can be run well as a federal polity ... there was a lot of *bonhomie* and goodwill, which was important.[64]

Other bureaucratic colleagues praised the prime minister for being a quick study, initiating dialogue with various political groups and grasping the essential political dynamics of the region.[65] In many ways, Deve Gowda's actions in the Northeast resembled his efforts in Jammu & Kashmir, but more positively. The high-level visit illustrated, once again, his apparent determination "to prove that he was capable of being a first-class PM", with Deve Gowda often comparing himself to US president Bill Clinton—a regional governor with national talent.[66]

The media expressed cautious optimism.[67] Providing adequate jobs for alienated youth was a perennial concern. Yet the economic package resembled previous attempts by the Centre to promote regional development, while few observers believed the proposed high-level committee could finish their work within the stipulated time-frame. Additionally, some Bodo leaders found their meetings with Deve Gowda "extremely frustrating".[68] Perhaps most importantly, promising to repeal the IMDT Act was a major concession to the AASU, but created unintended problems for the AGP, which had to appear not to oppose indigenous groups' demands for its repeal and yet protect the rights of minority communities that had become an important electoral base for the party. Hence the prime minister and home minister backtracked, stating the Government would revisit the issue.[69] Similarly, the Home Ministry raised the prospects of "major amendments" to the AFSPA, which granted immunity to the army and various paramilitary forces, but failed to push anything through.[70] In short, these various initiatives comprised well-intentioned but hasty steps that signaled a commitment to the region, whose implementation ultimately required a longer time horizon for the Government.

Growing political threats

Alas, developments in the Bihar fodder scam, and efforts by the Congress(I) to regroup, posed growing risks. On 4 October, the Comptroller and Auditor General (CAG) had confirmed a conspiracy to defraud the Bihar state treasury of hundreds of crores of rupees. Its report indicted senior bureaucrats in the state animal husbandry department and implicated several government ministers, including the JD chief minister, Lalu Yadav.[71] The same day, CBI joint director U.N. Biswas contended that his superior, Joginder Singh, had amended his original report, which stated that Lalu Yadav had participated in a conspiracy, stalled inquiry into the affair and failed to act on the CBI's recommendations.[72] Purportedly, Singh removed the accusation because of inadequate evidence.[73] His explanation failed to persuade the Patna High Court, however, which revoked his supervisory powers over the investigation and ordered all future reports to be submitted without prior vetting.[74]

Fears of tampering and prime ministerial interference drew in the Supreme Court, which directed the Government in mid-October to submit all relevant documents to ascertain Biswas' charges.[75] In early November, the apex judiciary restored to Singh full responsibility over the investigation, but instructed him not to make controversial decisions unilaterally.[76] The modified order stirred debate. On the one hand, it ensured that no agency exceeded its proper jurisdiction. However, reports that Deve Gowda had met Chief Justice Aziz Mushabber Ahmadi late at night on 10 October fuelled speculation of a cover-up.[77] According to a senior JD politician, "[n]obody surmised anything early on. The media was very anti-Lalu, very hostile in any case ... given the social equations [anti-OBC bias] in the country."[78] It was hard to verify. In any event, on 13 November, the Patna High Court summoned Singh following "re-attempts to present a truncated report", and his order that all progress reports should be "presented to the High Court after it is vetted by the legal division of the CBI and also my approval".[79]

Two competing developments suggested themselves. On the one hand, concern mounted that Deve Gowda had used his authority to protect Lalu Yadav, damaging the standing of the prime minister's office, which his efforts in Jammu & Kashmir and the Northeast had strengthened. On the other, heightened judicial vigilance checked the growing politicization of the central investigative branch of the police with the advent of multiparty Union governments. The second trend would prove stronger.

Discontent within the Congress(I) posed a greater external threat to the Government. The imposition of Article 356 in Uttar Pradesh foiled the party's hopes of capturing state power in a crucial arena, demonstrating the limits of its parliamentary veto. Accordingly, the Congress(I) started to reorder its apparatus and reconsider its parliamentary strategy. In late October, Sitaram Kesri displaced several Rao loyalists within the party organization.[80] Subsequent moves intimated larger designs. On 1 November, the CWC criticized the conduct of the United Front, while encouraging Sonia Gandhi to become a member of the party.[81]

Inviting the heir of the Nehru-Gandhi dynasty to bolster its fortunes, despite her minimal political experience, underscored the level of desperation. Yet the decision sent a clear signal. On 5 November, the MPVC rejoined the Congress(I).[82] The practical loss to the Government was negligible: one parliamentary seat. Moreover, Madhavrao Scindia's volte-face was unsurprising, given the poor electoral showing of his formation in recent by-elections. But his reintegration raised the possibility that other anti-Rao factions would follow. And it indicated that Kesri, who summoned Scindia without consulting Rao to counter more powerful Congressmen such as Sharad Pawar and K. Karunakaran, had ambitions of his own.[83]

Attempts to lure the most significant rebel unit failed. On 11 November, G.K. Moopanar maintained the independence of the TMC.[84] The party enjoyed an advantageous position within the United Front. It controlled four Union ministries, the most important being Finance, giving it considerable political clout. Its leading members enjoyed unprecedented power—something they felt would be severely circumscribed under the Congress(I).[85] The failure to entice the TMC temporarily halted, according to observers, Kesri's aspirations of taking over the Government.[86]

Smaller renegade groups gradually returned, however. On 17 November, Kesri revoked the suspension of M.L. Fotedar, Sheila Dixit and K. Natwar Singh, three senior leaders of the AIIC(T).[87] Sis Ram Ola and Satpal Maharaj, its ministerial representatives, refused to relinquish their posts. But the vast majority of the party complied with their leaders' decision. S. Bangarappa, leader of the one-man KCP, rejoined the Congress(I) as well.[88] The dissidents' return added merely four seats to its parliamentary tally. But their return signaled Sonia Gandhi's growing political weight.

Mending subcontinental relations

Despite these growing political concerns, diplomatic relations with China and Bangladesh witnessed progress. A historic visit to New Delhi by President Jiang Zemin, the first by the People's Republic's head of state, led to several bilateral statements. The most important was an agreement to implement confidence-building measures—exchanging national maps, reducing troop deployments and establishing better military-level communication—regarding their disputed Himalayan boundary.[89] Cross-border tension along the Line of Actual Control had abated since bilateral relations normalized in the late 1980s under Rajiv Gandhi, and the Peace and Tranquility Agreement in 1993, which pledged to resolve all boundary disputes through consultation.[90] China sidestepped Indian concerns over its reported nuclear assistance to Pakistan. President Zemin urged "peaceful bilateral consultations" to settle the conflict in Kashmir during his following trip to Pakistan, however, which China had maintained since 1993.[91] In short, the visit was a choreographed step in the context of improving Sino-Indian relations.

The Government played a more decisive role in improving relations with its smaller neighbors, however. In late November, it ratified the Mahakali Treaty with Nepal, renewed free access for Nepalese exports and supported its desire to build a road link with Bangladesh.[92] In early December, the United Front signed the historic Ganga Waters Accord, providing Bangladesh a comparatively greater share of the Ganga during its annual lean season for 30 years. Crucially, the Accord guaranteed Bangladesh a minimum flow of 35,000 cusecs per day for eight ten-day periods between 1 January and 31 May every year, which almost doubled the 18,000 cusecs it had previously received. In contrast, India would receive an analogous level for seven ten-day periods. In doing so, the Accord addressed Bangladesh's longstanding needs.[93] It was a major political achievement.

Key informants disagreed over what factors enabled the breakthrough. Many senior mandarins stressed the external factor: the election of Sheik Hasina in June 1996, the daughter of Bangladesh's first prime minister Sheik Mujib-ur-Rahman, who had sought refuge in India following her father's assassination in 1975.

Hasina came to power ... that was the key development.[94]

It had nothing to do with the coalition in India. The conditions suited Bangladesh ... Hasina kept the Islamists at bay ... [she] was upset at the blemish on her father's name.[95]

According to them, "there was a change in attitude in Bangladesh ... more reasonable ... less hostile", following the defeat of the Bangladesh Nationalist Party.[96] Preparatory foreign-secretary level visits in July and August had laid the groundwork. The Government oversaw a deal whose fruition was "a function of time".[97]

Other government officials emphasized its adroit decisions and conciliatory approach, however. I.K. Gujral undertook a three-day visit to Bangladesh in early September, reportedly the first by an Indian foreign minister since 1971.[98] Moreover, he asked the West Bengal chief minister, Jyoti Basu of the CPI(M), to undertake a five-day mission given the direct stake his state had. One senior diplomat downplayed it, remarking that

[it] was the victory of a weak Centre. Basu was sent to Bangladesh without any authority to bargain, [only] to soften and appease Hasina ... His role gave a cushion.[99]

Others, however, pointed to his "critical role of statesmanship":[100]

The dispute was very complicated. There was total deadlock ... Jyoti Basu played a great role. Without him, it would never have worked ... It was a stroke of genius by I.K. Gujral [to send him] ... of course, the Ministry of External Affairs will never accept it.[101]

Indeed, observers on both sides credited Basu's "catalytic initiative".[102]

The Ganga Waters Accord raised several questions. According to critics, it failed to account for the substantially lower flow of water on the Indian side in the post-1988 period, partly caused by Bihar and Uttar Pradesh drawing water before the Ganga reached the Farraka dam in Bangladesh, or to consider the danger of siltation in Calcutta and flooding in West Bengal due to the rising water levels of the Bhagirathi-Hooghly river.[103] The high-level meeting also failed to resolve other issues: the need to build transit facilities in Bangladesh for Indian goods and a gas pipeline to serve their respective energy requirements, and an agreement to return the Chakma refugees to Bangladesh. According to self-described realists, the Accord disproportionately recompensed Bangladesh.[104]

These criticisms warranted attention. They overlooked several geopolitical considerations, however. The Government had deliberately avoided linking concessions on the Ganga to other items to pre-empt opposition to the deal in Bangladesh.[105] The Accord also put India on stronger ground to pursue security cooperation in the Northeast. For the United Front, the strategic calculus of subcontinental relations extended beyond the narrow vantage of zero-sum gain. Even skeptical government officials recognized "the sense that we [India] can now afford to be good-neighborly".[106] It was a disposition widely shared within the coalition:

> We felt, "let us sacrifice something", even though Bengal and Bihar may suffer in bad years. For small matters we should not fight ... Basu has the problem, so Gujral suggested him ... The Congress would not have solved the Farraka [dispute]. It has an imperial attitude and wants problems to be alive.[107]

In sum, by recognizing an opportune moment and delegating power to key political figures that had the incentive, clout and experience to strike a deal, the governing coalition operationalized a more federal approach to subcontinental foreign policy. In early January, Gujral unilaterally removed all non-tariff barriers on exports from, and offered financial assistance to, Sri Lanka, stating that "we need neighbors who are developing at least as fast as we are to avoid imbalances which feed dissatisfaction and political problems".[108] The United Front displayed political imagination and good judgment in seizing the moment.

The fall of Narasimha Rao

Unfortunately, diplomatic feats could neither suppress growing political discontent within the governing coalition nor halt political re-alignments within the Congress(I). The prime minister took several well-intentioned but misguided decisions to display boldness in policy matters. Deve Gowda proposed to grant statehood to Uttarakhand without consulting the Left, forcing him to withdraw the offer. Colleagues also questioned his political motivations. The TDP chief minister Chandrababu Naidu criticized the Prime Minister for the slow pace and inadequate level of central assistance after a cyclone hit Andhra Pradesh in early November.[109] Colleagues in the JD objected to his proposal to include Jats and Sainis within the ambit of reservations for OBCs.[110] And the DMK and TMC criticized Deve Gowda for visiting their arch political rival, Jayalalitha. Finally, others alleged personal improprieties. The CPI(M) Politburo member Sitaram Yechuri rebuked him for taking his extended family to a G-15 summit in Zimbabwe. None of these criticisms entailed serious clashes. Yet they depicted a prime minister negligent towards, even dismissive of, the political sensitivities of his principal coalition partners.

More serious differences within the JD, and between the Left and the pro-liberalization lobby in the UF, had also materialized. On 11 December, the Patna High Court reprimanded the CBI director for not expediting the fodder scam inquiry, even though a simultaneous CAG report verified the personal involvement of the Bihar chief minister on various counts.[111] The JD had skirted the issue at its national political meeting in late November. The Bihar unit was one of its two remaining state-level bases. But the Left, which had already rallied against "corruption and maladministration", called a day-long *bandh* (strike) in the state on 18 December.[112] Lalu Yadav's growing political vulnerability formed a growing political fissure.

Economic policy caused further political disagreements. The revamping of the Foreign Investment Promotion Board by the industry minister in late October, and proposals by the finance minister in late November to open the insurance sector to domestic private capital and raise the ceiling on disinvestment of PSUs to 74 per cent, antagonized the Left.[113] Reportedly, Chidambaram promised to halve the price of basic food grains under the public distribution system (PDS). Yet his pledges failed to satisfy them entirely. The CPI(M) Politburo reiterated the need to establish a fixed venue and an agreed *modus operandi* for the Steering Committee, which it viewed as the main decision-making body for sensitive political questions.[114] Indeed, the CPI high command invited "the people's intervention" to pressure the Government to abide by the CMP.[115] The Left began to act as the internal loyal opposition of the governing coalition in economic affairs.

Still, the Congress(I) remained its greatest threat. The gradually shifting terms of power within the party marginalized Rao. Kesri displaced many of his loyalists in the Youth Congress in late November, promoting figures close to Sharad Pawar, K. Karunakaran and Madhavrao Scindia, given their respective clout in Maharashtra, Kerala and Madhya Pradesh.[116] The embattled former prime minister claimed that attempts to replace him required two-thirds support in the CPP. Yet the election of many anti-Rao candidates to its executive tilted the balance.[117] On 18 December, the CWC authorized Kesri to demand his resignation. Rao stepped down the following day. His exit gave the Congress(I) an opportunity to adhere to the one-man-one-post principle that ostensibly created many dissidents within its ranks. In early January, the party endorsed Kesri, the interim president of the AICC, as leader of the CPP. The integrity of principle had succumbed to the stratagems of power.

10

REFORM AMID CRISIS (JANUARY–APRIL 1997)

The United Front entered the new year facing significant challenges. It had had a relatively unsuccessful winter parliamentary session.[1] The joint select committee charged with reformulating the Women's Bill was unable to do so. The proposed Lok Pal bill, which excluded the prime minister's office from its purview, drew widespread opposition. And the governing coalition had failed to redeem its pledge to establish review committees regarding judicial reform, freedom of information and Centre-state relations. In addition, the formal elevation of Sitaram Kesri introduced a new dynamic into its relations with the Congress(I). On the one hand, certain conditions protected the Government. The former chief minister of Maharashtra, Sharad Pawar, supported Kesri in exchange for becoming the parliamentary floor leader of the Congress(I). More significantly, the former extracted a pledge from the latter not to topple the Government. Many regional chiefs believed the party needed time to reintegrate former dissidents and strengthen its organization. The Congress(I) had to face elections in Punjab, await a resolution in Uttar Pradesh and respond to the first major Union budget of the Government. Thus Pawar declared that it had an interest in ensuring political stability and preparing adequately for elections in the medium-term, presumably at least a year away, if not more.[2]

Mitigating factors existed, however. The first concerned Kesri himself. The new Congress(I) leader, seen by many within his party as "an old man in a hurry", had the prerogative to decide its stance vis-à-vis the United Front.[3] His desire for greater political influence and deference complicated matters. On 16 January, the prime minister offered to constitute a joint pre-budget panel with the Congress(I). Reportedly, Deve Gowda said either Pawar or Kesri could determine their respective nominees, enjoining the latter to dismiss it.[4] A few days later the TMC chief G.K. Moopanar proclaimed that only Sonia Gandhi could reintegrate the party. He even claimed that a section of it sought to join the Government.[5] Both incidents emphasized Kesri's perceived weakness.

Political self-preservation also drove him, however. On 19 January, the CBI grilled Kesri after a former Congress(I) advisor close to Rao petitioned the Delhi High Court to investigate his assets.[6] Sources in the CBI acknowledged the lack of evidence. Hence the Congress(I)'s spokesperson, V.N. Gadgil, accused the prime minister's office of pressing the probe.[7] Rumors also circulated that it had pressured the Intelligence Bureau to implicate Kesri in a 1993 murder inquiry. The disclosures suggested that Deve Gowda was trying to "find some dirt under [Kesri's] finger nails". According to one senior investigator,

> the case against Mr Kesri cannot be justified on purely professional grounds; the one thing one can say is that the Prime Minister is being very poorly advised to go after Mr Kesri. These advisors must be his worst enemies: they are needlessly hastening the end of the Government.[8]

It was a prescient diagnosis. The prime minister, who had allegedly misused the central police agencies to protect Rao, now used them to harass his successor.

The second factor that had altered circumstances was the Congress(I)'s political dynasty. On 21 January, the Government received secret documents from Switzerland relating to the Bofors scandal. Its prime suspects included Ottavio Quattrocchi, an Italian businessman with close ties to Rajiv and Sonia Gandhi, Win Chadha, who managed the Indian front company for Bofors, and the Hinduja brothers, whose international business holdings enabled their considerable political clout.[9] Rao had launched contentious litigation in 1991 to delay further investigation. But the inquiry now resumed. On 30 January, the CBI Director, Joginder Singh, set up a special investigative unit to pursue the case, vowing to process the relevant papers by the end of March. On 11 February, the agency identified Quattrocchi, Chadha and several members of their respective families for receiving kickbacks amounting to Rs. 16 crore between May 1986 and March 1987.[10] Kesri proclaimed that attempts to "politically blackmail" the Congress(I) would fail.[11] Nevertheless, the inquiry threatened to renew scrutiny of the ties between Quattrocchi and the Gandhi family. The party had reasons to fear revelations that might further stain its already disgraced image, and damage the reputation of its potential dynastic heirs.

Deteriorating electoral fortunes gave it further reason to reconsider its parliamentary strategy. On 9 February, the BJP-SAD(Badal) alliance captured an overwhelming majority in the Punjab assembly elections, routing the incumbent Congress(I).[12] The victorious alliance championed Sikh pride, claiming that only the SAD could restore peace to the state. The electorate punished the Congress(I), which only retained 14 of its 87 assembly seats, for its widely perceived corruption. It was a regional verdict. Nonetheless, the resounding defeat exposed the party's declining electoral prowess. On 16 February, the CWC reversed its hitherto "unconditional" support for the United Front. Many commentators dismissed the move as bluster.[13] Nevertheless, it signified deteriorating relations, casting a shadow over the upcoming parliamentary budget session. On 12 February, the governing coalition formed an eleven-member standing committee that would meet at the prime minister's office before all scheduled meetings of the larger Steering Committee.[14] The CPI(M) declared its continuing external support to the Government a few days later. But the party reiterated its opposition to further economic liberalization, galvanizing media attention in the run-up to the 1997–98 Union budget.[15]

Advancing economic liberalization

The first budget in July 1996 was an interim exercise designed to signal the Government's general orientation. The second was a better measure of its plans. The governing coalition confronted more difficult circumstances on various fronts.[16] First, core economic sectors exhibited recessionary tendencies. Agriculture had contracted; industry required greater levels of investment; export growth in both sectors had collapsed. Second, the fiscal situation required attention. In addition to servicing debt obligations, the Government faced the cost of the impending Fifth Central Pay Commission Report, concerning the revision of pay scales, allowances and benefits for central government employees, which traditionally shaped state-level and private sector negotiations as well. The cost of these three items alone equaled total expected revenues projected for the fiscal year. Lastly, continued economic growth required adequate investment in depleted infrastructural services.

Differences over priorities and strategy divided the United Front, however. The prime minister, finance minister and industry minister appeared receptive to demands for further investment, trade and financial sector reforms from domestic industry and big business in their pre-budget meetings.[17] The TDP chief minister, who had reduced food subsidies and inaugurated plans to build a major technology park, backed such initiatives. Indeed, Chidambaram had reportedly allowed Naidu to pursue further reforms in Andhra Pradesh by tapping World Bank loans, setting a precedent.[18] In contrast, their Left colleagues continued to criticize the direction of reform. Communist party officials emphasized the need to control the rising cost of basic food grains by increasing supply, devise restructuring plans for sick PSUs and make primary education a fundamental constitutional right during Steering Committee meetings in mid-January. They also demanded a parliamentary bill to secure a minimum wage for agricultural laborers, improve working conditions in the countryside and provide crop insurance to small farmers.[19] The *Approach Paper to the Fifth Five Year Plan (1997–2002)* also focused on increasing employment through small-scale sector development. Indeed, reports that Manmohan Singh had privately approved of the Government's economic orientation reinforced critiques that its leading economic ministers remained in thrall to the policy framework he had introduced in 1991.[20]

The debate within the United Front over the role of private domestic capital and foreign investment reflected the divergent views of its respective electoral bases. In general, approximately 40 per cent of the electorate took no clear view, a significant plurality. Their relative lack of awareness heightened the relative autonomy of the political class to shape public opinion and make policy decisions (see Tables 10.1 and 10.2). Nevertheless, significant differences existed. Despite the rhetoric of *swadeshi*, supporters of the BJP and its allies supported the entry of foreign private capital to the greatest extent, with approximately 42 per cent in favor. Their counterparts in the third force, who represented approximately 34 per cent of its base as a whole, were the least. The question of privatization marshaled greater cross-party opposition, even amongst the relatively privileged constituents of the BJP, with only 27 per cent voicing support. Again, the social base of the United Front was the least supportive, at 20 per cent.

Table 10.1: Desirability of Unrestricted Foreign Investment, 1996

Category	All Alliances									
	Third Front		Congress(I)+		BJP+		Other		Total	
	N	I	N	I	N	I	N	I		
Disagree	29	22.2	34.4	23.6	27.4	20.9	9.2	19.7	22	
Don't Know	30.6	43.9	31.6	40.5	26.1	37.3	11.7	46.8	41.2	
Agree	26.5	33.9	31.3	35.9	32.8	41.8	9.4	33.5	36.8	
Total	28.8	100.0	32.1	100.0	28.8	100.0	10.3	100.0	N=8295	

Category	Third Front					
	National Front		Left Front		Federal Front	
	N	I	N	I	N	I
Disagree	8.9	23.1	7.8	19.9	12.3	23.2
Don't Know	8.8	43	8.1	38.4	13.8	48.5
Agree	7.8	33.9	9.8	41.6	9	28.3
Total	8.4	100.0	8.7	100.0	11.7	100.0

Note: N represents the national vote share given to each alliance. I reflects the internal vote share within each alliance.
Source: CSDS National Election Survey, 1996.

Table 10.2 Desirability of Privatization, 1996

| Category | All Alliances ||||||||||
| | Third Front || Congress(I)+ || BJP+ || Other || Total |
	N	I	N	I	N	I	N	I	Total
Disagree	28.5	34.6	34.1	37.2	28.2	34.2	9.2	31.1	34.9
Don't Know	30.8	45.5	30.4	40.2	26.3	38.7	12.5	51.6	42.5
Agree	25.4	19.9	32.1	22.6	34.7	27.2	7.8	17.2	22.6
Total	28.8	100.0	32.1	100.0	28.8	100.0	10.3	100.0	N=8295

| Category | Third Front ||||||
| | National Front || Left Front || Federal Front ||
	N	I	N	I	N	I
Disagree	9.4	28.2	10.9	43.9	8.2	34.0
Don't Know	13.8	50.2	8.6	42.1	8.5	42.6
Agree	11.2	21.7	5.4	14.1	8.8	23.4
Total	11.7	100.0	8.7	100.0	8.4	100.0

Source: CSDS National Election Survey, 1996.

REFORM AMID CRISIS (JANUARY–APRIL 1997)

Real divisions existed within the latter, however. A strong plurality of Left supporters, exceeding 40 per cent, opposed unrestricted foreign investment and the privatization of PSUs. In contrast, regional parties' constituents supported both policies to a slightly greater extent (28 per cent to 23 per cent), even though critics of privatization comprised an important grouping (34 per cent to 23 per cent). In short, the growing political disagreement within the United Front reflected deeper social differences.

Thus what ultimately mattered was the question of power, in particular, the political strategy of the Left. In early January, the CPI(M) chief minister of West Bengal, Jyoti Basu, declared that his party had committed "a historical blunder":

> [I had argued that] since we have the experience, we know these people, we can keep them together for as long as possible. If we were there we would see that the programmes would be somewhat carried out, much better than what they [our UF partners] would do if we were not there in the government.... Our argument was: this cannot last five years. If we are there, much more than the others we can make them accept some policies, put them before the country, whatever the limits are. You can't remove every obstacle, that is not possible: but we could do something for self-reliance, for the countryside, for panchayats, all that we can push through. Anti poverty programmes: it [sic] is there but it does not reach the people.... But it is a political blunder. It is a historical blunder.... We do not accept many of their policies, they do not accept many of ours. But the minimum programme was there, and we could have implemented it much better than others. Because we have the experience, nothing more, nothing personal.[21]

His remarks ignited renewed debate. The Politburo, citing Engels' caution that communists risked being junior partners in bourgeois dominated coalitions as the experience of France in 1848 had demonstrated, framed the matter as "settled and closed".[22] Basu failed to explain himself further. But his statement warranted reflection. By calling the decision a "blunder", the venerable chief minister underscored its political consequences, namely the inability of the CPI(M) to shape government policies. Formal ministerial participation could not guarantee political clout as the marginalization of the CPI home minister in preceding months had shown. Yet exercising formal political authority, especially from the prime minister's office, might have given the CPI(M) a chance to exert real power and bolstered the CPI as well. By characterizing the blunder as "historic", Basu implied that it represented a rare opportunity, which might not materialize again. His judgment was political in a classic realist sense: rooted in context, dispassionate in spirit and comprehensive in scope, and attentive to the practicalities, strategic options and actual possibilities of the historical moment. As with every political judgment, Basu had to distinguish the foreseen, foreseeable and unforeseeable. Yet his interpretation was judicious. Subsequent events proved him right.

Indeed, proponents of liberalization burnished their credentials with few hindrances. In late January, Chidambaram announced a series of reforms—allowing foreign institutional investors to invest in dated government securities, liberalizing external commercial borrowing guidelines and scrapping the existing RBI ceiling on foreign exchange remittances—that opened the financial sector to greater private sector involvement. These various measures satisfied apex business organizations, who demanded capital market liberalization, at the risk of exposing the economy to greater external shocks.[23] In early February, the TDP Minister of State for Commerce, B.B. Ramaiah, eased import tariffs on 161 categories of consumer goods, placating the interests of industrial and finance capital and the country's urban middle classes. In late February, Deve Gowda unveiled reforms to the PDS. The Government increased its subsidy by Rs. 2,400 crore. But it also sought to target the poorest of the poor, dividing those above and below the official poverty line by introducing a dual price mechanism for rice and wheat, and capping food provisions at ten kg per family per month.[24] Supporters of the revamped PDS, called the Targeted Public Distribution System (TPDS), claimed that cross-subsidization would allow states to reach the most vulnerable sections of the population:

> The TPDS made a conscious link between poverty and PDS lifting. It was novel in this regard ... [it would] eliminate subsidies to the 10 per cent who lived above the poverty line.[25]

Some commentators praised the increase in funds. Yet critics voiced concern that targeting failed to address problems of misidentification (due to the illegibility of and fluctuation in poor households' incomes in the informal sector), vulnerability (families above the official poverty line suffered absolute deprivation and could easily fall below it due to exogenous shocks) and corruption (basic food grains often went to better off urban dwellers and found their way onto the open market).[26] They were right. But the CPI(M)'s lack of formal cabinet authority, and the fact that it would not exercise its veto over the Government

The 1997–98 Union budget

On 25 February, the finance minister presented the *Economic Survey 1997*, which estimated GDP growth at 6.8 per cent in 1996–97.[27] On 28 February, he unveiled the 1997–98 Union budget, which prescribed a neo-liberal growth-oriented strategy to spur economic productivity, generate employment and reduce mass poverty. The budget projected a reduction in the fiscal deficit from 5 per cent of GDP in 1996–97 to 4.5 per cent in 1997–98, largely through estimated tax receipts. In particular, Chidambaram argued that lower tax rates would yield higher returns, appealing to the logic of the so-called Laffer curve.[28] Hence the Budget reduced direct taxes (income and corporate), lowering peak rates between 5 and 10 per cent on average, and abolished many specific levies.[29] Concomitant reductions to indirect tax rates (customs and excise), and on tariffs for luxury consumer durables, mass consumption items and industrial goods, integrated the economy closer to global norms. The finance minister partly offset these reductions by expanding the tax net amongst urban dwellers, the service sector and previous tax evaders, pledging to direct three-quarters of the recovered funds to the states and allot the remainder for basic minimum services and other infrastructural needs, and raising selective rates for retail traders and postal services. He also satisfied a long-standing demand of the states, accepting the recommendation of the Tenth Finance Commission to form a single divisible pool of taxes and expand the states' gross share to 29 per cent. Indeed, during its tenure the United Front reversed the decline in central transfers to the states relative to GDP and central and state revenues that had characterized the post-1991 period.[30]

Projecting greater revenue buoyancy, Chidambaram increased social sector spending to Rs. 15,707 crore, allotting Rs. 3,300 crore for basic minimum services, an additional Rs. 500 crore to NABARD, and an increase of Rs. 7,958 crore for the Central plan with a focus on rural development, employment and poverty alleviation. The Budget also announced the elimination of ad hoc Treasury Bills, a short-term debt instrument used by the Centre to impose fiscal restraint, and granted more autonomy to the Reserve Bank of India (RBI) in setting monetary policy. However, total current expenditure—on debt-related interest payments (Rs. 68,000 crore), defense (Rs. 26,713 crore) and various agricultural subsidies (Rs. 17,130 crore)—remained high.

In addition, the finance minister relaxed various restrictions and unveiled new initiatives regarding trade, investment and finance. He raised the investment ceiling for foreign institutional investors, Non-Resident Indians (NRIs) and venture capital funds in company equity and revised the Companies Act and other tax-related measures. Chidambaram partially liberalized the non-banking financial sector, allowing private domestic capital to provide health insurance and enter joint ventures in pension funds. He also permitted investments in overseas joint ventures up to US$15 million, and pledged to replace the Foreign Exchange Regulatory Act (FERA) and explore the feasibility of full capital account convertibility. Finally, the finance minister announced supply-side incentives to spur private capital formation in infrastructure. The budget accepted the recommendations of the India Infrastructure Report, which declared tax holidays and lowered royalty payments in the petroleum sector, and provided further incentives in telecommunications.

In marked contrast, Chidambaram introduced few reforms relating to agriculture, labor or the public sector, key domains of second-stage reforms. He raised the rate of return on pensions for public sector workers and abolished controls on rice, milling and ginning. Reforms to the small-scale industry allowed greater investment and improved the formula for disbursing funds. But the budget essentially maintained the 896 items on the protected list. Finally, given its failure to raise Rs. 5,000 crore through disinvestment, a target that had been set in the 1996 budget, the Finance Minister announced steps to strengthen profitable state enterprises—identifying the so-called *navaratnas* (nine jewels)—by delegating greater monetary autonomy to their boards and increasing non-Plan loans. In other words, the Government proposed privatizing only a handful of PSUs recommended by the Disinvestment Commission. The financial burden of sick companies remained.

Boosters of liberalization in the media applauded Chidambaram for delivering "a dream budget".[31] The latter belied the view that a disparate governing coalition, with professedly social democratic credentials, would diminish the pace of further economic reform. Several factors accounted for it. First, the 1997–98 budget largely announced measures concerning "elite politics".[32] Second, although the pre-budgetary process encompassed many competing interests, the finance minister and, to a lesser extent, the prime minister enjoyed customary prerogative over its design.[33] Discontented groups could seek redress during the post-budget meetings and crucial parliamentary debates that would ensue. Yet ministerial authority gave them both the crucial power to set the agenda. Third, their ideological beliefs and political incentives had great import. Senior government bureaucrats contrasted Chidambaram's "ideological commitment" and his "desire to outshine Manmohan [Singh]", his predecessor, with the "pragmatic" disposition of Deve Gowda, who believed in the "necessity" of further economic liberalization.[34] Indeed, some observers claimed that Rao advised Deve Gowda to "unleash" Chidambaram during the parliamentary budget session to win the approval of the country's metropolitan elites.[35] Finally, proponents of liberalization within the United Front outweighed its detractors, led by the TDP, DMK and TMC:

> Nobody dreamt of it. It set a benchmark [on fiscal policy], [from] which nobody would dare to retreat.[36]
>
> It sent out good signals ... which were necessary at the time.[37]

The orientation of the regional bloc attested to the influence of regional capitalists within their ranks as well as the incentives generated by India's federal market economy post liberalization.

The JD and SP privately supported the budget too, although less enthusiastically. The maintenance of high subsidies and limited trade liberalization within agriculture satisfied the intermediate proprietary castes that dominated both parties.[38] The slow pace of public sector disinvestment, with proposed salary increases for central government employees, mollified urban constituencies. Measures to widen the tax base mirrored earlier proposals by the NDC. A few JD leaders opposed lower taxes. For the most part, however, their colleagues displayed little interest.[39] High economic policy and questions of basic material redistribution failed to excite their passions in contrast to the politics of recognition based on status and identity.

As expected, the Left disapproved of various aspects of the 1997–98 Union budget. The CPI national secretary, D. Raja, criticized the finance minister for making "excessive" concessions to industry and big business and failing to allocate sufficient capital for public goods.[40] The CPI(M) general secretary, H.S. Surjeet, endorsed the budget but indicated a desire to amend its tax proposals.[41] Privately, many senior figures in both parties asserted that "[we] didn't expect it to be so bad. It was very pro-rich ... lollipops were handed to them [big business and the socially privileged]".[42] Indeed, the CPI(M) released a statement praising Deng Xiaoping for creating "socialism with Chinese characteristics" and defending the public sector under a "people's democratic dictatorship".[43] Given its displeasure, many expected the party to seek amendments in the post-budget discussions and parliamentary debates that lay ahead. Yet the CPI(M)'s unwillingness to topple the Government defanged its veto power. A Basu prime ministership would have given the party—even with Chidambaram in the finance ministry—a better chance to shift the growth-oriented economic policy framework onto a more socially inclusive path.

The BJP finance spokesperson, Yashwant Sinha, rebuked Chidambaram for presenting a "capital market and foreign institutional investor-friendly" that courted the "approval of Davos".[44] The Congress(I) voiced little reaction in parliament. In private, however, its leading figures appraised it negatively:

It was a deceptive budget ... [made] in a context of gloom ... which in the end was a disaster. It was based on short-term optimism, to cheer up the animal spirits. The arithmetic projections didn't add up ... it was a case of irrational exuberance, mistaken by assumption.[45]

Indeed, Chidambaram's economic projections begged many questions. His fiscal strategy presumed strong revenue buoyancy, which presupposed economic growth reaching 7 per cent in the coming fiscal year through high exports and greater industrial output. Senior finance officials defended these expectations.[46] But critics on the Left knocked the budget.[47] They criticized the amnesty given to previous tax evaders and questioned the assumption that lower tax rates would yield higher government receipts. They also noted that, despite increasing in absolute terms, social sector spending represented a smaller percentage of GDP, leading them to deride the budget as "inequitable" for its "giveaway" tenor.[48] Lastly, they doubted the sufficiency of tax cuts and easier credit to stimulate rapid growth across the economy without adequate public investment in physical infrastructure. The lower propensity of (foreign) private capital to invest in large-scale infrastructure projects, given their long gestation periods and uncertain returns, cast doubt on the underlying strategy. Agriculture faced more severe challenges due to persisting barriers to domestic agro-trade and inadequate private funding. The budget promised higher rural credit, greater funds for infrastructural development and bank capitalization, and additional capital shares to NABARD. Yet these various schemes, which collectively increased absolute expenditures, masked an actual decline of spending in terms of total Central Plan outlay. Indeed, even ardent neoliberals voiced concern. The IMF Managing Director, Michel Camdessus, praised Chidambaram for pursuing a "bold but credible [budget] strategy" in early March. Yet he conceded, "anything reasonable at the time of formulation can be frustrated by unforeseen expenditure and revenue shortfalls".[49]

In the end, strong neoclassical convictions, situational demands and sectional pressures shaped the 1997–98 Union budget. In part, its fiscal strategy reflected economic constraints:

most non-Plan resources ... [were] already committed, so politics is about rationing resources. The revenue side offers clear political scope for action.[50]

Debt-related interest payments, defense spending and various agricultural subsidies accounted for most of these resources. India's trend rate of economic growth had picked up slightly, from 5 per cent in the 1980s to 6 per cent by the late 1990s. But central government revenue as a percentage of GDP remained very low. Simply put, the Centre lacked the resources to spend. In part, the fiscal strategy exposed political demands. It was difficult to "scale back" high agricultural subsidies to dominant rural interests, powerfully represented by the JD and major regional formations in the United Front, and food subsidies, which benefited state-level administrations of various hues and disproportionately those run by the Left.[51] Given these factors, wooing foreign private investment and stimulating economic growth through tax-based incentives made political sense. Yet the Finance Minister's strategy, according to one key associate, "[presumed] political stability and investor confidence".[52] Chidambaram advanced economic liberalization further than many commentators had anticipated when the United Front came to power. But the "dream budget", relying on private sector growth, was a gamble.

A political setback in Uttar Pradesh

Paradoxically, a political crisis instigated by the Congress(I), whose interests came under threat from two directions, shaped the fate of the budget. The first entailed ongoing criminal investigations. On

11 March, the CBI declared that it had adequate evidence against Rao and twenty-one other suspects in the JMM pay-off case.[53] Several days later Sudhir Mahato claimed that Rao had bribed the smaller regional party to defeat the 1993 no-confidence motion against the Congress(I).[54] Simultaneously, the CBI interrogated Arjun Singh, a former Union minister of state and associate of the Nehru-Gandhi family, regarding his alleged role in the Bofors scandal.[55] Although routine, the inquest recast light on Sonia Gandhi, who many hoped would enter the political fray.

The second threat concerned the party's fortunes in key electoral arenas. On 24 February, the home minister had stated in parliament that Uttar Pradesh was "heading towards anarchy, chaos and destruction".[56] Gupta's blunt declaration may have betrayed personal frustration at his relative powerlessness. It was an astonishing statement, however, given that such matters comprised his brief. It underscored the home minister's tendency to speak like a member of the Opposition as well as the breakdown in collective ministerial responsibility. The BJP demanded an inquiry. The Speaker, P.A. Sangma, directed the Government to establish an advisory parliamentary committee to investigate.[57]

But further indiscretions undercut its position. The controversial Uttar Pradesh Governor, Romesh Bhandari, repudiated Gupta's charge. Moreover, he proclaimed that law and order had improved under his tutelage, invoking the support of the prime minister.[58] Deve Gowda failed to intercede. Mulayam Singh Yadav spoke more frankly, claiming, "whatever I could not finish during my chief ministership is now being done by Bhandari".[59] On 5 March, Sangma reproved the Council of Ministers for violating collective responsibility and admitted a motion by the BJP to recall the governor.[60] On 10 March, Gupta criticized Bhandari for breaking protocol. The home minister also artfully deflected attention, blaming the BJP for communalizing the politics of Uttar Pradesh in the wake of Ayodhya.[61] Citing political deadlock, the Government extended President's rule for another six months on 13 March.[62]

But the impasse suddenly broke. On 19 March, the BJP and BSP staked a claim to form a government on the basis of a quasi-consociational power-sharing arrangement. The two parties agreed to alternate the chief ministership every six months, assign the assembly speaker post to the BJP and share an equal number of ministerial seats. They also established a senior monitoring panel to oversee political administration.[63] On 21 March, Mayawati was sworn in as the chief minister. The desire for office, and the two parties' shared enmity towards the SP, facilitated a seemingly incongruous alliance. Genuine differences persisted. The Sanskritic vision of the *Ramjanmabhoomi* movement championed by the BJP clashed with the desire of the BSP to empower the *dalitbahujan*. Moreover, kept out of the negotiations in New Delhi, Kalyan Singh questioned whether Mayawati would relinquish the chief ministership according to schedule.[64] The first attempt by the parties to share power had failed in 1995. That said, the deal illustrated their pragmatism and willingness to fashion a novel pact in order to capture state office. Mayawati immediately expanded the Ambedkar Village Program, loosening criteria for incorporating settlements and raising its budget, and transferred many senior administrative officials appointed by the SP. More controversially, the BSP chief directed all state funds for school building towards the Program.[65] Her actions simultaneously revealed a profound distrust of the impartiality of the state and yet confidence that it was the only effective agency of social change.

Domestic events coincided with significant external developments. In late February, the Government had agreed to resume foreign secretary-level negotiations with Pakistan after a hiatus of three years. The election in mid-February of the new prime minister, Nawaz Sharif, who adopted a relatively conciliatory posture towards India, provided an opening. The United Front, which stated its "willingness for wide-ranging and comprehensive talks on all issues of mutual concern", also deserved credit.[66] On 2 March, the Government accepted Sharif's proposal to prepare for a prime ministerial summit on the condition that "some progress [be achieved] on the core issue of Jammu and Kashmir" alongside movement on cultural and economic matters, as India had customarily demanded.[67]

The main objective was simply to revive high-level communication. On 20 March, the Pakistani authorities released 38 children apprehended two years earlier while traveling on Indian shipping boats. In return, I.K. Gujral exempted young and elderly Pakistani citizens from reporting to the police while in India, waived visa fees for artists and students, and opened more religious sites to Pakistani tourists.[68] The circumspection of these early gestures underscored their largely symbolic value.[69] Still, commentators acknowledged the revival of high-level dialogue. According to a senior government official, "the purpose of the first round [of negotiations] was simply to lay down the subjects".[70] Both delegations exchanged formal statements on the first day. The foreign secretaries claimed "some positive movement" on the second.[71] Unforeseen events stalled further progress, however. On 30 March, Sitaram Kesri withdrew the Congress(I)'s parliamentary support to the Government, staking his claim to rule.

The rise of Inder Kumar Gujral

Complex political motivations inspired his maneuver. Officially, Kesri presented two reasons. First, recent events had undermined the "secular logic" of the previous Congress(I) stance. Disagreements with the United Front had allowed the BJP to advance in various by-elections across the country:[72] in the Punjab assembly elections, where the governing coalition and the Congress(I) worked at cross-purposes; and most importantly in Uttar Pradesh, where the governing coalition failed to support the chief ministerial ambitions of Mayawati, allowing the BJP to strike a deal with the BSP. Second, Kesri highlighted the "deteriorating law and order situation", "the drift in the economy", and "the lack of cohesive functioning of the Government".[73] In short, genuine political concerns imperiled the country.

It was difficult to accept these charges at face value. Jyoti Basu claimed that several Congressmen accused Deve Gowda of being "communal" but "[t]hey could not give me one instance ... they said a face saving thing."[74] The United Front had displayed political disunity and administrative negligence in Uttar Pradesh and failed to cooperate with the Congress(I) in Punjab. Its most recent endeavors regarding economic liberalization and India-Pakistan relations articulated a sense of purpose, however. Invoking the principle of secularism failed to mask basic political interests. Several leaders in the governing coalition, especially on the Left, accused the Congress(I) of withdrawing support simply to capture office: "Congress people cannot reconcile themselves to being out of power".[75] Successive electoral defeats had exposed the party's organizational weakness and political unpopularity. These debacles also showed its inability to bend the United Front towards other designs. Indeed, one CPI(M) official claimed the Congress(I) was "anxious to join the Government" in order to halt its deteriorating position.[76] Indeed, Kesri had reportedly failed to justify his actions to the President, Shankar Dayal Sharma.[77]

REFORM AMID CRISIS (JANUARY–APRIL 1997)

Many focused on his private aspirations. Some claimed he coveted the prime ministership.[78] Others suggested lesser ambitions:

Kesri felt threatened because he did not like Deve Gowda being close to Rao … It was purely personal.[79]

Kesri felt he was being improperly consulted [by Deve Gowda].[80]

Kesri was the cause of the downfall … He had an aversion to Rao, and by extension to Deve Gowda … which led to his wild allegations against him [Deve Gowda].[81]

In other words, pride and insecurity drove Kesri to withdraw parliamentary support. The prime minister exacerbated the situation by not courting him and spurning intermediaries' efforts to placate him.[82] In doing so, Deve Gowda failed to take Kesri's measure.

The prime minister's alleged disregard was neither benign nor complete, however. Key observers claimed that he

was obsessed with Kesri … [saying] for instance, that Kesri wanted the cabinet secretary and principal secretary [to the Prime Minister] removed. But Kesri didn't even know the cabinet secretary's name.[83]

Indeed, Kesri complained to Indrajit Gupta that

this man, the prime minister, is treating us like an enemy and wants to destroy my party and I could not sit quietly and tolerate it. Ever since I became president of the Congress he seems to think I am a potential enemy or rival of him, which I am not.[84]

Many quarters blamed the Prime Minister for practicing a "politics of blackmail" and intimidating the Congress(I) leader, including an investigation into his personal assets and alleged role in a 1993 murder case, as well as most of his CWC associates.[85] Several figures also accused Deve Gowda of encouraging the pro-Rao faction to challenge Kesri's leadership at a CPP meeting on 17 March, and of trying to split the party by tempting the Maharashtra faction led by Sharad Pawar with possible ministerial posts.[86] Reportedly, the prime minister pursued these activities with his personal coterie of Karnataka politicians and compliant police officials: "by the time the others clued in, it was too late … [CPI(M) general secretary] Surjeet found out only ten days before Kesri pulled the plug".[87] Assailed by Deve Gowda on multiple flanks, the CWC authorized Kesri, who purportedly "was willing to wait for another year", to withdraw outside support in due course:

There was poor coordination between us. We were suffering from harassment, arrests, interference, neglect … Kesri's angst was a secondary reason …

we had no obligation to support the UF ... We cannot liquidate ourselves voluntarily."⁸⁸

In sum, the prime minister had many reasons to distrust the Congress(I) and its interim leader, who had intimated his prime ministerial ambitions. However, most Congressmen saw little reason to topple the Government immediately. In the end, Deve Gowda became convinced that Kesri was conspiring with others in the United Front to oust him, only to later admit he was wrong.⁸⁹

But the Congress(I) failed to grasp the reins of power. Tactical mistakes were partly to blame. On 29 March, K. Karunakaran publicly declared that G.K. Moopanar of the TMC had agreed to become prime minister with the party's support, leading to Kesri's gambit the next day.⁹⁰ Yet the CWC presumed he would first explore how many parties would support the move. The interim president failed to initiate the necessary discussions and weigh the odds, causing his colleagues to protest.⁹¹ "Calculations are for intellectuals—politicians work differently", Kesri remarked, invoking an essential feature of political judgment: the need to act without precisely knowing the possibilities and risks of various options.⁹² But his failure to appraise his political context, dispassionately assessing his practical chances of luring sufficient defections, betrayed an expectation that temptations of power would suffice. The debacle exemplified the hazards of imprudence. The art of managing an unwieldy coalition government, and deciding how to supplant it, required far better political judgment.

Ultimately the United Front stood firm. On 30 March, the Steering Committee agreed to face a confidence vote in parliament, backing Deve Gowda the next day. Subsequent attempts by the Congress(I) to woo potential turncoats failed. The TDP rebuffed its principal opponent. The DMK also spurned invitations to switch political sides, pressing the TMC to resist similar entreaties as well.⁹³ Both parties resisted similar overtures from the BJP, which considered propping up Deve Gowda in an attempt to gain *de facto* control.⁹⁴ On 9 April, the Steering Committee rejected the Congress(I)'s demand to lead the Government. "Kesri miscalculated that fear of the BJP would lead non-Left coalition partners [to] surrender".⁹⁵

On 11 April, the Deve Gowda ministry lost the confidence of parliament, 190 votes to 338. Not a single member of the United Front, or faction thereof, crossed the floor. It was a striking display given the perceived opportunism of many of its leaders and the manner in which previous incarnations of the Janata *parivar* had fallen out of office. On 21 April, the Government convened a special parliamentary meeting to approve the 1997–98 Union budget. The finance minister had already secured parliamentary approval to maintain spending commitments for the first two months of the financial year.[96] Nevertheless, the decision illustrated how parties that opposed elements of the budget prioritized stability, even the Left. Kesri's ill-fated ploy facilitated the passage of Chidambaram's controversial "dream budget" without further debate. But the sense of instability generated by the crisis simultaneously diminished the prospects that its underlying fiscal calculus would succeed.

Unable to break the United Front, the Congress(I) demanded it appoint a new prime minister. V.P. Singh orchestrated the governing coalition during the impasse, reportedly advising its leaders not to comply with the Congress(I), which might encourage further demands.[97] But he failed to persuade. On 13 April, the Steering Committee—"afraid to face an election"—forced Deve Gowda to step down.[98] Many senior officials praised the willingness of the Prime Minister to consult widely before taking a decision on most issues. Indeed, Deve Gowda himself frequently blamed his poor Hindi for causing misunderstandings, reportedly saying "if I knew this language, I could handle the Lalus and Mulayams easily".[99] Notwithstanding a handful of confidants, however, his previous handling of various matters had undermined wider political trust.

On 19 April, the United Front named I.K. Gujral as its parliamentary leader. Three factors informed his selection. First, most of its leaders "knew the Congress would strike again ... only everyone knew that it could not go to elections right away".[100] Hence few desired such political responsibility. Indeed, according to a senior JD leader, Gujral "was informed in the middle of the night ... [he was] overwhelmed with concern and depression".[101] It was safe to anticipate a short tenure for the country's twelfth prime minister. Second, Gujral was relatively weak. The external affairs minister was neither closely affiliated with, nor in command of, a particular caste grouping or regional base. Whatever political influence he enjoyed stemmed from his policy expertise. Gujral was bound to dominant party bosses in the JD: in particular, to Lalu Yadav, who had engineered his election to the Rajya Sabha from Bihar. Indeed, the prime minister designate's subordination fomented speculation over how investigations into the ongoing fodder scam would unfold. Third, crosscutting pressures within the

United Front thwarted other nominees. The self-proposed nomination of G.K. Moopanar evoked opposition from two quarters: the CPI(M) and Karunanidhi, his rival from the DMK.[102] Moopanar reacted by removing his four TMC ministers from the Government. Another nominee was Mulayam Singh Yadav, supported by the CPI(M), which feared the BJP in Uttar Pradesh.[103] But Sharad Yadav and Lalu Yadav, his competitors for the leadership of the OBCs in the Gangetic plains, threatened to defect. Regional party bosses reiterated their earlier disapproval. And the untimely absence of H.S. Surjeet, in Moscow for political meetings, deprived Mulayam Yadav of his principal CPI(M) ally during the negotiations.[104] Hence the leaders of the Steering Committee, determined to "neutralize everyone", selected Gujral at the behest of Jyoti Basu and V.P. Singh.[105] On 21 April, I.K. Gujral became India's thirteenth prime minister, winning a vote of confidence the following day.

PART III

THE FALL OF THE THIRD FORCE

11

THE DECLINE OF THE UNITED FRONT (MAY 1997–MARCH 1998)

The willingness of the Congress(I) to exercise its parliamentary veto changed the strategic calculus of the United Front. To promote stability, President Shankar Dayal Sharma urged both sides to set up a coordination panel. Establishing a proper institutional channel would allow information to flow, lessen the potential for individual political gaffes and heighten the possibility of negotiation and compromise. But neither side accepted his recommendation. On the one hand, "Kesri did not want it because of factions in the Congress".[1] A coordination panel would reduce his personal autonomy. On the other, the regional bloc and the Left disliked the idea of establishing "open relations" given their state-level rivalries with the Congress(I).[2] Kesri's abrupt maneuver had widened the gulf.

Indeed, it raised the question of his position within the party, which had to conduct organizational elections as mandated by the Election Commission. In early May, the AICC revealed that Sonia Gandhi had joined its Delhi unit in late March.[3] Concomitantly, the CBI director Joginder Singh charged Rajiv Gandhi and several other Congressmen in the Bofors scam. Many parties rebuked Singh for not informing parliament, raising suspicion of high-level collusion involving the prime minister's office.[4] Kesri formally became the Congress(I) president in early June after receiving the support of thirteen of the 20-member CWC and packing the party's electoral college with loyalists.[5]

Nevertheless, observers speculated that Sonia's formal entry would force Kesri to appear beholden to the heir of the Nehru-Gandhi family.[6] The subsequent exoneration of figures close to the dynasty that had been implicated in the *hawala* scandal, such as Madhavrao Scindia, N.D. Tiwari and Arjun Singh, accompanied her gradual political ascent.[7]

The crisis reworked internal relations within the United Front too. The governing coalition had stood firm against the Congress(I). Expectations that it would strike again shortened the Government's time horizon, however. The likelihood of an early general election encouraged its multiple leaders to identify and pursue their perceived core interests with greater vigor. Circumstances now required greater political leadership at the helm, to manage the disagreements and conflicts that were more likely to erupt, than when the governing coalition formed. Misfortune had a tendency to strike in such conditions. I.K. Gujral lacked an independent political base, making it hard to impose political bargains. The new prime minister, despite his formal constitutional authority, faced a very difficult job.

The formation of the Rashtriya Janata Dal

The prime minister began his tenure with little fanfare. Gujral made some minor changes to the Council of Ministers, dropping the Minister for Food and Civil Supplies, D.P. Yadav, at the behest of the Bihar chief minister Lalu Prasad Yadav. He also appointed two JD associates, S. Jaipal Reddy and Maqbool Dar as minister of information and broadcasting and minister of state for home, respectively, and restored the status of the four TMC ministers after their party rejoined the Government. According to one JD colleague, Gujral's "first imperative was to maintain continuity".[8] Another claimed that Deve Gowda counseled against expanding the Council of Ministers, saying it would generate a spiral of demands.[9] The move may have also constituted an attempt to retain influence. Lastly, many figures concurred, saying Gujral "was not a decisive leader temperamentally", and lacked the "ability and the mandate to prize open the coalition struck in May '96".[10]

Events tested his mettle immediately. On 5 May, the prime minister designated Bhabani Sengupta as officer on special duty in his office. Clamor broke out in the Lok Sabha the following day. Both the Congress(I) and the BJP protested against the "revisionist" views of

THE DECLINE OF THE UNITED FRONT (MAY 1997–MARCH 1998)

the new appointee regarding Indo-Pakistani relations. Sengupta had previously counseled demilitarizing the Siachen glacier and recognizing the Line of Control dividing Kashmir.[11] Yet many observers had advocated that such measures, which contravened official state policy, would help resolve the conflict. Perhaps most importantly, they extended the spirit of the Gujral doctrine. But the prime minister, having just been appointed by more powerful colleagues, failed to defend his decision. Sengupta resigned. The episode illustrated Gujral's tendency to avoid conflict.

This became clear in his response to the crisis in the JD. On 27 April, the CBI director had recommended the indictment of 56 persons in the fodder scam in Bihar, including Lalu Yadav, Minister of State for Rural Areas and Employment C.P. Verma and several other high-ranking officers.[12] The previously compliant investigative officer, whom the Supreme Court reprimanded in early April for poor conduct, now seemed reluctant to protect the Government. But Yadav refused to quit office, stating that "a mere chargesheet does not establish the guilt of the person".[13] He disparaged the central police agency, saying that a BJP-CBI nexus existed to depose him. More significantly, the beleaguered chief minister questioned the powers of the Bihar governor, A.R. Kidwai, to arrest him, contending that only the state legislative assembly could sanction it.[14]

Yadav's actions lent themselves to various interpretations. Some commentators associated "a loosening of earlier administrative protocols and a steady erosion of the institutional insulation of the decision-making process in public administration and economic management" with the rise of the OBCs, as well as "a certain nonchalance in … rampant corruption … seen as a collective entitlement in an amoral game of group equity".[15] Others underscored how political leaders increasingly sought to evade the wider institutional rules, procedures and norms that circumscribed executive power in India's democratic regime. The vertical accountability bestowed by popular electoral mandates, according to Yadav, trumped the separation of powers and checks and balances that provided horizontal accountability in modern representative democracies.[16] The prime minister assured parliament that anyone found guilty would be charged.[17] The chorus of disapproval grew in early June. The Minister of Railways, Ram Vilas Paswan of the JD, and Chandrababu Naidu, the TDP chief minister of Andhra Pradesh, implored Yadav to step down.[18] The CPI Home

Minister, Indrajit Gupta, backed their calls.[19] Yet the Bihar chief minister disregarded their entreaties, suspending detractors within his ministry and provoking street demonstrations in Bihar to shore up his position.[20] The Delhi High Court appointed a retired judge, N.C. Kochhar, to supervise the upcoming JD organizational election, which had to be completed by mid-June. The justices also requested two senior figures in the party, Madhu Dandavate and S. Jaipal Reddy, to assist in the process.[21] Yadav convinced the court to rescind its directive. But his luck turned when Governor Kidwai authorized the CBI to prosecute everyone involved in the fodder scam. On 23 June, after the CBI decided to book all suspects in the case, the prime minister requested his colleague to relinquish the chief ministership.

But Yadav resisted. On 19 June, the Supreme Court instructed Dandavate and Jaipal Reddy to hold the JD's organizational elections by 3 July. The Bihar chief minister, alleging that his opponent, Sharad Yadav, had stacked the party's national council, persuaded the apex court to allow Raghuvansh Prasad Singh to supervise the poll with Dandavate instead.[22] But R.P. Singh's attempt to reopen the list of voters renewed political misgivings. On 28 June, the Supreme Court designated Dandavate as the sole referee. Two days later, Prime Minister Gujral discharged Joginder Singh on grounds of "incompetence" and "exhibitionism", appointing R.C. Sharma as his successor.[23]

Officially, the deposed CBI director had broken protocol by speaking to the media. His earlier conduct under Deve Gowda attracted scrutiny too. Partisan motives appeared to trigger his sudden reassignment, however:

> Gujral asked him [Joginder Singh] to bail Lalu out. So [he] asked the prime minister to give the order in writing ... [Gujral] took it badly ... I don't blame him. His job was on the line.[24]

It was hard to verify what happened. But Singh had finally pursued the Bihar chief minister. His dismissal intensified charges of deliberate partisan obstruction, compelling the Patna High Court to demand an explanation.[25] Indeed, the Supreme Court would eventually issue a landmark ruling in December 1997 seeking to enhance the autonomy of the CBI, reversing the single directive principle that placed it under prime ministerial control.[26] But changing its leadership failed to save the JD, which proceeded to elect Sharad Yadav as its president. On 5 July, Lalu Yadav formed the Rashtriya Janata Dal (RJD) with the support of more than a third of the JD in parliament, indicating his willingness to accept Congress(I) support.[27]

Reactions ranged widely. The BJP asked President Sharma to impose Article 356. The Congress(I) was more equivocal. The break-up of the JD offered potential dividends. The Left demanded Yadav's resignation, asking the prime minister to discharge RJD cabinet ministers on grounds of inadequate probity. Senior JD politicians, seeking to diminish the Bihar satrap, endorsed the ultimatum.[28] Members of the regional bloc disagreed, however, saying that only Lalu Yadav had been indicted. The DMK chief minister of Tamil Nadu, M. Karunanidhi, proposed accepting external parliamentary support from the RJD until the investigation was complete. The prime minister reiterated that anyone indicted of a crime should relinquish their post.[29]

But these admonishments had little effect. On 13 July, Lalu Yadav promoted various members of the JMM and the Jharkhand Area Autonomous Council (JAAC) into his state-level ministry, and advocated statehood for Jharkhand, which he had previously opposed.[30] The RJD secured a majority of votes in the assembly amidst scenes of disorder two days later following the suspension of opposition MLAs.[31] On 25 July, after the CBI sanctioned his arrest, Lalu Yadav installed his wife Rabri Devi as the new chief minister of Bihar. His brazen attempt to maintain rule by proxy succeeded. On 28 July, the new RJD ministry secured its second confidence vote with the help of the Congress(I).[32] Indeed, despite surrendering to the police, Yadav stalled court proceedings against him until the autumn on grounds of poor health. The loss of his formidable political machine in Bihar, combined with the expulsion of Ramakrishna Hegde from the Karnataka state unit of the party in June 1996, severely truncated the reach of the JD.

The brash political ambitions that increasingly characterized the JD damaged related policy initiatives too. On 20 May, the last day of the parliamentary budget session, the Government reintroduced the Women's Bill. A majority of the parliamentary joint select committee had approved it, rejecting demands for an OBC quota. But the soon to be anointed president of the JD, Sharad Yadav, galvanized objections against its passage.[33] Disorder erupted on the floor, disgracing parliamentary norms, and forcing the Bill's withdrawal. On 3 June, the prime minister attempted to redeem the setback, inducting four women ministers of state: Renuka Chowdhury of the TDP, Jayanthi Natarajan of the TMC, and Ratnamala Savanoor and Kamala Sinha of the

JD. The impasse over the Women's Bill showed how the original socialist vision of tackling the inequalities of caste and gender simultaneously had atrophied during the 1990s, providing ammunition to critics of the politics of Mandal. Yet the failure of parties that had previously supported the measure to defend it exposed their hypocrisy as well. Indeed, it would take more than a decade before another Union government would attempt to pass the bill again.

Advancing bilateral relations with Pakistan and Nepal

Bilateral relations in the subcontinent witnessed greater movement. The Sengupta incident had exhibited the conservatism of the foreign policy apparatus and political establishment, constraining heterodox initiatives. Nonetheless, the prime minister sought to normalize Indo-Pakistani relations through a multi-track approach. In mid-May, Gujral met Prime Minister Sharif during the Maldives summit of the South Asian Association for Regional Cooperation (SAARC). The two leaders agreed to establish working groups on all outstanding bilateral issues. They also proposed to reactivate a high-level political hotline, to defuse potential military clashes, and to sign an electricity purchase agreement.[34] Significantly, neither side abandoned longstanding nuclear-related activities or their extant positions. Gujral defended the presence of Indian troops in Jammu & Kashmir to combat the insurgency. Reportedly, he also privately ordered the production of tritium, necessary to ignite thermonuclear devices.[35] Sharif called for a plebiscite on Kashmiri self-determination, inviting external intervention.[36] Pakistan advanced its covert nuclear programme, while India moved several Prithvi medium-range ballistic missiles to its western border.[37] Still, the resumption of dialogue broke the stalemate. It was partly due to the initiative of both prime ministers. The media highlighted their shared Punjabi heritage and personal *bonhomie*.[38] Gujral invested considerable political capital into securing tangible achievements. Sharif withstood immense domestic pressure from the Pakistani army, which dictated national security policy, by agreeing to explore issues beyond Kashmir.[39] Geopolitical considerations provided further incentives.

Some observers attributed the resumption of high bilateral negotiations to economic liberalization and a weak multiparty government in New Delhi, eager to strike a more accommodating posture vis-à-vis Pakistan.[40] Yet one of Gujral's first acts upon becoming prime minister was to shut down the covert organizational capabilities of the Research and Analysis Wing (RAW) in Pakistan, the principal arm of India's foreign intelligence apparatus. Given its secrecy, the media failed to report it. The decision would eventually breed controversy within the national security establishment, limiting its ability to retaliate against Pakistani-sponsored attacks on Indian soil, which gradually escalated.[41] However, terminating such clandestine activities reflected the spirit of the Gujral doctrine, which sought to reduce the scope for conflict by making unilateral concessions and recognizing the sovereignty of its neighbors. The United Front deserved much credit for resuming high-level talks with Pakistan and taking such a courageous move.

The prime minister extended concessions to other neighbors as well.[42] On 5 June, Gujral and his Nepalese counterpart, Lokendra Bahadur Chand, signed a trade pact allowing cross-border private investment in electricity generation. They also agreed to instruments of ratification of the Mahakali Treaty, which established principles for water sharing along the river; strengthened the 1996 Treaty of Trade, which granted duty-free access to all Nepalese exports on a non-reciprocal basis; and agreed to revisit the 1950 Indo-Nepal Treaty of Peace and Friendship, given its bias towards India.[43] Consensus on other issues, such as the Kalapani border dispute, proved more elusive. Several mandarins attributed these accomplishments to the diplomatic machinery: Gujral "merely ratified" proposals that were "in the pipeline".[44] But others stated that while "another [Indian] government [might have found it difficult] to say no", the prime minister's special interest and ability to persuade his ministerial colleagues made an important difference.[45] In short, Gujral took advantage of opportunities that presented themselves, exploiting his room for maneuver to the fullest possible extent.

Indeed, the second round of foreign secretary-level negotiations between India and Pakistan in late June established joint working groups on eight outstanding issues, including Jammu & Kashmir; terrorism and drug trafficking; economic cooperation and promotion of friendly exchanges in various fields; and boundary-related and resource sharing disputes regarding the Siachen glacier, Sir Creek and Wullar Barrage.[46] Additional gestures included the relaxation of visa requirements for selected Pakistani groups and release of approximately 400 fishermen previously detained for crossing maritime boundaries.[47] Both sides sought to conciliate hawkish quarters through political grandstanding.[48] Yet the basis of dialogue had been established:

The second round [concerned] working on mechanisms and modalities ... Our stand was, "we urge you to a broader relationship... Let's move from easier to harder issues".[49]

The Government failed to make headway in the third round, however. The sticking point, unsurprisingly, was Kashmir. The Pakistani delegation wanted the disputed region to receive separate treatment while their Indian counterpart sought to integrate its discussion with other issues. The fiftieth anniversary of Partition might have provided an opportunity to address its legacy with greater introspection, equanimity and imagination. Official state nationalism proved stronger in both countries, alas. By the end of August, heavy cross-border artillery shelling resumed along the Line of Control, causing many casualties. According to a senior government official,

Pakistan rejected our stand [to pursue multi-track negotiations]. So they wanted to create an international scandal ... they negotiated while encouraging cross-border terrorism.[50]

Nevertheless, both sides agreed to hold further talks in Islamabad by the end of the year.[51] Observers credited Gujral for pushing to normalize bilateral relations and resisting pressures from more hawkish quarters in the atomic energy establishment.[52] Unfortunately, domestic political events in India would soon undercut his position, thwarting further progress.

The deceleration of economic reform

The economy was the last realm to witness concerted policy action. In early June, the prime minister had reorganized the Council of Ministers, assigning berths to recently inducted ministers, shuffling portfolios and introducing new faces. The restructuring sought to boost critical sectors, especially regarding infrastructure, suffering relative neglect.[53] The fiscal calculus of the 1997–98 Union budget banked on resurgent industrial growth. Yet quarterly economic indicators had revealed a revenue shortfall, despite the RBI cutting interest rates in April to offset shortfalls in investment, industrial production and capital market activity.[54] In mid-June, the finance minister launched the Voluntary Income Disclosure Scheme (VIDS), providing a six month amnesty to individuals and companies willing to pay arrears at rates of 30 and 35 per cent, respectively, waiving interest charges.[55] Gujral constituted an export promotion board to lessen delays caused by inter-ministerial wrangling.⁵⁶ And in July the Government granted complete financial and operational autonomy to nine PSUs in the energy sector capable of competing in the global market.⁵⁷

Several factors shaped its disinvestment strategy. A senior government official claimed that

> disinvestment had different meanings in various contexts ... The UF was divided over several questions. What percentage? Which market? When? ... And do we take the Disinvestment Commission Report as final or as a recommendation?⁵⁸

Senior CPI politicians opposed privatization on principle, but also for practical reasons:

> We were opposed the disinvestment of profitable PSUs ... those that were sick ... their viability must be examined ... a set of options must be given.⁵⁹

> We are opposed to it. It amounts to privatization ... as for sick units, one or two can be considered, but who is going to buy them?⁶⁰

Others cited poor market conditions:

> The Government couldn't get its act together ... the markets were low.⁶¹

> You must fetch an adequate price ... it is not possible [prudent] to sell profit-making units.⁶²

Given these factors, even proponents of liberalization urged caution:

> We wanted to avoid scams ... it cannot be done in a stroke ... The public sector is a sacred cow. There are many vested interests ... Yet you can't throw money into a bottomless pit ... so we were cautious.⁶³

Accordingly, the Council of Ministers authorized the privatization of only 3 of the 48 PSUs recommended by the Disinvestment Commission.

The reluctance to tackle the issue foreshadowed difficulties regarding other second-stage reforms. In early July, the prime minister endorsed raising the price of several petroleum-based products to tackle the rising oil pool deficit. But pressure by the Left prevented any decision.⁶⁴ In early August, the finance minister withdrew the Insurance Regulatory Authority Bill, which sought to allow private investment in the sector. His retreat was partly due to bad luck: many Congress(I) MPs were at a plenary session in Calcutta.⁶⁵ But it was largely due to opposition from the Left, which objected entirely, and the BJP, which protested the entry of foreign private actors.⁶⁶ According to a senior bureaucrat,

The BJP [had] privately agreed to the bill ... Jaswant Singh [a BJP politician] met Chidambaram [and indicated his support] ... yet the party rejected it on the floor of the house ... and accused Gujral of selling out the country. It was sheer populism.[67]

Purportedly, Chidambaram had gambled on the BJP to outflank the Left, and lost.

Nonetheless, he pressed ahead. On 25 August, the Council of Ministers approved the Foreign Exchange Management Act (FEMA), which sanctioned transactions on the capital account over a staggered period.[68] Financial market turmoil in South-East Asia eventually hindered the initiative, demonstrating the prudence of capital controls, as the Left had consistently argued. Yet the introduction of FEMA constituted another incremental step towards greater external liberalization. And on 1 September, the Government tackled the oil pool deficit, which had surpassed Rs. 18,000 crore, by raising the prices of liquefied natural gas, diesel and petrol between 5 and 20 per cent. The Government vowed to abolish the administered price mechanism, which cross-subsidized various products, within two years.[69] Proponents criticized it for not acting sooner. Cabinet ministers and government bureaucrats involved in the decision defended the delay, however:

We were for rational economic management ... [but] extensive consultation is part of the democratic exercise ... we cannot bulldoze our partners.[70]

The CPI(M) gradually reconciled themselves [to the decision] since we explained it to them.[71]

Forging political agreement in a diverse multiparty government was inevitably time-consuming. But the decision exacerbated internal differences. The CPI(M) General Secretary H.S. Surjeet castigated the Government for not implementing the "pro-people" policies of the CMP, inciting fellow travelers to thwart further "anti-people" measures.[72]

The growing political rift within the United Front erupted over the Fifth Central Pay Commission Report, the most significant economic policy decision of Gujral's prime ministership. On 18 July, the Council of Ministers had accepted the recommendations of the empowerment committee of secretaries, which advocated substantial pay raises for elite public servants. But the latter rejected the majority of proposals to restructure administrative services—such as instituting a six-day week and increasing the age of retirement to 65—and to cut the number of central government employees by 30 per cent.[73] In early August, trade unions representing lower-tier employees campaigned for greater parity, forcing the Government to postpone the Report's implementation. A senior labor official claimed that

> [t]he Fifth Pay Commission's scope was beyond its terms of reference since it included optimizing the size of government. But it had no mandate to do that ... an administrative reforms commission is required for structural changes to take place.[74]

On 5 September, the country's major public sector unions announced they would launch an indefinite strike on 24 September. Various senior politicians attended the rallies. The Left had always held that its parties could simultaneously pursue "partial participation" in coalition governments and foment mass action.[75] Yet the CPI had joined the Council of Ministers. The prime minister appointed Indrajit Gupta to lead the Government in negotiations with the joint consultative machinery. It was a costly decision. On 11 September, the ministerial delegation acceded to the unions' demands. Amongst other things, it agreed to increase across-the-board basic pay by 40 per cent, rejecting possible disinvestment and the abolition of 3.5 lakh vacant posts. The estimated cost of the settlement was Rs. 18,350 crore, approximately 50 per cent greater than originally budgeted, forcing the Centre to immediately to raise various taxes.[76] The financial burden would increase over the long run.

Several government officials involved in the negotiations exonerated the United Front, claiming that any Union administration would have succumbed to such concerted pressure from the unions.[77] A senior Congress(I) official agreed: "It was a total disaster but had nothing to do with the UF ... [it was] a conspiracy of the entire political class ... competitive populism".[78] The terms of reference of the Report, which exceeded its customary purview, created further difficulties. Specific political factors exerted additional pressure, however. First, by accepting complaints against the original recommendations in February 1997, the Government encouraged the staff-side to press further claims:

> The initial reaction was general acceptance without too much change, leading to five or six meetings ... but some members in the Cabinet became overenthusiastic. They wanted to give special recognition to various groups: Gupta to the police, Mulayam [Singh Yadav] to the army and so on ... [which] signaled pliability. So the staff side [of the Joint Consultative Machinery] felt they could make a better deal ... the whole damn edifice is interlocked. One change and

it all cracks ... Giving in to their [staff-side] demands without enforcing the reforms ... was a major mistake.[79]

Acquiescing to these early demands generated a strong domino effect. Second, Gujral selected Gupta as head of the ministerial delegation, which comprised several leading communists and socialists. Reportedly, the former ordered the latter to avoid a strike "at any cost":

Gujral's calculations of choosing Gupta ... to compromise him ... backfired.[80]

Gujral [was] an amiable person ... eminently reasonable. But the cost [of these qualities in] a prime minister is terrible ... small-time temporary prime ministers do not have the psychological guts [sic] to face an all-India strike.[81]

The prime minister's injunction constrained the Government's hand. But delegating negotiating authority to the home minister in a bid to restrain the unions betrayed poor judgment.

Finally, many government officials blamed Gupta, who encouraged the unions to agitate; Chidambaram, who purportedly left the negotiations after registering his disapproval; and the Group of Ministers as a whole, which conceded numerous demands without extracting any concessions in return:

Gupta was still psychologically in the Opposition ... He was trapped by his ideological lineage ... and high integrity quotient. [He] obliquely attacked Chidambaram, who irresponsibly stopped attending meetings, in front of the unions.[82]

Chidambaram was completely isolated ... so he disassociated [himself] from the Cabinet decision.[83]

It was a theatre of the absurd ... a model of how not to conduct the negotiations. The accountability measures and non-pay related matters were dealt with first ... and simply shelved.[84]

Facing cross-party trade union opposition over a controversial report would have been difficult under any circumstances. G.K. Moopanar reportedly ordered Chidambaram to abandon the negotiations to avoid putting the TMC into an awkward position.[85] But psychological miscalculations, fear of opposition and the abdication of collective political responsibility worsened the outcome. The implementation of the Fifth Pay Commission Report, which unduly rewarded public sector employees, represented an important political victory for communists and socialists in the United Front. But its "self-indulgently generous" terms placed an enormous burden on central government finances as well as the states, drastically constraining much needed investments in basic social capabilities and physical infrastructure.[86] The increasingly bleak prospects of the governing coalition emboldened its constituents to pursue their perceived interests with greater alacrity. Their unity seemed to be reaching its nadir.

The vicissitudes of Centre-state relations

The installation of the Rabri Devi ministry in Bihar reignited political debate regarding the functioning and autonomy of state legislative assemblies. On 17 June, the ISC standing committee had accepted the Finance Commission's proposal to allocate 29 per cent of the central tax revenues to the states.[87] The measure was in keeping with steps the United Front had taken the previous summer, such as transferring many centrally sponsored schemes and granting more autonomy over developmental planning to the states. The standing committee also addressed Article 356. According to a senior government official, most of its members supported its continuation, saying that "without [Article] 356 you [the Centre] couldn't have saved Jammu & Kashmir or managed the Northeast".[88] The justification, in light of the flawed assembly elections held in Kashmir the previous year, was telling. However, the quorum also discussed how to prevent its misuse:

The problem of law and order [was] not a sufficient pretext ... [Article] 356 [could] only be imposed if the state government is colluding with internal insecurity [sic].[89]

The assembled chief ministers agreed several provisos, such as requiring the Centre to notify the state of its intentions after receiving a full report from the governor, allowing the state seven days to reply, and requiring the Centre to muster the approval of two-thirds of the Lok Sabha before dismissing a state-level ministry.[90] But their 8 July meeting ended in deadlock:

There was debate on several issues. The Lok Sabha might be out of session ... and sometimes the immediacy of the situation might demand it.[91]

The prime minister, unable to resolve these differences at the full ISC meeting on 17 July, shelved the issue until its next scheduled meeting in November.[92] Given its composition, the desire of the United Front to promote greater state autonomy was unsurprising.

In any event, political developments elsewhere soon revived the debate. Tensions had surfaced within the ruling BJP-BSP coalition in

Uttar Pradesh. In early September, Mayawati reluctantly gave up her chief ministership after completing six months in office, as previously agreed. Kalyan Singh took over on 21 September.[93] However, he reorganized the state ministry, reneging a vow to maintain its portfolio distribution. Singh had opposed expanding the Ambedkar Village Scheme and implementing the SC/ST (Prevention of Atrocities) Act on grounds of *samajik samarasata* (social integration).[94] On 18 October, BSP ministers boycotted a cabinet meeting, which reversed many decisions that Mayawati had taken.[95] The party revoked its support for the Singh ministry two days later, requesting Governor Bhandari to dissolve the assembly. Bhandari asked Singh to test his strength on the floor, triggering defections from the BSP, JD and especially the Congress(I).[96] On 21 October, the newly inducted BJP chief minister won a legislative majority, marred by reports of violence.[97] The United Front recommended President's rule.

The collapse of the BJP-BSP government in Uttar Pradesh failed to surprise most observers. The BSP had formed and abandoned tactical alliances with great ease in preceding years. Shifts in relations of power, real or perceived, usually led to their demise. Kalyan Singh had opposed the power-sharing arrangement from the start, accepting its terms simply to block the SP. But Mayawati had used the chief ministership to channel extensive patronage to her *Dalit* base: constructing theme parks, expanding village development schemes and erecting prominent statues of B.R. Ambedkar. She had also transferred many state-level officers from other castes and tried to enforce laws protecting *Dalits* and *Adivasis* from upper-caste violence. Reportedly, she authorized most of these decisions without seeking cabinet approval, exacerbating latent conflicts with the BJP.[98] The willingness of the BSP to make expedient alliances over time demonstrated its belief that maneuvers to gain short-term advantage trumped long-term agreements, and that wielding government power enhanced its electoral prospects.

Nonetheless, the decision to impose Article 356 provoked dissension within the Council of Ministers.

> It was a very long Cabinet meeting. There was a lot of disagreement … recommending it was a fraud on the Constitution.[99]

> The proceedings [were] bringing shame to everyone … [the] Cabinet was unanimous after a prolonged session … but Gujral, Indrajit Gupta and [Murasoli] Maran disagreed.[100]

THE DECLINE OF THE UNITED FRONT (MAY 1997–MARCH 1998)

Senior bureaucrats maintained that Mulayam Singh Yadav, whose political interests lay overwhelmingly in Uttar Pradesh, swayed deliberations.

Gujral was very indecisive... He had avoided taking decisions in national politics. He didn't know national politics. And he had no experience of being a state [level] politician... He was mortally afraid of Lalu [Yadav] and Mulayam [Yadav]. So he never said what he wanted in Cabinet ... sat around and waited for a consensus to happen ... would let the dominant man take over.[101]

Unlike the CPI, the CPI(M) was a stalwart ally of the SP and backed the decision, demonstrating its willingness to exploit discretionary executive powers at the Centre to block Hindu nationalists in the states.

The newly inducted President, Kocheril Raman Narayanan, surprised everyone, however, by asking the Government to reconsider its decision. A professional diplomat who had risen to the vice presidency of the republic, Narayanan had been elected with overwhelming cross-party support in July 1997.[102] He would have had to abide if the Government stuck to its position. According to a 1976 constitutional amendment, the President could not refuse such a request twice.[103] But the Council of Ministers demurred. On 22 October, the Gujral ministry rescinded its recommendation.[104]

The reversal startled many quarters. Few expected President Narayanan to exercise his prerogative so firmly. The emergence of national coalition governments enabled the Election Commission, Supreme Court and Presidency—hitherto more passive regulatory political institutions—to check perceived abuses of power by the executive and legislature. Yet Narayanan's intervention demonstrated that rules could only shape the capacity and incentive of these institutions to exercise their mandated constitutional authority. Rather, particular office-holders had to judge how to use the powers at their disposal, which ultimately determined their efficacy.

The Congress(I), labeling the decision "unfortunate", indicated "no immediate move" to topple the Government.[105] Mulayam Yadav declared his readiness to ally with any party, including the BSP, against the BJP. Yet his sudden political amenability came too late. In late October, Kalyan Singh inducted 93 members into his cabinet, making it the largest ministry ever formed in the state, and ordered high-level inquiries into Mayawati's decision to build the Ambedkar Udyan Complex on land belonging to the Lucknow Development Authority.[106]

Neither a pliant governor nor central rule could stop the BJP from crafting alliances through political horse-trading. The failure to defuse the animosity between Mulayam Yadav and Mayawati, and address the conflicts of interest dividing their parties' organizational ranks and social bases, had worked to the detriment of the third force as a whole.

The fall of the United Front government

The setback in Uttar Pradesh occurred amidst increasing tensions within the United Front. In mid-October, Chidambaram had criticized the Agriculture Minister, Chaturanan Mishra of the CPI, for not spending his budgetary allocation for the first quarter of the fiscal year. Mishra retaliated, accusing the Finance Ministry of promoting anti-farmer policies and limiting agricultural credit, and of vetoing several of his earlier proposals, including a pilot scheme for crop insurance. Indeed, he threatened to resign. The prime minister persuaded Mishra otherwise.[107] But the incident revealed growing resentment between Chidambaram and the Left. In late October, the CPI(M) criticized the Minister for Information & Broadcasting, S. Jaipal Reddy, for using an ordinance to amend the Broadcasting Corporation of India (Prasar Bharati) Bill, originally introduced by the National Front. Reddy had notified the law in late September, which had languished for years under the Congress(I), to widespread approval. But his colleagues on the Left believed the amendments, which limited parliamentary oversight of All India Radio and Doordarshan and state restrictions on private advertising, threatened the public mission of the original act.[108] The party also complained that Gujral had neglected domestic policy issues—even though he had dispatched Chidambaram to recent ASEAN meetings and Murasoli Maran to meet the new Labour government in Britain—leading the prime minister to cancel his visit to the G-15 summit in Malaysia.[109] Finally, in early November, Karunanidhi threatened to provide "outside support" to the Government because of it's allegedly "scant regard for Tamil Nadu's views and requests on the Cauvery issue".[110] The cause of his fulmination was clear. The Centre had framed a draft scheme in April 1997 establishing a regulatory body, the Cauvery River Water Authority, to implement the 1991 interim award. However, an all-party conclave convened by the JD Chief Minister J.H. Patel stated that his ministry would await the final Tribunal award given the scheme's bias against Karnataka. Indeed, it would take fifteen years before another chief ministerial meeting would occur.[111] These public spats exposed the growing political vulnerability of the United Front.

An external threat stiffened its unity, however. On 9 November, the media leaked excerpts from the Jain Commission, constituted in 1991 under the direction of Justice Milind Chand Jain, to investigate Rajiv Gandhi's assassination.[112] Its interim findings caused a stir. On the one hand, the Report indicted the DMK for tacitly supporting the LTTE and the National Front government of V.P. Singh for removing security protection for the former prime minister. On the other, it exonerated the previous Congress(I) governments of Indira and Rajiv Gandhi as well as the prior AIADMK state government of M.G. Ramachandran on grounds that all three administrations had engaged the LTTE "for self-defense purposes" prior to its conflict with the IPKF. Karunanidhi derided the interim findings as "old propaganda".[113] He had good reason. Rajiv Gandhi had consistently encouraged the DMK to maintain contact with the LTTE. The alleged difference between the parties was a hollow distinction.

Reactions within the Congress(I) differed. The principal aim of the party, formally adopted at its Calcutta plenary session in August 1997, was to form a "viable and stable one-party government".[114] Significantly, Kesri had rejected the idea that "coalitions are here to stay", contending that "the Congress itself is the most successful coalition".[115] Yet he also reportedly concurred with Rao and Pawar that joining the Government was in the party's long-term interest. In mid-November, Kesri met Gujral, confirming the Government's stability. The prime minister promised to table the interim Commission report at the start of the winter parliamentary session on 19 November. An understanding had been reached.

Senior Congressmen close to the Nehru-Gandhi family took another view, however. On 11 November, Jitendra Prasad asked the Government to expedite the Report. Three days later Arjun Singh attacked Kesri for besmirching the memory of their slain former leader. Other dynastic followers, such as K. Karunakaran, Pranab Mukherjee and K. Vijayabhaskara Reddy, demanded the Government expel the DMK. Reports later surfaced that Prasad and Singh had hired a team of lawyers to link the DMK to the LTTE. "Blissfully unaware" of these machinations, Kesri gave Gujral time to respond.[116] Yet pressure mounted within the CPP. On 16 November, the Congress(I) president met Sonia Gandhi. It was unclear what transpired. But Kesri and Pawar quickly changed their tune. On 19 November, speaking at a function commemorating Indira Gandhi's birth anniversary, Kesri urged Sonia to lead the party in order to "salvage the situation".[117] The CWC sent a letter to Gujral commanding the expulsion of the DMK.

The directive suggested a cynical ploy: to exploit the tragedy of Rajiv Gandhi's assassination to create popular support and divide the United Front. Several members of the party high command also believed that electoral alliances with the SP in Uttar Pradesh and RJD in Bihar, combined with discontent towards non-Congress(I) governments in Rajasthan, Maharashtra and Karnataka, would help them in an early general election. Kesri apparently resisted. But the dynastic political cabal outmaneuvered him, creating a groundswell of pressure in the Congress(I).

Arjun Singh fancied himself as champion of Rajiv and Sonia ... though there was no love lost during Rajiv's life ... he pressured Kesri into withdrawing support.[118]

[The Report] became an instrument of power struggle in the party ... Arjun Singh said, "We will come to power" ... as did most Congressmen ... [who] felt they were in striking distance.[119]

The ultimatum caused some disagreement in the United Front. Mulayam Singh Yadav was reluctant to support the DMK at the Government's expense. State-level considerations—an expected alliance between the SP and the Congress(I) in Uttar Pradesh—informed his stance. G.K. Moopanar also remained conspicuously silent. The TMC had left the Congress(I) in opposition to Rao and its tie-up with the AIADMK. The prospect of Sonia Gandhi at the helm created a more awkward position for Moopanar, however, given the dynastic character of the party. The Congress(I) looked poised to exploit, as it had vis-à-vis Moraji Desai in 1979 and V.P. Singh in 1990, the internal conflicts of the latest Janata dispensation.

But the United Front bucked the past. On 20 November, the Government tabled a formal response to the Jain Commission Report, dismissing its allegations.[120] Realizing its folly, the Congress(I) disrupted parliamentary scrutiny of the Report. Its anti-DMK brigade suggested various compromises, such as appointing a high judicial bench or an eminent public figure to review the Report and make a binding recommendation. It even mooted the possibility of changing the Council of Ministers and naming either Moopanar or Mulayam

THE DECLINE OF THE UNITED FRONT (MAY 1997–MARCH 1998)

Yadav as prime minister. Yet the United Front rejected every proposal, belying the view of many Congressmen that it would cave in.[121] On 24 November, Gujral expressed his willingness to discuss other possibilities. The Speaker of the Lok Sabha, P.A. Sangma, suggested the formation of a national government.[122] Forty first-time MPs beseeched President Narayanan not to dissolve parliament.[123] The fiasco impelled the Congress(I), which now simply asked the governing coalition to remove cabinet ministers from the DMK until their party was absolved of wrongdoing, to recover the situation.[124] But the Steering Committee rebuffed the suggestion. "The Congress are like the Bourbons," claimed a senior JD leader, "they learn nothing, forget nothing".[125]

On 28 November, the party withdrew its support to the seven-month-old Gujral ministry, instigating a second political crisis. The prime minister resigned, yet refrained from recommending the dissolution of parliament, in case a compromise might be reached. The BJP and its allies, as well as the Congress(I), registered their respective claims to rule. Yet their endeavors proved futile. Not a single member of the United Front, despite their well-known differences and recent political conflicts, offered to support either bid. On 4 December, President Narayanan dissolved the eleventh Lok Sabha, compelling the country to face its twelfth general election.

India's twelfth general election

The capacity of the United Front to withstand the blunderbuss tactics of the Congress(I) was impressive. Few expected it. Chandrababu Naidu, the convener of its Steering Committee, proclaimed:

> Right from the beginning, the thirteen-party alliance has been united despite heavy odds. It is true that there are differences in the UF about how we look at certain issues. But we never allowed our differences to unsettle our basic relationship. Ours is a real coalition.[126]

Naidu clearly exaggerated. Contrary to declarations that it was an "electorally anomalous formation", however, the governing coalition had stuck together against tremendous odds.[127] The TDP chief predicted that it would gain a parliamentary majority in the India's twelfth general election.

Nonetheless, the alliance confronted three steep challenges.[128] Its leadership, a perennial concern of the third force, was the first. Formally, the prime minister headed the coalition. But his refusal to dismiss two RJD ministers from the Government, at the behest of his colleagues, provoked much displeasure. The Steering Committee eventually forced their expulsion on 3 January, showing that Gujral could not lead the United Front into battle. In fact, lacking an independent political base, he approached the SAD to support a bid to reclaim his old Jalandhar constituency in Punjab. During his tenure, Gujral had committed the Centre to sharing the cost that Punjab had incurred fighting Sikh extremists in the late 1980s and early 1990s. His overture caused much dismay: the SAD was still aligned with the BJP. But the travails of Lalu Yadav, Gujral's erstwhile patron, narrowed his options. The instinct for political self-preservation trumped wider solidarities, leaving a vacuum of leadership.

The second obstacle facing the United Front was the JD. On 15 December, Naveen Patnaik broke up its 43-member unit in Orissa, which he had led following the death of his father, Biju Patnaik, in April 1997. The EC recognized the 29-member Biju Janata Dal (BJD). Various factors, all regionally inflected, caused the split.[129] Organizationally, the JD had suffered from deep internal rifts in Orissa, which the patriarch's overwhelming presence had historically contained. Socially, the party represented a diverse cross-section of groups, privileging the aspiring intermediate Khandayats, and winning the support of Brahmins and *karans*. But it encompassed few minority communities, in contrast to the JD in Karnataka and especially in Bihar. Ideologically, Biju Patnaik was an ardent modernizer, welcoming economic liberalization and courting foreign investment to develop basic utilities, mineral resources and heavy industries. Indeed, he had lambasted V.P. Singh for notifying the Mandal Commission Report without specifying economic criteria, and had only reserved 12 per cent of government posts for OBCs in Orissa in 1993. Lastly, personal equations had soured at the top. The younger Patnaik alleged that his father, who reportedly compared his colleagues to "a cluster of lobsters fighting against each other", had been marginalized after the United Front coalesced.[130] In short, the demise of the longstanding chief minister allowed his heir to form a distinct political formation. The tripartite foundations of the JD had imploded during the Government's tenure: first with the exit of Ramakrishna Hegde in Karnataka in June 1996, followed by Lalu Yadav in Bihar one year later. The break-up in Orissa dealt a massive blow, questioning the ability of the party to survive.

THE DECLINE OF THE UNITED FRONT (MAY 1997–MARCH 1998)

The third challenge to the United Front was formulating a common electoral strategy amidst many conflicting pressures. First, its members confronted ambiguous electoral incentives. The FPTP electoral system encouraged them to consolidate their respective vote shares vis-à-vis the Congress(I) and the BJP. However, the logic of plurality rule in tight multi-cornered fights reduced the threshold of winning, encouraging parties to contest alone if they perceived an advantage in doing so. Second, the federal party system produced diverse state-level pressures. Many constituents faced dissimilar rivals in their respective home states. The Congress(I) remained the principal adversary of the TDP and AGP in Andhra Pradesh and Assam, respectively, and of the Left in West Bengal, Kerala and Tripura. Its reckless decision to attack the DMK, and topple two JD prime ministers within eight months, roused considerable enmity. However, both the SP and the TMC expressed greater willingness to cooperate with the Congress(I) in Maharashtra and Tamil Nadu, respectively. The emergence of a BJP-led ministry in Uttar Pradesh, and the entry of Sonia Gandhi into national politics, had altered their incentives. Moreover, several partners—primarily the SP, what remained of the JD, and the parties of the Left—competed in more than one state. Maximizing their chances of success required seat adjustments across the third force. Yet such negotiations inevitably provoked disagreements over relative electoral strength.

In the end, the United Front resolved its differences in only a handful of states. The coalition released a joint policy declaration in the middle of February, stressing the dangers of militant Hindu nationalism and the need for improved federal governance, alongside the separate political manifestoes of its constituents. Greatest cooperation occurred in West Bengal, Kerala and Tripura, where the Left maintained its traditional alliance; in Andhra Pradesh, where the TDP offered a few seats to the CPI, CPI(M) and JD; and in Assam, where the AGP contested with the Left and JD. Crucially, a single party or bloc enjoyed unquestioned primacy in all these states, lessening rival claims.

Greater political disagreement arose in Tamil Nadu. The DMK and TMC divided the majority of seats between themselves, rebuffing the demands of the JD, CPI(M) and CPI, which collectively bid for ten. The former two parties also rejected entreaties by smaller regional forces—the Pattali Makkal Katchi (PMK), Puthiya Tamilagam and the Thirunavukkarasu faction of the AIADMK—that had partly sought to revive Dravidian ideology.[131] The DMK and TMC highlighted the weak organizational presence of the JD in Tamil Nadu, a reasonable claim. Its leaders spurned the Left on more contentious grounds, however. The TMC recalled how the CPI(M) blocked Moopanar from the prime ministership in April 1997. The DMK opposed the CPI(M) and JD for supporting its erstwhile faction, V. Gopalaswamy's MDMK, in the eleventh general election. Ultimately the TMC and DMK allotted just one seat to the CPI.

The pursuit of partisan electoral advantage inflicted greatest damage elsewhere, however, including several decisive political arenas. In Uttar Pradesh, the SP refused to support the JD in sixteen constituencies the latter had fought in 1996 because it had only won two, and rejected the CPI's bid as well. In Bihar, the rump JD competed against the CPI, CPI(M) and SP, as well the Communist Party of India (Marxist-Leninist) (CPI-ML), after spurning their proposed seat adjustments. Hence the extant allies separately contested the Jan Morcha, an expedient multiparty alliance led by their former colleague, Lalu Yadav. The JD contested virtually alone in Karnataka, while in Maharashtra the SP joined the Congress(I) and two factions of the Republican Party of India (RPI), sidelining the JD and the Left. Even in Rajasthan, where they were independently weak, the constituents of the United Front could not agree to share different seats. Finally, the MGP and UGDP both pursued local supremacy in Goa. In sum, an exceedingly local calculus of political survival undermined the cohesion of the recently dethroned alliance. The most distinctive manifestation of the third force in modern Indian democracy had rapidly drifted apart.

The Congress(I) faced serious obstacles as well.[132] Formally, Sitaram Kesri remained its leader. Mounting personal disaffection compelled several MPs, including former Union ministers, to defect however. Far more damaging, Kesri divested the authority of Mamata Banerjee to determine whom the Congress(I) should field in West Bengal. The populist maverick reacted by forming the All-India Trinamool Congress (AITC) in mid-December. It was the latest example of the Congress(I) high command attempting to micro-manage its state units from afar. Kesri's power rapidly ebbed away. On 28 December, it received a further jolt: Sonia Gandhi declared she would campaign for the party. Although a novice, her decision galvanized the rank-and-file. Reportedly, Sonia Gandhi joined the campaign to protect her dynastic inheritance. The interim Jain Commission Report encouraged her intervention. Her decision to campaign marked a significant political moment for the Congress(I).

THE DECLINE OF THE UNITED FRONT (MAY 1997–MARCH 1998)

But the reappearance of the Nehru-Gandhi dynasty could not resuscitate the party's immediate prospects on its own. Electoral alliances and seat adjustments decisively shaped political fortunes in a deeply regionalized federal party system. Yet the Congress(I) struck few. This was partly due to structural reasons. The party was the main electoral rival of many state-based formations across the country. Hence it was difficult to strike electoral agreements without making significant political concessions. But inept political management, its ham-fisted conduct over the Jain Commission and an unwillingness to defer to increasingly assertive forces in the regions also robbed the Congress(I) of potential electoral allies. The party largely maintained traditional electoral alliances in West Bengal, Kerala and most of the Northeast. It agreed limited seat adjustments with the BKKP in Uttar Pradesh, and had a tacit understanding with the SP in Uttar Pradesh and in Maharashtra alongside factions of the RPI. Finally, the Congress(I) collaborated with the United Communist Party of India (UCPI) in Tamil Nadu. However, its limited attempts to establish or renew pacts in other arenas either failed or broke down. In Tamil Nadu, the AIADMK and its allies joined forces with the BJP, a massive setback that reportedly surprised Kesri, wounding his standing. The embattled Congress(I) president wooed the RJP of Shankar Singh Vaghela in Gujarat, and the BSP in Madhya Pradesh, Punjab and Rajasthan, and to a lesser extent in Uttar Pradesh. Yet his efforts bore little fruit. Indeed, although the party allied with the Jan Morcha in Bihar, the RJD chief Lalu Yadav offered it relatively few seats.

In sum, the Congress(I) had much to lose. The party grossly misjudged its ability to split the United Front. The entry of Sonia Gandhi arrested its declining prospects. Yet winning national power required shrewd coalition building—a task that exceeded the willingness, capacities and grasp of its high command. The Congress(I) remained hostage to an earlier conception of modern Indian democracy, unwilling to accept the rise of parties, interests and agendas that undermined its presumption to rule. It was the start of a long political winter.

The BJP faced a tough strategic dilemma. On the one hand, it had to stitch together a broad electoral alliance in order to maximize its chances at the polls. Otherwise, the party would replay its debacle after the eleventh general election. Unlike the Congress(I), the BJP was not the principal electoral rival of many parties outside the Hindi heartland, enabling a diverse multi-party coalition in theory. The party could only grasp this possibility, however, by devising a credible political strategy. The prospect of national power might entice a few state-based formations. But many parties opposed its militant Hindu nationalism. The BJP could only win new allies by moderating its official agenda. At the same time, it could not alienate the leaders and foot-soldiers of the *sangh parivar*. Moreover, hardliners within the party constantly pressured its leadership to fulfill their demands. In short, the advent of national coalition politics deepened the oscillation between strategies of moderation and militancy that had historically marked Hindu nationalist politics.[133]

In the end, the BJP exploited the disarray of the United Front and the Congress(I).[134] First, the party renewed existing alliances with the SAP in Bihar and Uttar Pradesh, HVP in Haryana, Shiv Sena in Maharashtra and SAD in Punjab. Next, it struck an agreement with the AIADMK in Tamil Nadu in mid-December, pre-empting the overtures of the Congress(I). The historic openness of the AIADMK to Sanskritic culture and Hindu revivalism intimated possible affinities.[135] The media paid closer attention to charges against Jayalalitha, however, for purportedly acquiring government land below market prices in the early 1990s. The tie-up provoked speculation that the former chief minister had extracted a pledge from the BJP to frustrate legal proceedings against her if the alliance captured national power. It also permitted a wider anti-DMK alliance with the Janata Party, Tamilaga Rajiv Congress (TRC), PMK and MDMK.

Lastly, the BJP pursued various arrangements with other state-level parties. By the middle of January the party had fashioned a pact with the BJD in Orissa. It struck less momentous, yet still noteworthy, alliances with the Jantantrik faction of the BSP and Loktantrik Congress in Uttar Pradesh, and the TDP (NTR) in Andhra Pradesh. Finally, the BJP courted Lok Shakti in Karnataka, which had been launched by Ramakrishna Hegde, and the AITC of Mamata Banerjee in West Bengal. Significantly, all of these state-level formations opportunistically joined hands with the BJP to avenge parties to which they formally belonged.

Nevertheless, ideological tensions persisted. Mamata Banerjee, fearing the loss of key Muslim voters, refused to acknowledge her pact with the BJP. The chiefs of the AIADMK, Lok Shakti and BJD also emphasized their secular credentials. National power was a necessary, but insufficient, political motivation. Ultimately, pragmatism triumphed.

THE DECLINE OF THE UNITED FRONT (MAY 1997–MARCH 1998)

Desperate for support, the BJP focused its campaign on Atal Bihari Vajpayee, touting his leadership and the need for national political stability. Most importantly, the party agreed to shelve its most controversial measures: the abrogation of Article 370 regarding Jammu & Kashmir, establishment of a uniform civil code and construction of a Ram temple in Ayodhya. The decision allowed the BJP to buttress its position in several traditional strongholds in the north and west, while enabling the party to enter more difficult arenas, most notably in Karnataka, Andhra Pradesh and Tamil Nadu in the south. It was a strategic political judgment decisively shaped—but not determined—by the incentives generated by plurality-rule elections and parliamentary cabinet government in a deeply regionalized federal party system.

The BJP captures power

India's twelfth general election in February–March 1998 produced yet another deeply fractured verdict. Thirty-nine parties and six independents secured a presence in parliament.[136] All seven national parties, which together won 384 of a possible 543 seats, survived the ballot. Twenty state parties, of the thirty that contested, attained a further 100 seats. And eleven registered parties, many of which comprised recent breakaway factions, prevailed in forty-nine parliamentary constituencies. The dispersion of votes across a wider range of parties, surpassing the fragmentation of the eleventh general election, underscored the importance of building wide electoral coalitions for national power.

Indeed, alliances made all the difference. The BJP substantially increased its national vote share from 20.3 per cent in 1996 to 25.5 per cent. Yet its tally of seats, now 179, increased only marginally.[137] The party largely retained its presence in traditional strongholds in the north and west. However, the BJP significantly expanded its reach in the south and east, thanks to its coalition strategy, which paid striking dividends. The party's electoral allies, which collectively won 72 seats, captured 11.7 per cent of the all-India vote. Significantly, its newfound partners seized the majority of these gains. The AIADMK-led coalition in Tamil Nadu won 26 parliamentary seats. The BJD, Lok Shakti and Trinamool Congress collectively secured 19. Lastly, the former partners of the BJP—the SAP, SAD, Shiv Sena and HVP—captured 27. The combined strength of the BJP-led coalition, 251 seats, came within striking distance of the 272 necessary to secure a parliamentary majority (see Table 11.1).

The Congress(I) suffered a terrible defeat. Its parliamentary tally, of 141 seats, was simply one greater than before. Far more significantly, the national vote share of the party declined to 25.9 per cent, its lowest to date. The Congress(I) gained electoral support in constituencies that Sonia Gandhi had visited. Her ability to draw large crowds, according to observers, had saved the party from further losses. Yet the overall verdict proved very damning. The RJD, determined "not to allow the issue of corruption to be used as an instrument of political vendetta," managed to ward off the BJP-SAP alliance with the help of the JMM and Congress(I) and dominate the JD, CPI and CPI(M) in Bihar.[138] But the larger alliance failed to deliver. Its combined tally, of 165 seats, deflated its pretensions to rule.

The most ignominious result, however, concerned the recently deposed United Front. Its overall strength crashed from 173 to 97 parliamentary seats. The national vote share of the coalition, 22.2 per cent, was almost 6 per cent lower than in 1996. Collectively, the Left, SP and TDP captured 78 parliamentary seats, mostly in their respective strongholds. Indeed, the only party to increase its strength was the SP, which now had 20 parliamentary seats. But it was the stunning diminution of the JD, whose presence fell from 46 to 6, that largely accounted for the overall defeat. The rise of the BJD, RJD and Lok Shakti, which together won 29 seats and 4 per cent of the national vote, approximated the 4.9 per cent decline registered by the JD. The collapse of the DMK-TMC combine in Tamil Nadu, whose parliamentary tally fell from 37 to 9, contributed to the debacle as well. Their unwillingness to agree seat adjustments with extant coalition allies and smaller parties from the state, combined with the BJP's appeal for national stability, led to a massive anti-incumbency swing. Finally, the AGP failed to retain a single constituency in Assam. The perceived economic failures of its state administration, during which the party reversed its former demand to withdraw the army, led to a resounding defeat.[139]

On 6 March, the Steering Committee reviewed its options.[140] The CPI(M) General Secretary H.S. Surjeet, stressing the danger of *Hindutva*, advised its constituents to provide external parliamentary support to the Congress(I). The Left had staunchly attacked the latter during the campaign. The main political enemy, however, remained the BJP. The SP chief Mulayam Yadav, given his state-level understandings with the Congress(I), supported the proposal. The TMC, given its severe electoral losses, expressed its amenability as well. The new elec-

THE DECLINE OF THE UNITED FRONT (MAY 1997–MARCH 1998)

Table 11.1: Twelfth General Election, 1998

	Votes	Seats
BJP+		
BJP	25.47	179
AIADMK	1.84	18
SAP	1.77	12
BJD	1	9
SAD	0.82	8
WBTC	2.43	7
SHS	1.78	6
PMK	0.42	4
LS	0.69	3
MDMK	0.44	3
HVP	0.24	1
JP	0.12	1
NLP	0.04	0
NTRTDP (LP)	0.1	0
Total	37.16	251
INC+		
INC	25.88	141
RJD	2.71	17
RPI	0.37	4
UMFA	0.01	1
SJP(R)	0.32	1
KEC(M)	0.1	1
JMM	0.07	0
IUML	0	0
BKKP	0	0
Total	29.46	165
UF		
CPI(M)	5.18	32
SP	4.95	20
TDP	2.78	12
CPI	1.75	9
DMK	1.45	6
JD	3.25	6
RSP	0.55	5
TMC	1.41	3
FBL	0.33	2
JKN	0.15	2

AGP	0.29	0
MAG	0.02	0
UGDP	0.04	0
Total	22.15	97

Source: Election Commission of India.

toral scenario in Tamil Nadu even forced the DMK, despite the Congress(I)'s ham-fisted tactics, to consider supporting the proposal. The rapidly shifting electoral landscape pushed many leaders to contemplate an option that had seemed beyond the pale three months earlier.

However, analogous state-level calculations and partisan resentment pushed their associates towards the other side. The Congress(I) remained the principal electoral opponent of the TDP in Andhra Pradesh. Chandrababu Naidu declared its equidistance from the party and the BJP. The AGP, having suffered a humiliating loss to the Congress(I) in Assam, was also unsupportive. The JD, having lost two prime ministers to the machinations of the Congress(I), resisted the idea too. Even the JKN, which had sought since the late 1980s to maintain good relations with and leverage greater aid from New Delhi for Jammu & Kashmir, had opened negotiations with the BJP. In short, differing threat perceptions and political circumstances in their regions polarized the remnants of the United Front.

Ultimately the dilemma caused its demise. In order to avoid the imbroglio of May 1996, President Narayanan requested the contending parties to supply evidence that a viable coalition could be formed.[141] On 15 March, after receiving letters from over 250 MPs, he appointed A.B. Vajpayee as prime minister. The BJP inducted a 42-member ministry a few days later.[142] Critically, the TDP chief assured Narayanan of his neutrality. Yet Naidu changed his stance. On 23 March, he successfully ensured the election of his colleague, G.M.C. Balayogi, as parliamentary Speaker, which granted the latter the authority to determine the timing of business and legitimacy of defections and splits.[143] The sudden *volte-face* generated acrimony in the United Front. Reportedly, the CPI(M) Politburo tried to expel the TDP at a Steering Committee meeting after Naidu indicated his neutrality, provoking objections from the AGP and DMK. The TDP chief retorted that he was "insulted" for not being invited to the meeting, despite being its official convener, complaining that his opposition towards the Congress(I) fell on deaf ears.[144] Severe political constraints dictated his hand. His intense rivalry with the latter, combined with the simultaneous rise of the BJP and the prospect of state assembly elections in less than two years in Andhra Pradesh, convinced Naidu to abandon ship. On 28 March, the Vajpayee ministry won a parliamentary majority, sealing the fate of the United Front.

The outcome exposed a naïve faith that India's regional parties were inherently secular.[145] The Shiv Sena combined nativist regional appeals with militant Hindu chauvinism in Maharashtra. The AIADMK had been funding temple endowments and "allying with prominent Hindu priests" in Tamil Nadu.[146] Meanwhile other parties, such as the SAP and HVP, had allied with the BJP even before the latter had shelved its most controversial proposals.

Nonetheless, not every regional formation exhibited such pusillanimity. The BJP moderated its agenda at the behest of its newfound allies. Reportedly, the TDP warned the BJP it would withdraw parliamentary support if the latter "passed any policies that harmed minorities" or undermined the "secular fabric" of the country.[147] Whether they would contain the politics of *Hindutva* remained a vital question. Yet federal coalition politics had constrained the space available to militant national ideologies. In addition, the unwillingness of the TDP to support the Congress(I) was understandable. The regionalization of the federal party system had created dissimilar electoral fields. The CPI(M), SP and RJD dominated their respective home states. Indeed, the rise of the Trinamool Congress had bolstered the Left in West Bengal. The TDP, in contrast, was less secure. Short-term political survival, the most basic drive in any politics, compelled its decision. Finally, Naidu had good reason to distrust his chief state-level rival, which had made few overtures to potential regional allies, unlike the BJP. The high-handedness of the Congress(I), despite its steadily deteriorating fortunes, betrayed a failure to grasp the evolving dynamics of India's post-Congress polity.

12

THE DISSOLUTION OF THE THIRD FORCE (1998–2012)

The outcome of the twelfth general election reconfigured the politics of the third force.[1] The United Front had formally come apart. The TDP, its most important regional partner since the 1980s, abandoned its ranks. Several regional components of the JD, its longstanding fulcrum, had splintered into separate parties during the tenure of the alliance. Consequently the mantle of leadership passed to the Left Front, especially the CPI(M), the most relentless opponent of the BJP. Its primary focus was to contain the dangers of *Hindutva* and disallow the *sangh parivar* from consolidating its newfound power.

The BJP faced a strategic dilemma. On the one hand, the party had to make programmatic compromises and share power within a diversity of parties in order to capture national power. Maintaining ideological purity would have ensured political defeat. Hence it conceded several important ministries and parliamentary offices to smaller regional allies in the governing coalition, named the National Democratic Alliance (NDA) in May 1998. Moreover, its National Agenda for Governance omitted key Hindutva proposals, including the establishment of a Uniform Civil Code, repeal of Article 370 of the Constitution and construction of a *mandir* in Ayodhya. The BJP distanced itself from the Gujral doctrine by advocating "peaceful relationship [sic] with all neighbors on a reciprocal basis". But the Agenda failed to mention its longstanding pledge to reclaim territories lost to China and Pakistan.² The new BJP president, Kushabhau Thakre, proclaimed:

It may be difficult for the party to go beyond the confines of the National Agenda because a coalition government functions under many compulsions. We have to be realistic about the situation and our workers must realise working as part of a coalition, the BJP can only do this much and no more."³

The "politics of social exclusion" had become "self-limiting in the domain of representative democracy" due to "the compulsions of coalition politics".⁴

Nonetheless, the BJP could not ignore the *sangh parivar*, which ultimately distinguished its politics and mobilized crucial support on the streets. Numerous political concessions frustrated politicians in the party close to the RSS, with many forced to accept junior cabinet posts, who lamented that "although we are the largest party in the ruling coalition at the Centre, we also seem to be the weakest".⁵ The oscillation between strategies of moderation and militancy that had historically shaped the Hindu nationalist movement acquired a new dynamic amidst the pressures of national coalition politics.⁶ The BJP placated the *sangh parivar* by pursuing key objectives through tactics of stealth and obfuscation; seeking to frame the terms of debate, and the wider historical discourse that partly constituted such debates, to its own partisan advantage; and selectively justifying acts of intimidation and violence by hardline followers on the ground. By "offering a weighted, and ambiguous, mix of appeals," the party legitimized several Hindu nationalist claims, gradually pushing the center of gravity to the right.⁷

The first National Democratic Alliance, 1998–1999

The NDA began its tenure with a momentous decision. In early May, the Vajpayee ministry tested India's nuclear devices at Pokhran. The extension in 1995 of the NPT into perpetuity, and mounting international pressure on India to sign the CTBT or face sanctions after it came into force in 1999, encouraged the decision.⁸ Yet conducting a test was a longstanding desire of the *sangh parivar*. The BJP claimed the tests demonstrated Hindu pride, helping the party to outflank the Opposition as well as its new coalition allies. Significantly, none of the latter had been briefed in advance. National security questions remained the purview of the highest executive powers. Yet reports that even the Defense Minister, George Fernandes of the SAP, had been informed of the decision after it had been approved by senior RSS figures emphasized the limits of power-sharing within the NDA. The Congress(I) demanded credit for developing the nuclear option, reminding the media that Narasimha Rao had almost tested in 1995. I.K. Gujral, who had opposed the tests, vouched that his ministry would have taken the step if circumstances had demanded.[9] The CPI(M) would later castigate the BJP for its "pro-imperialist policy" in the wake of secret talks between US Secretary of State Strobe Talbott and his Indian counterpart, Jaswant Singh.[10] Yet few parties questioned the decision on either strategic or ethical grounds amidst the nationalist euphoria that erupted over Pokhran II.[11]

That said, Atal Bihari Vajpayee reached out to his counterpart, Nawaz Sharif. Indeed, the prime minister undertook a historic bus journey to Pakistan in February 1999, signing the Lahore Declaration, which committed both countries to avert nuclear conflict. Several commentators praised Vajpayee personally for the undertaking, evoking his endeavors during the Janata Party administration in the late 1970s. Others argued that only an eminent leader of BJP could have taken such a risk. Yet the pursuit of rapprochement also relied upon the foundations laid by I.K. Gujral. The free trade agreement signed by India and Sri Lanka in December 1999, the first ever linking both sides, reflected similar preparatory groundwork too.[12] The NDA reiterated its commitment to the Gujral doctrine under Jaswant Singh, its external affairs minister, until July 2002.[13]

Indeed, observers praised the party for managing the governing coalition relatively well, often justifying decisions by citing the National Agenda for Governance. The coordination committee of the NDA met frequently in New Delhi and in the states as well, in contrast to the United Front. And the BJP called for greater coalition *dharma* (principles of discipline, order and harmony) in February 1999, establishing a code of conduct for all party meetings, chief ministers' conferences and Union cabinet deliberations. In doing so, the party "attempted to transform partners into stakeholders" of the NDA.[14]

Nonetheless, hardliners sought to expand *Hindutva* in other realms. Some proved unsuccessful. Attempts by the Minister for Human Resources Development, the RSS stalwart Murli Manohar Joshi, to rewrite school textbooks, introduce new curricula and appoint educational officials sympathetic to Hindu nationalist views encountered stiff opposition from secular opponents and smaller regional allies.[15]

Campaigns of intimidation, pressure and violence orchestrated by the *sangh parivar* against Christians in the tribal belts of Bihar, Madhya Pradesh and Gujarat proved harder to contain, however, culminating in the desecration of churches and several heinous murders of priests and nuns. The VHP defended the atrocities as "expressions of the anger of patriotic Hindu youth against anti-national forces" against the conversion efforts of Christian missionaries.[16] National coalition politics, which compelled the BJP to shelve its pledge to build a Ram temple in Ayodhya, introduce a UCC and repeal Article 370, limited the damage militant Hindu nationalists could inflict in New Delhi. But these constraints redirected their animus towards other religious minorities in the peripheries, either unilaterally or in connivance with the BJP, which failed to rein them in. Regional secular allies expressed dismay. Yet they proved unwilling to topple the government, portending further violence.

The BJP's first spell of national power ended in April 1999, resurrecting the possibility of another United Front. The failure of the NDA to last had nothing to do with attempts to introduce *Hindutva* into high politics or everyday social life. Myopic power moves bore the blame. Reportedly, J. Jayalalitha, the mercurial leader of the AIADMK, pressured the BJP to obstruct criminal investigations against her and dismiss the state-level ministry of her arch rival, the DMK.[17] The prime minister failed to oblige, leading the AIADMK—his largest coalition ally—to withdraw its support. The BSP chief Mayawati, seeking to avenge the loss of her ministry in September 1997 at the hands of the BJP, followed suit. The first NDA lasted thirteen months. The RSP and AIFB promoted a Third Front with Jyoti Basu as its prime minister-designate.[18] But Sonia Gandhi dismissed the idea: "We are not ready to support a Third Front, Fourth Front or whatever it is called. We will not give our support to anybody else".[19] Frustrated, Basu declared "a communist cannot become prime minister of India", invoking Jagjivan Ram's famous lament in 1979.[20] Yet Basu's own party had stymied the opportunity during the United Front. The alignment of forces had decisively changed. Critics accused the "domesticated" Left, bereft of powerful mass organizations with a national presence, of being afraid of taking responsibility as a "caretaker administration". "The passions of youth", said one, "have become the lust of old men".[21]

The CPI and CPI(M), maintaining that it was too dangerous to remain "equidistant" vis-à-vis the BJP, pledged to back the Congress(I).[22] The Left's shifting coalition strategy gave the latter another chance to recapture national power at the head of a diverse multiparty alliance. The party faced genuine obstacles. Despite occupying the center of gravity, which theoretically bestowed it with maximum political flexibility, the Congress(I) was "less coalitionable".[23] Simply put, the party remained the main electoral rival of various political formations in the east and south of the country, making it harder to forge power-sharing alliances. Everything depended on striking innovative compromises. Yet none transpired. Jyoti Basu upbraided both sides:

> Mulayam Singh said: I cannot support the Congress government. Then I asked him: Why did you vote against the BJP? He said: the alternative is you. (It is ridiculous, that I become Prime Minister.) I said: Why should the Congress accept me? Those days are over, no more there. Then Pranab Mukherjee, Arjun Singh and others came to my house and said: We shall form the government on our own. I asked: how can you ... The Congress had also made the mistake. They would not have a coalition government.... Then Sonia Gandhi rang me up and said the same thing. My working committee has already taken a decision, either we form the government or nothing happens. We cannot support the alternative suggested. I said: very good. Then I do not know why you people voted again the BJP because [it] is now saying rightly that they are so irresponsible that threw us out but could not form a government.... not only Mulayam Singh, even the RSP and the Forward Bloc ... also opposed and they could not give me a reason why they voted against the BJP, but then opposed the Congress forming a government.[24]

Short-term electoral considerations played a role. The SP feared losing the support of Muslims and OBCs to the Congress(I) and the BJP, respectively.[25] Hence Mulayam Yadav reportedly struck an informal understanding with Kalyan Singh, the OBC leader of the BJP, whose state-level ministry had countenanced the destruction of the Babri masjid. Taking Muslim support for granted was a precarious decision. Yet Yadav also criticized Sonia Gandhi's "foreign" origins, a grievance that simultaneously inspired the Maratha strongman Sharad Pawar to form the Nationalist Congress Party (NCP), which subsequently joined the third force. Individual political ambition insidiously allowed Hindu nationalist claims to spread. The rebuff to the Congress(I) simultaneously belittled, however, its presumption to rule. Indeed, the party refused to concede anything. Declaring its opposition to national coalition building "unless absolutely necessary" in Pachmarhi in September 1998, the party refused to grasp the exigencies of power in a post-Congress polity.[26]

The campaign for the thirteenth general election in September–October 1999 sowed the dissolution of the Third Front. The NDA returned to power with a comfortable majority, winning over 40 per cent of the vote and 300 parliamentary seats. The Congress(I)-led coalition came second, with 34 per cent and 137 constituencies, with 23 of the latter due to its allies. The ten parties now comprising the Third Front experienced a humiliating defeat, winning less than 16 per cent of the national vote and 77 parliamentary seats (see Table 12.1).

Many commentators argued that nationalist fervor secured victory for the BJP. The party reacted to the incursion of irregular Pakistani troops in Kashmir during the summer of 1999, provoking the so-called Kargil war, by claiming that "cross-border terrorism" embodied India's external "Muslim threat".[27] The charge had clear political reverberations. Clever state-level alliances and the party's shrewd decision to contest fewer constituencies in areas of strength, however, ultimately facilitated its greater winning percentage.[28] Indeed, the self-destruction of the third force on two fronts bolstered the fortunes of the BJP.

The relentless implosion of the Janata *parivar* was the first. In June 1998, Mulayam Singh Yadav and Lalu Prasad Yadav temporarily joined hands to form the Rashtriya Loktantrik Morcha in north India. But the RJD eventually contested the general polls with the Congress(I). The BSP chief Kanshi Ram contemplated an alliance with Mulayam Yadav. But Mayawati, arguing that Muslims were abandoning the SP given its tacit pact with Kalyan Singh amidst the growing vertical cleavages within the OBCs, courted the most backward castes.[29] More significantly, intense factional rivalries in the JD—pitting J.H. Patel against Deve Gowda in Karnataka, and Sharad Yadav and Ram Vilas Paswan against Lalu Yadav in Bihar—inflicted severe damage. In July 1999, Patel, Sharad Yadav and Paswan joined hands with the Lok Shakti and SAP to form the Janata Dal (United) (JD(U)), allying with the NDA. The three leaders opposed Sonia Gandhi. Yet personal rivalries in the regions—all three desired to regain political office—compelled their exit. Together, the JD(U) and BJD secured 31 parliamentary seats. The rump of the original party, rechristened the Janata Dal (Secular) (JD(S)) under H.D. Deve Gowda, managed just one.[30] The inability of its highest ranks to resist the entreaties of the BJP by sharing power amongst themselves in the states or respond to the challenges of further economic liberalization destroyed the legatee of India's socialist tradition.

Second, the former regional stalwarts of the third force entered or maintained tactical alliances with the BJP. Seeking to exploit the deba-

THE DISSOLUTION OF THE THIRD FORCE (1998–2012)

Table 12.1: Thirteenth General Election, 1999

	Votes	Seats
NDA		
BJP	23.75	182
TDP	3.65	29
JD(U)	3.1	21
SHS	1.56	15
DMK	1.73	12
BJD	1.2	10
AITC	2.57	8
PMK	0.65	5
INLD	0.55	5
MDMK	0.44	4
JKN	0.12	4
SAD	0.69	2
MADMK	0.11	1
MSCP	0.06	1
SDF	0.03	1
HVC	0.07	0
AC	0.02	0
Total	40.3	300
INC+		
INC	28.3	114
AIADMK	1.93	10
RJD	2.79	7
RLD	0.37	2
MUL	0.23	2
KCM	0.1	1
RPI(A)	0.13	1
RPI	0.13	0
Total	33.98	137
Third Front		
CPI(M)	5.4	33
SP	3.76	26
NCP	2.27	8
CPI	1.48	4
RSP	0.41	3
FBL	0.35	2
JD(S)	0.91	1
TMC	0.56	0

AGP	0.32	0
RPI	0.14	0
Total	15.6	77

Source: Election Commission of India.

cle of the AIADMK, the DMK brokered a deal with the PMK and MDMK, with all three joining the NDA. Reportedly, the Left antagonized M. Karunanidhi by reaching out to AIADMK, leading him to proclaim that "Jayalalitha's corruption is more dangerous than communalism".[31] Many would disagree. Ultimately, these state-level parties inverted the strategic judgment their national counterparts had previously made, entering seemingly incongruous alliances in New Delhi to protect their turf in the states. The new Tamil combine won 21 seats. Lastly the TDP maintained its formal political neutrality vis-à-vis the BJP in exchange for retaining dominance in Andhra Pradesh, posting its highest ever tally of 29 parliamentary seats.[32] In short, these erstwhile members of the United Front together contributed 52 MPs to the NDA, enabling it to become the first political formation to muster a parliamentary majority since 1989.

The second National Democratic Alliance, 1999–2004

The NDA, which resumed office in October 1999, survived a full parliamentary term.[33] It partly reflected changed arithmetic: no smaller coalition partner could topple the administration on its own.[34] Yet the willingness of the BJP to accommodate, for instrumental and substantive reasons, allies' political claims and wider social demands clearly helped. The new Vajpayee ministry reallocated cabinet portfolios eleven times during its tenure, signaling its maneuverability and resilience.[35] The BJP made several overtures to appease its expanding social bloc. The party elected Bangaru Laxman, the first Dalit to become its organizational president, in August 2000.[36] It preserved the traditional Haj subsidy for Muslims, appointed well-known representatives to the National Minorities Commission and improved financial allocations to Wakf boards.[37] And the NDA promoted greater regionalization.[38] In September 2000, the Government created three new states, Chhattisgarh, Jharkhand and Uttarakhand. Bureaucratic reasons and political incentives gave strong impetus: the highly populous states of Madhya

Pradesh, Uttar Pradesh and Bihar generated administrative challenges; the BJP had strong ties with *Adivasi* communities and high-caste groups that varyingly dominated the electoral landscape in each new state. Nevertheless, it was a historic decision. More significantly, the NDA supported the principle of a greater Nagaland and established a Bodo Territorial Council between 2001 and 2003, despite the nationalist federal vision of the BJP. Finally, and perhaps most surprising, the Government continued to pursue better relations with Pakistan. Vajpayee met General Pervez Musharraf in Agra in July 2001 and, despite subsequent respective attacks on the Kashmir state assembly and national parliament by the Jaish-e-Mohammed and Lashkar-e-Taiba, militant Islamist groups with bases in Pakistan, held assembly elections in Jammu & Kashmir in October 2002, the first relatively fair polls since 1983.[39]

Nonetheless, the aspirations unleashed by the second democratic upsurge remained alive, even if the prospects of a third force became increasingly elusive:

> The third space, occupied by various non-Congress, non-BJP formations, has not shrunk in any significant way. What has declined, of course, is the vision and organizational capacity of those wanting to create a Third Front in national politics.[40]

If anything, the politics of its erstwhile protagonists diverged further. In part, it reflected growing economic disparities and inter-state rivalries. In August 2000, the TDP chief minister Chandrababu Naidu accused the 11th Finance Commission of encouraging fiscal profligacy by rewarding high population-low economic growth states like Bihar and Uttar Pradesh as well as Assam and West Bengal, while penalizing the low population-high economic growth performance of Gujarat, Maharashtra, Andhra Pradesh, Karnataka and Tamil Nadu.[41] Significantly, the last three represented the home states of ardent economic liberalizers in the United Front: the TDP, DMK and the southern wing of the JD. Prior NDA actions, which had made grants conditional upon states' fiscal performance without consulting the ISC, created the situation.[42] Yet Naidu's reaction spoke volumes. Styling himself as a CEO devoted to "e-governance", the TDP chief minister had successfully courted investments from leading multinational corporations in Hi-Tec City, with ambitions to expand into finance and biotechnology aimed at leapfrogging stages of industrial transformation.[43] The self-promotion was only partly true. Although West Bengal gained a higher share of central tax revenues, it had recorded the fifth fastest rate of per capita growth in state domestic product in the 1990s; Kerala received less, despite scoring the sixth fastest rate.[44] Moreover, the TDP had received approximately Rs. 40,000 crores in grants and loans from New Delhi during the NDA for power, food and rural development, exploiting national coalitions to cement its rivalry vis-à-vis the Congress(I) in Andhra Pradesh.[45] Yet while population growth had declined in Andhra Pradesh in the 1990s, its trend economic growth mirrored the previous decade, while state finances remained poor and agrarian distress persisted.[46] Nonetheless, the Commission agreed to grant supplementary funds to reforming states, in contrast to previous awards. The centrifugal tendencies stoked by economic liberalization reinforced the growing political clout of the states in the third electoral system, withering the third space in the 1990s.

Indeed, further evidence mounted. Uttar Pradesh witnessed a sharp economic deceleration, rising government expenditure and severe financial constraints, creating a level of indebtedness similar to Rajasthan and Madhya Pradesh, and levels of social under-development akin to Bihar.[47] Declining public investment compelled fierce competition for scarce private capital, which now accounted for three-quarters of gross fixed investment, as the states of the Union submitted to regulation by the Centre and domestic and international credit ratings agencies.[48] Some cooperation occurred. A proposal to create a uniform sales tax, first raised by the West Bengal chief minister Jyoti Basu in 1993, lessened beggar-thy-neighbor dynamics. State finance ministers agreed in 2002 to adopt a uniform central value added tax as well.[49] Horizontal inter-state competition remained the norm, however, testing the compensatory mechanisms of India's federal political economy. Growing regional inequalities strained political ties between the former protagonists of the third force, which had only recently begun to demand more equitable Centre-state relations against the unitary visions of their national party rivals.

The more significant cleavage dividing the former third force, however, was the failure of its ostensibly secular forces to challenge growing communal violence. Many observers believed that federal coalition politics, and the centrist institutional logic of India's democratic regime, would compel the BJP to restrain Hindu nationalist militancy. It was a reasonable judgment to make. The anti-Muslim pogrom that shook Gujarat in March 2002 tested it savagely, however.[50] On

27 February, a train returning from Ayodhya caught fire in the town of Godhra. Fifty-eight Hindus died. Over the next three days, Hindu nationalist foot-soldiers rampaged across the state, wrecking Muslim establishments, killing hundreds of civilians and displacing thousands of citizens. The BJP state ministry, led by chief minister and stalwart former RSS worker, Narendra Modi, failed to stop the violence. The NDA vacillated in dispatching the army, despite evidence mounted that his ministry abetted the carnage. Perhaps most damagingly, Prime Minister Vajpayee appeared to justify the reprisal, invoking the threat of pan-Islamic terrorism in the wake of the September 11, 2001 attack on the United States:

> Wherever there are Muslims, they do not want to live in peace with others. Instead of living peacefully, they want to propagate their religion by creating terror in the minds of others.[51]

Modi subsequently characterized the perpetrators of the fire in Godhra as "terrorists", jeopardizing national security, which resonated with growing anti-Muslim rhetoric and closer Indo-Israeli strategic ties.[52] The TDP distanced itself by not claiming the Speaker's post after G.M.C. Balayogi died in plane crash in early March.[53] In mid-April, Chandrababu Naidu demanded Modi's resignation, backed by the Trinamool Congress, Lok Janshakti Party (LJP) and JD(U). All four parties boycotted the subsequent meeting called by the NDA central committee. Yet not a single ally extracted a major political resignation from the BJP, even though Muslims constituted a sizeable electoral constituency in many regions. Only Ram Vilas Paswan, the minister for coals and mines from the LJP, resigned.[54] Indeed, some ventured that Naidu exploited the situation by extracting more Central resources to quell growing discontent in his state, eventually supporting the NDA during a no-confidence motion in August 2003.[55] The DMK abandoned the ruling alliance by the end of the year. Yet it was the abuse of the Prevention of Terrorism Act, 2002, by its state-level rival the AIADMK, not the events in Gujarat, which precipitated its departure in the run-up to the fourteenth general election.

The survival of the NDA highlighted the paradoxical effects of the federal party system in an era of diverse coalition governments. Nominally secular parties resisted joining the Opposition given divergent political stances vis-à-vis the Congress(I) in their respective states. Yet it also reflected a brutally cynical assessment: the cost of remaining in the NDA against the potential electoral backlash they would face in the regions. Godhra constituted their moment of reckoning. Good political judgment demanded responsibility for the consequences of action where, at some point, an absolute ethical value transcended the calculus of survival.[56] The collective failure of several erstwhile members of the United Front to leave the NDA belied the professed secular ideals of the third force.

The first United Progressive Alliance, 2004–2009

Most commentators believed the NDA, the first non-Congress administration on course to survive a full parliamentary term since independence, would retain power. Touting a slogan of "India Shining", the alliance called for an early general election in April–May 2004 to capitalize on seemingly propitious conditions. Politically, the BJP looked strong. Atal Bihari Vajpayee had become the country's leading politician. The party's victories in three state assembly elections in December 2003, on a platform stressing economic development presaged further success. Economically, a very good monsoon and 8 per cent aggregate GDP growth on the basis of rising exports and low inflation in the previous fiscal year underscored the re-branding of India as a successful globalizing economy.[57] Internationally, the Prime Minister's assiduous overtures to General Pervez Musharraf to resolve the conflict in Kashmir earned popular support. Hence few believed that Sonia Gandhi could lead the Congress(I), its dispirited party-workers and newfound allies in the United Progressive Alliance (UPA) back to power.

The downfall of the NDA stunned most observers. Its final seat count fell from 299 to 189, with the BJP crashing from 182 to 138. In contrast, the Congress(I) raised its tally from 112 to 145, despite its declining vote share, which fell from 28.3 to 26.4 per cent. Far more importantly, its UPA allies captured another 76 parliamentary seats, bringing their cumulative tally to 221 (see Table 12.2).

Many observers read the result as a protest against economic liberalization, which had exacerbated social inequalities between classes, sectors and regions, and a rejection of militant Hindu nationalism.[58] Following the polls the TDP, AIADMK and TC left the NDA, citing the electoral backlash against Godhra. Their belated defense of secularism underscored the limits of pursuing militant *Hindutva* in India's state nation.[59] But it was a complex verdict. The NDA retained a majority of seats in historic bastions such as Rajasthan, Madhya

THE DISSOLUTION OF THE THIRD FORCE (1998–2012)

Table 12.2: Fourteenth General Election, 2004

	Votes	Seats
UPA		
INC	26.44	145
TRS	0.6	5
IND(INC)	0.16	1
RJD	2.39	24
LJP	0.66	4
NCP	1.78	9
JMM	0.41	4
PDP	0.07	1
MUL	0.19	1
KCM	0.05	0
JDS	0.05	0
RPI	0.04	0
RPI(A)	0.09	1
PRBP	0.06	0
DMK	1.81	16
MDMK	0.43	4
PMK	0.56	6
Total	36.43	221
NDA		
BJP	22.16	138
TDP	3.04	5
JD(U)	1.94	8
IND(BJP)	0.18	1
IFDP	0.07	1
SHS	1.77	12
BJD	1.3	11
SAD	0.9	8
AIADMK	2.19	0
AITC	2.06	2
MNF	0.05	1
SDF	0.04	1
NPF	0.18	1
Total	35.88	189
Third Front		
SP	4.31	36
RLD	0.61	3
CPI(M)	5.66	43

CPI	1.32	9
RSP	0.43	3
FBL	0.35	3
KEC	0.09	1
IND(Left)	0.08	1
JD(S)	1.47	3
Total	14.32	102

Source: Election Commission of India.

Pradesh and Gujarat. Yet the demobilization of its traditional high-caste urban supporters and affiliates in the *sangh parivar*, dissatisfied with the ideological compromises the BJP had pursued, cost the party in vital states such as Uttar Pradesh.[60] The Congress(I)'s slogan of *aam aadmi* (common man) captured the specter of vulnerability in the rural hinterlands and far-flung peripheries of the country. Yet the party had acquired its new electoral base, disproportionately representing socially marginalized groups, essentially by default. Short-term political tactics rather than a transformational social agenda facilitated its success.[61] In the end, a deft coalition strategy resurrected the Congress(I)'s electoral fortunes. The party recognized the primacy of the PDP in Jammu & Kashmir, RJD in Bihar and DMK in Tamil Nadu. However, it remained the dominant partner vis-à-vis the NCP and Telengana Rashtra Samiti (TRS) in Maharashtra and Andhra Pradesh, respectively. Moreover, the Congress(I) struck indirect electoral agreements with the CPI(M) in Andhra Pradesh and Tamil Nadu, but contested the latter directly in West Bengal and Kerala. In contrast, the BJP committed several mistakes in alliance making, imploding the wider social bloc it had shrewdly cultivated.

Compared to 1999, the Congress(I) had become "more coalitionable".[62] The position of the party in different state-level coalitions took into account local electoral strength, however. Its capacity to exploit these opportunities, moreover, required strategic reorientation as well as political finesse. These changes began in 2002, when the Congress(I) accepted a secondary role vis-à-vis the PDP in the Jammu & Kashmir assembly election, overriding its state unit to convey that "we treat our junior partners with equal respect" and "show that we are the only true national alternative to the NDA government".[63] Reaping the dividends, the Congress(I) unveiled a more accommodative stance at its

THE DISSOLUTION OF THE THIRD FORCE (1998–2012)

Shimla party session in July 2003, reversing its earlier Pachmarhi declaration. The party leadership had finally submitted to the logic of the third electoral system. In addition, the Congress(I) had skillfully refused to name Sonia Gandhi as its prime ministerial candidate during the 2004 campaign, saying it was a decision for the UPA as a whole. Gandhi's subsequent endorsement of Manmohan Singh, following xenophobic attacks against her by the BJP, adroitly undermined the latter. Her gesture simultaneously allowed erstwhile rivals, particularly the NCP and DMK, to join the incipient coalition government without losing face over past imbroglios. The NCP's poor showing in the 1999 general election, and its acquisition of the Union agricultural ministry and anxiety over the upcoming October 2004 state assembly polls in Maharashtra, helped of course. Promising the DMK that it would repeal POTA and recognize Tamil as a classical language, and awarding the party seven cabinet berths in New Delhi, helped the Congress(I) mend relations after the ignominious Jain Commission debacle in 1998. The party's maneuvers revealed a new strategic outlook, greater political skill and sober learning. In the end, its improved fortunes evinced better political judgment.

Nonetheless, the Congress(I) could not grasp power without the communist parties' support, which captured a historic 61 parliamentary seats. The Left Front mobilized a diverse electoral coalition to win more than half of the popular vote in West Bengal, its highest level since 1989, with the CPI(M) posting its best performance since 1980.[64] Despite their state-level electoral agreements, the Left remained noncommittal during the campaign. Indeed, it expected the Congress(I) to adopt the "pro-imperialist neoliberal policies" of the BJP.[65] Many prominent artists, intellectuals and activists from the broader Indian left implored the CPI and CPI(M) to join the government:

> [At this] "historical juncture, which calls for creative and constructive initiative ... the left can undertake the task only as part of the government and not by supporting it from outside. The latter ... is a negative approach, which the people are likely to interpret as shirking the responsibility. To usher in changes on secular and democratic lines the left has to use the possible access to power, even if there are certain political risks ... After all the quality of the life of the people in West Bengal and Kerala would not have been possible without the left wielding power."[66]

Indeed, the former prime minister H.D. Deve Gowda proclaimed,

> there is no such concept as a third front. Now, we have the Congress-led UPA government at the Centre and the erstwhile National Democratic Alliance

headed by the BJP. It is time we stopped harping on the third front. We have to forget about evolving alternatives to the UPA government".[67]

The CPI(M) helped to draft a Common Minimum Programme (CMP), mirroring the agenda of the United Front government a decade earlier, some observers noted.[68] The party agreed to participate in a National Advisory Council (NAC) and Coordination Committee to discuss policy and political matters, respectively. The party allowed Somnath Chatterjee, one of its most seasoned parliamentarians, to assume the post of Speaker. Yet its Politburo and Central Committee, vowing to push for the implementation of the CMP and oppose further neoliberal reform, offered external parliamentary support.[69]

Ideologically, it was a consistent move. The CPI(M) had always believed that effective power required a majority, or at least a plurality of seats, in any cabinet. The wider leadership of the Left, moreover, displayed considerable integrity in an era of growing political corruption and socioeconomic inequality.[70] The Congress(I) accounted for 52 per cent of total assets held by members of the fourteenth Lok Sabha, despite representing only 27 per cent of the latter, and had the largest number of *crorepatis* (individuals worth ten million rupees). Yet the traditional vanguard of the second democratic upsurge had sullied its egalitarian aspirations too. An extraordinarily high percentage of MPs from the SP enjoyed disproportionate material assets and faced serious criminal indictments too. The ranks of the BSP and RJD exhibited similar profiles. In stark contrast, the accumulated wealth of Left parliamentarians comprised less than 2 per cent of the total, averaging Rs. 24 lakh per MP.

Nonetheless, the CPI(M)'s refusal to participate in government begged the question of whether it would ever acquire a plurality of power. Indeed, the CPI General Secretary A.B. Bardhan proclaimed that resurrecting even a Third Front without Congress(I) support was impossible.[71] Sympathetic critics claimed the Left had lost another opportunity to become a genuinely all-India formation, or to put employment generation, agrarian reform and decentralized planning on the national policy agenda and showcase the integrity and ability of its leading members.[72]

According to some, the capture of power by the UPA signified the "closure" of the third electoral system, indicating the "saturation" of the second democratic upsurge and the "domestication" of policy choices in New Delhi.[73] The emergence of a bipolar electoral contest at the Centre, with state-based parties oscillating between the Congress(I) and the BJP, extinguished the possibility of radicalism. Rising socioeconomic inequalities ultimately overpowered the possibilities of the second democratic upsurge. Lower-caste, communist and regional parties had failed to construct a viable third force in modern Indian democracy.

Yet several things had changed. First, unlike its heyday, the Congress(I) could no longer easily absorb its new allies. Many of these state-based parties, which had formed after 1989, lacked the explicit regionalist orientation of the early third force. Still, collectively they garnered more than 50 per cent of the national vote, determining their national rivals' fortunes.[74] Moreover, the Congress(I) showed less desire to destabilize rival formations and a greater willingness to address the needs of historically subordinate groups.[75] Second, the newly appointed government endorsed the CMP, rechristening it the National Common Minimum Programme (NCMP). The newly established NAC, led by Sonia Gandhi, would oversee its implementation. And the Congress(I) established a Coordination Committee between the UPA and the Left, foreshadowing greater collective decision-making.[76] Finally, declining rates of anti-incumbency in the states partly reflected growing state revenues and greater social expenditures, aided by higher economic growth and tax reforms.[77] The new fiscal situation broadened programmatic opportunities. The tasks facing the UPA, according to the newly elected prime minister Manmohan Singh, were to "focus on the poor, the rural, the agricultural sector. We have to provide water, schools, health facilities, jobs for the youth and a favorable environment for businesses and industry to flourish".[78] "[L]ife is never free of contradictions," he continued, "… we will have to find a practical way".[79]

The Left won many battles during the first two years of the UPA. On the one hand, the Front frustrated greater economic liberalization in New Delhi. Its parties opposed the disinvestment of profitable state enterprises, entry of foreign direct investment into the retail sector and external financial liberalization. They argued that greater foreign investment in aviation, insurance and telecommunications had to augment productive capacity, technological capabilities and formal sector employment. And the Left stymied efforts by the Planning Commission to ask foreign consultants to evaluate social programmes.[80] On the other hand, the UPA introduced several landmark acts that heralded a new welfare architecture. The Right to Information Act (RTI), 2005, mandated government agencies to release information regarding their activities to individual citizens upon request in a timely manner. The National Rural Employment Guarantee Act (NREGA), 2005, granted adult members of every rural household the right to demand 100 days of wage-employment from the state. And the Scheduled Tribes and Other Traditional Forest Dwellers (Recognition of Forest Rights) Act, 2006, gave tribal communities the right to own and use traditionally cultivated land and to protect and conserve forests. The efforts of committed social activists and progressive intellectuals, directed by Sonia Gandhi, deserved immense credit for the formulation of these acts. Yet concerted Left pressure helped their passage in parliament.

Indeed, its assertive posture crystallized at the 18th party congress of the CPI(M) in April 2005. According to Prakash Karat, who replaced the 89 year old General Secretary H.S. Surjeet, the CPI(M) was the "only party which can rally all the left, democratic and progressive forces to carry forward the struggle for a left and democratic alternative".[81] Despite the "steadily developing ... economy ... and all round progress" of China, "the biggest socialist country", American imperialism had advanced with the re-election of George W. Bush. Hence the political tactical line of the party was to "to advance the struggle against communalism, the pro-big business economic policies and imperialism":

> The Party has extended support to the UPA government with the clear understanding ... [that] it depends on its willingness to implement the pro-people measures in the Common Minimum Programme and how it maintains the unity of the UPA coalition... To accomplish this, the Party will build mass movements and conduct struggles... It should act as the sentinel of the people's interests.

Simultaneously, however, the CPI(M) pledged

> to build a third alternative ... by conducting joint movements and campaigns. A viable alternative will emerge only if there is a common policy framework at least on some major issues. Experience has shown that this cannot be just an electoral alliance. Left unity and strengthening of the Left is essential if we are to successfully rally the other democratic forces.

The ritual invocation of the party as the sole leader of the broader Indian left, and its depiction of China as socialist, held little sway beyond its most loyal ranks. Nonetheless, some commentators rightly perceived a generational shift, its new leadership increasingly discomfited by the high coalition deal-making that Surjeet had pursued since the mid 1990s.[82] The CPI(M) sought once again to build a united-front-from-below.

THE DISSOLUTION OF THE THIRD FORCE (1998–2012)

The Left faced a growing quandary, however. The LDF had managed to secure rising per capita incomes and lower absolute poverty in Kerala.[83] Its achievements partly reflected the lagged effects of earlier social investments in basic human capabilities. Yet it was largely because the alliance maintained social expenditures despite lower public investment, declining central transfers and volatile commodity prices. Indeed, the introduction of participatory development planning, inaugurated in 1996, stimulated higher participation and greater public expenditure on locally perceived needs as well as economic productivity.[84] In short, the Left had radicalized the so-called Kerala model of development against tremendous odds.

In West Bengal, however, the Left Front embraced greater economic liberalization, especially after Buddhadeb Bhattacharya succeeded Jyoti Basu in 2002. The state administration unveiled *Agribusiness Vision 2010*, drafted by the multinational consultancy McKinsey, which recommended contract farming, relaxing labor regulations and exempting prospective agribusiness from the Land Ceiling Act.[85] It courted foreign capital and public-private-partnerships in manufacturing, software and urban industrial development, while conducting "lockouts, retrenchments and closures" of failing public sector enterprises.[86] Crucially, the promotion of industrialization through compulsory acquisition of fertile land ignited violent conflicts between the party, its electoral rivals and small rural landholders. The government acquired approximately 120,000 acres between 2001 and 2006, displacing more than 2.5 million landless peasants and allowing a resurgent Naxalite campaign to intervene, led by the Communist Party of India (Maoist).[87] Bhattacharya defended the strategy, declaring, "Marxism is a science, not a dogma. It will have to keep pace with changing times".[88] But the deaths of several local inhabitants fighting attempted land acquisition in Nandigram in January 2007 galvanized protests across the state, halting the process. In the summer of 2008, Mamata Banerjee of the Trinamool Congress organized demonstrations in Singur, accusing the Left government of acquiring poor farmers' land on behalf of Tata without adequate compensation. The ensuing violence persuaded the company to relocate its plant to Gujarat.[89] Finally, basic social opportunities had stagnated in relative terms.[90] The middling performance of West Bengal in health and education—especially amongst women, lower-caste groups and Muslims—put the state behind Tamil Nadu, Punjab, Maharashtra, Gujarat and Haryana, despite their more pro-market orientations. Evidence mounted of increasing pauperization, landlessness and hunger in the interior. Supporters reiterated the inherent difficulty of pursing autonomous development within India's federal constitutional regime. Yet Kerala pursued genuine decentralization, as they conceded, despite equal constraints. Indeed, populist inter-party competition allowed Karnataka, Tamil Nadu and Andhra Pradesh significantly to outperform West Bengal in human development, suggesting deeper sociological causes peculiar to southern India.[91]

Disappointed observers accused the Left Front, which championed a "Chinese reformist strategy" and decimated the opposition during the May 2006 assembly polls, of presuming that it therefore did not have to consult local *panchayats*.[92] Defensive party intellectuals blamed the Centre for supporting corporate industrialization under a neoliberal dispensation.[93] To absorb relatively unskilled labor into capital-intensive manufacturing required safety nets, critical investments in education and training, and gradual economic integration.[94] Disillusioned former comrades argued that rampant politicization had sown corruption, incompetence and clientelism in the local *panchayats* and state bureaucracy.[95] Tactical mistakes exacerbated popular grievances. Ultimately, three decades of uninterrupted Left rule had ossified into a domineering regime.

Yet these episodes simultaneously raised strategic questions regarding the national coalition strategy of the Left. Efforts continued episodically to create another third force. Most remained haphazard. They also pitted several former constituents against each other. In April 2006, V.P. Singh and Raj Babbar revived the Jan Morcha in Uttar Pradesh, a popular front that Singh had launched in the late 1980s en route to the prime ministership. Revolving around a new party, the Jan Dal, Babbar and Singh targeted the *"goonda raj"* (rule of thugs) of the SP, whose Special Economic Zone (SEZ) policy and "brokers and fixers" exploited vulnerable women, peasants and laborers and backward castes.[96] Attracting support from eighteen political entities, including the LJP, RJD and Left Front, the new formation contested the SP in the May 2007 assembly elections. The BSP dethroned the SP, expanding its social base to win 206 of 403 seats, a stunning legislative majority. The Jan Morcha, which captured just one seat, would eventually merge with the Congress(I).[97] Similarly, in August 2007, eight parties formed the United National Progressive Alliance (UNPA), proclaiming their equidistance from the Congress(I) and the BJP. The alliance, comprising the AIADMK, INLD, MDMK, Kerala Congress(T) and Jharkhand Vikas Morcha, included only two parties from the erstwhile United Front: the TDP and the SP. The membership of the UNPA was fluid; nothing distinguished its policy agenda. In fact, every constituent had previously supported the NDA, save the SP. Little united them beyond their desire to regain power. The ostensible principles of the third force, premised on secular politics and egalitarian development, had slowly dissipated.

A pitched confrontation over external affairs galvanized the Left, however, which still enjoyed a veto over the UPA. Significantly, the former supported the latter with the precondition that it "maintain[ed] the independence of India's foreign policy", a clear attempt to limit closer ties with the United States that had developed under the NDA.[98] In July 2005, Manmohan Singh and George W. Bush had announced their intention to forge a "global partnership" exempting India from the NPT since it was "a responsible state with advanced nuclear technology".[99] In exchange, India agreed to separate its civilian and military reactors and place them under IAEA safeguards. The statement came on the heels of the New Framework for the US-India Defense Relationship. The landmark agreement, which foresaw greater military collaboration in security operations, missile defense and nuclear nonproliferation as well as protecting international commerce, provided the basis for a long-term alliance.[100] In March 2006, Singh and Bush formalized the deal, recognizing India as a "great power and key strategic partner" in the wake of growing bilateral trade and investment and deepening military cooperation and technological exchange.[101] In June, the US Congress passed the Henry J. Hyde United States-India Peaceful Atomic Energy Cooperation Act, circumventing the restrictions of the 1954 Atomic Energy Act and providing further momentum. Critics feared American strategic designs. The first test was Iran, a signatory to the NPT, on grounds that it was covertly seeking to militarize its nuclear capabilities. The larger strategic concern was China. In September 2005 and February 2006, India backed two IAEA resolutions, referring Iran to the UNSC. Tensions with Iran, which had opposed the 1998 nuclear tests in India and pursued clandestine nuclear relations with Pakistan, exerted pressure in New Delhi.[102] Yet the censures in Vienna reinforced fears on the Left. Similarly, Indian foreign officials sought to mollify their Chinese counterparts, despite continuing multidimensional assistance to Pakistan and establishing new commercial ports and naval bases around the Indian Ocean, declaring that both countries were "too big to contain each other or be contained by any other country".[103] Nevertheless, India participated in US-promoted joint naval exercises with Australia, Japan and Singapore in September 2007, aggravating Chinese suspicions.[104] Crucially, the Government refused to disclose to parliament the terms of the IAEA safeguards agreement, which it had signed in August 2007. The Congress(I) leadership simply maintained that it superseded the Hyde Act, which expressed confidence that India would increasingly align with the United States and authorized Washington to suspend transferring nuclear fuel, technology and equipment if New Delhi violated the terms of the deal.

The ambiguities surrounding the deal and its various pieces ignited a political storm over the following year. The BJP criticized the so-called 123 Agreement, saying it restricted India's strategic options and imposed huge costs.[105] Prime Minister Singh defended the deal, pointing to strong continuing relationships with Russia, France and Iran.[106] Indeed, he argued that closer Indo-US ties would enhance multipolarity in the global order.[107] Yet most American officials, facing skeptics at home, argued it would buttress US geopolitical dominance and expand sales of nuclear and military technology.[108] The UPA defended its stance, agreeing to discuss the safeguards agreement with parliament before finalizing negotiations. But in early July the IAEA proceeded to circulate a draft of the agreement to its board of governors. The Congress(I) justified the move, saying the Government had discussed the issue in parliament but that only cabinet members were privy to the "privileged document".[109] On 8 July 2008 the Left Front, calling the move a "shocking betrayal", withdrew its external parliamentary support.[110]

The opposition of the CPI(M), notwithstanding its steadfast opposition to the United States and support for China, was genuine. It raised serious concerns over whether India could avail of corrective measures if any member of the Nuclear Suppliers Group (NSG) reneged on its commitment to supply nuclear fuel. The political difficulties of the BJP, which had laid the genesis of the deal and whose leadership experienced deep rifts after its surprising loss in 2004, emboldened the communists. Arguably, given the contention within the broader Left Front over economic liberalization in West Bengal, taking an anti-imperialist stance made political sense too. Indeed, the CPI(M) had issued approximately 300 political statements during the tenure of the UPA.[111] The impasse opened the possibility of a reinvigorated third force.

Yet the CPI(M) failed to construct an alternative. Indeed, the party found itself aligned with its existential adversary, the BJP. The SP rescued the Congress(I), a party that it had opposed for as long as it had allied with the Left, enabling the Government to survive: 275 votes to 256 with ten abstentions. It was a highly controversial vote of confidence. Twenty-eight MPs defied their respective party whips. Far more significantly, the BJP alleged gross impropriety, displaying bundles of cash on the Lok Sabha floor allegedly given by the UPA in exchange for support.[112] Refuting the allegations, Mulayam Yadav justified his about-face by saying that former president A.P.J. Kalam rectified his misapprehensions of the civil-nuclear deal, and by claiming that communalism was more dangerous than imperialism.[113] Yet the capture of power by BSP in Uttar Pradesh the previous year likely exerted greater pressure. On 22 July, Prime Minister Singh questioned his rivals' political judgments, mocking their intentions and refusal to accept responsibility for the consequences of action:

The Leader of Opposition, Shri L.K. Advani, has chosen to use all manner of abusive abjectives to describe my performance. He described me as the weakest Prime Minister, a *nikamma* [good for nothing] PM, and of having devalued the office of PM. To fulfill his ambitions, he has made at least three attempts to topple our government. But on each occasion his astrologers have misled him ... At his ripe old age, I do not expect Shri Advani to change his thinking. But for his sake and India's sake, I urge him at least to change his astrologers so that he gets more accurate predictions of things to come....

Our friends in the Left Front should ponder over the company they are forced to keep because of the miscalculations of their General Secretary.... Our Left colleagues should tell us whether Shri L.K. Advani is acceptable to them as a Prime Ministerial candidate. Shri L.K. Advani should enlighten us if he will step aside as Prime Ministerial candidate of the opposition in favour of the choice of UNPA....

The moral of the story is that political parties should be judged not by what they say while in opposition but by what they do when entrusted with the responsibilities of power.[114]

On 18 August, the IAEA board of governors approved the safeguards agreement. On 6 September, the NSG granted India a waiver, removing previous restrictions on its civil nuclear trade. On 8 October, the US Congress passed the US-Indian Nuclear Cooperation Approval and Non-Proliferation Enhancement Act. Critics as well as skeptics questioned whether the Indian establishment, given its growing ideological admiration for the US, could resist strategic overtures by the latter.[115] Subsequent events provided evidence to both sides of the debate. India continued reducing oil imports from Iran. The proposed gas pipeline linking the two countries via Pakistan stalled. And critics feared that India had begun accommodating US positions in multilateral negotiations over global trade and climate change.[116] That said, although signatories to the NPT resented its special dispensation, India eventually signed nuclear energy pacts with Britain, Russia and France. Despite calling for stronger enforcement of traditional non-proliferation treaties, the subsequent Obama administration acquiesced to India processing spent nuclear fuel under IAEA safeguards.[117] Indeed, pressure by the Left and BJP compelled the UPA ultimately to pass the Civil Liability for Nuclear Damages Bill, 2010, extending responsibility for potential accidents to overseas nuclear suppliers.[118] The ground-breaking provision jeopardized foreign cooperation and provoked much criticism abroad. Nonetheless, it underscored Indian sovereignty. And while the United States won several military contracts in India, it also lost several to European counterparts.[119] In short, the Indo-US civil nuclear deal portended a closer long-term alliance between the two countries, while raising serious concerns over the cost and safety of nuclear energy development. Nonetheless, by allowing greater multipolarity in certain domains, it challenged the static zero-sum conception of power harbored by the Left. The reconfigured UPA survived its full parliamentary term.

The Fifteenth General Election and the fall of the Left Front, 2009–2012

Buoyed by its recent success, the Congress(I) entered the fifteenth general election in April–May 2009, asserting greater independence. Significantly, the party entertained discrete state-level pacts where necessary, rejecting the idea of a national alliance. "Coalitions mean positive support from all sides," declared Sonia Gandhi, "but working in a coalition does not mean we lose our political space… Such a coalition cannot be at the cost of the revival of the Congress, particularly in states where its base has been eroded".[120] In particular, the high command focused its attention on recapturing the Hindi heartland, rebuff-

THE DISSOLUTION OF THE THIRD FORCE (1998-2012)

ing the SP, LJP and RJD in Bihar and Uttar Pradesh. The Congress(I)'s strategic turn created the possibility of mobilizing another third force. Despite its pledge in 2005 to build a united-front-from-below, exhibiting programmatic unity, the CPI(M) forged another united-front-from-above. The alliance encompassed several leading figures from its avatar in the mid 1990s: the Left Front in Kerala, Tripura and West Bengal; the TDP and TRS, formed in 2001 to seek statehood for Telangana, in Andhra Pradesh; and the BJD and JD(S) in Orissa and Karnataka, respectively. It also fielded new members, the most important being the BSP in Uttar Pradesh and an AIADMK-led alliance in Tamil Nadu. Finally, former chief minister Bhajan Lal defected from the Congress(I), forging the Haryana Janhit Congress (HJC). Conspicuously absent from the latest manifestation of the third force, however, were its socialist protagonists from the Gangetic plains. The SP, LJP and RJD, which rescued the UPA during the July confidence vote, ventured a Fourth Front.

It was a shambles. The inability of its constituents to choose a leader, and their shifting declarations and state-level disagreements, created a specter of instability.[121] The JD(S) aligned with the Left in Karnataka, but competed against it in West Bengal and Kerala, and kept talking with the Congress(I). The BJP, promising to grant statehood to Telangana, persuaded the TRS to break from the TDP-led "grand alliance" during the campaign.[122] The BJD abandoned the NDA on the eve of polling. Conflicting perceptions vis-à-vis the BJP over their relative electoral strength, and anti-Christian violence by Hindu extremists in the Kandhamal region in 2008, influenced its volte-face. But the BJD subsequently tied up with the NCP, part of the UPA, underscoring its exceedingly regional strategy. Even the CPI(M) appeared tactically split: Prakash Karat intimidating that it might accept Congress(I) support, Jyoti Basu venturing that his party might join the government and Sitaram Yechury saying the latter depended on the concrete situation.[123] "[The] difficulty ... for the Third Front," remarked a leading observer, "is that it is hard to know who is in and who is out on any given day".[124]

The UPA vanquished its rivals, returning to office with 262 parliamentary seats and 36 per cent of the national vote, a significant improvement against 2004. The Congress(I) captured 206 constituencies, 61 more than in 2004, winning 28.6 per cent of ballots cast. In contrast, the NDA attained 159 constituencies and 24.1 per cent, a massive collapse. The BJP suffered its worst defeat since 1989, taking 116 seats and 18.8 per cent of the vote. Lastly, the Third Front drew approximately 20 per cent of the vote but just 77 parliamentary seats, its worst tally ever (see Table 12.3).

The verdict revealed the failure of proponents of the third force to deepen popular democratic mobilization, and recognize the long-term nexus between economic development and social empowerment, in their traditional regional strongholds.[125] The successors of the Janata *parivar* experienced momentous setbacks. In Bihar, Nitish Kumar of the JD(U) castigated the governance failures and poor development record of the RJD and LJP. Championing the slogan *vikaas nahin, sammaan chahiye* ("we need dignity, not development"), Lalu Yadav had reduced communal violence, even registering improvements in female literacy and infant mortality.[126] But his tenure depleted wider socioeconomic opportunities, state capacity and the rule of law, betraying the promise of "social justice". In contrast, Kumar had empowered Mahadalits, lower OBCs and poorer Muslim communities, who had suffered relative exclusion, and revived the *panchayats* with reservations for lower OBCs and women.[127] The RJD crashed, saving only four parliamentary constituencies, compared to 24 in 2004, while the LJP simply blanked. In Uttar Pradesh, the BSP, aligning with the third force for the first time, increased it tally to 21 seats, becoming the third largest formation in the Lok Sabha. Yet the party had contested 503 constituencies, more than any other contestant, encouraged by its sweeping victory in the 2007 assembly elections. Its overt personalization of power and inability to attract further upper-caste and Muslim voters, both relatively satisfied with the UPA, constrained its expansion.[128] The SP won 23 seats, 13 less than in 2004. By associating with former BJP chief minister Kalyan Singh, the party alienated Muslim voters, allowing the Congress(I) to recover ground. The traditional regional proponents of the third force, moreover, suffered greater losses in the south. Narrow-minded intrigues by Deve Gowda vis-à vis the Congress(I) and the BJP over the years generated widespread opposition, undermining the JD(S) in Karnataka.[129] The TDP sought to recover its former presence in Andhra Pradesh, offering new welfare schemes and aligning with the Left. But the ardent liberalizing bent of the former prior to 2004, which the latter had strenuously opposed, dented the credibility of both.[130] Finally, and most significantly, the Left witnessed a historic defeat. In Kerala,

THE DISSOLUTION OF THE THIRD FORCE (1998–2012)

Table 12.3 Fifteenth General Election, 2009

	Votes	Seats
UPA		
INC	28.56	206
JKN	0.12	3
MUL	0.2	2
KCM	0.1	1
NCP	2.04	9
DMK	1.83	18
AITC	3.19	19
JMM	0.21	2
RPI	0.12	0
VCK	0.18	1
IND(INC)	0.11	1
Total	36.66	262
NDA		
BJP	18.81	116
AGP	0.43	1
JD(U)	1.42	20
INLD	0.31	0
SS	1.51	11
NPF	0.2	1
SDF	0.04	1
RLD	0.43	5
TRS	0.01	2
AD	0.96	4
Total	24.12	161
Third Front		
BSP	6.17	21
TDP	2.51	6
AIADMK	1.67	9
BJD	1.59	14
JD(S)	0.01	3
MDMK	<0.01	1
PMK	0.01	0
HJC	<0.01	1
CPI(M)	5.33	16
CPI	1.43	4
RSP	0.38	2
FBL	0.32	2

KEC	0.08	0
IND(Left)	0.07	0
Total	19.58	79
Fourth Front		
SP	3.43	23
LJP	<0.01	0
RJD	1.27	4
Total	4.7	27

Source: Election Commission of India; CSDS National Election Study, Appendix I, *Economic and Political Weekly*, 44, 39 (26 September 2009): 203; Yogendra Yadav and Suhas Palshikar, "Between fortuna and virtu: explaining the Congress' ambiguous victory in 2009," *Economic and Political Weekly*, 44, 39 (26 September 2009): 34.

despite the achievements of participatory development planning, the Left endured a serious rout. High-level rifts over corruption, and allying with the communalist People's Democratic Party (PDP), tarnished its credentials.[131] In West Bengal, recent agrarian conflicts in Singur and Nandigram exposed the increasingly "smug", "high-handed" and "supercilious" underbelly of the Left's "party society", delivering a massive electoral defeat.[132] Indeed, although their national vote share fell by only 0.3 per cent, the communist parties' rate of support declined by 6 per cent in their bastions.[133]

Hence many commentators posited that India's fifteenth general election gave national parties a mandate to pursue secular politics, better governance and wider socioeconomic development. The CPI(M), glossing its losses by saying voters sought national political stability, partly agreed. The enhanced parliamentary strength of the UPA, championing its various progressive measures, encouraged such interpretations. Indeed, the JD(S) and BSP offered to support the UPA after the verdict was declared. The strength of the third force, its historically more distinctive claims now widely shared, had reached its political nadir.

The reality was more complex, however.[134] Supporters of the UPA credited its efforts to improve political governance and social welfare.[135] The fortunes of the NDA declined for the third successive election, exposing its deteriorating social base, geographical spread and ideological appeal. Legatees of the Janata *parivar* that simply emphasized social justice along narrow caste lines fared badly. Indeed, the tenure of the UPA saw levels of anti-incumbency, which undercut three-quarters of all state governments between 1989 and 1999, fall to 46 per cent between 2004 and 2008.[136] Yet the federal party system had not renationalized. The collective seat ratio of the Congress(I) and BJP had substantially rebounded. Yet the combined relative vote share, in long-term secular decline, remained less than 48 per cent.[137] Conversely, state parties together accounted for approximately 29 per cent of the overall popular vote, a constant ratio since 1999. 70 per cent of the electorate professed loyalty to their region before the nation.[138] Moreover, the Congress(I) improved its vote share by 2 per cent since 2004 by contesting 27 extra seats. Yet the party seized less than 30 per cent nationally. Its improved parliamentary strength reflected smart alliances in key states and the rise of several new regional formations, such as the Maharashtra Navnirman Sena in Maharashtra and Praja Rajyam Party in Andhra Pradesh, which disproportionately hurt its state-level adversaries. In contrast, the BJP lost the BJD during the campaign and failed to recover erstwhile allies such as the TDP, Trinamool and the AIADMK. The Congress(I)'s domination of the so-called third space, in short, owed as much to strategy as to fortune.

The CPI(M) attributed its massive defeat to three related factors. The poorer social majority, while crediting the UPA for the progressive social policies the Left had ensured, opposed the BJP. The urban middle classes and big bourgeoisie, which benefitted immensely from primitive accumulation and external liberalization, reinforced the Congress(I). Lastly, "[the] call for an alternative secular government [without programmatic unity] ... could not be believed by the people." One part of the CPI(M) leadership vowed to overcome "wrong trends" within the party—the factional infighting in Kerala and backlash against industrialization and social welfare deficits in West Bengal—by pursuing a "rectification campaign" to restore organizational discipline and ideological morale, while another underscored efforts to reunify the socialists and the communists.[139] Nonetheless, party intellectuals stood defiant. Highlighting the resurgence of the Latin American left, they claimed that embracing neoliberal development in India would pauperize the countryside and push the basic classes towards Islamic fundamentalism or Maoist nihilism, annihilating the true modern Left. "Anti-imperialism is not a product of the Left's imagination; it arises from the objective conditions faced by the people."[140] Quoting Lenin, the CPI(M) declared it was impossible to fight for socialism

> without making mistakes, without retreats, without numerous alterations to what is finished or wrongly done. Communists who have no illusions, who do not give way to despondency, and who preserve their strength and flexibility to "begin from the beginning" over and over again ... are not doomed.[141]

Yet the end came first. In May 2011, the communist Left simultaneously lost elected office in the two states that had defined its trajectory. In Kerala, the Congress(I)-led UDF defeated the LDF, the former winning 72 assembly seats while the latter retained 68. It was the narrowest of victories: less than 1 per cent of the popular vote separated the rival blocs.[142] In West Bengal, however, the Left Front endured an ignominious defeat.[143] Led by the maverick populist Mamata Banerjee, the Trinamool Congress swept to power, winning 39 per cent of the popular vote and 184 of 294 state assembly seats. The Left kept just 60. To some extent, it was a misleading result. The communist parties, retaining a plurality of support amongst the early beneficiaries of land reform, still carried 41 per cent of ballots cast. Nevertheless, small farmers and sharecroppers, as well as Muslims, swung to the Trinamool in large numbers.[144] Some blamed the deradicalization of the Left, beginning in the mid-1980s and accelerating post-liberalization.[145] Many others claimed that its parties had to face new economic dynamics. The communist Left, having persevered for an astonishing 34 years, had finally succumbed to defeat.

Purists in the CPI(M) stood firm. On the one hand, they distinguished "the practical policies of state governments" from the party, which "embodie[d] a theory". Conflating the two was an "inversion of reason".[146] On the other, merely reforming neoliberalism amounted to "empiricisation" and "the small change of politics", "uninformed by the project of transcending capitalism". Doing so would destroy the distinctiveness of the Left, which remained the sole custodian of basic class interests, which "no coalition of *reformist* forces, no matter how well-meaning and serious, can possibly *replace*".[147] In April 2012, the CPI(M) steeled its conviction at the 20th party congress in Kozhikode, re-electing General Secretary Prakash Karat. Renouncing future attempts to create a third front, the party undertook to build Left unity by focusing on peasants and workers and *Dalits* and *Adivasis*, limiting joint action with other parties to questions of federalism and secularism. The latest experiment of the third force, amidst rising inequalities and financial crises, was a shambles. It justified the change of tack.

Yet the grounds for idealizing a future communist society, and the strategies necessary to achieve it, provoked genuine debate. "The CPI(M) has always believed in applying Marxism-Leninism to the concrete conditions of India," proclaimed Karat, "[we] have never tried to emulate models abroad."[148] Yet other leading Politburo members, such as Sitaram Yechury, called for a model of socialism rooted in Indian conditions.[149] The absence of former chief minister Buddhadeb Bhattacharya, ostensibly due to poor health, intimated further dissent.[150] Ultimately the smaller communist parties expressed historic resignation.[151] Relinquishing a post he had held since 1996, the CPI General Secretary Ardhendu Bhushan Bardhan confessed, "I do not claim the Left is an alternative. We too drew a blank ... [the] idea of a programmatic front is not yet clear. We will have to go through many struggles". He humbly acknowledged that "I could not spread the party in the Hindi heartland. It is futile to think of any radical or social change in Indian politics without the heartland." Ultimately, Bardhan revisited the pivotal years of the 1990s, culminating in the United Front: "It was a historic blunder. The fact that India would have had a communist prime minister, maybe for one-and-a-half or two years ... That Obama phrase, 'Yes we can', it is a meaningful phrase." Many of his comrades undoubtedly blanched. Yet Bardhan, echoing Jyoti Basu, had it right. Imagining the possibilities of the broader Indian left meant judging plausible futures, at a given historical moment, against the mistakes of the past.

CONCLUSION

The politics of constructing a Third Front in modern Indian democracy, from its roots in the Janata Party (1977–80) and crystallization during the National Front (1989–91) to its maturation as the United Front (1996–98), has always been a struggle. Many observers blame the myopic personal ambitions of individual party leaders, bereft of any substantive agenda, simply desiring power for its own sake. Others emphasize the multiplicity of forces—distinct ideological traditions, competing policy goals and divergent social interests—underlying these inherently unwieldy coalitions. Yet most assessments, whatever their explanations, have been harsh. "The Janata had few real political designs, and it was incapable of implementing them. Survival was of the essence."[1] "What had begun as a bold experiment in party building, political realignment and national reconstruction along Gandhian lines ended in shambles."[2] Likewise, commentators lamented how the National Front spent "long days assuaging hurt feelings, imagined slights, political differences".[3] The coalition "had no machinery worth its name ... to build bridges ... dilute differences or even agree to disagree".[4] Finally, the United Front was a "patchwork creature of doubtful longevity", according to critics, encouraging an "almost palpable sense of a nation adrift with no cohesiveness or sense of purpose".[5] Indeed, even sympathetic observers claimed its predominantly lower-caste leadership took "an entirely instrumental view of the Union", lacked "a coherent view of Indian identity and operate[d] with more restricted horizons".[6] The rise and fall of the third force resembles the chronicle of a death foretold.[7]

Yet its obvious political failures, as some noted, concealed complex countercurrents. "[The] obvious weaknesses of the Janata system ...

tend to conceal the significance of new beginnings.... What the Janata Party ... accomplished during the brief span of two years was not self-evident when it assumed joint responsibility of democratic restoration and developmental reconstruction".[8] The National Front, tumultuous in office, attracted little praise. Yet some commentators found the United Front "refreshing", articulating the demands and sentiments of groups historically ignored by the Congress(I) and BJP, while others thought its constituents "managed to reconcile their differences with a fair degree of success ... a remarkable achievement [given the challenges] of growing economic liberalization and political federalization of the governance system".[9] Analyzing the trajectory of the Third Front in proper historical context requires a more nuanced perspective. And doing so helps to situate national coalition politics in India in comparative historical perspective, revealing its distinctive causal dynamics, which suggest wider conceptual, theoretical and methodological implications.

Indeed, the Janata Party, National Front and United Front differed in several important respects. First, each coalition encompassed a progressively wider constellation of parties, groups and interests. The Janata Party, led by the "squabbling gerontocratic triumvirate" of Morarji Desai, Charan Singh and Jagjivan Ram, coalesced to restore parliamentary democracy after Emergency rule.[10] It represented the merger of four distinct parties, all based in northern India, symbolizing the ascent of rich capitalist farmers in the Gangetic plains. The National Front suffered analogous tensions at the helm between V.P. Singh, Chandrasekhar and Devi Lal. Clashing personal ambitions and deeper partisan disputes spoiled the plot. Nonetheless, it was a seven-party coalition, incorporating ascendant regional formations from the east and south. Its minority parliamentary status left it vulnerable to the communist Left and Hindu right. Crosscutting tensions between the "*kisan* politics" of the erstwhile BLD and the "quota politics" of the socialists, and between the latter and *Hindutva*, precipitated its demise.[11] Finally, the United Front was a minority governing coalition encompassing fifteen parties, representing states across the country. Given the number and diversity of constituents in the alliance, which relied on the external parliamentary support of the CPI(M) and Congress(I), its leadership established a steering committee to manage inter-party relations. And the active government participation of the TDP, AGP and DMK, whose electoral fortunes had risen significantly

CONCLUSION

since the late 1980s, signified the arrival of non-Hindi-speaking regions in New Delhi. Hence the United Front represented the growing political ambitions of intermediate social forces and formerly regionalist parties in modern Indian democracy.

Second, the politics of the third force witnessed a gradual rapprochement with the communist Left as it grew increasingly antagonistic towards the Hindu right. The RSS and Jan Sangh played integral roles, respectively, in the JP Movement and Janata Party. Several eminent figures, most notably Atal Bihari Vajpayee, acquired executive power. But tensions over the membership of Jan Sangh parliamentarians in the RSS eventually led the Socialists to break away. In contrast, the National Front accepted parliamentary support from the Left Front as well as the BJP. But rising political tensions over *Hindutva*—committed to introducing a Uniform Civil Code, abrogating the special constitutional status of Jammu & Kashmir and building a Ram temple in Ayodhya—and the implementation of the Mandal Commission Report tore the coalition apart. The formation of the United Front, involving the crucial government participation of the CPI and outside support from the CPI(M), prevented the BJP from seizing national power. It represented the most secular avatar of the third force.

Finally, each of these multiparty experiments displayed greater unity. Competing prime ministerial aspirations in New Delhi and deeper partisan battles in the states plagued the Janata. The Congress(I) exploited its rivalries, temporarily propping up Charan Singh, before recapturing power. The party similarly exploited high-level intrigues during the National Front, briefly supporting Chandrasekhar, only to abandon him a few months later. Hence many commentators expected a similar fate to befall the United Front. Despite its manifold internal tensions, however, its government only fell after the Congress(I) withdrew outside support. The need to carry many state-based allies encouraged it to share power better than its predecessors.[12] Yet the willingness of the United Front to rebuff successive Congress(I) demands while in office surprised many observers.

In short, viewing the Janata Party, National Front and United Front as ill-fated experiments whose politics never change obscures as much as it reveals. The precise reasons for their respective achievements and shortcomings in economic policy, Centre-state relations and foreign affairs, lie embedded in the narrative that comprises the heart of this book. Nevertheless, several continuities, ruptures and trends become clearer in hindsight.

Ideologically, the Janata Party championed a neo-Gandhian outlook, vowing to eliminate mass poverty through political and economic decentralization in the countryside. Apart from introducing food-for-work schemes for the poorest and local development projects, however, it made little progress on these fronts. If anything, the government began to liberalize the economy, simplifying industrial licensing and corporate taxes, boosting economic growth, while greatly expanding agricultural subsidies that disproportionately benefited intermediate proprietary castes. The rise of the latter, facilitated by the Green Revolution, militated against a broader coalition of Dalits, small peasants and landless laborers. Similarly, the National Front promised greater employment, welfare and decentralization, exemplified by the Approach Paper to the Eighth Five Year Plan. Its maiden budget introduced employment guarantee schemes and earmarked greater funds for rural development. Escalating agricultural subsidies and further industrial deregulation proved more significant legacies, however, leaving critical shortfalls in health and education largely unaddressed. Accelerating fiscal and external deficits, precipitated by the Congress(I) and exacerbated by the first Iraq war, ignited a full-blown economic crisis. Finally, the United Front continued to liberalize industry, trade and investment regimes, while lowering tax rates and increasing public spending that rewarded the rising middle classes. The devolution of economic decision-making to the regions increased the chances that a minority governing coalition of state-based parties would advance the reforms. Economic liberalization by the Centre sowed divisions among the states, between Third Front constituents and within many of its parties too. But it was not a foregone conclusion. Advocates of liberalization had feared a Third Front government in New Delhi. The capture of key cabinet posts by liberal economic reformers from the regions, matched by the refusal of the CPI(M) to participate in government, made a difference. The United Front failed to embrace privatization, agricultural liberalization and labor reforms. Yet both the National Democratic Alliance and United Progressive Alliance found it difficult to advance these second-stage reforms too. In short, the radical economic rhetoric and well-formulated social democratic plans of the Janata Party, National Front and United Front largely failed to materialize. The progressive campaign promises and government pronouncements of all three coalitions might simply have been empty rhetoric and cheap talk. Yet the relatively disappointing implementation of

CONCLUSION

pro-poor policies reflected, more importantly, the failure of leftists within each administration to capture, maintain and exercise sufficient power vis-à-vis coalition partners that held rival material interests and political views. And these shortcomings exposed deeper ideologies of power, elaborated further below, understood most broadly as a set of ideas regarding its nature, purposes and limits.

Efforts to strengthen parliamentary democracy and Centre-state relations fared better in varying degrees. The Janata Party repealed Emergency rule, restoring the autonomy of parliament, the judiciary and state legislative assemblies, constraining executive power. The coalition enhanced the institutional standing of the Opposition in the Lok Sabha. And greater participation by the states in the NDC, and the first free and fair elections in Jammu & Kashmir, burnished its pro-federalist credentials. The Janata Party abused its historic mandate, alas, by dismissing nine state governments run by the Congress(I). Yet it largely improved the constitutional machinery and balance of powers governing India's federal parliamentary democracy. The record of the National Front was more contradictory. It established two important mechanisms, the Inter-State Council and Cauvery Water Disputes Tribunal, to improve Centre-state relations. The coalition also sought to check executive power by introducing Prasar Bharati, which guaranteed greater freedom for the press. Yet its vow to resolve deadly conflicts in Punjab and Jammu & Kashmir, despite early acts of goodwill towards opposition parties from both states that had participated in the regional conclaves of the 1980s, foundered against rising home-grown insurgencies. Cross-border support from Pakistan abetted the latter. But the National Front unleashed a severe military response, stoking greater violence in both regions, undermining Indian democracy for several years.

Centre-state relations improved during the United Front, which vowed to promote 'cooperative federalism'. The coalition pledged 'maximum autonomy' to Jammu & Kashmir and re-engaged the Northeast, directing central funds and scheduling high-level visits to both regions, heralding new beginnings. But it failed to ensure free assembly elections in Kashmir or reach out to secessionist forces, despite calls for both from within the JD, forfeiting an important political opportunity to the next administration. The United Front handled water-sharing disputes, between Karnataka, Andhra Pradesh and Tamil Nadu, with mixed success. The coalition brought together the relevant parties, but political manipulation exacerbated longstanding tensions. Yet neither previous Congress(I) administrations nor subsequent multi-party governments resolved the issues at stake, underscoring their intractability. Indeed, it took fifteen years before the chief ministers from the relevant states met again to address the problem. Similarly, the United Front exploited mitigating circumstances to employ national power for partisan ends, imposing President's rule against BJP state governments on two occasions, despite its constituents' professed demands against such interventions. That said, the coalition sought to reverse the decline in central transfers to the states that characterized the post-liberalization era, while revitalizing languishing federal institutions. It held more sessions of the full ISC and its standing committee in eighteen months—albeit without granting either independent statutory authority—than the National Democratic Alliance and United Progressive Alliance managed to do over six years.[13] In doing so, the United Front began to redress the asymmetries of power and resources that had historically damaged Centre-state relations.

Finally, all three governments pushed foreign policy in new directions, a realm where few observers expected much headway. The Janata Party pursued greater non-alignment vis-à-vis the United States and the Soviet Union, while putting a moratorium on nuclear testing. The coalition re-engaged China as well as Pakistan, making high-level visits to both countries, ending a long and difficult hiatus. And it implemented a 'good neighbor policy', granting Bangladesh a greater share of the Ganga waters and acceding to longstanding Nepalese requests for separate trade and transit agreements, undoing some of the hardships unilaterally imposed by the Congress(I). The National Front promoted subcontinent relations further, ending the blockade against Nepal imposed by the Congress(I) in the late 1980s, and withdrawing the increasingly contentious IPKF from Sri Lanka, alleviating bilateral strains. Finally, many presumed the lower-caste, communist and regional parties that comprised the United Front would prove unable to develop a national perspective out of their ostensibly provincial horizons. Yet the coalition belied such fears. It withstood concerted international pressure to renounce India's nuclear option. More distinctively, the United Front promoted bilateral agreements vis-à-vis its smaller neighbors on the basis of non-reciprocity, exemplified by the Ganga Waters Treaty Accord, employing state-level leaders to achieve breakthroughs in a manner that has yet to be surpassed.[14] And

CONCLUSION

the coalition resumed bilateral dialogue with Pakistan. Some observers contended that an unwieldy minority government, facing the need to bolster economic growth and gain credit abroad, was bound to pursue such initiatives.[15] Yet the United Front went much further, covertly reducing foreign intelligence operations in Pakistan. It would prove to be a controversial decision. But it was hardly the sign of a weak Centre. Arguably, all these initiatives helped the National Democratic Alliance subsequently to pursue deeper bilateral negotiations. The presence, expertise and interest of I.K. Gujral proved critical. Yet it was not simply because every foreign minister enjoyed greater discretion in coalition governments.[16] The 'Gujral doctrine' represented the broader political orientation of the third force, tempering India's traditionally overbearing posture in the subcontinent, in contrast the Congress(I) and even the BJP. The divergent regional perspectives that tested its unity in New Delhi, paradoxically, inspired its conciliatory spirit and enhanced its capacity in foreign affairs. Indeed, given the tremendous odds facing the coalition, it was surprising that it accomplished anything at all. The United Front embodied a more progressive version of the "federal nationalist" principle and asymmetric dynamics that have unevenly shaped the Indian "state-nation".[17]

Despite their differences, achievements and shortcomings, however, the Janata Party, National Front and United Front failed to endure. No single overarching theoretical explanation, either the lust for power or irreconcilable social, ideological or policy differences, accounts for their common fate. The leadership of the Janata, given the immense popular goodwill and rare parliamentary majority enjoyed by the coalition, bears tremendous responsibility for its downfall. Indeed, the myopic political machinations that marred its tenure encouraged observers to explain the dynamics of subsequent anti-Congress experiments in New Delhi in roughly similar terms. In general, however, complex interaction effects, involving structure, agency and conjuncture, shaped distinct outcomes.

The vicissitudes of third force politics, and successive governing coalitions since 1989, owe much to the compound tripartite logic of FPTP elections in a progressively regionalized federal parliamentary democracy. From the beginning, India's federal system institutionalized distinct regional cleavages, making it hard for opposition parties to create horizontal alliances across state boundaries. The Congress party, successor to the nationalist movement, found it easier. The reorganization of states into distinct linguistic-cultural zones in the 1950s and 1960s, however, encouraged opposition parties to mobilize electoral support in local idioms of caste, region and language.[18] Their rise undermined the old Congress system by 1967. The state-level coalition governments that emerged failed spectacularly. Shortsighted power struggles, substantive partisan differences and the difficulty of building cross-state alliances caused their demise. Yet the consolidation of power by Indira Gandhi in the 1970s, culminating in the Emergency (1975–77), could not prevent forces of regionalization gathering pace. Indeed, centralization generated powerlessness in New Delhi.[19] Plurality-rule elections fostered the emergence of distinct party systems in the states in which two parties or blocs competed for power in the vernacular. The gradual inability of the Congress(I) to maintain a dominant presence across the Union created a federal party system marked by "multiple bipolarities" to surface in the late 1980s, increasing parliamentary fragmentation in New Delhi.[20] Complex state-level rivalries, combined with a parliamentary system of cabinet government that divided executive authority and required a majority to remove sitting administrations, enabled successive minority coalition governments to form.[21] The rise of lower-caste, regional and communist parties, which decisively shaped the struggles over Mandal, *mandir* and market of the 1990s, was partly a long-term consequence of these complex processes. The struggles of the National Front crystallized the politics of the third force vis-à-vis the Congress(I) and BJP. The United Front represented its maturation as an idea.

However, the same regime dynamics that encouraged successive coalition governments to arise paradoxically tested their ability to survive. Distinct electoral incentives in the states, with parties contesting local issues against heterogeneous rivals in desynchronized polls over time, hindered durable alliances at the Centre. The disproportional effects of FPTP—where small vote swings led to massive seat changes—generated mistrust within particular coalition governments.[22] The regionalization of many parties' social bases, their tendency to fragment into increasingly narrow segments and relatively high levels of electoral volatility added further difficulties. And the introduction of structural economic reforms in the 1990s deepened these centrifugal tendencies. Inter-state competition over private capital generated rising economic disparities. Partisan differences over the pace and scope of reform began to widen. In sum, India's macro-democratic regime, despite its relative institutional stability, generated very high political uncertainty after 1989. Party leaders confronted an intensely competitive federal party system, where marginal electoral swings, tight electoral races and multiparty blocs determined the balance of power.[23] Indeed, the minority parliamentary status of virtually every coalition government in New Delhi meant they always faced "election time", impeding the development of inter-elite trust that required longer "historical time".[24] Sustaining a diverse multiparty administration in such circumstances, especially an alliance of state-based parties seeking to forge a Third Front, became exceedingly hard.

Indeed, few established democracies pose comparable difficulties. First, despite its FPTP electoral system, which normally yields two-party systems and single-party majority governments, India has witnessed continued parliamentary fragmentation since 1989. Hung parliaments, minority governments and power-sharing executives arise in other plurality-rule regimes. Yet they rarely become the norm. Polities where governing coalitions regularly arise, such as in continental Europe, mostly contest elections under PR. Second, the disproportional effects of FPTP normally encourage parties to stick to their policy aims and to form durable majority alliances. Yet successive governing coalitions in India have seen parties adjust their professed preferences and shift political alliances with relative ease between elections in contrast to the west European scene. Factional disputes and weak party organizations, critics rightly emphasize, encourage these differences. Indeed, regardless of their stated views on social, economic and foreign policy, many party leaders enter 'unholy alliances'. Yet the latter also reveals the cross-cutting structural dynamics, generating an unparalleled degree of bargaining complexity, of the Indian party system after 1989.[25] Third, several European polities with a history of national coalition governments—Austria, Germany, Switzerland—are federal democracies as well. Yet their governing coalitions largely comprise national party organizations that span the polity. The frequency of minority coalition governments in Canada, which has a FPTP electoral system and strong regional proclivities like India, offers a more revealing comparison. With the exception of Quebec and the recently institutionalized self-governing indigenous territories of the north, none of its provinces are organized along distinct linguistic-cultural lines. Put differently, India exhibits greater affinities with diverse ethnofederal states in other parts of the world.[26] Yet few of them are consolidated representative democracies. Indeed, many have collapsed as states. In short, the number and range of parties that have secured parliamentary representation and participated in government in India—with largely state-based parties receiving approximately 50 per cent of the national vote—constitutes a difference in order of magnitude that distinguishes its coalition politics decisively.

The crosscutting pressures engendered by India's democratic regime failed to determine the fate of the third force vis-à-vis rival governing coalitions, however. Plurality-rule elections, parliamentary cabinet government and the regional dynamics of the federal party system created incentives, opportunities and constraints that party leaders had to appraise, especially when the rules presented multiple possible outcomes. Parliamentary cabinet government, compared to the winner-takes-all logic of most presidential systems, supported executive power sharing. The succession of different governing coalitions in New Delhi since 1989 makes this clear. But forging political agreement and enforcing collective discipline posed challenges, especially in unwieldy minority administrations, beholden to external parliamentary support. Moreover, the chief protagonists did not always play by the rules of the game. The weak organizational coherence and poor ideological moorings of many parties allowed individuals and factions to undermine collegial responsibility. Party leaders had to devise power-sharing formulas and conflict resolution mechanisms to accommodate contending interests. A number of practices and techniques—from local pacts, friendly electoral contests and joint state-level campaigns to national electoral alliances, common minimum programmes and high-level steering committees—facilitated bargaining, negotiation and compromise. The greater cohesiveness of the United Front vis-à-vis the Janata Party and National Front demonstrated their usefulness. The even greater durability of the National Democratic Alliance and United Progressive Alliance, even if both enjoyed more propitious conditions, underscored their importance.

These empirical nuances offer several insights for the comparative study of coalition politics. First, they suggest that power-oriented explanations often elide important distinctions, namely what drives actors to seek high office. Personal vanity, partisan interest and the desire to represent distinct social groups comprise genuinely different motives. Grasping these distinctions requires greater conceptual discrimination. Conversely, a diversity of policy goals, social interests and

CONCLUSION

political ideologies shaped the Janata Party, National Front and United Front, not to mention the National Democratic Alliance and United Progressive Alliance. Simply put, despite the ubiquity of manifold power struggles during their respective tenures, only a single minimum winning coalition emerged after 1989. Hence the premise that parties seek office, policy and votes, and view these goals as clearly defined, mutually exclusive and exogenously determined, overstates the polarity between them. Moreover, it evades the questions of strategy and tactics—what to prioritize, with whom and to what extent, and how—that bedevil actual coalition bargaining. Comprehending these debates through singular theoretical paradigms becomes very difficult.

Second, the proclivity of comparativists to study the formation and demise of coalitions at the expense of actual government decision-making, and to construe such moments as independent of prior events and future expectations, makes it hard to answer such questions. Indeed, the ontological presuppositions that underlie these methodological choices mask real dilemmas. Parsimonious theoretical models, which posit linear causal mechanisms to elucidate composite outcomes, maximize explanatory leverage. Yet they risk misconstruing significant facts and messy realities that bear on the outcome. Studying the formation, performance and demise of coalition experiments as interconnected processes, and analyzing the concatenation of actors, structures and contingencies that produce various outcomes, frequently reveal important configurative effects. Yet temporal contingencies and complex causal chains make theoretical generalization difficult. Clear trade-offs exist. The predominance of parsimonious theoretical models and relatively static large-N studies of coalition politics in comparative scholarship, however, provides a strong incentive to pursue analytical case studies that trace causal processes to a greater degree.

Finally, despite representing an extreme case, the study of national coalition politics in India offers potential comparative insights. Since the 1980s, the number of single-party majority governments in advanced industrialized democracies in western Europe and beyond has significantly declined, comprising less than a third of all governing formations.[27] The vast majority of these polities employ PR regimes. Nonetheless, the conditions that historically enabled two-party dominance in several Westminster-style democracies have waned too, encouraging greater parliamentary fragmentation over the last decade.[28] In Britain, the growing electoral popularity of the Liberal Democrats,

Scottish National Party, Greens and United Kingdom Independence Party has challenged the overwhelming post-WWII preeminence of Labor and the Conservatives, whose combined vote share has declined from approximately 97 percent in 1951 to roughly 65 percent in 2010. Many observers expect the 2015 general election to produce another hung parliament, with Labor and the Conservatives facing dissimilar challengers in different regions, increasing the likelihood that both parties will cobble together distinct regional majorities rather than a national mandate.[29] Similarly, despite their wildly oscillating fortunes, the rise of the Bloc Quebecois, Reform Party and Canadian Alliance, and New Democratic Party in Canada has significantly lowered the collective vote share of the Liberals and Progressive Conservatives since the late 1980s. The two dominant parties, the latter having integrated its regional conservative rivals, continue to alternate in power over long stretches. Moreover, both continue to seek parliamentary majorities, viewing coalition arrangements with extreme suspicion. Nonetheless, despite the recent parliamentary majority of the Conservatives, the succession of minority federal governments between 2004 and 2011 and growing electoral volatility reveal new political dynamics. The relative significance and underlying causes of these emergent trends—the devolution of political authority and economic resources, emergence of new social dynamics and regional forces, decline of trust in established institutional arrangements—attract much debate in each country.[30] Yet all these factors make it harder for traditionally dominant parties to enjoy a balanced national presence. The potential consequences stimulate growing attention too. Strikingly, many comparativists venture that regionalization correlates with poorer democratic governance, greater political instability and potentially the breakdown of national unity.[31] The structure and character of federal parliamentary democracy in India, given its distinctive linguistic-cultural states and tremendous socioeconomic diversity, makes it hard to infer simple generalizations. However, the composite record of the third force in domestic policy matters, and its relatively imaginative efforts in foreign affairs, belies the assumption that regional parties are inherently destabilizing forces. And the vicissitudes of Third Front governments suggest theoretical insights and practical lessons for other polities, especially Westminster-style parliamentary democracies, where processes of regionalization and demands for federalism may fragment national electoral mandates in the future.

CONCLUSION

Ultimately, strategies, tactics and choices matter in politics. The imperative of good political judgment challenged every party in India, including the Congress(I) and the BJP, the latter adjusting to the demands of power sharing during the third electoral system earlier than the former. Yet it tested the socialists and communists, the two principal axes of the broader Indian left that inspired the Third Front, to a greater degree.

Social democracy has always required a political coalition. The "red-green" alliance between workers and farmers that emerged in northern Europe in the mid-twentieth century remains the paradigmatic model.[32] On the surface, communists and socialists in India appeared to be natural political representatives of these two historical protagonists. Yet they faced less propitious conditions. First, the relative lateness and political compromises of state-led industrialization produced a very small class of organized sector workers against a massive reservoir of poor informal labor, whose magnitude in the countryside far exceeded the circumstances facing any society in twentieth-century Europe. Moreover, the politics of *Bharat* that sought to project rural society as a class unto itself vis-à-vis metropolitan elites since the late 1960s euphemized divergent class interests.[33] The rural propertied castes that benefited from *zamindari* abolition and the Green Revolution opposed programmes uplifting the rural poor. The strident opposition of Charan Singh and his successors to egalitarian social programmes in the countryside underscored these cleavages. Socialists in interwar Europe had to address conflicts between small farmers and agricultural labor as well. Land reform and higher agricultural wages threatened the material interests and social prestige of propertied classes everywhere. Yet their capacity to organize urban workers and rural producers into a social democratic bloc often presumed that non-socialist parties had already mobilized rural labor. Conversely, socialist parties' attempts to radicalize the latter alienated propertied classes, presaging the road to fascism.[34] Second, democracy preceded industrialization in India. The advent of universal adult franchise in a poor agrarian society, beset with destitution and inequality on a massive scale, was an unprecedented historical experiment. Yet it decoupled the mutually reinforcing struggles for political liberties and socioeconomic rights that had developed in most European social democracies.[35] Finally, the vexed relationship between class and caste and relative importance given to political representation versus material redistribution divided socialists and communists in India, a predicament that had no significant correlate in European social democracy. In short, the struggle to forge a coalition uniting the broader Indian left confronted unique historical challenges from the start. Daunting material conditions demanded an innovative political vision.

The socialists emphasized the primacy of caste for understanding the disparities of power, status and wealth in Indian society. Since the 1960s, its leaders sought to justify, implement and extend a policy of reservations for backward castes in legislative assemblies, educational institutions and state bureaucracies, and to alter the discourse and landscape of the public sphere through symbolic acts that challenged upper-caste dominance. Rhetorically, they also advocated the right to health and education and, following Gandhi, self-sufficient small-scale development in the countryside. In contrast, the communist Left deployed a classic Marxist idiom of class struggle, championing labor mobilization, land reform and agricultural investment, and local self-government. In other words, the socialists pursued a politics of recognition based on lower-caste identities, while the communists promoted economic redistribution according to material class interests. The first addressed the degradations of caste, which violated the respect, dignity and self-worth of its most downtrodden citizens. The second tackled the structures of exploitation that shaped their relations of production. Together, these two traditions provided a critical foundation for progressive democratic politics.[36]

The socialists and communists struggled to find "an ideological chain of equivalence", however, capable of synthesizing their respective conceptual insights into a transcendent emancipatory politics.[37] Indeed, the achievements and shortcomings of each movement inversely mirrored the other. On the one hand, class-based cleavages—which divided the SSP and PSP in the 1950s, socialists and the BLD in the 1970s, and the SP and BSP in the 1990s—undermined solidarity in the Janata *parivar*. On the other, the upper-caste dominated echelons of the parliamentary communists reflected its deep resistance to caste-based empowerment, especially in West Bengal.[38] Similarly, the socialists neglected to tackle the structural relations binding land, capital and labor, or to expand the social capabilities of and economic opportunities for their most disadvantaged constituents. Yet the communists failed to recognize how social humiliation and inadequate political self-representation comprised distinct manifestations of social injustice,

CONCLUSION

irreducible to material class exploitation.[39] Ultimately, building a broad political coalition that integrated the strengths of both political movements proved difficult. Their respective struggles for ascendancy, beginning with attempts to unite the CSP and CPI during the anti-colonial movement, made it tough to bridge the divide. Their leaders, ideologies, strategies, policies and bases of support differed. Socialists and communists have remained distinct political forces ever since.

What has received less scholarly attention, however, is the notion of power that informed major sections of both political movements. In particular, their respective party leaders frequently embraced a fixed, indivisible and zero-sum concept of power. Crucially, its scope differed. On the one hand, the socialists instrumentally joined hands with many other forces, most controversially the Hindu right. Yet they struggle to share power amongst themselves, unleashing destabilizing cycles of domination, insubordination and defiance, which have proven self-liquidating for its elites. On the other, the Left Front successfully forged communist alliances over several decades in the regions. Yet the CPI(M), its leading party since the 1960s, refused to share power with non-communist parties in New Delhi, revealing a deeply moralistic, instrumental and majoritarian understanding of power. To put it in Gramscian terms, the socialists pursued too many self-destructive "wars of maneuver", allowing rivals to cannibalize its ranks, while the communists have mounted a steadfast "war of position", imprisoning them in particular bastions.[40]

As the most popular non-Congress force in the 1950s, the socialists led the call for opposition unity, sharing power with a variety of formations. Ram Manohar Lohia believed that collaboration would attenuate partisan differences. Yet non-Congressism, which manifested in the SVD coalitions in the late 1960s, the Janata Party a decade later and the National Front in the late 1980s, stretched too far. On the one hand, high-level disagreements over the Congress caused the socialists repeatedly to split. On the other, by allying with the BLD and Jan Sangh, genuine leftists in their ranks found themselves practically overwhelmed. Power sharing with Hindu conservatives and rich peasant proprietors failed to sway either group towards socialist ideology. Hence the conflicts that wracked the Janata Party and the National Front.

Still, the bewildering history of splits and mergers in the Janata *parivar* since the late 1970s requires further scrutiny. Some observers contend that caste-based politics, given the inherently local character of particular *jati* groups, inevitably generates conflict. The classification and enumeration of castes by the state, which began during the colonial regime, further encourages specific leaders to direct sectional benefits to their own communities.[41] Amazingly, as early as the 1960s, Lohia foresaw the danger that Yadavs and Chamars, the "colossi" amongst OBCs and *dalits* in northern India, groups that later would dominate the JD and its successors and BSP in Uttar Pradesh and Bihar, would corner the gains of power: "They do not change the social order ... but merely cause a shift in status and privileges ... sectional elevation ... disgusts all the other lower castes and enrages the high caste."[42] Others argue that federalism prevented the development of horizontal caste solidarities. The JD had distinct social bases in Bihar, Karnataka and Orissa by the mid 1990s: predominantly lower-caste in the first, rural propertied interests in the second, and a cross-section of social groups in the third. The subsequent formation of the RJD, JD(S) and BJD in each state reflects, to some extent, the regionalization of the party. Finally, electoral changes and political developments created additional pressures. The Anti-Defection Law and emergence of national coalition politics in the 1980s created powerful incentives for factions to maximize their political influence by breaking away.[43] All three factors carry weight. Yet each has its limits. Caste-based appeals have reshaped a social order defined by hierarchy, ordained status and ritually defined roles to one based on plurality, negotiated status and politically defined positions.[44] The Congress had distinct social bases and dissimilar electoral rivals in the regions too.[45] Finally, the predecessors of the Janata *parivar* suffered many splits long before the 1980s.

Ultimately, it was a failure of leadership. Many socialists attributed their reluctance to share power and recognize a single authority figure to a lack of experience, organization and time:

Those who have experience with power ... money, privilege, status ... learn how to negotiate power. Intermediates lack sophistication ... just like a hungry man who wants to gulp his food at once.[46]

[No] experiment of the Janata [lasts] long enough at the Centre to develop a hierarchy ... a leadership to distribute power among personalities and social forces. Visible power struggles compound the problem ... [and] generate their own logic and momentum.[47]

There is no long association of struggle together ... just a short period of cohesion. So [the Janata forces] are like a flash flood [at the Centre] ... leaving no mark thereafter.[48]

CONCLUSION

Yet an enduring Gandhian skepticism towards parties shares blame too. As some acknowledged, the socialists' neglected the "intensive work" of party building: Jayaprakash Narayan "strayed into the Sarvodaya path, which he later admitted was a blind alley", compelling its forces to rely on "imported Congressmen".[49] The respective prime ministers of the Janata Party and National Front—Morarji Desai and V.P. Singh—as well as their principal nemeses—Charan Singh and Jagjivan Ram during the former and Devi Lal and Chandrasekhar during the latter—began their careers with Congress. H.D. Deve Gowda and I.K. Gujral, who led the United Front, did as well. Narayan had hoped that a "new composite elite", untouched by the "evils of self-aggrandizement, dynastic succession, and family enrichment", would slowly emerge.[50] The socialist parties' ascent broadened Indian democracy by vernacularizing its politics and transforming the meaning of caste from *jati* to *samaj*.[51] Yet the politics of martial virtue and state patronage, which leveled traditional hierarchies between forward and backward castes, descended into a morally dubious *goonda raj* in the Gangetic plains. Worse, it showed how renegotiated caste identities could nonetheless allow hierarchical practices to persist.[52] The narrow strategic judgments of the socialists, reluctant to construct strong party organizations, stabilize power hierarchies amongst themselves and tackle material conflicts on the ground, prevented the emergence of a genuine *dalitbahujan* formation.

The greater stability and ideological consistency of the communist Left, which acquired leadership of the third force in the late 1990s, posed a striking historical contrast. Yet the choices of its two main formations, the political judgments that informed their strategic line and tactical maneuvers at key historical junctures, also generated intense debate. Indeed, the question of contesting the ballot, and whether and how to share power with other parties, has been an existential matter for parliamentary communists across the world in the twentieth century.

In western Europe, communist parties frequently entered diverse multiparty coalitions during national crises, allying with moderate parties in order to supplant the socialists as the leading progressive front, promising major gains for their constituents. Yet the perception of crisis often led their leadership to accept conservative policy measures pushed by their newfound coalition allies. Workers suffered disproportionately, causing substantial electoral losses, instigating relative decline.[53] The great social upheavals of the late 1960s, coinciding with renewed labor militancy and growing communist party membership in Italy, France and Spain, crystallized new hopes. In 1977, the Partito Comunista Italiano (PCI), Parti Communiste Français (PCF) and Partido Comunista de España (PCE) agreed to pursue a broader social coalition through democratic politics, while distancing themselves from the Soviet Union in varying degrees, inaugurating so-called Eurocommunism. The most famous experiment was the "historic compromise" of the PCI. Winning an unprecedented 34 per cent of the popular vote in the 1976 Italian general election, amidst a sense of mounting national crisis, the party supported the Christian Democrat-led ruling coalition. Its strategic rationale was twofold: to overcome the *conventio ad excludendum* (an unwritten pact to keep the PCI out of power) and to protect itself against elimination (even an electoral majority, as the overthrow of Allende in Chile in 1971 demonstrated, could not rule out a military coup or civil war against the communists). Power-sharing was inevitable.

But it proved one-sided. Facing renewed hostility from the socialists, and unable to prevent severe austerity, organized labor blamed the PCI for imposing "sacrifices without compensation". A massive electoral backlash ensued in 1979, inaugurating a decade of declining party membership and electoral support, amidst labor demobilization.[54] The PCF took a more orthodox line. Conceptualizing social pluralism as a "bloc, a concentration of practically all major interests ... around [its] leadership," the party persisted in believing that "the problems of the country reside[d] uniquely in the fact that the Communists [were] not in power".[55] Yet it suffered falling membership and electoral popularity too.[56] Some observers highlighted the inherent limitations of liberal capitalist democracy to explain these historic failures.[57] Others blamed the communist parties' Stalinism, unable to grasp the character and potential of the social revolts of the late 1960s or to resolve the atomization and heterogeneity of late industrial society in the 1970s and 1980s, leading them to defend existing structures.[58] Hence its leaders continued to see multiparty alliances primarily as means to power rather than instruments of change.[59]

Analyzing the opportunities, strategies and fortunes of the CPI and CPI(M) against the record of Eurocommunism yields several insights. By definition, the "rightist" CPI has always proved more open to power-sharing with non-communist parties, causing the split in 1964. The political judgments of its leadership have been erratic. The party disastrously misread the politics of the mid-1970s, simply labeling the JP Movement fascist, only to support the tyranny of Emergency rule. Yet the CPI also demonstrated a capacity to learn, supporting reformists in the USSR and PRC in the 1980s, and participating in the United Front government in the 1990s. The party struggled to achieve its goals. Nonetheless, it was an historic attempt to steer the most coherent manifestation of the Third Front. If anything, the isolation of the CPI vis-à-vis other communist parties at both junctures refocuses attention towards the strategy, tactics and judgments of its larger partner.

The CPI(M) consistently refused to participate in every governing coalition in New Delhi.[60] Indeed, ever since the first socialist entreaties in the 1930s, leftists communist leaders have rarely joined coalitions they could not dominate. Forsaking national office, in order to maintain the political autonomy and "accumulated moral hegemony" of the party, has always been a price worth paying. Given the opportunism that frayed the Janata *parivar*, such remarkable self-restraint deserves admiration. Yet the judgments underlying its strategic vision grew more contestable over time. The CPI(M) declined to join the Janata Party in the late 1970s, given the participation of the RSS and Jan Sangh, displaying considerable prudence. The party acknowledged that it had missed an opportunity to radicalize the JP Movement beforehand, questioning its extant tactical line. Nonetheless, the prominence of the BLD in the Janata Party, in conjunction with the Jan Sangh in the states, would have overwhelmed the CPI(M) just as it had captured power in West Bengal. Similarly, despite protestations that it was a natural ally, the party had good reason to offer external parliamentary support to the National Front a decade later. V.P. Singh disallowed the BJP from entering government. Yet it was partly due to pressure by the Left. Moreover, the BJP shared power—if precariously—with the JD in various states. The growing militancy of *Hindutva*, symbolized by the *Ramjanmabhoomi* movement, required staunch opposition. In short, the CPI(M) tactical line evinced good political judgment.

But preventing its chief minister of West Bengal, Jyoti Basu, from becoming prime minister of the United Front in the late 1990s made less sense. Sharing executive power in a large governing coalition necessarily involved ambiguities, constraints and risks. The CPI(M) would have lacked a majority of seats in the Council of Ministers. Indeed, the national electoral popularity of the party was a shadow of what the PCI enjoyed in Italy in the mid-1970s when the latter struck its 'historic compromise'. Yet the PCI was never offered cabinet participation, let alone the prime ministership. Joining government, given that it would have enjoyed sizeable representation alongside the CPI, afforded opportunities. Prime ministers lack de facto control over ministries, especially in large coalition governments, yet the capacity of parties to make or break governments in parliamentary systems typically enhances the benefits of office.[61] Formal ministerial authority furnished the power to set the agenda, to draft, shape and push legislative proposals in cabinet, and to delay or veto them before parliament had a chance to deliberate if necessary.[62] Historically, many Left parties have sought to shape government agendas through outside pressure in order to maintain their ideological distinctiveness, especially in polities where elections are competitive and decisive. Yet such a strategy usually presumed considerable policy influence through parliamentary oversight and strong, decentralized and proportional legislative committees, features that India's parliamentary democracy lacked in the late 1990s.[63] Moreover, power is a function of complex strategic behavior in large multiparty coalitions: the greater the number of players, the more strategic action matters, diminishing the salience of simple legislative strength.[64] Being the single largest party, as the BJP and Congress(I) respectively discovered in the National Democratic Alliance and United Progressive Alliance, cannot determine every important outcome. And ministerial office does not ensure real clout. Deve Gowda and Gujral circumvented the CPI Home Minister Indrajit Gupta on key appointments under his remit, demonstrating the limits of formal executive power. Yet *de jure* authority, as the Fifth Pay Commission demonstrated, more often mattered. The longstanding complaints of discrimination by the Centre towards the CPI(M) during the Congress era, and acknowledgement that such practices lessened during other dispensations, suggested as much. A Basu prime ministership would have altered the balance of power in the Council of Ministers. The formation of the United Front constituted a political opening, creating new pathways, amplifying the significance of agency. It demanded the exercise of good political judgment, to foresee the likely consequences of particular actions at the moment of decision, even if these uncertainties made it harder to judge well.

Leading a diverse minority government, beholden to a historic rival, posed obvious risks. The Congress(I) would have likely toppled a Basu ministry prematurely too. Redirecting capitalist development from

CONCLUSION

New Delhi, in a federal parliamentary democracy undergoing economic liberalization, threw up enormous challenges as well. Critical junctures do not guarantee political change. Given these risks, the CPI(M)'s decision not to participate in government appeared supremely rational. Protecting its vote base and tenure in office in West Bengal, Kerala and Tripura vis-à-vis the Congress(I), while seeking to exercise policy influence at the Centre, was a shrewd political stance. Indeed, the party deployed its parliamentary veto to slow the pace and scope of liberalization in New Delhi during the United Front as well as the United Progressive Alliance after 2004, helping to push new social entitlements during the latter to ameliorate the inequalities unleashed by uneven capitalist growth. The CPI(M) simply tried 'have its cake and eat it' too.

But ostensibly clever instrumental reasoning, allied with high-minded theoretical consistency, can impede good political judgment in intricate strategic environments. On the one hand, the party could not dictate policy agendas from outside. Sympathetic observers blamed the United Front for not making the Steering Committee its primary decision-making body. Granting every party a veto over fundamental issues would have avoided unnecessary policy conflicts and enabled more egalitarian measures to emerge.[65] By arrogating the right to lead without bearing formal political responsibility, however, the CPI(M) weakened its stand. Few believed the party would withdraw external parliamentary support during the United Front. On the other, the compulsion to liberalize pressured every state in the 1990s, causing increasing divergence between the policies of CPI(M)-led government in West Bengal vis-à-vis the stance of its Politburo in New Delhi. It became impossible to square the contradictions and fend off charges of hypocrisy. Put differently, the party's short-term instrumentality evoked the historic predicament of "rational choice Marxists" in interwar Europe, struggling to legislate socialism through parliamentary democracy. The need to win broad electoral support and political allies, and failure to answer the crises of capitalism, enabled Keynesian social democrats to win the day.[66] In the end, the CPI(M) played a two-level game, stranding it between government and opposition. The party courageously walked out of the United Progressive Alliance over its opposition to the Indo-US civil nuclear deal. But the field of play by then had changed decisively against it.

According to some, rapid political change encourages actors to rely on existing templates, whereas changes that persist over longer durations tend to facilitate political learning.[67] Yet ideologies are powerful impediments. The steadfast refusal of the CPI(M) to reconsider its basic tactical line, in light of changing political circumstances, suggested an understanding of politics beholden to the "textbook solutions" offered by "high theory".[68]

Moreover, the Stalinism that pervaded a majority of the CPI(M) leadership, the belief that it was the only party that could rally all left, progressive and democratic forces, made it "incapable of learning the lessons of the past or of strategically orienting [its own cadre]. Ideological and programmatic mistakes were not directly confronted; errors were sidestepped, not overcome."[69] Yet the problem, arguably, lay even deeper. The theory of strategy-and-tactics that inspired the CPI(M), which privileged a traditional understanding of class relations and presumed that laws governed their dynamics, evaded the task of judgment, of making fine-grained assessments of specific historical contexts and moments of change, in order to grasp the possibilities and limits of autonomous political action in specific strategic situations.

Suffice to say, neither a Basu prime ministership during the United Front nor formal cabinet participation in the United Progressive Alliance would have engendered radical transformations. Notwithstanding its denunciation of capitalism and US imperialism, which prevented it from castigating similar developments in the PRC or from emulating the path taken by the Partido dos Trabalhadores in Brazil,[70] the CPI(M) became a social democratic party long ago.[71] Only a revolutionary political organization that promoted anti-capitalist structural reforms, practiced internal democracy with proportional representation for historically subordinate classes, and allied with socialist movements across the subcontinent could offer a more radical alternative.[72]

Yet the belief in a pristine socialist future, capable of avoiding the mistakes of the past, remains an article of faith for many too. "The future is certain," as the old Soviet joke has it, "it is only the past that is unpredictable".[73] Counterfactuals are hard to evaluate in a world of probability. Multiple causal forces interact, in a non-linear fashion, to shape historic outcomes.[74] Nonetheless, highlighting the scope for agency during periods of transition, particularly the importance of good political judgment, requires observers to exercise the latter as well: to analyze the foreseeable consequences of different political options by reconstructing, as carefully as possible, the historic contexts that gave them meaning. We do not have to rewrite Indian history

CONCLUSION

drastically to envision a plausible alternative to the present. The United Front constituted a rare political opening for the CPI(M) to consolidate a progressive alliance, however short-lived, and to introduce egalitarian measures that might have galvanized its ancillary front organizations, buttressed egalitarian movements and earned wider electoral support in new domains. Tactical maneuvers rarely "change the structure of the world", but they frequently allow political actors to "profit from its moment of fragility".[75] Renouncing the chance to lead a national coalition government, with all its inevitable flaws, was a missed historical opportunity. A more imaginative grasp of the real possibilities available at the time, a determination to identify political opportunities amidst structural constraints, would have enabled likeminded allies amongst the socialists and the communists to advance the cause of a third force to a greater degree.

The idea, despite its waning political fortunes since the late 1990s, nevertheless persisted. The advent of the sixteenth general election in 2014 revived the ambition. Most observers expected the demise of the United Progressive Alliance. High-level tussles over power and policy between Prime Minister Manmohan Singh and the Congress(I) president, Sonia Gandhi, had paralyzed decision-making during its second term in office. High inflation, industrial stagnation and rapid economic deceleration ensued, amidst revelations of high-level corruption, overshadowing the rights-based welfare agenda the government had introduced. Public dissatisfaction mounted.

The spring of 2012 presaged a new political order. Intriguingly, it recalled the ferment that enabled the Janata Party to emerge in the late 1970s. On the one hand, an array of forces, including Gandhian social activists, urban middle classes and Hindu nationalist groups, channeled their discontent into the India Against Corruption movement, demanding the introduction of a Lok Pal to restore political integrity in public life. On the other, the CPI(M), criticizing the composition and ideology of the anti-corruption movement, called for a multiparty alliance of left and democratic forces at its 20th congress in Kozhikode, pledging to defend sovereignty and democracy, federalism and secularism, and people's livelihoods and rights. The party believed that bringing together rivals of the Congress(I) and BJP, which collectively had almost 300 parliamentary seats, could enable yet another "third alternative" to form the next Union government.[76]

The subsequent formation of the Aam Aadmi Party (AAP), a neo-Gandhian formation that emerged from the anti-corruption movement to win a plurality of seats in the 2013 Delhi assembly elections, altered the field of battle. Relations between the older communist vanguard and upstart socialist party remained prickly. In February 2014, the CPI(M) announced an eleven-party bloc in the Lok Sabha, comprising several longstanding protagonists of the third force, and obliquely invited the AAP to join. The latter dismissed its latest incarnation of the third force as "usual, predictable and boring", a "substitute" to the ruling establishment, not an "alternative".[77] It proved astute. Within days the AIADMK and BJD, wishing to maintain political flexibility, distanced themselves from the bloc. But the massive electoral wave for the BJP, with each survey forecasting greater numbers for the party, compelled its opponents to reconsider their stances. Indeed, some erstwhile members of the third force, notably the MGP in Goa, LJP in Bihar and TDP in Andhra Pradesh, whose leaders had previously castigated Modi for the anti-Muslim pogrom in 2002, joined the National Democratic Alliance.[78] Thus in the final phase of the campaign the Left, Congress(I) and the AAP declared their willingness to unite following the verdict.[79]

The BJP routed its rivals. The party expanded its presence in every major state of the Union, winning 31 per cent of the national vote and 282 seats in the Lok Sabha, becoming the first party to win a parliamentary majority since 1984. The electoral turnout, which saw 66 per cent of eligible voters cast a ballot, was the highest since independence. Yet the consolidation of votes and seats amongst relatively fewer parties, and the increasing margins of victory at the constituency level, suggested a new electoral phase.[80] Several regional parties—especially the BJD in Orissa, AIADMK in Tamil Nadu and TC in West Bengal—swept their respective state arenas. But the need to share national power, the distinctive feature of the third electoral system, had been contained for the first time since 1989.

The politics of the broader Indian left, wherever its forces may lie, now confronts its greatest test. The Congress(I), retaining only 44 of 206 seats, faces existential crisis. The communist Left, reduced to twelve parliamentary seats, does as well. The stunning electoral sweep of the BJP in Uttar Pradesh and Bihar, at the expense of the SP, BSP and JD(U), underscores the limitations of their extant political strategies.[81] The terrible misjudgment of the AAP, relinquishing its electoral goodwill by prematurely resigning its minority administration in Delhi, inflicted unnecessary self-defeat. Each of these parties, to be sure,

CONCLUSION

retains electoral popularity. The disproportional effects of FPTP, in a divided oppositional field, over-rewarded the BJP. Moreover, despite its dominance of the Lok Sabha, the National Democratic Alliance lacks a majority of seats in the Rajya Sabha. It will have to earn wider cross-party support to pass legislation.[82] Nevertheless, the BJP earned a significant popular mandate, revealing widespread desire for leadership, stability and growth. The United Progressive Alliance, which neglected to foster the conditions for greater investment, employment and production alongside its pioneering welfare schemes, paid the price.

Yet opponents of the present BJP-led government may soon have reason to regroup. Its pro-business economic strategy of prioritizing capital accumulation and infrastructural development, if zealously pursued, risks enhancing social inequalities and human dispossession. The phenomenal amount of money spent by the party during the campaign, supplied by titans of finance and industry, underscores the nature of the new ruling alliance.[83] The presumption that modernization and growth will inevitably promote greater social opportunities begs obvious questions, given the severe market failures neoliberal economic policies have produced around the world in recent years. Moreover, critics rightly fear the BJP may weaken the checks and balances of India's democratic regime, fanning nationalistic sentiment while suppressing political dissent. Signs of intolerance and acts of intimidation during the campaign did not augur well. A new dispensation demanding political loyalty could further undermine institutional autonomy. Comparisons with the centralization of power under Indira Gandhi, and its calamitous effects, deserve considerable attention. Finally, the new prime minister made several efforts to articulate a new political discourse during the campaign, frequently expressing his ambition to transform economic development into a mass social movement. But evidence of massive communal polarization in northern India and the dearth of minorities in the BJP in parliament underscore the persistence of Hindu nationalist beliefs among many sections of the *sangh parivar*. Indeed, several high-level appointments in the new administration desire a more aggressive posture against civic organizations and neighboring states.[84] For the remnants of the contemporary Indian left to resurrect themselves amidst such adversity will require astute political leadership, committed to building durable party organizations and connecting with progressive social movements, to forge a new electoral coalition. Devising an imaginative strategic vision of how to promote social equality without jeopardizing economic dynamism, and constructing transparent, representative and accountable political institutions without undermining their capacity to implement decisions, are equally difficult balancing acts. The ability of the broader Indian left to forge social democracy anew will shape the life-chances of multitudes in the world's largest democracy.

NOTES

INTRODUCTION

1. Most contemporary scholarship of the Indian left exclude its socialist parties, given their weak organizational structures and heterodox ideological programmes, focusing solely on the parliamentary and non-parliamentary communists. However, it is important to examine both traditions, whatever their advantages and shortcomings. A significant exception remains Paul R. Brass and Marcus F. Franda (eds), *Radical Politics in South Asia* (Cambridge: MIT Press, 1973).
2. India is a federal polity: a Union of 29 states, as of today, and seven union territories. In 1996, there were 25 states. The national government in New Delhi is frequently called the Centre; states, with their own capitals and legislative assemblies, are sometimes called the regions.
3. See Yogendra Yadav, "Reconfiguration in Indian politics: state assembly elections, 1993–1995," in Partha Chatterjee (ed.), *State and Politics in India* (New Delhi: Oxford University Press, 1997), pp. 177–208; and idem, "The third electoral system," *Seminar*, 480 (August 1999): 14–20.
4. Ramesh Thakur, "A changing of the guard in India," *Asian Survey*, 38, 6 (June 1998), p. 616.
5. For example, see Paul R. Brass, *The Politics of India since Independence*, 2nd edition (Cambridge: Cambridge University Press, 1994), p. 337.
6. E. Sridharan, "Principles, power and coalition politics in India: lessons from theory, comparison and recent history," in D.D. Khanna and Gert W. Kueck (eds.), *Principles, Power and Politics* (New Delhi: Macmillan India, 1999), p. 284.
7. For example, see T.V. Sathyamurthy, "Impact of Centre-state relations on Indian politics: an interpretative reckoning, 1947–84," in Partha Chatterjee (ed.), *State and Politics in India* (New Delhi: Oxford University Press, 1998), p. 252.
8. To invoke Gabriel García Márquez, *Chronicle of a Death Foretold* (New York: Vintage, 2003).

9. For example, see Baldev Raj Nayar, "Policy and performance under democratic coalitions," *Journal of Commonwealth and Comparative Politics*, 37, 2 (July 1999): 22–56.
10. Important exceptions include Lloyd I. Rudolph and Susanne Hoeber Rudolph, *In Pursuit of Lakshmi: the political economy of the Indian state* (Chicago: University of Chicago Press, 1987), pp. 172–177; and Mahendra Prasad Singh, "India's National Front and United Front coalition governments: a phase in federalized governance," in Mahendra Prasad Singh and Anil Mishra (eds), *Coalition Politics in India: problems and prospects* (New Delhi: Manohar, 2004), pp. 75–99.
11. Editorial, "Perils: real and fanciful," *Economic and Political Weekly*, 33, 10 (7 March 1998), p. 491.
12. On the concept of "federal nationalism" and "state-nation", respectively, see Balveer Arora, "Political parties and the party system: the emergence of new coalitions," in Zoya Hasan (ed.), *Parties and Party Politics in India* (New Delhi: Oxford University Press, 2004), pp. 504–33, and Alfred Stepan, Juan Linz and Yogendra Yadav, *Crafting State-Nations: India and other multinational democracies* (Baltimore: Johns Hopkins University Press, 2011).
13. Yadav, "The third electoral system."
14. See Sridharan, "Principles, power and coalition politics in India."
15. Sridharan, "Principles, power and coalition politics in India."
16. In general, see Herbert Kitschelt, "Party systems," in Carles Boix and Susan Stokes (eds), *The Oxford Handbook of Comparative Politics* (New York: Oxford University Press, 2009), pp. 530–1.
17. For example, see Peter A. Hall and Rosemary Taylor, "Political science and the three new institutionalisms," *Political Studies*, 44, 5 (December 1996): 936–57; Wolfgang Streeck and Kathleen Thelen (eds), *Beyond Continuity* (Oxford: Oxford University Press, 2004); and Mark Blyth, "Structures do not come with an instruction sheet: interests, ideas, and progress in political science," *Perspectives on Politics*, 1, 4 (December 2003): 695–706.
18. Tim Büthe, "Taking temporality seriously: modeling history and the use of narratives as evidence," *American Political Science Review*, 96, 3 (September 2002): 481–94.
19. Paul Pierson, *Politics in Time: history, institutions and social analysis* (Princeton: Princeton University Press, 2004).
20. See J.W.N. Watkins, "Ideal types and historical explanation," in Alan Ryan (ed.), *The Philosophy of Social Explanation* (Oxford: Oxford University Press, 1973).
21. Geoffrey Hawthorn, *Plausible Worlds: possibility and understanding in history and the social sciences* (Cambridge: Cambridge University Press, 1991), xi.
22. For example, see Robert H. Bates, Avner Grief, Margaret Levi, Jean-Laurent Rosenthal and Barry R. Weingast, *Analytic Narratives* (Princeton: Princeton University Press, 1998); Doug McAdam, *Political Process and the Development of Black Insurgency, 1930–1970* (Chicago: University of Chicago Press, 1982); and Theda Skocpol, *States and Social Revolutions: a comparative analysis of France, Russia and China* (Cambridge: Cambridge University Press, 1979).

NOTES

23. Andrew Abbott, "From causes to events," in Andrew Abbott, *Time Matters* (Chicago: University of Chicago Press, 2001), p. 186; Paul Roth, "Narrative explanations," in Michael Martin and Lee C. McIntyre (eds), *Readings in the Philosophy of Social Science* (Cambridge: MIT Press, 1995), p. 709.
24. A classic account remains Gosta Esping-Andersen, *The Three Worlds of Welfare Capitalism* (Princeton: Princeton University Press, 1990).

1. THE PARADOXES OF INDIA'S COALITION POLITICS

1. Many early studies of European coalition politics analyzed specific cases. Their impact on recent theoretical debates is slight, however. See Vernon Bogdanor (ed.), *Coalition Government in Western Europe* (London: Heinemann Educational Books, 1983); and Geoffrey Pridham (ed.), *Coalitional Behaviour in Theory and Practice* (Cambridge: Cambridge University Press, 1986).
2. Bent Flyvbjerg, *Making Social Science Matter* (Cambridge: Cambridge University Press, 2001), pp. 66–7.
3. See the articles of Paul R. Brass, "Party systems and governmental instability in Indian states" and "Coalition politics in north India," in Paul R. Brass, *Caste, Faction and Party in Indian Politics, Volume One: faction and party* (New Delhi: Chanakya Publications, 1984), pp. 19–62 and 97–135, respectively.; Bruce Bueno de Mesquita, *Strategy, Risk and Personality in Coalition Politics: the case of India* (Cambridge: Cambridge University Press, 1975); and Subrata K. Mitra, *Governmental Instability in Indian States: West Bengal, Bihar, Uttar Pradesh and Punjab* (New Delhi: Ajanta Publications, 1978).
4. For example, see Saral K. Chatterji (ed.), *The Coalitional Government: a critical examination of the concept of coalition, the performance of some coalitional governments and the future prospects of coalition in India* (Madras: The Christian Literature Society, 1974); Manju Verma, *The Coalition Ministries in Punjab* (Patiala: Shivalik Printing House, 1978); and E.J. Thomas, *Coalition Game Politics in Kerala* (New Delhi: Intellectual Publishing House, 1985).
5. In particular, see Bidyut Chakrabarty, *Forging Power: coalition politics in India* (New Delhi: Oxford University Press, 2006).
6. In particular, see Csaba Nikolenyi, *Minority Governments in India: the puzzle of elusive majorities* (London: Routledge, 2010).
7. In particular, see E. Sridharan, "Principles, power and coalition politics"in India: lessons from theory, comparison and recent history," in D.D. Khanna and Gert W. Kueck (eds), *Principles, Power and Politics* (New Delhi: Macmillan India, 1999), pp. 270–291; "The fragmentation of the Indian party system: seven competing explanations," in Zoya Hasan (ed.), *Parties and Party Politics in India* (New Delhi: Oxford University Press, 2002), pp. 478–504; "Coalitions and party strategies in India's parliamentary federation," *Publius*, 33, 4 (Autumn 2003): 135–52; "Unstable parties and unstable alliances: births, splits, mergers and deaths of parties, 1952–2000," in Mahendra Prasad Singh and Anil Mishra (eds), *Coalition Politics in India: problems and prospects* (New Delhi: Manohar, 2004), pp. 43–74; and "Why are multi-party minority governments

viable in India? theory and comparison," *Commonwealth & Comparative Politics*, 50, 3 (July 2012): 314–43.
8. See Ian Shapiro, "Problems, methods and theories in the study of politics, or: what's wrong with political science and what to do about it," *Political Theory*, 30, 4 (August 2002): 596–620.
9. Shapiro, "Problems, methods and theories in the study of politics," pp. 600–605.
10. Michael Laver and Norman Schofield, *Multiparty Government: the politics of coalition in Europe* (Oxford: Oxford University Press, 1990), p. 215.
11. There are some prominent exceptions. Michael Laver and Kenneth A. Shepsle (eds), *Cabinet Ministers and Parliamentary Government* (Cambridge: Cambridge University Press, 1994) examines the role of cabinet ministers. Gregory M. Luebbert, *Comparative Democracy: policymaking and governing coalitions in Europe and Israel* (New York: Columbia University Press, 1986) focuses on party leaders.
12. Michael Laver, "Legislatures and parliaments in comparative context," in Barry Weingast and Donald A. Wittman (eds), *The Oxford Handbook of Political Economy* (New York: Oxford University Press, 2008), pp. 121–40; and Kaare Strøm and Benjamin Nyblade, "Coalition theory and government formation," in Carles Boix and Susan C. Stokes (eds), *The Oxford Handbook of Comparative Politics* (New York: Oxford University Press, 2009), pp. 782–802.
13. Sridharan, "Unstable parties and unstable alliances," p. 44.
14. Terence Ball, "Party," in Terence Ball, James Farr and Russell Hanson (eds), *Political Innovation and Conceptual Change* (Cambridge: Cambridge University Press, 1989), pp. 155–76.
15. Nikolenyi, *Minority Governments in India*, p. 72.
16. Adam Ziegfeld, "Coalition government and party system change: explaining the rise of regional political parties in India," *Comparative Politics*, 45, 1 (October 2012): 69–87.
17. See Peter Mair, "Introduction," in Peter Mair (ed.), *The West European Party System* (Oxford: Oxford University Press, 1990), pp. 1–25.
18. For example, see Moshe Maor, *Parties, Conflicts and Coalitions in Western Europe: organisational determinants of coalition bargaining* (London: Routledge/LSE, 1998); and Laver, "Legislatures and parliaments in comparative context".
19. Akhtar Majeed, "Coalitions as power-sharing arrangements," in Akhtar Majeed (ed.), *Coalition Politics and Power Sharing* (New Delhi: Manak Publications PVT Ltd., 2000), p. 14. In general, see James Manor, "India's chief ministers and the problem of governability," in Philip Oldenburg (ed.), *India Briefing* (New York: M.E. Sharpe, 1995), pp. 47–75.
20. Seminal power-maximization theories include William H. Riker, *The Theory of Political Coalitions* (New Haven: Yale University Press, 1962); William Gamson, "A theory of coalition formation," *American Sociological Review*, 26, 4 (June 1961): 373–82; and Lawrence C. Dodd, *Coalitions in Parlia-*

mentary Government (Princeton: Princeton University Press, 1976). Dodd argues that social cleavages may constrain parties' abilities to pursue office, however.
21. Gregory M. Luebbert, "Coalition theory and government formation in multiparty democracies," *Comparative Politics*, 15, 2 (January 1983), p. 238.
22. Strøm and Nyblade, "Coalition theory and government formation," pp. 786–88.
23. Joseph A. Schumpeter, *Capitalism, Socialism and Democracy* (London: Routledge, 1994).
24. Seminal policy-realization theories include Robert Axelrod, *Conflict of Interest: a theory of divergent goals with applications to politics* (Chicago: Markham Publishing Company, 1970); and Abram de Swann, *Coalition Theories and Cabinet Formations: a study of formal theories of coalition formation applied to nine European parliaments after 1918* (New York: Elsevier Scientific Publishing Company, 1973).
25. Luebbert, "Coalition theory and government formation in multiparty democracies," p. 239.
26. Anthony Downs, *An Economic Theory of Democracy* (New York: Harper, 1957), p. 28.
27. Downs, *An Economic Theory of Democracy*, pp. 30–31.
28. Downs, *An Economic Theory of Democracy*, p. 159.
29. Sridharan, "Principles, power and coalition politics in India, pp. 270–291.
30. Majeed, "Coalitions as power-sharing arrangements," pp. 10–11.
31. Paul R. Brass, "Leadership conflict and the disintegration of the Indian socialist movement: personal ambition, power and policy," in Paul R. Brass, *Caste, Faction and Party in Indian Politics, Volume One: faction and party* (New Delhi: Chanakya Publications, 1984), pp. 169 and 181–186.
32. Sridharan, "Principles, power and coalition politics in India."
33. Majeed, "Coalitions as power-sharing arrangements," p. 25.
34. Respectively, see Kanchan Chandra, *Why Ethnic Parties Succeed: patronage and ethnic head counts in India* (Cambridge: Cambridge University Press, 2004), pp. 220–1; and Andrew Wyatt, "The limitations on coalition politics in India: the case of electoral alliances in Uttar Pradesh," *Commonwealth & Comparative Politics*, 37, 2 (July 1999): 1–21.
35. Max Weber, "The profession and vocation of politics," in Peter Lassman and Ronald Speirs (eds), *Weber: political writings* (Cambridge: Cambridge University Press, 1994), p. 353.
36. Zoya Hasan, "Introduction," in Zoya Hasan (ed.), *Parties and Party Politics in India* (New Delhi: Oxford University Press, 2004), pp. 21–33.
37. Luebbert, "Coalition theory and government formation in multiparty democracies," p. 243.
38. See Luebbert, "Coalition theory and government formation in multiparty democracies," p. 244, and Gregory Luebbert, "A theory of government formation," *Comparative Political Studies*, 17, 2 (July 1984), pp. 237–43.
39. See Wolfgang C. Müller and Kaare Strøm, "Political parties and hard choices."

In Wolfgang C. Müller and Kaare Strøm (eds), *Policy, Office or Votes? How political parties in Western Europe make hard decisions* (Cambridge: Cambridge University Press, 1999), pp. 1–35

40. Müller and Strøm, "Political parties and hard choices," p. 26.
41. Bernard Williams, "Realism and moralism in political theory," in Geoffrey Hawthorn (ed.), *In the Beginning Was the Deed: realism and moralism in political argument* (Princeton: Princeton University Press, 2005), p. 12.
42. Guillermo O'Donnell, "Delegative democracy," in Larry Diamond and Marc F. Plattner (eds), *The Global Resurgence of Democracy*, 2nd edition (Baltimore: Johns Hopkins University Press, 1996), p. 97.
43. Müller and Strøm, "Political parties and hard choices," pp. 22–23.
44. See Juan J. Linz, "The perils of presidentialism" and "The virtues of parliamentarism," in Diamond and Plattner (eds), *The Global Resurgence of Democracy*, 2nd edition (Baltimore: The John Hopkins University Press, 1996), pp. 124–43 and 154–62, respectively; and Scott Mainwaring and Matthew S. Shugart, "Juan Linz, presidentialism, and democracy: a critical appraisal," *Comparative Politics*, 29, 4 (July 1997): 449–71.
45. The following discussion relies on Laver and Schofield, *Multiparty Government*, pp. 144–163.
46. Arend Lijphart, *Patterns of Democracies: government forms and performance in thirty-six countries* (New Haven: Yale University Press, 1999), p. 97. The rest of the paragraph summarizes Strøm and Nyblade, "Coalition theory and government formation," pp. 788–794.
47. Daniel Diermeier, Hulya Eraslan and Antonio Merlo, "Coalition government and comparative constitutional design," *European Economic Review*, 46, 4 (2002): 893–907.
48. Strøm and Nyblade, "Coalition theory and government formation," p. 796.
49. Strøm and Nyblade, "Coalition theory and government formation," p. 797. Earlier statistical analyses found that "country effects" outweighed the significance of cross-national patterns. See M. Franklin and T. Mackie, "Reassessing the importance of size and ideology for the formation of governing coalitions in parliamentary democracies," *American Journal of Political Science*, 28 (1984): 94–103, and B. Grofman, "The comparative analysis of coalition formation and duration: distinguishing between-country and within-country effects," *British Journal of Political Science*, 19, 2 (April 1989): 291–302.
50. Debate persists over whether random critical events have a uniform impact across coalition governments irrespective of time, or if their vulnerability increases the longer they stay in office. See E.C. Browne and D.W. Gleiber, "An 'events' approach to the problem of cabinet stability," *Comparative Political Studies*, 17, 2 (July 1984): 167–97; Gary King, James Alt, Nancy Burns and Michael Laver, "A unified model of cabinet dissolution in parliamentary democracies," *American Journal of Political Science*, 34, 3 (August 1990): 846–71; A. Lupia and K. Strøm, "Coalition termination and the strategic timing of parliamentary elections," *American Political Science Review*, 89, 3 (September 1995): 33–50; and D. Diermeier and R. Stevenson, "Cabinet sur-

vival and competing risks," *American Journal of Political Science*, 43, 4 (October 1999): 1051–68.
51. See Maurice Duverger, *Political Parties: their organization and activity in the modern state* (New York: Wiley, 1963).
52. Sridharan, "The fragmentation of the Indian party system, 1952–1999," pp. 478–504.
53. David Samuels, "Separation of powers," in Carles Boix and Susan C. Stokes (eds), *The Oxford Companion to Comparative Politics* (New York: Oxford University Press, 2009), pp. 709–10.
54. Nikolenyi, *Minority Governments in India*, pp. 5–6.
55. Nikolenyi, *Minority Governments in India*, pp. 11–12.
56. Sridharan, "Why are multi-party minority governments viable in India?" pp. 328–332.
57. Ziegfeld, "Coalition government and party system change," pp. 69–87.
58. Thomas Blom Hansen and Christophe Jaffrelot, "Introduction," in Thomas Blom Hansen and Christophe Jaffrelot (eds), *The BJP and the Compulsions of Politics in India*, 2nd edition (New Delhi: Oxford University Press, 2001), pp. 1–22.
59. Yogendra Yadav, "Reconfiguration in Indian politics," in Partha Chatterjee (ed.), *State and Politics in India* (New Delhi: Oxford University Press, 1997), pp. 177–208. E. Sridharan coined the term "multiple bipolarities".
60. Sridharan, "Why are multi-party minority governments viable in India?" pp. 328–32.
61. Alfred Stepan, Juan Linz and Yogendra Yadav, *Crafting State-Nations: India and other multinational democracies* (Baltimore: Johns Hopkins University Press, 2011), pp. 17–22 and 54–56.
62. Pradeep K. Chibber and John R. Petrocik, "The puzzle of Indian politics: social cleavages and the Indian party system," *British Journal of Political Science*, 19, 2 (April 1989): 207.
63. See Alistair McMillan, "The BJP coalition: partisanship and power-sharing in government," in Katharine Adeney and Lawrence Sáez (eds), *Coalition Politics and Hindu nationalism* (London: Routledge, 2005), pp. 13–35.
64. I owe this insight to E. Sridharan.
65. Nikolenyi, *Minority Governments in India*, pp. 152–153.
66. James Chiriyankandath, "'Unity in diversity?' coalition politics in India (with special reference to Kerala)," *Democratization*, 4, 4 (Winter 1997), p. 34.
67. See Herbert Kitschelt, "Party systems," in *The Oxford Handbook of Comparative Politics*, pp. 530–1.
68. Peter A. Hall and Rosemary Taylor, "Political science and the three new institutionalisms," *Political Studies*, 44, 5 (December 1996), p. 940.
69. O'Donnell, "Delegative democracy," p. 96.
70. Gerald Berk and Dennis Galvan, "How people experience and change institutions: a field guide to creative syncretism," *Theory and Society*, 38, 6 (2009), p. 544; and Adam D. Sheingate, "Political entrepreneurship, institutional change, and American political development," *Studies in American Political Development*, 17, 2 (Fall 2003): 185–9.

71. McMillan, "The BJP coalition," p. 16. More generally, see Lloyd I. Rudolph and Susanne Hoeber Rudolph, *In Pursuit of Lakshmi: the political economy of the Indian state* (Chicago: University of Chicago Press, 1987), pp. 95–8.
72. In general, see Wolfgang Streeck and Kathleen Thelen (eds), *Beyond Continuity* (Oxford: Oxford University Press, 2004), and Devesh Kapur and Pratap Bhanu Mehta (eds), *Public Institutions in India: performance and design* (New Delhi: Oxford University Press, 2008).
73. Karen Orren and Stephen Skowroneck, *The Search for American Political Development* (Cambridge: Cambridge University Press, 2004).
74. See Mark Blyth, "Structures do not come with an instruction sheet: interests, ideas, and progress in political science," *Perspectives on Politics*, 1, 4 (December 2003): 695–706.
75. Rogers M. Smith, "Science, non-science and politics," in Terrence J. McDonald (ed.), *The Historic Turn in the Human Sciences* (Ann Arbor: University of Michigan Press, 1996), p. 145.
76. The following paragraph summarizes Gene D. Overstreet and Marshall Windmiller, *Communism in India* (Berkeley: University of California Press, 1959), pp. 5–12.
77. Sumanta Banerjee, "Strategy, tactics, and forms of political participation among left parties," in T.V. Sathyamurthy (ed.), *Class formation and political transformation in post-colonial India* (New Delhi: Oxford University Press, 1999), pp. 205–6.
78. See Eric C. Browne and John Dreijmanis (eds), *Coalition Governments in Western Democracies* (London: Longman, 1982); Kaare Strøm, Wolfgang C. Müller and Torbjörn Bergman (eds), *Cabinets and Coalition Bargaining: the democratic life cycle in western Europe (*Oxford: Oxford University Press, 2008); and Catherine Moury, *Coalition Government and Party Mandate: how coalition agreements constrain ministerial action* (London: Routledge, 2013).
79. K.K. Kailash, "Middle game in coalition politics," *Economic and Political Weekly*, 42, 4 (27 January 2007): 307–17.
80. It partly reflects the assumption that formal political institutions adequately settle power equations within the executive. But it is also because most comparativists seek to explain the formation and demise of specific coalition governments as discrete events in large-N studies, paying less attention to how coalition governments actually work, which are hard to model using current statistical techniques. See Strøm and Nyblade, "Coalition theory and government formation," p. 799.
81. Kailash, "Middle game in coalition politics," pp. 307–317.
82. Majeed, "Coalitions as power-sharing arrangements," p. 6.
83. In general, see Gretchen Helmke and Steven Levitsky, "Informal institutions and comparative politics: A research agenda," *Perspectives on Politics*, 2, 4 (December 2004), p. 730.
84. Kailash, "Middle game in coalition politics," pp. 309 and 314.
85. Ruth Berins Collier and David Collier, *Shaping the Political Arena: critical junctures, the labor movement, and regime dynamics in Latin America* (Princeton: Princeton University Press, 1991), p. 27.

86. Ira Katznelson, "Periodization and preferences: reflections on purposive action in comparative historical social science," in James Mahoney and Dietrich Rueschemeyer (eds), *Comparative Historical Analysis in the Social Sciences* (Cambridge: Cambridge University Press, 2003), pp. 274 and 283.
87. I do not analyze any differences within this broad tradition here. For further elaboration, see Sanjay Ruparelia, "The role of judgment in explanations of politics," Political Methodology Working Paper 29 (May 2010), Committee on Concepts & Methods, International Political Science Association, www.concepts-methods.org.
88. Geoffrey Hawthorn, "Pericles' unreason," in Richard Bourke and Raymond Geuss (eds), *Political Judgment: essays for John Dunn* (Cambridge: Cambridge University Press, 2009), pp. 225–6.
89. See John Dunn, *The Cunning of Unreason: making sense of politics* (London: HarperCollins, 2001).
90. Niccolò Machiavelli, *The Prince*, edited by Quentin Skinner and Russell Price (Cambridge: Cambridge University Press, 2008), p. 85.
91. Raymond Geuss, *Philosophy and Real Politics* (Princeton: Princeton University Press, 2008), p. 98.
92. Isaiah Berlin, "Political judgment," in Henry Hardy (ed.), *The Sense of Reality: studies in ideas and their history*, with an introduction by Patrick Gardner (London: Pimlico, 1997), p. 46.
93. Geuss, *Philosophy and Real Politics*, pp. 1–18 and 31–2.
94. Raymond Geuss, "Political judgment in historical context," in Raymond Geuss, *Politics and the Imagination* (Princeton: Princeton University Press 2010), pp. 11–4.
95. Berlin, "Political judgment," p. 50.
96. See Alexander L. George and Andrew Bennett, *Case Studies and Theory Development in the Social Sciences* (Cambridge: MIT Press, 2005), p. 145; and Charles C. Ragin, "Turning the tables: how case-oriented research challenges variable-oriented research," in Henry E. Brady and David Collier (eds), *Rethinking Social Inquiry: diverse tools, shared standards* (New York: Rowman & Littlefield Publishers Ltd., 2004), p. 134.
97. John Dunn, "Unger's Politics and the appraisal of political possibility," in John Dunn, *Interpreting Political Responsibility: essays 1981–1989* (Cambridge: Cambridge University Press, 1990), pp. 170–5.
98. Jon Elster, "Rational choice history," *American Political Science Review*, 94, 3 (September 2000): 685–95.
99. Weber, "The profession and vocation of politics," p. 353.
100. Luebbert, "Coalition theory and government formation in multiparty democracies," p. 242.
101. Two notable examples include Bueno de Mesquita, *Strategy, Risk and Personality in Coalition Politics*, and Nikolenyi, *Minority Governments in India*. Mesquita argues that party leaders that share power successfully over the long term demonstrate a need for achievement, possess an organizational capability to assess risk and pursue a mixed strategy of competition and cooperation vis-à-vis other actors.

102. M.J. Akbar, "Interview with Jyoti Basu," *The Asian Age*, 2 January 1997.
103. "An action is rational ... if it meets *three optimality requirements*: the action must be optimal [choosing the best means to realize one's desires], given the beliefs; the beliefs must be as well supported as possible, given the evidence [agents must be Bayesian, i.e. they update their beliefs in light of new evidence]; and the evidence must result from an optimal investment in information gathering." Jon Elster, *Explaining Social Behavior: more nuts and bolts for the social sciences* (Cambridge: Cambridge University Press, 2007), p. 191.
104. Elster, *Explaining Social Behavior*, pp. 125 and 206–13.
105. James C. Scott, *Seeing like a State: how certain schemes to improve the human condition have failed* (New Haven: Yale University Press, 1998), p. 327.
106. See Anna Gryzmala-Busse, "Time will tell? temporality and the analysis of causal mechanisms and processes," *Comparative Political Studies*, 44, 9 (December 2010): 1267–97.
107. See Edgar Kiser and Howard T. Welser, "The microfoundations of analytic narratives," *Sociologica* 3 (November–December 2007): 1–19.
108. Daniel Diermeier, "Coalition government," in Barry M. Weingast and Donald A. Wittman (eds), *The Oxford Handbook of Political Economy* (New York: Oxford University Press, 2008), p. 162–79.
109. Adam Przeworksi, *Capitalism and Social Democracy* (Cambridge: Cambridge University Press, 1991), pp. 1–2.
110. John Ferejohn, "External and internal explanation," in Ian Shapiro, Rogers M. Smith and Tarek E. Masoud (eds), *Problems and Methods in the Study of Politics* (New York: Cambridge University Press, 2004), p. 162.
111. See Daniel Kahneman, "Nobel lecture: maps of bounded rationality: psychology for behavioral economics," *American Economic Review*, 93, 5 (2003): 1449–75.
112. Przeworksi, *Capitalism and Social Democracy*, p. 240.
113. Daniel Carpenter, "Commentary: what is the marginal value of 'analytic narratives'?" *Social Science History*, 24, 4 (Winter 2000): 653–67.
114. Kurt Weyland, *Bounded Rationality and Policy Diffusion: social sector reform in Latin America* (Princeton: Princeton University Press, 2007), pp. 5–11.
115. Margaret Levi, "An analytic narrative approach to puzzles and problems," in *Problems and Methods in the Study of Politics*, p. 220.
116. Charles Tilly and Robert E. Goodin, "It depends," in Robert E. Goodin and Charles Tilly (eds), *The Oxford Handbook of Contextual Political Analysis* (New York: Oxford University Press, 2008), p. 17.
117. Of course, rationality properly understood—the use of reason to scrutinize our ends as well as the means we select to secure them—shares many features of good judgment. See Amartya Sen, *Rationality and Freedom* (Cambridge: Harvard University Press, 2004), pp. 3–64.
118. Peter A. Hall and Rosemary Taylor, "Political science and the three new institutionalisms," *Political Studies*, 44, 5 (December 1996), p. 939.

119. James D. Fearon, "Counterfactuals and hypothesis testing in political science," *World Politics*, 43, 2 (January 1991): 169–95.
120. Margaret Levi, "A model, a method and a map: rational choice in comparative and historical analysis," in Mark Irving Lichbach and Alan S. Zuckerman (eds), *Comparative Politics: rationality, culture and structure* (New York: Cambridge University Press, 1999), p. 34.
121. Levi, "A model, a method and a map," p. 21.
122. David Collier, James Mahoney and Jason Seawright, "Claiming too much: warnings about selection bias," in Henry E. Brady and David Collier (eds), *Rethinking Social Inquiry: diverse tools, shared standards* (Lanham, MD: Rowman and Littlefield, 2004), p. 93.
123. Ian Shaprio, "Problems, methods and theories in the study of politics, or what's wrong with political science and what to do about it," *Political Theory*, 30, 4 (August 2002), pp. 614–5.
124. Gary King, Robert O. Keohane and Sidney Verba, *Designing Social Inquiry: scientific inference in qualitative research* (Princeton: Princeton University Press, 1994), pp. 34–74.
125. James Mahoney, "After KKV: the new methodology of qualitative research," *World Politics*, 62, 1 (January 2010), pp. 125–31.
126. Scott, *Seeing like a State*, p. 390, fn. 37.
127. John Gerring, *Case Study Research: principles and practices* (Cambridge: Cambridge University Press, 2007), pp. 172–85.
128. The following draws on Timothy Garton Ash, *History of the Present: essays, sketches and dispatches from Europe in the 1990s* (London: Penguin, 1999), xvii-xix.
129. See Atul Kohli, Peter Evans, Peter J. Katzenstein, Adam Przeworski, Susanne Hoeber Rudolph, James C. Scott and Theda Skocpol, "The role of theory in comparative politics: a symposium," *World Politics*, 48, 1 (October 1995), p. 14.
130. Exceptions include Jean Blondel and Ferdinand Müller-Rommel (eds), *Governing Together: the extent and limits of joint decision-making in western European cabinets* (London: Macmillan, 1993), Müller and Strøm, *Policy, Office or Votes?* and Moury, *Coalition Government and Party Mandate*.
131. See Charles Taylor, "Neutrality in political science," in Alan Ryan (ed.), *The Philosophy of Social Explanation* (Oxford: Oxford University Press, 1973), pp. 139–71.
132. Bent Flyvbjerg, *Rationality and Power: democracy in practice* (Chicago: University of Chicago Press, 1998), p. 7.
133. Geoffrey Hawthorn, *Plausible Worlds: possibility and understanding in history and the social sciences* (Cambridge: Cambridge University Press, 1991), p. 121.

2. THE ROOTS OF THE BROADER INDIAN LEFT (1934–1977)

1. See Rajni Kothari, "The Congress 'system' in India," *Asian Survey*, 4, 12

(December 1964): 1161–73; and W.H. Morris-Jones, *Politics Mainly Indian* (Madras: Orient Longman, 1978), pp. 196–232. For a synthesis of their views, see James Manor, "Parties and the party system," in Atul Kohli (ed.), *India's Democracy: an analysis of changing state-society relations* (Princeton: Princeton University Press, 1988), pp. 63–71.

2. Atul Kohli, *The State and Politics in India: the politics of reform* (Cambridge: Cambridge University Press, 1987), pp. 52–61.
3. Manor, "Parties and the party system," p. 65.
4. Kothari, "The Congress 'system' in India," p. 1162.
5. Stuart Corbridge and John Harriss, *Reinventing India: liberalization, Hindu nationalism and popular democracy* (Cambridge: Polity Press, 2000), pp. 20–67.
6. See Sunil Khilnani, *The Idea of India* (New York: Farrar Straus Giroux, 1999), pp. 28–42 and 166–79, and Partha Chatterjee, *A Possible India: essays in political criticism* (New Delhi: Oxford University Press, 1997), pp. 228–63.
7. Kothari, "The Congress 'system' in India," p. 1163.
8. Pradeep K. Chibber and John R. Petrocik, "The puzzle of Indian politics: social cleavages and the Indian party system," *British Journal of Political Science*, 19, 2 (April 1989): 191–210.
9. The Congress never exercised political hegemony in West Bengal, Kerala, Tamil Nadu and Punjab, or dominance in Orissa, Rajasthan, Madhya Pradesh and Assam. See Yogendra Yadav and Suhas Palshikar, "From hegemony to convergence: party system and electoral politics in the Indian states, 1952–2002," *Journal of Indian School of Political Economy*, XV, 1&2 (January–June 2003): 12–13.
10. Vivek Chibber, *Locked in Place: state-building and late industrialization in India* (Princeton: Princeton University Press, 2003), pp. 110–126.
11. Kohli, *The State and Poverty in India*, pp. 61–64.
12. As the reader will note, the vote shares of the Hindu Right and the Socialist Left in 1977 are missing, due to difficulties calculating reliable data. Hence the table only provides the corresponding values of the Congress and the Communist Left.
13. Kohli, *The State and Poverty in India*, p. 60. See Ronald J. Herring, *Land to the Tiller: the political economy of agrarian reform in South Asia* (New Haven: Yale University Press, 1983).
14. Chatterjee, *A Possible India*, pp. 12–35.
15. H. M. Rajeshekara, "The nature of Indian federalism: a critique," *Asian Survey*, 37, 3 (March 1997): 245–54.
16. Myron Weiner, *Party Politics in India: the development of a multi-party system* (Princeton: Princeton University Press, 1957), pp. 6 and 26.
17. Corbridge and Harriss, *Reinventing India*, pp. 46–9.
18. See Weiner, *Party Politics in India*, pp. 19–33.
19. Madhu Limaye, *Janata Party Experiment: an insider's account of opposition politics: 1977–1980, volume two* (New Delhi: B.R. Publishing, 1994), pp. 541–3. See also D.L. Sheth "Ram Manohar Lohia on caste in Indian politics," in

Ghanshyam Sheth (ed.), *Caste and Democratic Politics in India* (London: Anthem Press, 2004), pp. 79–99.
20. Ramachandra Guha (ed.), *The Makers of Modern India* (Cambridge: Harvard University Press, 2011), pp. 353–67.
21. Weiner, *Party Politics in India*, p. 29.
22. Paul R. Brass, "Political parties and the radical left in South Asia," in Paul R. Brass and Marcus F. Franda (eds), *Radical Politics in South Asia* (Cambridge: MIT Press, 1973), p. 36.
23. Javeed Alam, "Communist politics in search of hegemony," in Partha Chatterjee (ed.), *Wages of Freedom: fifty years of the Indian nation-state* (New Delhi: Oxford University Press, 1998), p. 185.
24. Gene D. Overstreet and Marshall Windmiller, *Communism in India* (Berkeley: University of California Press, 1959), p. 473.
25. Zoya Hasan, "The prime minister and the left," in James Manor (ed.), *Nehru to the Nineties: the changing office of the prime minister of India* (London: Hurst, 1994), p. 208.
26. Achin Vanaik, *The Painful Transition: bourgeois democracy in India* (London: Verso, 1990), p. 177. According to Vanaik, the CPI failed to be anti-imperialist and anti-capitalist in a "consistent, principled manner while retaining tactical flexibility".
27. Aditya Nigam, "Communist politics hegemonized," in Partha Chatterjee (ed.), *Wages of Freedom: fifty years of the Indian nation-state* (New Delhi: Oxford University Press, 1998), p. 214.
28. Sumanta Banerjee, "Strategy, tactics, and forms of political participation among left parties," in T.V. Sathyamurthy (ed.), *Class Formation and Political Transformation in Post-Colonial India* (New Delhi: Oxford University Press, 1999), p. 209.
29. See Patrick Heller, *The Labor of Development: workers and the transformation of capitalism in Kerala, India* (Ithaca: Cornell University Press, 1999), pp. 53–87.
30. Limaye, *Janata Party Experiment*, p. 542.
31. Unless otherwise stated, the following relies on Overstreet and Windmiller, *Communism in India*, pp. 156–69, and Sumit Sarkar, *Modern India 1885–1947* (New Delhi: Macmillan Indian Limited, 1986), pp. 331–43.
32. Sarkar, *Modern India*, p. 247.
33. Chibber, *Locked in Place*, pp. 110–126; Corbridge and Harriss, *Reinventing India*, pp. 46–49.
34. Guha, *The Makers of Modern India*, p. 242.
35. Overstreet and Windmiller, *Communism in India*, pp. 508–527.
36. A.G. Noorani, "Communist memories," *Frontline*, 3 December 2011.
37. Noorani, "Communist memories," and "Of Stalin, Telangana & Indian revolution," *Frontline*, 17 December 2011.
38. E.M.S. Namboodiripad, "The Left in India's freedom movement and in free India," *Social Scientist*, 14, 8/9 (August–September 1986), p. 10.
39. Francine R. Frankel, *India's Political Economy 1947–2004: the gradual revolution*, 2nd edition (New Delhi: Oxford University Press, 2006), pp. 65–6.

40. Overstreet and Windmiller, *Communism in India*, p. 328.
41. Overstreet and Windmiller, *Communism in India*, pp. 526–7.
42. Heller, *The Labor of Development*, pp. 71–4.
43. T.J. Nossiter, *Marxist State Governments in India: politics, economics and society* (London: Pinter Publishers, 1988), p. 179.
44. Heller, *The Labor of Development*, p. 85.
45. Quoted in Guha, *The Makers of Modern India*, p. 382.
46. Overstreet and Windmiller, *Communism in India*, pp. 359–364.
47. Guha, *The Makers of Modern India*, pp. 370 and 460.
48. See Brass, "Political parties and the radical left in South Asia," pp. 9–19.
49. Weiner, *Party Politics in India*, pp. 30–31; Lewis P. Fickett, Jr., "The Praja Socialist Party of India–1952–1972: a final assessment," *Asian Survey*, 13, 9 (September 1973), p. 827.
50. Lewis P. Fickett, Jr., "The major socialist parties of India in the 1967 election," *Asian Survey*, 8, 6 (June 1968), p. 490. According to Fickett, the PSP and SSP directly contested 43 parliamentary constituencies and 299 state assembly districts in the fourth general elections in 1967, which likely cost them 9 and 41 seats in each arena, respectively.
51. Paul Brass, "Leadership conflict and the disintegration of the Indian socialist movement: personal ambition, power and policy," in Paul R. Brass, *Caste, Faction and Party in Indian Politics, Volume One: faction and party* (New Delhi: Chanakya Publications, 1984), pp. 169 and 181–6. Fickett describes the socialists' leaders as "*prima donnas*, each espousing his own kind of political salvation, each indulging in the fruitless ideological abstractions ... and each unwilling to compromise with the others." Fickett, "The Praja Socialist Party of India," pp. 829–32. See also Christophe Jaffrelot, *India's Silent Revolution: the rise of the low castes in north Indian politics* (London: Hurst/New Delhi: Permanent Black, 2003), p. 307.
52. Quoted in Guha, *The Makers of Modern India*, pp. 358.
53. Jaffrelot, *India's Silent Revolution*, pp. 260–1. By 1967, four-fifths of the leadership of the PSP was college educated, three quarters were upper caste and more than nine-tenths were greater than 40 years old. See Fickett, "The Praja Socialist Party of India," pp. 827–9. Brass contends that caste was a significant factor for the break-up of the socialists in Bihar, but less so in Uttar Pradesh. See "Leadership conflict and the disintegration of the Indian socialist movement," pp. 171–81.
54. Fickett, "The Praja Socialist Party of India," pp. 829–32. That said, Fickett also agrees that the party "neglected a considerable portion of its natural constituency".
55. See Brass, "Political parties and the radical left in South Asia," pp. 9–19.
56. Marcus F. Franda, *Radical Politics in West Bengal* (Cambridge: MIT Press, 1971), pp. 54–6; Nigam, "Communist politics hegemonized," p. 214.
57. Unless noted, the following summarizes Franda, *Radical Politics in West Bengal*, pp. 82–113.
58. Hasan, "The prime minister and the left," pp. 210–4.

59. Franda, *Radical Politics in West Bengal*, p. 109.
60. Franda, *Radical Politics in West Bengal*, p. 216.
61. Brass, "Political parties and the radical left in South Asia," pp. 40–1.
62. Maurice Duverger, *Political Parties: their organization and activity in the modern state* (New York: Wiley, 1963), p. 217.
63. Kothari, "The Congress 'system' in India," p. 1165.
64. Manor, "Parties and the party system," pp. 66–73; and Yogendra Yadav, "Reconfiguration in Indian politics: state assembly elections, 1993–1995," in Partha Chatterjee (ed.), *State and Politics in India* (New Delhi: Oxford University Press, 1997), p. 191.
65. See Francine R. Frankel, "Middle classes and castes in Indian politics," in Atul Kohli (ed.), *India's Democracy: an analysis of changing state-society relations* (Princeton: Princeton University Press, 1988), pp. 225–62; Lloyd I. Rudolph and Susanne Hoeber Rudolph, *In Pursuit of Lakshmi: the political economy of the Indian state* (Chicago: University of Chicago Press, 1987), pp. 335–46.
66. Francine R. Frankel, "Conclusion," in Francine R. Frankel and M.S.A. Rao (eds), *Dominance and State Power in India: decline of a social order, Volume II* (New Delhi: Oxford University Press, 1990), pp. 482–518.
67. The Congress share of seats in parliament, however, remained approximately 54 per cent. Ramachandra Guha, *India after Gandhi: the history of the world's largest democracy* (New York: HarperCollins, 2007), p. 420.
68. Myron Weiner, "The 1971 elections and the Indian party system," *Asian Survey*, 11, 12 (December 1971), p. 1156.
69. The first was the National Conference of Jammu & Kashmir. See Narendra Subramanian, *Ethnicity and Populist Mobilization: political parties, citizens and democracy in south India* (New Delhi: Oxford University Press, 1999), p. 30.
70. Subramanian, *Ethnicity and Populist Mobilization*, p. 31; Guha, *The Makers of Modern India*, p. 223.
71. See Jaffrelot, *India's Silent Revolution*, pp. 166–71.
72. Alfred Stepan, Juan Linz and Yogendra Yadav, *Crafting State-Nations: India and other multinational democracies* (Baltimore: Johns Hopkins University Press, 2011), p. 124.
73. The 1950 Constitution had stipulated that English would no longer serve as the main language across the Union after 15 years. See Stepan, Linz and Yadav, *Crafting State-Nations*, pp. 127–30.
74. Heller, *The Labor of Development*, pp. 85 and 49.
75. Heller, *The Labor of Development*, pp. 174–5.
76. Heller, *The Labor of Development*, p. 49.
77. Paul R. Brass, "Coalition politics in north India," in Paul R. Brass, *Caste, Faction and Party in Indian Politics, Volume One: faction and party* (New Delhi: Chanakya Publications, 1984), pp. 120–4.
78. Bidyut Chakrabarty, *Forging Power: coalition politics in India* (New Delhi: Oxford University Press, 2006), pp. 68–71.
79. Franda, *Radical Politics in West Bengal*, pp. 124 and 146.

80. Guha, *India after Gandhi*, pp. 427–32.
81. Jaffrelot, *The Silent Revolution*, pp. 31–47.
82. Limaye, *Janata Party Experiment*, pp. 541–3.
83. Chakrabarty, *Forging Power*, p. 107.
84. Heller, *The Labor of Development*, pp. 76–80.
85. Nossiter, *Marxist State Governments in India*, pp. 87–94.
86. Franda, *Radical Politics in West Bengal*, pp. 225–30.
87. Chakrabarty, *Forging Power*, pp. 68–71.
88. Franda, *Radical Politics in West Bengal*, pp. 184–90 and 196–201.
89. Franda, *Radical Politics in West Bengal*, pp. 150–170.
90. Nossiter, *Marxist State Governments in India*, pp. 133–136.
91. Bruce Bueno de Mesquita, *Strategy, Risk and Personality in Coalition Politics: the case of India* (Cambridge: Cambridge University Press, 1975), p. 113.
92. E. Sridharan, "Unstable parties and unstable alliances: births, splits, mergers and deaths of parties, 1952–2000," in Mahendra Prasad Singh and Anil Mishra (eds), *Coalition Politics in India: problems and prospects* (New Delhi: Manohar, 2004), pp. 68–70.
93. See Paul R. Brass, "Party systems and governmental instability in Indian states" and "Coalition politics in north India" in Paul R. Brass, *Caste, Faction and Party in Indian Politics, Volume One: faction and party* (New Delhi: Chanakya Publications, 1984), pp. 19–62 and 97–135; Chakrabarty, *Forging Power*, pp. 71–7.
94. Guha, *India after Gandhi*, pp. 427–32.
95. Brass, "Leadership conflict and the disintegration of the Indian socialist movement," pp. 182–5.
96. Limaye, *Janata Party Experiment*, p. 542.
97. Franda, *Radical Politics in West Bengal*, pp. 232–8.
98. Franda, *Radical Politics in West Bengal*, pp. 201–3.
99. Fickett, "The Praja Socialist Party of India," p. 827; Subramanian, *Ethnicity and Populist Mobilization*, p. 227.
100. Subramanian, *Ethnicity and Populist Mobilization*, pp. 204–21.
101. Hasan, "The prime minister and the left," pp. 217–8.
102. Kumaramangalam also influenced the passage of the 24th, 25th and 26th amendments to the Constitution, which granted parliament the authority to dilute fundamental rights, restrict private property and abolish the privy purses. Rudolph and Rudolph, *In Pursuit of Lakshmi*, pp. 107–8.
103. Myron Weiner, "India's new political institutions," *Asian Survey*, 16, 9 (September 1976): 898–901.
104. I thank Phil Oldenburg for pointing this out to me.
105. Weiner, "The 1971 elections and the Indian party system," p. 1163.
106. In August 1971, the remnants of the PSP formed a new Socialist party, but it split yet again by December 1972. See Brass, "Leadership conflict and the disintegration of the Indian socialist movement," pp. 160–1.
107. Weiner, "The 1971 elections and the Indian party system," p. 1161.

108. Limaye, *Janata Party Experiment*, p. 543.
109. Atul Kohli, *Democracy and Discontent: India's growing crisis of ungovernability* (Cambridge: Cambridge University Press, 1990), p. 16. Unless noted, the following paragraph summarizes his thesis.
110. Guha, *India after Gandhi*, pp. 189–209.
111. Weiner, "The 1971 elections and the Indian party system," pp. 1155–8. Weiner qualifies their collective rise in several ways, however.
112. Yadav and Palshikar, "From hegemony to convergence," p. 27. The KHAM social coalition comprised Kshatriyas, Harijans (dalits), Adivasis and Muslims.
113. See Oliver Mendelsohn, "The collapse of the Indian National Congress," *Pacific Affairs*, 51, 1 (Spring 1978), pp. 47–8.
114. Guha, *India after Gandhi*, p. 476.
115. See Bipan Chandra, *In the Name of Democracy: JP movement and the Emergency* (New Delhi: Penguin Books, 2003), pp. 94–124.
116. Frankel, *India's Political Economy*, p. 652. The following relies on Guha, *India after Gandhi*, pp. 476–490.
117. Corbridge and Harriss, *Reinventing India*, p. 87.
118. Guha, *India after Gandhi*, pp. 490–514. Rajni Kothari and Kohli largely agree. See, respectively, *Politics and the People: in search of a humane India*, volume I (New Delhi: Ajanta, 1989), p. 234; and *Democracy and Discontent*, p. 251.
119. Christophe Jaffrelot, *The Hindu Nationalist Movement and Indian Politics: 1925 to the 1990s* (New Delhi: Penguin Books, 1999), p. 272.
120. By 1973, the DMK had increased reservations for OBCs in public institutions to 51 per cent. See Subramanian, *Ethnicity and Populist Mobilization*, pp. 52, 76 and 204–21; and more generally, M.S.S. Pandian, "Stepping outside history? new dalit writings from Tamil Nadu," in Partha Chatterjee (ed.), *Wages of Freedom: fifty years of the Indian nation-state* (New Delhi: Oxford University Press, 1998), pp. 292–310.
121. Corbridge and Harriss, *Reinventing India*, p. 87.
122. Chandra, *In the Name of Democracy*, pp. 55–6.
123. Transcript of interview with Madhu Dandavate, Acc. No. 788, Oral History Division, Nehru Memorial Museum and Library, New Delhi, pp. 139–40.
124. Jaffrelot, *The Hindu Nationalist Movement and Indian Politics*, pp. 266.
125. Chandra, *In the Name of Democracy*, p. 32.
126. Hasan, "The prime minister and the left," p. 224.
127. Weiner, "India's new political institutions," pp. 898–901.
128. Guha, *India after Gandhi*, p. 484.
129. Chandra, *In the Name of Democracy*, p. 188.
130. His expected succession failed to happen due to a plane crash in June 1980. After thirty years of silence, the Congress(I) blamed Sanjay for the most controversial aspects of the Emergency. See *Times of India*, 29 December 2010.

3. THE JANATA PARTY (1977–1980)

1. See *Janata Party Election Manifesto 1977*.
2. "Free India from the dictators," in Arun Shourie (ed.), *Institutions in the Janata Phase* (Bombay: Popular Prakash Private Ltd, 1980), pp. 5–8.
3. See Table 2 in William G. Vanderbok, "The tiger triumphant: the mobilization and alignment of the Indian electorate," *British Journal of Political Science*, 20, 2 (April 1990), p. 242.
4. The Election Commission (EC) distinguishes four types of contestants: as members of national, state and registered (unrecognized) parties; and as Independents. To acquire accreditation as a state party, parties had to be engaged in political activity for five continuous years, and have at least one member for every twenty-five or thirty members of the Lok Sabha (LS) and state legislative assembly (SLA)/union territory (UT), respectively, in the state they contested, or have greater than 4 per cent of the total valid votes polled for either the LS or the SLA/UT in the state they contested. To be a national party, a party had to be recognized as a state party in four or more states. Parties that registered with the EC, but failed to meet the criteria of state or national parties, were designated "unrecognized".
5. The latter had been 22 per cent in 1952. See Christophe Jaffrelot, *India's Silent Revolution: the rise of the low castes in north Indian politics* (London: Hurst & Co./New Delhi: Permanent Black, 2003), p. 311, and Ramachandra Guha, *India after Gandhi: the history of the world's largest democracy* (New York: HarperCollins, 2007), p. 529.
6. Myron Weiner, "The 1977 parliamentary elections in India," *Asian Survey*, 17, 7 (July 1977): 625.
7. Election Commission of India, *Statistical Report on General Elections, 1977, to the Sixth Lok Sabha: Volume I*, pp. 85–8 and 93–4; Bipan Chandra, *In the Name of Democracy: JP movement and the Emergency* (New Delhi: Penguin Books, 2003), pp. 248–9.
8. Guha, *India after Gandhi*, pp. 521–522.
9. Myron Weiner, "Congress restored: continuities and discontinuities in Indian politics," *Asian Survey*, 22, 4 (April 1982): 341.
10. Francine R. Frankel, "Conclusion," in Francine Frankel and M.S.A. Rao (eds), *Dominance and State Power in Modern India: decline of a social order, volume II* (New Delhi: Oxford University Press, 1990), pp. 510–1.
11. Francine R. Frankel, *India's Political Economy 1947–2004: the gradual revolution*, 2nd edition (New Delhi: Oxford University Press, 2006), p. 571.
12. Madhu Limaye, *Janata Party Experiment: an insider's account of opposition politics: 1977–1980, volume two* (New Delhi: B.R. Publishing, 1994), p. 544.
13. Oliver Mendelsohn, "The collapse of the Indian National Congress," *Pacific Affairs*, 51, 1 (Spring 1978), p. 59.
14. Transcript of interview with Shri Madhu Dandavate, Acc. No. 788, Oral History Division, Nehru Memorial Museum and Library, New Delhi, pp. 104–105.

15. Weiner, "The 1977 parliamentary elections in India," p. 620.
16. The Election Commission could not grant the Janata Party legal recognition because the merger was de facto. Hence its need to campaign under the BLD election symbol. See David Butler, Ashok Lahiri and Prannoy Roy, *India Decides: elections 1952–1995*, 3rd edition (New Delhi: Books and Things, 1995), p. 24.
17. The other prominent Congressmen to join Ram were H.N. Bahuguna and Nandini Satpathy. See Weiner, "The 1977 parliamentary elections in India," p. 620.
18. Alfred Stepan, Juan Linz and Yogendra Yadav, *Crafting State-Nations: India and other multinational democracies* (Baltimore: Johns Hopkins University Press, 2011), pp. 92–3.
19. Paul R. Brass, "The Punjab crisis and the unity of India," in Atul Kohli (ed.), *India's Democracy: an analysis of changing state-society relations* (Princeton: Princeton University Press, 1988), p. 176.
20. Leftists in the CPI(M), such as Hare Krishna Konar, objected for principled reasons. They also thought pursuing such tactics might compel the party's leftists to join CPI(ML). See Marcus F. Franda, *Radical Politics in West Bengal* (Cambridge: MIT Press, 1971), pp. 204–8 and 216–7.
21. Transcript of interview with Shri Jyoti Basu, Acc. No. 781, Oral History Division, Nehru Memorial Museum and Library, New Delhi, p. 118.
22. Aditya Nigam, "Communist politics hegemonized," in Partha Chatterjee (ed.), *Wages of Freedom: fifty years of the Indian nation-state* (New Delhi: Oxford University Press, 1998), p. 220.
23. Chandra, *In the Name of Democracy*, p. 219.
24. Chandra, *In the Name of Democracy*, pp. 32 and 56–57.
25. See Zoya Hasan, "The prime minister and the left," in James Manor (ed.), *Nehru to the Nineties: the changing office of the prime minister of India* (London: Hurst, 1994), p. 224.
26. Harkishen Singh Surjeet, "The present political situation and the evolution of our tactics in perspective," *The Marxist*, 8, 4 (October–December 1990), pp. 9–10.
27. Guha, *India after Gandhi*, p. 520.
28. See James Manor, "Parties and the party system," in Atul Kohli (ed.), *India's Democracy: an analysis of changing state-society relations* (Princeton: Princeton University Press, 1988), p. 75.
29. The latter included the Utkal Congress, Rashtriya Loktantrik Congress, Kisan Mazdoor Party, and Punjabi Khetibari Zamindari Union. Terence J. Byres, "Charan Singh, 1902–87: an assessment," *The Journal of Peasant Studies*, 15, 2 (1988): 146.
30. Lloyd I. Rudolph and Susanne Hoeber Rudolph, *In Pursuit of Lakshmi: the political economy of the Indian state* (Chicago: University of Chicago Press, 1987), p. 170.
31. Limaye, *Janata Party Experiment*, p. 544.
32. Jaffrelot, *India's Silent Revolution*, p. 318.

33. Jaffrelot, *India's Silent Revolution*, p. 305.
34. Letter from Charan Singh to Jayaprakash Narayan, 1 June 1976, in Dhirendra Sharma (ed.), *The Janata (People's) Struggle: the finest hour of the Indian people—with underground documents, resistance literature and correspondence relating to the advent of the Janata Party* (New Delhi: Sanjivan Press, 1977), pp. 300–4.
35. Letter from Charan Singh to Ashok Mehta, 27 September 1976, in Sharma, *The Janata (People's) Struggle*, p. 312.
36. Byres, "Charan Singh," p. 164.
37. The following summarizes Christophe Jaffrelot, *The Hindu Nationalist Movement and Indian Politics: 1925 to the 1990s* (New Delhi: Penguin Books, 1999), pp. 279–81.
38. See Csaba Nikolenyi, *Minority Governments in India: the puzzle of elusive majorities* (London: Routledge, 2010), p. 38; and Rudolph and Rudolph, *In Pursuit of Lakshmi*, pp. 167–8. The respective shares of parliamentary seats are approximate figures.
39. Since the constituent parties agreed to contest the elections under the BLD, however, official records of the Janata Party government do not distinguish their prior affiliations.
40. Jaffrelot, *The Hindu Nationalist Movement and Indian Politics*, p. 283.
41. Atal Bihari Vajpayee, "All responsible for the Janata crisis," in *Ideals or Opportunism?* (Janata Party Publication, 1979), p. 9.
42. Granville Austin, *Working a Democratic Constitution: a history of the Indian experience* (New Delhi: Oxford University Press, 2000), p. 401.
43. Vajpayee, "All responsible for the Janata crisis," p. 9.
44. Byres, "Charan Singh," p. 145; Austin, *Working a Democratic Constitution*, pp. 401–402.
45. Guha, *India after Gandhi*, p. 523.
46. James Manor, "Introduction," in James Manor (ed.), *Nehru to the Nineties: the changing office of the prime minister of India* (London: Hurst, 1994), p. 7.
47. Guha, *India after Gandhi*, pp. 531–534.
48. Frankel, *India's Political Economy*, p. 579.
49. The Janata government eventually created special courts, in May 1979, to adjudicate the findings of the Shah Commission Reports. But the premature collapse of the Government, and a negative subsequent ruling by the Supreme Court, allowed the proponents of the Emergency to evade political responsibility.
50. Rudolph and Rudolph, *In Pursuit of Lakshmi*, pp. 173–174.
51. Madhu Dandavate, "Crisis rooted in summit power politics," in *Ideals or Opportunism?* (Janata Party Publication, 1979), p. 21.
52. "1977 mein janata mein jo amootpurva utsah pragat hua tha vah bhi keval satta-parivartan ke liye nahi tha, balki ek naye samaj ke nirman ki akansha ka ghotak tha ... is avsar ko apni vyaktigat mahattvakansha ya aagrah ke khatir kho dena ek prakar se janata ke sath vishvasghat hoga. Hum in choti-moti baton mein uljhe rahe ... to desh mein tanashahi pravrittiyon our takaton ko

phir se ubharne ka bhi khatra hai ... uske anoorup hamein sabit hona chahiye." Letter from Jayaprakash Narayan to Chandrasekhar, 9 May 1978, Janata Party Papers, Subject File 311, Nehru Memorial Museum and Library, New Delhi.
53. Rudolph and Rudolph, *In Pursuit of Lakshmi*, pp. 164–5.
54. Guha, *India after Gandhi*, p. 537.
55. Nivedita Menon and Aditya Nigam, *Power and Contestation: India since 1989* (Hyderabad: Orient Longman, 2008), p. 6.
56. Austin, *Working a Democratic Constitution*, p. 403.
57. Bidyut Chakrabarty, *Forging Power: coalition politics in India* (New Delhi: Oxford University Press, 2006), p. 109; Stuart Corbridge and John Harriss, *Reinventing India: liberalization, Hindu nationalism and popular democracy* (Cambridge: Polity Press, 2000), p. 90.
58. Austin, *Working a Democratic Constitution*, pp. 571–2.
59. Guha, *India after Gandhi*, pp. 527–8.
60. The most prominent of these provisions include the right of Jammu & Kashmir to concur on all national legislation barring defense, foreign affairs and communications, and to promulgate a separate constitution, rules for citizenship (necessary for holding land) and titles for its officials. However, the Congress(I) violated many of these rights from 1953 to 1968, repeatedly imprisoning Sheikh Abdullah and issuing 48 presidential orders. See Stepan, Linz and Yadav, *Crafting State-Nations*, pp. 111–3. Several commentators have criticized the Shimla Accord, saying it underscored Abdullah's submission to Indian rule and Congress dominance. See Sumantra Bose, *The Challenge in Kashmir: democracy, self-determination and a just peace* (New Delhi: Sage, 1997), pp. 88–90.
61. Guha, *India after Gandhi*, p. 484.
62. See Sten Widmalm, "The rise and fall of democracy in Jammu and Kashmir: 1975–1989," in Amrita Basu and Atul Kohli (eds), *Community Conflicts and the State* (New Delhi: Oxford University Press, 2000), p. 152.
63. Pranab Bardhan, *The Political Economy of Development in India, expanded edition with an epilogue on the political economy of reform in India* (New Delhi: Oxford University Press, 2012), p. 82.
64. Austin, *Working a Democratic Constitution*, p. 404.
65. See Ashis Nandy, "Federalism, the ideology of the state and cultural pluralism," in Nirmal Mukherji and Balveer Arora (eds), *Federalism in India: origins and development* (New Delhi: Vikas Publishing House, 1992), pp. 27–41.
66. Ghanshyam Shah, "The prime minister and the 'weaker sections of society'," in James Manor (ed.), *Nehru to the Nineties: the changing office of the prime minister of India* (London: Hurst, 1994), p. 252.
67. Rekha Saxena, *Situating Federalism: mechanisms of intergovernmental relations in Canada and India* (New Delhi: Manohar, 2006), pp. 265–266.
68. See Jyotirindra Das Gupta, "The Janata phase: reorganization and redirection in Indian politics," in Zoya Hasan (ed.), *Parties and Party Politics in India* (New Delhi: Oxford University Press, 2004), pp. 357–62.

69. See Shah, "The prime minister and the 'weaker sections of society'," p. 252; Chakrabarty, *Forging Power*, pp. 118–22.
70. Brass, "Political parties and the radical left in South Asia," pp. 72–3.
71. A. Appadorai and M.S. Rajan, *India's Foreign Policy and Relations* (New Delhi: South Asian Publishers, 1985), pp. 597–8.
72. Appadorai and Rajan, *India's Foreign Policy and Relations*, pp. 603–5.
73. Achin Vanaik, *The Painful Transition: bourgeois democracy in India* (London: Verso, 1990), p. 265.
74. Appadorai and Rajan, *India's Foreign Policy and Relations*, p. 607.
75. Achin Vanaik, *India in a Changing World* (New Delhi: Orient Longman, 1995), p. 34.
76. The agreement gave the state of West Bengal 20,500 cft (cubic feet) in lean seasons, compared to 34,500 cft for Bangladesh, leading the former to protest that ports in Calcutta would silt up. The two countries also discussed border disputes, the vexed question of illegal migrants and trade diversification. See Vanaik, *The Painful Transition*, p. 265; Appadorai and Rajan, *India's Foreign Policy and Relations*, p. 611; Menon and Nigam, *Power and Contestation*, p. 192, fn 10.
77. See Jaffrelot, *The Hindu Nationalist Movement*, p. 285.
78. John Whelpton, *A History of Nepal* (Cambridge: Cambridge University Press, 2005), p. 155.
79. George Perkovich, *India's Nuclear Bomb: the impact of global proliferation* (Berkeley: University of California Press, 1999), pp. 199–221.
80. Vipin Narang and Paul Staniland, "Institutions and worldviews in Indian foreign security policy," *India Review*, 11, 2 (2012): 78.
81. See Baldev Raj Nayar, "India and the super powers: deviation or continuity in foreign policy?" *Economic and Political Weekly*, 12, 30 (July 23, 1977): 1185–9.
82. Appadorai and Rajan, *India's Foreign Policy and Relations*, pp. 615–9.
83. Desai was willing to forego it. See Nayar, "India and the super powers," p. 1189; Nalini Kant Jha, "Coalition governments and India's foreign policy," in Mahendra Prasad Singh and Anil Mishra (eds), *Coalition Politics in India: problems and prospects* (New Delhi: Manohar, 2004), p. 304.
84. See Baldev Raj Nayar and T.V. Paul, *India in the World Order: searching for major-power status* (Cambridge: Cambridge University Press, 2003), pp. 186–7.
85. Appadorai and Rajan, *India's Foreign Policy and Relations*, p. 609.
86. Atal Bihari Vajpayee, *New Dimensions of India's Foreign Policy*, with a foreword by M.C. Chagla (New Delhi: Vision Books, 1979), p. 38.
87. Compare Sumit Ganguly, "The prime minister and foreign and defence policies," in James Manor (ed.), *Nehru to the Nineties: the changing office of the prime minister of India* (London: Hurst, 1994), p. 154; C. Raja Mohan, *Crossing the Rubicon: the shaping of India's new foreign policy* (New York: Palgrave, 2004), p. 264; and Nalini Kant Jha, "Coalition governments," p. 302.

88. Guha, *India after Gandhi*, pp. 531–534.
89. The Congress government appointed the Kalelkar commission on 29 January 1953 and tabled its report in parliament on 3 September 1956. However, parliament only discussed the report in November 1965. See Jaffrelot, *India's Silent Revolution*, pp. 221–229.
90. Zoya Hasan, *Quest for Power: oppositional movements and post-Congress politics in Uttar Pradesh* (New Delhi: Oxford University Press, 1999), p. 146.
91. Byres, "Charan Singh," p. 162.
92. Ashutosh Varshney, *Democracy, Development and the Countryside: urban-rural struggles in India* (Cambridge: Cambridge University Press, 1995), pp. 104–5.
93. Atul Kohli, *Democracy and Discontent: India's growing crisis of ungovernability* (Cambridge: Cambridge University Press, 1990), p. 312.
94. For data, see Rudolph and Rudolph, *In Pursuit of Lakshmi*, pp. 240–2.
95. Bardhan, *The Political Economy of Development in India*, pp. 65–8.
96. Vijay Joshi and I.M.D. Little, *India: macroeconomics and political economy, 1964–1991* (New Delhi: Oxford University Press, 1994), pp. 146–53, 164 and 216.
97. Frankel, *India's Political Economy*, p. 572.
98. Frankel, *India's Political Economy*, p. 577.
99. Chandra, *In the Name of Democracy*, p. 139.
100. Atul Kohli, *The State and Poverty in India: the politics of reform* (Cambridge: Cambridge University Press, 1987), pp. 89–91.
101. Corbridge and Harriss, *Reinventing India*, p. 92.
102. Jaffrelot, *The Hindu Nationalist Movement*, p. 285.
103. "Choudhary Charan Singh ... stated his firm belief that no RSS Volunteers can join the new party and [sic] member of the new party can join the RSS." See Minutes of meeting of four Opposition parties, 8 July 1976, in Sharma, *The Janata (People's) Struggle*, p. 305.
104. Vanaik, *The Painful Transition*, p. 147.
105. Austin, *Working a Democratic Constitution*, p. 467.
106. Jaffrelot, *The Hindu Nationalist Movement*, p. 308.
107. Transcript of interview with Dandavate, p. 129.
108. Guha, *India after Gandhi*, 535–6.
109. Rudolph and Rudolph, *In Pursuit of Lakshmi*, pp. 176–7. In
110. James Manor, "The prime minister and the president," in James Manor (ed.), *Nehru to the Nineties: the changing office of the prime minister of India* (London: Hurst, 1994), p. 131.
111. Paul R. Brass, "Chaudhuri Charan Singh: an Indian political life," *Economic and Political Weekly*, 28, 39 (25 September 1993): 2088–90.
112. Kohli, *The State and Poverty in India*, p. 196.
113. See *Janata Party Election Manifesto 1980*.
114. I thank James Manor for sharing this information with me.
115. Vajpayee, "All responsible for the Janata crisis," pp. 9, 12.
116. Chandra Shekhar, "Betrayal of Janata Party," in *Ideals or Opportunism?* (A Janata Party Publication, 1979), pp. 1–9.

117. "*Bagh ujad gaya hai.*" See transcript of interview with Madhu Dandavate, p. 135.
118. See James Manor, "The electoral process amid awakening and decay: reflections on the Indian general election of 1980," in Peter Lyon and James Manor (eds), *Transfer and Transformation: political institutions in the new commonwealth* (Leicester: Leicester University Press, 1983), pp. 97–104.
119. See *Janata Party Election Manifesto 1980*.
120. See Indian National Congress(I), *Election Manifesto 1980*.
121. Harold A. Gould, "The second coming: the 1980 elections in India's Hindi belt," *Asian Survey*, 20, 6 (June 1980): 596–8.
122. Weiner, "Congress restored," p. 340.
123. Susanne Hoeber Rudolph and Lloyd I. Rudolph, "The centrist future of Indian politics," *Asian Survey*, 20, 6 (June 1980): 594.
124. Manor, "The electoral process amid awakening and decay," p. 99.
125. Gould, "The second coming," p. 615.
126. Gould, "The second coming," p. 598.
127. See Paul R. Brass, "Congress, the Lok Dal, and the middle-peasant castes: an analysis of the 1977 and 1980 parliamentary elections in Uttar Pradesh," in Paul R. Brass, *Caste, Faction and Party in Indian Politics, Volume Two: election studies* (New Delhi: Chanakya Publications, 1984), pp. 162–204.
128. Weiner, "Congress restored," p. 347.

4. THE RISE OF THE REGIONS (1980–1989)

1. *Janata Party Election Manifesto 1980*, p. 36.
2. Achin Vanaik, *The Furies of Indian Communalism: religion, modernity and secularization* (London: Verso, 1997), p. 45.
3. The Congress(U) became the Congress(S) after the Maharashtra leader Sharad Pawar became its president in 1981. Walter K. Andersen, "India in 1981: stronger political authority and social tension," *Asian Survey*, 22, 2 (February 1982): 127.
4. Andersen, "India in 1981," p. 127.
5. See James Manor, "Parties and the party system," in Atul Kohli (ed.), *India's Democracy: an analysis of changing state-society relations* (Princeton: Princeton University Press, 1988), pp. 62–99.
6. Myron Weiner, "Congress restored: continuities and discontinuities in Indian politics," *Asian Survey*, 22, 4 (April 1982): 339–47.
7. Deepak Nayyar, "Economic development and political democracy: interaction of economics and politics in independent India," *Economic and Political Weekly*, 33, 49 (5 December 1998), pp. 3125–6.
8. Atul Kohli, "Politics of economic growth in India, 1980–2005—Part I: the 1980s," *Economic and Political Weekly*, 41, 13 (1 April 2006): 1251–9.
9. Assam accounted for approximately 60 per cent of the timber, tea and oil harvested by the country. Yet its royalties were a fraction of the total revenue that New Delhi accrued from these resources. Central resource transfers remained

poor too. The reorganization of states in the northeast, which reduced the size of "greater Assam", intensified these grievances. See Jyotirindra Das Gupta, "Ethnicity, democracy and development in India: Assam in a general perspective," in Atul Kohli (ed.), *India's Democracy: an analysis of changing state-society relations* (Princeton: Princeton University Press, 1988), pp. 144–69; and Subir Bhaumik, "North-east India: the evolution of a post-colonial region," in Partha Chatterjee (ed.), *Wages of Freedom: fifty years of the Indian nation-state* (New Delhi: Oxford University Press, 1998), p. 317.

10. See Sanjib Baruah, *India Against Itself: Assam and the politics of nationality* (Philadelphia: University of Pennsylvania Press, 1999), pp. 115–38.
11. Andersen, "India in 1981," p. 128. The CPI had rejoined the parliamentary left and anti-Congress forces more generally after belatedly recognizing that Mrs Gandhi's administration was an "authoritarian, anti-democratic force". Achin Vanaik, *The Painful Transition: bourgeois democracy in India* (London: Verso, 1990), p. 180.
12. Vanaik, *The Painful Transition*, pp. 132–3. The following summarizes Atul Kohli, *Democracy and Discontent: India's growing crisis of ungovernability* (Cambridge: Cambridge University Press, 1990), pp. 288–93.
13. Ashutosh Varshney, *Democracy, Development and the Countryside: urban-rural struggles in India* (Cambridge: Cambridge University Press, 1995), p. 139.
14. Transcript of interview with Shri Jyoti Basu, Acc. No. 781, Oral History Division, Nehru Memorial Museum and Library, New Delhi, p. 124.
15. Kohli, *Democracy and Discontent*, p. 267.
16. Strike activity was twice the national average. Annual industrial growth slowed to 1.7 per cent between 1975 and 1985. Patrick Heller, *The Labor of Development: workers and the transformation of capitalism in Kerala, India* (Ithaca: Cornell University Press, 1999), pp. 176–7.
17. See Patrick Heller, "Degrees of democracy: some comparative lessons from India," *World Politics*, 52, 4 (2000): 484–519.
18. Polly Datta, "The issue of discrimination in Indian federalism in the post-1977 politics of West Bengal," *Comparative Studies of South Asia, Africa and the Middle East*, 1, 2 (2005), pp. 453–7.
19. Henry C. Hart, "Political leadership in India: dimensions and limits," in Atul Kohli (ed.), *India's Democracy: an analysis of changing state-society relations* (Princeton: Princeton University Press, 1988), p. 54.
20. Interview with senior regional politician, New Delhi, 18 February 2000.
21. Interview with senior government official, New Delhi, 7 February 2000.
22. Interview, New Delhi, 18 February 2000.
23. In August 1984, the governor of Andhra Pradesh, acting on behalf of Mrs Gandhi, appointed Shaskara Rao, who defected from the TDP, as chief minister without allowing Rama Rao to test his majority on the floor of the assembly. A popular backlash against the ploy, combined with the widespread popularity of the TDP's rice subsidy, helped the party win re-election in the 1985 assembly polls. Observers subsequently noted a growing personality cult and

centralization of power within the party, however. See Lloyd I. Rudolph and Susanne Hoeber Rudolph, "Redoing the constitutional design: from an interventionist to a regulatory state," in Atul Kohli (ed.), *The Success of India's Democracy* (Cambridge: Cambridge University Press, 2001), p. 146; Francine R. Frankel, "Middle classes and castes in Indian politics: prospects for political accommodation," in Atul Kohli (ed.), *India's Democracy: an analysis of changing state-society relations* (Princeton: Princeton University Press, 1988); pp. 247–8; and Ramachandra Guha, *India after Gandhi: the history of the world's largest democracy* (New York: HarperCollins, 2007), pp. 548–9.

24. James Manor, "Prologue: caste and politics in recent times," in Rajni Kothari (ed.), *Caste in Indian Politics*, 2nd edition (New Delhi: Orient Blackswan, 2010), xiv.

25. See H. M. Rajeshekara, "The nature of Indian federalism: a critique," *Asian Survey*, 37, 3 (March 1997): 245–54.

26. Granville Austin, *Working a Democratic Constitution: a history of the Indian experience* (New Delhi: Oxford University Press, 2000), pp. 534–5.

27. Unless noted, the following draws on Austin, *Working a Democratic Constitution*, pp. 541–6 and 610–1.

28. Guha, *India after Gandhi*, p. 561.

29. Ashutosh Varshney, "Contested meanings: India's national identity, Hindu nationalism, and the politics of anxiety," *Daedalus*, 122, 3 (Summer 1993): 227–61.

30. See Philip Oldenburg, "Pollsters, pundits and a mandate to rule: interpreting India's 1984 parliamentary elections," *Journal of Commonwealth and Comparative Politics* 26, 3 (1988): 296–317; and James Manor, "The Indian general election of 1984," *Electoral Studies*, 4, 2 (1985): 149–58.

31. See Indian National Congress (I), *Election Manifesto, Eighth General Elections-1984*.

32. Robert L. Hardgrave, Jr., "India on the eve of elections: Congress and the opposition," *Pacific Affairs*, 57, 3 (Autumn 1984): 414–8.

33. See the popular survey results of Table 6 in Oldenburg, "Pollsters, pundits and a mandate to rule," p. 308.

34. Manor, "The Indian general election of 1984," p. 150.

35. See Kohli, *Democracy and Discontent*, pp. 318–319.

36. The Agreement pledged to make Chandigarh the capital of Punjab by 26 January 1986, appoint a commission to determine which parts of the latter should be transferred to Haryana, refer the dispute over water-sharing between adjoining states to a tribunal and consider all remaining demands of the 1978 Anandpur Sahib Resolution under the auspices of the Sarkaria Commission. Paul R. Brass, "The Punjab crisis and the unity of India," in Atul Kohli (ed.), *India's Democracy: an analysis of changing state-society relations* (Princeton: Princeton University Press, 1988), p. 210.

37. The Accord granted migrants from the former East Pakistan that had settled in the region after 1 January 1966 the right to stay, but not to vote for ten years, and authorized the state government the right to deport those arriving after 25 March 1971. Baruah, *India Against Itself*, p. 139.

38. See Kohli, *Democracy and Discontent*, pp. 319–352.
39. See K.S. Krishnaswamy, I.S. Gulati and A. Vaidyanathan, "Economic aspects of federalism in India," in Nirmal Mukarji and Balveer Arora (eds), *Federalism in India: origins and development* (New Delhi: Vikas Publishing House, 1992), pp. 191–3; and Rekha Saxena, *Situating Federalism: mechanisms of intergovernmental relations in Canada and India* (New Delhi: Manohar, 2006), pp. 267–269.
40. The number of new "registered" parties contesting national elections increased from 33 in 1984 to 113 in 1989. The Election Commission classified the vast majority of the latter, 85 in total, as "unrecognized" due to the small vote shares. See Csaba Nikolenyi, *Minority Governments in India: the puzzle of elusive majorities* (London: Routledge, 2010), pp. 70–86.
41. See Niraja Gopal Jayal, *Democracy and the State: welfare, secularism and development in contemporary India* (New Delhi: Oxford University Press, 2002), pp. 101–50.
42. Christophe Jaffrelot, *The Hindu Nationalist Movement: 1925 to the 1990s* (New Delhi: Penguin Books, 1999), pp. 375–8.
43. Jaffrelot, *The Hindu Nationalist Movement*, pp. 363–75 and 399.
44. See Kohli, *Democracy and Discontent*, pp. 364–76.
45. The so-called IMDT Act put the onus of determining citizenship status on the accuser, required the latter to live within three kilometers of the accused, and allowed the latter to claim Indian citizenship on the basis of a ration card. It also did not apply to migrants that had arrived before 25 March 1971. See Baruah, *India Against Itself*, pp. 116–7 and 141–3; and Sanjoy Hazarika, *Strangers of the Mist: tales of war and peace from India's northeast* (New Delhi: Penguin, 1995), p. 156.
46. Yogendra Yadav and Suhas Palshikar, "From hegemony to convergence: party system and electoral politics in the Indian states, 1952–2002," *Journal of Indian School of Political Economy*, XV, 1&2 (January–June 2003): 28.
47. Balveer Arora, "India's federal system and the demands of pluralism: crisis and reform in the '80s," in Joyotpaul Chaudhari (ed.), *India's Beleaguered Federalism: the pluralist challenge* (Tempe: Arizona State University, 1992), pp. 13–5.
48. Seema Mustafa, *The Lonely Prophet: V.P. Singh* (New Delhi: New Age International, 1995), pp. 69–79.
49. The alleged bribes amounted to US \$37 million. Walter K. Andersen, "Election 1989 in India: the dawn of coalition politics?" *Asian Survey*, 30, 6 (June 1990): 530.
50. Mustafa, *The Lonely Prophet*, p. 120.
51. Mustafa, *The Lonely Prophet*, pp. 88–9.
52. T.J. Nossiter, *Marxist State Governments in India: politics, economics and society* (London: Pinter Publishers, 1988), p. 192.
53. E.M.S. Namboodiripad, "Opening Speech," 13th Congress of CPI(M), Calcutta, 25–30 December 1988, in A.M. Zaidi (ed.), *Annual Register of Indian Political Parties: proceedings and fundamental texts 1989: part one* (New Delhi: Indian Institute of Applied Political Research, 1992), p. 260.

54. Communist Party of India (Marxist), "Left Parties Appeal," New Delhi, 1 October 1989, in O.P. Rahlan (ed.), *Encyclopedia of Political Parties Post-Independence India, Volume 97: Samajwadi Janata Party and other smaller groups* (New Delhi: Anmol Publications, 2001), pp. 137–9.
55. Partha Chatterjee, *A Possible India: essays in political criticism* (New Delhi: Oxford University Press, 1997), p. 163.
56. Vanaik, *The Painful Transition*, p. 278.
57. L.K. Advani, "Opening speech," BJP National Executive, Udaipur, 3–5 March 1989, in A.M. Zaidi (ed.), *Annual Register of Indian Political Parties: proceedings and fundamental texts 1989: part one* (New Delhi: Indian Institute of Applied Political Research, 1992), pp. 118–9.
58. Andersen, "Election 1989 in India," p. 535.
59. Jaffrelot, *The Hindu Nationalist Movement*, pp. 378–81.
60. See Mustafa, *The Lonely Prophet*, pp. 88 and 119–31.
61. Mustafa, *The Lonely Prophet*, p. 121.

5. THE NATIONAL FRONT (1989–1991)

1. Election Commission of India, *Statistical Report on General Elections, 1989, to the Ninth Lok Sabha: Volume I* (New Delhi), p. 94.
2. See Richard Sisson, "India in 1989: a year of elections in a culture of change," *Asian Survey*, 30, 2 (February 1990): 114–5, 119 and 123.
3. See Table 10.4, Christophe Jaffrelot, *India's Silent Revolution: the rise of the low castes in north Indian politics* (London: Hurst & Co./New Delhi: Permanent Black, 2003), p. 351.
4. Sisson, "India in 1989," p. 119.
5. Csaba Nikolenyi, *Minority Governments in India: the puzzle of elusive majorities* (London: Routledge, 2010), p. 54.
6. Amrita Basu, "Parliamentary communism as a historical phenomenon: the CPI(M) in West Bengal," in Zoya Hasan (ed.), *Parties and Party Politics in India* (New Delhi: Oxford University Press, 2004), p. 341.
7. Achin Vanaik, *The Painful Transition: bourgeois democracy in India* (London: Verso, 1990), pp. 189–95.
8. Sisson, "India in 1989," p. 115.
9. Francine R. Frankel, "Middle classes and castes in Indian politics: prospects for political accommodation," in Atul Kohli (ed.), *India's Democracy: an analysis of changing state-society relations* (Princeton: Princeton University Press, 1988), pp. 245–6.
10. Unless otherwise noted, the following summarizes Yogendra Yadav, "Reconfiguration in Indian politics: state assembly elections, 1993–1995," in Partha Chatterjee (ed.), *State and Politics in India* (New Delhi: Oxford University Press, 1997), pp. 177–208.
11. Atul Kohli, "From majority to minority rule: making sense of the 'new' Indian politics," in Marshall M. Bouton and Philip Oldenburg (eds), *India Briefing, 1990* (Boulder: Westview Press, 1990), p. 4.

12. Peter A. Hall and Rosemary Taylor, "Political science and the three new institutionalisms," *Political Studies* 44, 5 (December 1996): 942. According to Hall and Taylor, critical junctures normally result from institutional changes or exogenous shocks. However, they may also occur due to cumulative political change. For a parallel theoretical argument, see Adam Ziegfeld, "Coalition government and party system change: explaining the rise of regional political parties in India," *Comparative Politics* 45, 1 (October 2012): 69–87.
13. National Front, "Constitution," in A.M. Zaidi (ed.), *Annual Register of Indian Political Parties: proceedings and fundamental texts 1990: part one* (New Delhi: Indian Institute of Applied Political Research, 1992), p. 466.
14. The last promise was the first point made by the Janata Dal, *The Peoples Charter—The Party Manifesto*, in O.P. Rahlan (ed.), *Encyclopedia of Political Parties Post-Independence India, Volume 96: Janata Dal proceedings* (New Delhi: Anmol Publications, 2001), p. 31.
15. National Front, *Manifesto: Lok Sabha Elections 1989*.
16. Kohli, "From majority to minority rule," pp. 21–23.
17. Jaffrelot, *India's Silent Revolution*, p. 336.
18. Seema Mustafa, *The Lonely Prophet: V.P. Singh* (New Delhi: New Age International, 1995), p. 101.
19. Mustafa, *The Lonely Prophet*, pp. 103, 106, 114.
20. Mustafa, *The Lonely Prophet*, p. 130.
21. Robert L. Hardgrave, Jr., "India on the eve of elections: Congress and the opposition," *Pacific Affairs* 57, 3 (Autumn 1984): 423.
22. Granville Austin, *Working a Democratic Constitution: a history of the Indian experience* (New Delhi: Oxford University Press, 2000), pp. 541–6 and 610–1; Aditya Nigam, "Communist politics hegemonized," in Partha Chatterjee (ed.), *Wages of Freedom: fifty years of the Indian nation-state* (New Delhi: Oxford University Press, 1998), pp. 233–234.
23. John Adams, "Breaking away: India's economy vaults into the 1990s," in Marshall M. Bouton and Philip Oldenburg (eds), *India Briefing, 1990* (Boulder: Westview Press, 1990), p. 97.
24. Sumanta Banerjee, "Strategy, tactics, and forms of political participation among left parties," in T.V. Sathyamurthy (ed.), *Class Formation and Political Transformation in Post-Colonial India* (New Delhi: Oxford University Press, 1999), p. 218.
25. Per capita Central Plan assistance to West Bengal between 1980–81 and 1984–85 was Rs. 132, compared to Rs. 214 nationally. Further, New Delhi delayed approving bills (for example, returning a "watered down" version of the state's Land Reform Act in 1987, fours years later: hence the adage that Congress' best weapon was the "sands of time"). Lastly, its average share of industrial licenses had declined since the 1970s, albeit erratically. See T.J. Nossiter, *Marxist State Governments in India: politics, economics and society* (London: Pinter Publishers, 1988), pp. 181–6, and Jùrgen Dige Pedersen, "India's industrial dilemmas in West Bengal," *Asian Survey*, 41, 4 (2001): 654.
26. Zoya Hasan, "The prime minister and the left," in James Manor (ed.), *Nehru*

to the Nineties: the changing office of the prime minister of India (London: Hurst, 1994), pp. 227–8.
27. See Basu, "Parliamentary communism as a historical phenomenon."
28. Patrick Heller, *The Labor of Development: workers and the transformation of capitalism in Kerala, India* (Ithaca: Cornell University Press, 1999), pp. 207–27.
29. The following summarizes Jaffrelot, *India's Silent Revolution*, pp. 336–40.
30. H.N. Bahuguna, its other leader, died in March 1989.
31. Kohli, "From majority to minority rule," pp. 8–9.
32. Mustafa, *The Lonely Prophet*, 85–99.
33. See Bidyut Chakrabarty, *Forging Power: coalition politics in India* (New Delhi: Oxford University Press, 2006), pp. 139–40.
34. Unless noted, the following draws on Adams, "Breaking away," pp. 93–4.
35. Richard Sisson and Munira Majmundar, "India in 1990: political polarization," *Asian Survey*, 31, 2 (February 1991): 104.
36. Interview with senior JD leader, New Delhi.
37. Sisson, "India in 1989," pp. 124–5.
38. James Manor, "Introduction," in James Manor (ed.), *Nehru to the Nineties: the changing office of the prime minister of India* (London: Hurst, 1994), p. 10.
39. Kohli, "From majority to minority rule," p. 23.
40. Sisson and Majmundar, "India in 1990," pp. 108–109.
41. See George Perkovich, *India's Nuclear Bomb: the impact of global proliferation* (Berkeley: University of California Press, 1999), pp. 303–17.
42. Rose, "India's foreign relations," pp. 73–74; Sisson and Majmundar, "India in 1990," pp. 104–106.
43. Balraj Puri, *Kashmir: towards insurgency* (Hyderabad: Orient Longman, 1995), pp. 72–75.
44. Christophe Jaffrelot, *The Hindu Nationalist Movement and Indian Politics: 1925 to the 1990s* (New Delhi: Penguin Books, 1999), pp. 413–414.
45. Perkovich, *India's Nuclear Bomb*, pp. 310–314.
46. Partha Chatterjee, *A Possible India: essays in political criticism* (New Delhi: Oxford University Press, 1997), p. 200; Sisson and Majmundar, "India in 1990," pp. 106–107.
47. See Annexure A in Balveer Arora and Douglas Verney (eds), *Multiple Identities in a Single State: Indian federalism in comparative perspective* (New Delhi: Konark, 1995), pp. 357–360.
48. Balveer Arora, "Political parties and the party system. the emergence of new coalitions," in Zoya Hasan (ed.), *Parties and Party Politics in India* (New Delhi: Oxford University Press, 2004), pp. 523–527.
49. Parvathi Menon, "In favor of talks," *Frontline*, 9 August 1996, pp. 110–111.
50. Sisson, "India in 1989," p. 124; Leo E. Rose, "India's foreign relations: reassessing basic policies," in Marshall M. Bouton and Philip Oldenburg (eds), *India Briefing, 1990* (Boulder: Westview Press, 1990), p. 75; M.V. Bratersky and S.I. Lunyov, "India at the end of the century: transformation into an Asian regional power," *Asian Survey*, 30, 10 (October 1990): 934.

51. Guha, *India after Gandhi*, p. 506.
52. Leo E. Rose, "India's foreign relations: reassessing basic policies," in Marshall M. Bouton and Philip Oldenburg (eds), *India Briefing, 1990* (Boulder: Westview Press, 1990), p. 70.
53. See Dhruba Kumar, "Managing Nepal's India policy?" *Asian Survey* 30, 7 (July 1990): 697–710.
54. Sumit Ganguly, "The prime minister and foreign and defence policies," in James Manor (ed.), *Nehru to the Nineties: the changing office of the prime minister of India* (London: Hurst, 1994), p. 156; Padmaja Murthy, "The Gujral doctrine and beyond," Mimeo, Institute for Defence Studies and Analysis, p. 3, available at: http://www.idsa-india.org/an-jul9–8.html.
55. Nalini Kant Jha, "Coalition governments and India's foreign policy," in Mahendra Prasad Singh and Anil Mishra (eds), *Coalition Politics in India: problems and prospects* (New Delhi: Manohar, 2004), p. 306.
56. See Achin Vanaik, *India in a Changing World* (New Delhi: Orient Longman, 1995), p. 35–7.
57. Paul Staniland, "Foreign policy making in India in the pre-liberalization and coalition era: unit level variables as determinants," in Amitabh Matoo and Happymon Jacob (eds), *Shaping India's Foreign Policy: people, politics and places* (New Delhi: Har-Anand Publications, 2010), p. 268.
58. See Ministry of Finance, Government of India, *1990–91 Union Budget Speech*.
59. Ashutosh Varshney, *Democracy, Development and the Countryside: urban-rural struggles in India* (Cambridge: Cambridge University Press, 1995), p. 143.
60. Planning Commission, Government of India, *Towards Social Transformation: Approach to the Eighth Five-Year Plan, 1990–95*.
61. BM, "Liberalisation lobby's attempted coup," *Economic and Political Weekly* 25, 23 (9 June 1990): 1237–8.
62. See Vijay Joshi and I.M.D. Little, *India: macroeconomics and political economy, 1964–1991* (New Delhi: Oxford University Press, 1994), pp. 62–5 and 181–6; and Amit Bhaduri and Deepak Nayyar, *The Intelligent Person's Guide to Liberalization* (New Delhi: Penguin, 1996), pp. 22–8.
63. Chatterjee, *A Possible India*, p. 202.
64. Harkishan Singh Surjeet, "The present political situation and the evolution of our tactics in perspective," *The Marxist* 8, 4 (October–December 1990), p. 15.
65. See Editorial, "What is different?" *Economic and Political Weekly* 25, 12 (24 March 1990): 579–80; Sanjaya Baru, "A quiet coup in the fisc," *Economic and Political Weekly* 25, 16 (21 April 1990): 887–92.
66. BM, "Precarious balancing act," *Economic and Political Weekly* 25, 13 (31 March 1990): 648–9; Pulapre Balakrishnan, "Union budget for 1990–91," *Economic and Political Weekly* 25, 16 (21 April 1990): 895.
67. Francine R. Frankel, *India's Political Economy 1947–2004: the gradual revolution*, 2nd edition (New Delhi: Oxford University Press, 2006), p. 688.
68. Jaffrelot, *India's Silent Revolution*, p. 321.
69. Cited by Guha, *India after Gandhi*, pp. 599–600.

70. Marc Galanter, "Pursuing equality in the land of hierarchy: an assessment of India's policies of compensatory discrimination for historically disadvantaged groups," in Rajeev Dhavan (ed.), *Law and Society in Modern India*, (New Delhi: Oxford University Press, 1997), pp. 185–208.
71. André Béteille, "Distributive justice and institutional well-being," in idem, *The Backward Classes in Contemporary India* (New Delhi: Oxford University Press, 1992), pp. 45–72.
72. Zoya Hasan, *Quest for Power: oppositional movements and post-Congress politics in Uttar Pradesh* (New Delhi: Oxford University Press, 1999), pp. 152–3.
73. Sudipta Kaviraj, "Democracy and social inequality," in Francine R. Frankel, Zoya Hasan, Rajeev Bhargava and Balveer Arora (eds), *Transforming India: social and political dynamics of democracy* (New Delhi: Oxford University Press, 2000), pp. 89–120.
74. Zoya Hasan, "Representation and redistribution: the new lower caste politics of north India," in Francine R. Frankel, Zoya Hasan, Rajeev Bhargava and Balveer Arora (eds), *Transforming India: social and political dynamics of democracy* (New Delhi: Oxford University Press, 2000), pp. 146–76.
75. In addition, it excluded Patels, Lingayats, Kammas, Reddys and Vellalas. Frankel, *India's Political Economy*, p. 688.
76. Chakrabarty, *Forging Power*, p. 141.
77. Mustafa, *The Lonely Prophet*, 158–165.
78. Austin, *Working a Democratic Constitution*, p. 452.
79. See Janata Dal, *The Peoples Charter*.
80. Hasan, *Quest for Power*, pp. 149–150.
81. Mahendra Prasad Singh, "India's National Front and United Front coalition governments: a phase in federalized governance," in Mahendra Prasad Singh and Anil Mishra (eds), *Coalition Politics in India: problems and prospects* (New Delhi: Manohar, 2004), pp. 83–4.
82. Mustafa, *The Lonely Prophet*, p. 174.
83. Jaffrelot, *India's Silent Revolution*, p. 255.
84. Mustafa, *The Lonely Prophet*, pp. 159–165.
85. Kancha Ilaiah, "Towards the dalitization of the nation," in Partha Chatterjee (ed.), *Wages of Freedom: fifty years of the Indian nation-state* (New Delhi: Oxford University Press, 1998), p. 282.
86. Ghanshyam Shah, "Social backwardness and the politics of reservation," in Ghanshyam Shah (ed.), *Caste and Democratic Politics in India* (London: Anthem Press, 2004), p. 306.
87. Hasan, *Quest for Power*, pp. 154–65.
88. Varshney, *Democracy, Development and the Countryside*, pp. 144–5.
89. A.G. Noorani, "The prime minister and the judiciary," in James Manor (ed.), *Nehru to the Nineties: the changing office of the prime minister of India* (London: Hurst, 1994), p. 114.
90. Rekha Saxena, *Situating Federalism: mechanisms of intergovernmental relations in Canada and India* (New Delhi: Manohar, 2006), pp. 271–2.
91. B.D. Dua, "The prime minister and the federal system," in James Manor (ed.),

NOTES   pp. [121–129]

Nehru to the Nineties: the changing office of the prime minister of India (London: Hurst, 1994), p. 41.
92. Lewis P. Fickett, Jr., "The rise and fall of the Janata Dal," *Asian Survey*, 33, 12 (December 1993): 1156–7.
93. *National Front Manifesto: Lok Sabha Elections, May 1991.*
94. Mustafa, *The Lonely Prophet*, pp. 128–129.
95. Fickett, Jr., "The rise and fall of the Janata Dal," pp. 1157–8.
96. Chatterjee, *A Possible India*, pp. 219–20.
97. Walter K. Andersen, "India's 1991 election: the uncertain verdict," *Asian Survey*, 31, 10 (October 1991): 985–7.

6. THE CRYSTALLIZATION OF THE THIRD FORCE (1991–1996)

1. *Bharatiya Janata Party (BJP) Election Manifesto 1996*, p. 7.
2. Sunil Khilnani, *The Idea of India* (New York: Farrar Straus Giroux, 1999), p. 151.
3. Amrita Basu, "The dialectics of Hindu nationalism," in Atul Kohli (ed.), *The Success of India's Democracy* (Cambridge: Cambridge University Press, 2001), p. 171.
4. Christophe Jaffrelot, *The Hindu Nationalist Movement and Indian Politics: 1925 to the 1990s* (New Delhi: Penguin Books, 1999), pp. 478–91 and 536–51.
5. Sudipta Kaviraj, "The general elections in India," *Government and Opposition*, 32, 1 (Winter 1997): 10.
6. See Rob Jenkins, *Democratic Politics and Economic Reform in India* (Cambridge: Cambridge University Press, 1999), pp. 119–71; Aseema Sinha, "The changing political economy of federalism in India: a historical institutionalist approach," *India Review*, 3, 1 (January 2004): 25–63.
7. See Ashutosh Varshney, "Mass politics or elite politics? India's economic reforms in comparative perspective," in Jeffrey Sachs, Ashutosh Varshney and Nirupam Bajpai (eds), *India in the Era of Economic Reforms* (New Delhi: Oxford University Press, 1999), pp. 8–13 and 27–33.
8. Interview with senior regional politician, New Delhi, 1 December 1998.
9. Jenkins, *Democratic Politics and Economic Reform in India*, pp. 172–207.
10. Stuart Corbridge and John Harriss, *Reinventing India: liberalization, Hindu nationalism and popular democracy* (Cambridge: Polity Press, 2000), pp. 159–69.
11. E. Sridharan, "Principles, power and coalition politics in India: lessons from theory, comparison and recent history," D.D. Khanna and Gert W. Kueck (eds), *Principles, Power and Politics* (New Delhi: Macmillan India, 1999), pp. 270–91.
12. *Indian National Congress(I) Election Manifesto 1996*, pp. 3–4.
13. *Congress(I) Election Manifesto 1996*, p. 20.
14. Kaviraj, "The general elections in India," p. 13.
15. Sukumar Muralidharan, "Heat and haze," *Frontline*, 3 May 1996, pp. 4–7.

16. *Congress(I) Election Manifesto 1996*, p. 13.
17. Vir Sanghvi, "The beginning of the end," *Seminar*, 461 (January 1998): 39–43.
18. Venkitesh Ramakrishnan, "Corruption at large," *Frontline*, 9 February 1996, pp. 4–11.
19. Interview, New Delhi, 30 January 2000.
20. Lloyd I. Rudolph and Susanne Hoeber Rudolph, "Redoing the constitutional design: from an interventionist to a regulatory state," in Atul Kohli (ed.), *The Success of India's Democracy* (Cambridge: Cambridge University Press, 2001), p. 134.
21. A.G. Noorani, "A gross case," *Frontline*, 9 February 1996, pp. 14–5.
22. See James Manor, "'Ethnicity' and politics in India," *International Affairs*, 72, 3 (July 1996): 459–75.
23. Interview with senior Congress(I) official, New Delhi, 30 January 2000.
24. Sanghvi, "The beginning of the end," p. 40.
25. The EC issued a series of directives in 1996. These included a ban on chief ministers from using official aircraft for electoral campaigning; on politicians from distributing complimentary rail passes; and on parties from fixing posters on public walls. Sudha Mahalingam, "Matters of conduct," *Frontline*, 3 May 1996, pp. 14–5.
26. The allegations concerned a conspiracy, with Rao's alleged involvement, to forge documents that indicated illegal offshore funds in the name of the son of former prime minister V.P. Singh. Prashant Bhushan, "Corruption and the law," *Frontline*, 9 February 1996, pp. 119–20.
27. On 5 February 1988, Mr Lakhubhai Pathak, a non-resident Indian (NRI) businessman living in the United Kingdom, filed a complaint against the controversial godman, Chandraswami, for defrauding him of approximately US $100,000. Chandraswami allegedly assured the former that his personal contacts with Rao would enable him to secure government contracts.
28. In January 1996, four JMM leaders—Shibu Soren, Suraj Mandal, Simon Mirandi and Shailendra Mahato—confessed to accepting bribes to vote for Rao's ministry in the July 1993 confidence vote.
29. Venkitesh Ramakrishnan, "Strategy slips," *Frontline*, 3 May 1996, pp. 7–9.
30. Pratap Bhanu Mehta, "Reform political parties first," *Seminar*, 497 (January 2001): 38–41.
31. Interview with senior AIIC(T) politician, New Delhi, 8 February 2000.
32. Interview with senior Congress(I) official, New Delhi, 8 October 1998.
33. Interview with senior Congress(I) official, New Delhi, 8 October 1998.
34. Interview with senior regional politician, New Delhi, 1 December 1998.
35. *Bharatiya Janata Party (BJP) Election Manifesto 1996*, pp. 10 and 64.
36. *BJP Election Manifesto 1996*, p. 8.
37. *BJP Election Manifesto 1996*, pp. 34–40.
38. *BJP Election Manifesto 1996*, p. 18.
39. *BJP Election Manifesto 1996*, pp. 42–5.
40. Jaffrelot, *The Hindu Nationalist Movement and Indian Politics*, pp. 536–9.
41. *BJP Election Manifesto 1996*, pp. 16–20.

42. *BJP Election Manifesto 1996*, pp. 21–4.
43. *BJP Election Manifesto 1996*, pp. 27–28.
44. For an analysis of the BJP in Maharashtra, see Thomas Blom Hansen, "The ethics of Hindutva and the spirit of capitalism," in Thomas Blom Hansen and Christophe Jaffrelot (eds), *The BJP and the Compulsions of Politics in India*, 2nd edition (New Delhi: Oxford University Press, 2001), pp. 309–13; and the sections on Rajasthan in Jenkins, *Democratic Politics and Economic Reform in India*.
45. Kaviraj, "The general election in India," p. 15.
46. *Janata Dal (JD) Election Manifesto 1996*, pp. 1 and 17–18.
47. *JD Election Manifesto 1996*, p. 14.
48. *JD Election Manifesto 1996*.
49. *JD Election Manifesto 1996*, pp. 3–5; Praveen Swami, "Aiming to cohere," *Frontline*, 3 May 1996, pp. 12–3.
50. *JD Election Manifesto 1996*, p. 5.
51. *JD Election Manifesto 1996*, pp. 14–15.
52. *JD Election Manifesto 1996*, pp. 6–7.
53. *JD Election Manifesto 1996*, pp. 22–3.
54. *JD Election Manifesto 1996*, pp. 7–9.
55. Harkishen Singh Surjeet, "The present political situation and the evolution of our tactics in perspective," *The Marxist*, 8, 4 (October–December 1990): 5 and 19.
56. *Communist Party of India (CPI) Election Manifesto 1996*, p. 14.
57. *CPI Election Manifesto 1996*, p. 4.
58. *Communist Party of India-Marxist (CPI(M)) Election Manifesto 1996*, p. 14.
59. *CPI(M) Election Manifesto 1996*, p. 1.
60. *CPI Election Manifesto 1996*, p. 7.
61. *CPI(M) Election Manifesto 1996*, pp. 12–13.
62. Patrick Heller, *The Labor of Development: workers and the transformation of capitalism in Kerala, India* (Ithaca: Cornell University Press, 1999), pp. 227–36.
63. See Richard Sandbrook, Marc Edelman, Patrick Heller and Judith Teichman, *Social Democracy in the Global Periphery: origins, challenges, prospects* (Cambridge: Cambridge University Press, 2007), pp. 65–92.
64. Abhijit Banerjee, Pranab Bardhan, Kaushik Basu, Mrinal Datta Chaudhuri, Maitresh Ghatak, Ashok Sanjay Guha, Mukul Majumdar, Dilip Mookherjee and Debraj Ray, "Strategy for economic reform in West Bengal," *Economic and Political Weekly*, 37, 41 (12 October 2002): 4203–18.
65. Jùrgen Dige Pedersen, "India's industrial dilemmas in West Bengal," *Asian Survey*, 41, 4 (2001): 657–658.
66. Sumanta Banerjee, "Strategy, tactics, and forms of political participation among left parties," in T.V. Sathyamurthy (ed.), *Class Formation and Political Transformation in Post-Colonial India* (New Delhi: Oxford University Press, 1999), p. 235; Sinha, "The changing political economy of federalism in India," p. 43.
67. Javeed Alam, "Communist politics in search of hegemony," in Partha

Chatterjee (ed.), *Wages of Freedom: fifty years of the Indian nation-state* (New Delhi: Oxford University Press, 1998), p. 180; Polly Datta, "The issue of discrimination in Indian federalism in the post-1977 politics of West Bengal," *Comparative Studies of South Asia, Africa and the Middle East*, 1, 2 (2005): 459.

68. Interview with senior regional politician, New Delhi, 12 November 1998.
69. Transcript of interview with Shri Jyoti Basu, Acc. No. 781, Oral History Division, Nehru Memorial Museum and Library, New Delhi, p. 196.
70. Yogendra Yadav and Suhas Palshikar, "From hegemony to convergence: party system and electoral politics in the Indian states, 1952–2002," *Journal of Indian School of Political Economy*, XV, 1&2 (January–June 2003): 30–2.
71. Jagdish Bhagwati, "The design of Indian development," in Isher Judge Ahluwalia and I.M.D. Little (eds), *India's Economic Reforms and Development: essays for Manmohan Singh* (New Delhi: Oxford University Press, 1998), p. 38.
72. James Manor, "Regional parties in federal systems," in Balveer Arora and Douglas Verney (eds), *Multiple Identities in a Single State: Indian federalism in comparative perspective* (New Delhi: Konark, 1995), p. 116.
73. Chatterjee, *A Possible India: essays in political criticism* (New Delhi: Oxford University Press, 1997), pp. 223–4.
74. Kaviraj, "The general election in India," p. 15.
75. Interview with CPI(M) Politburo member, New Delhi, 14 February 2000.
76. Quoted in Radhika Desai, "Forward march of Hindutva halted?" *New Left Review*, 30 (November–December 2004): 49.
77. Interview with CPI(M) Politburo member, New Delhi, 12 November 1998.
78. Interview with senior CPI official, New Delhi, 10 February 2000.
79. Interview, New Delhi, 23 October 1998.
80. Sitaram Yechury, "Caste and class in Indian politics today," *P. Sundarayya Memorial Lecture*, 1997, p. 3.
81. Communist Party of India (Marxist), "Review Report on 1996 General Elections, adopted by the Central Committee, July 27–29, 1996," p. 14.
82. Yechury, "Caste and class in Indian politics today," p. 11.
83. Yechury, "Caste and class in Indian politics today," p. 17.
84. Sudipta Kaviraj, "Marxism in translation: critical reflections on Indian radical thought," in Richard Bourke and Raymond Geuss (eds), *Political Judgment: essays in honor of John Dunn* (Cambridge: Cambridge University Press, 2009), pp. 172–200.
85. See E. Sridharan and Ashutosh Varshney, "Toward moderate pluralism: political parties in India," in Larry Diamond and Richard Gunther (eds), *Political Parties and Democracy* (Baltimore: Johns Hopkins University Press, 2001), pp. 231–2.
86. André Béteille, "The future of the Backward Classes: the competing demands of status and power," in André Béteille, *Society and Politics in India: essays in comparative perspective* (New Delhi: Oxford University Press, 1997), p. 157; and Myron Weiner, "The struggle for equality: caste in Indian politics," in

Atul Kohli (ed.), *The Success of India's Democracy* (Cambridge: Cambridge University Press, 2001), pp. 214–23.
87. Marc Galanter, "Pursuing equality in the land of hierarchy: an assessment of India's policies of compensatory discrimination for historically disadvantaged groups," in Rajeev Dhavan (ed.), *Law and Society in Modern India* (New Delhi: Oxford University Press, 1997), p. 206.
88. See Christophe Jaffrelot, *India's Silent Revolution: the rise of the low castes in north Indian politics* (London: Hurst/New Delhi: Permanent Black, 2003), ff. 409.
89. Kanchan Chandra, *Why Ethnic Parties Succeed: patronage and ethnic head counts in India* (Cambridge: Cambridge University Press, 2004), p. 181.
90. Chandra, *Why Ethnic Parties Succeed*, pp. 148–50.
91. Jaffrelot, *India's Silent Revolution*, pp. 372–5; Zoya Hasan, *Quest for Power: oppositional movements and post-Congress politics in Uttar Pradesh* (New Delhi: Oxford University Press, 1999), p. 161.
92. Sudha Pai, *Dalit Assertion and the Unfinished Democratic Revolution: the Bahujan Samaj Party in Uttar Pradesh* (New Delhi: Sage Publications, 2002), pp. 167–8.
93. Jaffrelot, *India's Silent Revolution*, pp. 409–13.
94. Yogendra Yadav, "The third electoral system," *Seminar*, 480 (August 1999), p. 18.
95. Lewis P. Fickett, Jr., "The rise and fall of the Janata Dal," *Asian Survey* 33, 12 (December 1993): 1152; John R. Wood, "On the periphery but in the thick of it: some recent Indian political crises viewed from Gujarat," in Philip Oldenburg (ed.), *India Briefing: Staying the Course* (Armonk: M.E. Sharpe, 1995), pp. 32–7.
96. Fickett, "The rise and fall of the Janata Dal," p. 1152; E. Sridharan, "Unstable parties and unstable alliances: births, splits, mergers and deaths of parties, 1952–2000," in Mahendra Prasad Singh and Anil Mishra (eds), *Coalition Politics in India: problems and prospects* (New Delhi: Manohar, 2004), p. 51.
97. Shafiuzzaman, *The Samajwadi Party: a study of its social base, ideology and programme* (New Delhi: A.P.H. Publishing Corporation, 2003), pp. 55–7.
98. Transcript of interview with Shri Madhu Dandavate, Acc. No. 788, Oral History Division, Nehru Memorial Museum and Library, New Delhi, pp. 276–7.
99. Samata Party, "Statement of Policy and Programmes," National Convention, New Delhi, 19 August 1994.
100. Jaffrelot, *India's Silent Revolution*, pp. 367 and 377–87; Atul Kohli, *Democracy and Discontent: India's growing crisis of ungovernability* (Cambridge: Cambridge University Press, 1990), p. 208.
101. See S. P. Sathe, *Judicial Activism in India: transgressing borders and enforcing limits*, second edition, with a foreword by Upendra Baxi (New Delhi: Oxford University Press, 2004), pp. 272–5.
102. Jaffrelot, *India's Silent Revolution*, pp. 187–99.

103. Interview, New Delhi, 13 October 1998.
104. Pranab Bardhan, "Sharing the spoils: group equity, development and democracy," in Atul Kohli (ed.), *The Success of India's Democracy* (Cambridge: Cambridge University Press, 2001), p. 227.
105. Interview, New Delhi, 16 February 2000.
106. Interview, New Delhi, 28 October 1998.
107. Sanjaya Baru, "Economic policy and the development of capitalism in India: the role of regional capitalists and political parties," in Francine R. Frankel, Zoya Hasan, Rajeev Bhargava and Balveer Arora (eds), *Transforming India: social and political dynamics of democracy* (New Delhi: Oxford University Press, 2000), pp. 207–31; Pranab Bardhan, *The Political Economy of Development in India, expanded edition with an epilogue on the political economy of reform in India* (New Delhi: Oxford University Press, 2012), pp. 124–5. Starting in the mid-1990s, forward castes and upper classes increasingly constituted the social base of the AGP, too. See Suhas Pulshikar, "Regional and caste parties," in Atul Kohli and Prerna Singh (eds), *Routledge Handbook of Indian Politics* (New York: Routledge, 2013), p. 99.
108. Interview with senior regional official, New Delhi, 7 February 2000.
109. Interview, New Delhi, 12 November 1998.
110. See Aditya Nigam, "India after the 1996 elections: nation, locality and representation," *Asian Survey*, 36, 12 (December 1996): 1157–70.
111. Mahendra Prasad Singh, "India's National Front and United Front coalition governments: a phase in federalized governance," in Mahendra Prasad Singh and Anil Mishra (eds), *Coalition Politics in India: problems and prospects* (New Delhi: Manohar, 2004), pp. 90–1.
112. N. Ram, "Elections, 1996," *Frontline*, 3 May 1996, p. 21.
113. Csaba Nikolenyi, *Minority Governments in India: the puzzle of elusive majorities* (London: Routledge, 2010), p. 59.
114. Swami, "Aiming to cohere," pp. 12–13.
115. S.K. Pande, "Hopes on Third Front," *Frontline*, 19 April 1996, p. 16; S. Viswanathan, "For a Third Front," *Frontline*, 9 February 1996, pp. 126–9.
116. Interview with senior JD official, New Delhi, 10 February 2000. The MDMK, which championed the rights of Sri Lankan Tamils, sought to revitalize the DMK's assertive populism. See Narendra Subramanian, *Ethnicity and Populist Mobilization: political parties, citizens and democracy in south India* (New Delhi: Oxford University Press, 1999), pp. 32 and 222.
117. Interview with senior JD official, New Delhi, 10 February 2000.
118. Achin Vanaik, *The Furies of Indian Communalism: religion, modernity and secularization* (London: Verso, 1997), pp. 297–8.
119. Yogendra Yadav, "Reconfiguration in Indian politics: state assembly elections, 1993–1995," in Partha Chatterjee (ed.), *State and Politics in India* (New Delhi: Oxford University Press, 1997), pp. 177–208.

7. THE FORMATION OF THE UNITED FRONT (MAY 1996)

1. Parties lost their financial deposits if they failed to secure more than one-sixth of the total valid votes polled in a given constituency.
2. James Chiriyankandath, "'Unity in diversity?' coalition politics in India (with special reference to Kerala)," *Democratization* 4, 4 (Winter 1997): 20.
3. Yogendra Yadav, "How India voted," *India Today*, 31 May 1996, p. 22.
4. The following analysis is based on data presented in Election Commission of India, *Statistical Report on General Elections, 1996, to the Eleventh Lok Sabha, Volume I* (hereafter *Election Commission Report 1996*), pp. 32–40.
5. Yadav, "How India voted," p. 22.
6. The *National Election Study, 1996, Post-Poll Survey*, was designed by V.B. Singh and Yogendra Yadav of the Centre for the Study of Developing Societies (CSDS), Delhi. It was the largest nationwide social scientific survey of political attitudes and opinions in India during the Eleventh General Election. Conducted between early June and mid July after the 1996 general election, 250 trained researchers interviewed 9457 respondents from a multi-stage random sample of 108 Lok Sabha constituencies. Each interview, translated into fifteen national languages, took between one and two hours. Each constituency comprised 216 state assembly segments and 432 polling booth areas. The composition of the interviewees included rural dwellers (75 per cent), women (49 per cent), the unlettered (42 per cent), Muslims (11 per cent), SCs/*dalits* (19 per cent) and STs/*Adivasis* (9 per cent). The failure of all respondents to answer every question led to some variation in the total number of respondents interviewed, but only marginally. See Yogendra Yadav, "The maturing of a democracy," *India Today*, 31 August 1996, pp. 28–43.
7. Anthony Heath and Yogendra Yadav, "The united colours of the Congress: social profile of Congress voters, 1996 and 1998," in Zoya Hasan (ed.), *Parties and Party Politics in India* (New Delhi: Oxford University Press, 2004), pp. 144–5.
8. Aditya Nigam, "India after the 1996 elections: nation, locality and representation," *Asian Survey*, 36, 12 (December 1996), ff. 1162.
9. *Election Commission Report 1996*, pp. 31–40.
10. Unless otherwise noted, the following statistics are from Yadav, "How India voted," p. 24.
11. Sudipta Kaviraj, "The general elections in India," *Government and Opposition*, 32, 1 (Winter 1997): 22.
12. The social profile of the three main formations employ bivariate, or cross-tabulated, analyses of data provided by the *National Election Study, 1996, Post-Poll Survey*. The limitation of such analyses—which duplicate the significance of factors that tend to overlap in reality—is well known. Nevertheless, the data reveals broad salient differences. See Oliver Heath, "Anatomy of BJP's rise to power: social, regional, and political expansion in the 1990s," in Zoya Hasan (ed.), *Parties and Party Politics in India* (New Delhi: Oxford University Press, 2002), p. 233.

13. I thank Stuart Corbridge for emphasizing this point. See also Tariq Thachil, *Elite Parties, Poor Voters: how social services win votes in India* (Cambridge: Cambridge University Press, forthcoming/October 2014).
14. E. Sridharan, "Unstable parties and unstable alliances: births, splits, mergers and deaths of parties, 1952–2000," in Mahendra Prasad Singh and Anil Mishra (eds), *Coalition Politics in India: problems and prospects* (New Delhi: Manohar, 2004), p. 60.
15. Lloyd I. Rudolph and Susanne Hoeber Rudolph, "Redoing the constitutional design: from an interventionist to a regulatory state," in Atul Kohli (ed.), *The Success of India's Democracy* (Cambridge: Cambridge University Press, 2001), p. 145.
16. Interview with senior Congress(I) official, New Delhi, 2 September 1998.
17. *Times of India*, 5 June 1996.
18. Interview with senior Congress(I) official, New Delhi, 2 September 1998.
19. Venkitesh Ramakrishnan, "Staying power," *Frontline*, 31 May 1996, pp. 17–8.
20. Interview with senior AIIC(T) politician, New Delhi, 12 October 1998.
21. Interview with senior Congress(I) official, New Delhi, 4 December 1998.
22. The AIIC(T) captured four parliamentary seats. S. Bangarappa, a factional opponent of the Rao coterie, led the KCP, which won only one of eleven contested seats. Ramakant D. Khalap, who led the MAG–also known as the Maharashtra Gomantak Party (MGP)–secured one of two parliamentary contests. See David Butler, Ashok Lahiri and Prannoy Roy, *India Decides: update 1996* (New Delhi: Books and Things, 1996), pp. 1–2.
23. Interview with senior regional politician, New Delhi, 7 February 2000.
24. Interview senior regional politician, New Delhi, 18 February 2000.
25. Interview senior regional politician, New Delhi, 1 December 1998.
26. Sukumar Muralidharan, "A new equilibrium," *Frontline*, 28 June 1996, pp. 4–7.
27. Aijaz Ahmad, *Lineages of the Present: ideology and politics in contemporary South Asia* (London: Verso, 2000), p. 218.
28. Interview with senior JD politician, New Delhi, 18 October 1998.
29. Interview with senior JD politician, New Delhi, 18 October 1998.
30. Interview with senior JD politician, New Delhi, 18 October 1998.
31. Interview with senior JD politician, New Delhi, 18 October 1998.
32. However, Basu conceded that his opposition to Ranadive "was a sort of a negative attitude." Transcript of interview with Shri Jyoti Basu, Acc. No. 781, Oral History Division, Nehru Memorial Museum and Library, New Delhi, pp. 58–69.
33. Transcript of interview with Jyoti Basu, pp. 151–2.
34. Sukumar Muralidharan, "An unstable set-up," *Frontline*, 31 May 1996, pp. 4–8.
35. E.M.S. Namboodiripad, *The Frontline Years: selected articles* (New Delhi: LeftWord Books, 2010), p. 106.
36. Interview with senior CPI(M) politician, Calcutta, 19 November 1998.
37. Interview with senior CPI(M) official, New Delhi, 8 February 2000.

38. Interviews with senior CPI(M) officials, New Delhi, 14 February 2000 and 12 November 1998, respectively.
39. Communist Party of India(Marxist), "Review of the 1999 Lok Sabha elections," adopted at the Central Committee meeting, 20–22 November 1999, p. 8.
40. Ahmad, *Lineages of the Present*, p. 234.
41. Interview with senior JD politician, New Delhi, 1 December 1998.
42. Interview, New Delhi, 2 September 1998.
43. Ahmad, *Lineages of the Present*, pp. 235–7.
44. Interview with senior Congress(I) official, New Delhi, 2 September 1998.
45. Interview, New Delhi, 12 November 1998.
46. Achin Vanaik, *The Furies of Indian Communalism: religion, modernity and secularization* (London: Verso, 1997), pp. 332–3.
47. See Nivedita Menon and Aditya Nigam, *Power and Contestation: India since 1989* (Hyderabad: Orient Longman, 2008), pp. 105–6; Aditya Nigam, "Logic of failed revolution: federalisation of CPI(M)," *Economic and Political Weekly*, 35, 5 (29 January 2000): 263–4.
48. Javeed Alam, "Communist politics in search of hegemony," in Partha Chatterjee (ed.), *Wages of Freedom: fifty years of the Indian nation-state* (New Delhi: Oxford University Press, 1998), p. 180.
49. Vanaik, *The Furies of Indian Communalism*, pp. 281–4.
50. Vanaik, *The Furies of Indian Communalism*, p. 357, fn 25.
51. T.J. Nossiter, *Marxist State Governments in India: politics, economics and society* (London: Pinter Publishers, 1988), pp. 188–96.
52. Interview with senior JD politician, New Delhi, 18 October 1998. Many other interviewees corroborated his claim independently.
53. Interviews with senior JD politician and senior regional politician, New Delhi, 1 December 1998 and 12 November 1998, respectively.
54. His caste status reflected the relative absence of indigenous Kshatriyas and Vaisyas in the south. James Manor, "Prologue: caste and politics in recent times," in Rajni Kothari (ed.), *Caste in Indian Politics* (New Delhi: Orient Blackswan, 2010), xx.
55. Interview with senior JD politician, New Delhi, 6 November 1998.
56. Interview with senior JD politician, New Delhi, 23 October 1998.
57. Interview with senior JD politician, New Delhi, 1 December 1998.
58. Interview with senior JD politician, New Delhi, 23 October 1998.
59. Interview with senior JD politician, New Delhi, 1 December 1998.
60. Interview with senior JD politician, New Delhi, 18 October 1998.
61. Parvathi Menon, "Road to the top," *Frontline*, 14 June 1996, pp. 13–4.
62. Interview with senior JD politician, New Delhi, 1 December 1998.
63. Interview with senior government official, New Delhi, 2 September 1998.
64. Interview with senior JD politician, New Delhi, 23 October 1998.
65. Interview, New Delhi, 1 October 1998.
66. Interview with senior JD politician, New Delhi, 18 October 1998.
67. See Venkitesh Ramakrishnan, "In the spoiler's role," *Frontline*, 31 May 1996, p. 138.

68. Interview with senior JD politician, New Delhi, 1 December 1998.
69. A.G. Noorani, "A grave lapse," *Frontline*, 14 June 1996, pp. 26–8.
70. V. Venkatesan, "A tale of two ordinances," *Frontline*, 19 April 1996, p. 30.
71. Interview with senior government official, New Delhi, 2 September 1998.
72. *Hindustan Times*, 20 May 1996.
73. See "Platform for secular democratic alternative," *New Age Weekly* (21–27 April 1996).
74. *Times of India*, 19 May 1996.
75. See Chapter III, Term No. (II), *Report of The Srikrishna Commission Appointed for Inquiry into The Riots at Mumbai during December 1992—January 1993 and the March 12, 1993 Bomb Blasts, Volume I*, by Hon. Justice B.N. Srikrishna, Mumbai High Court.
76. *Hindustan Times*, 23 May 1996.
77. Interview with senior JD politician, New Delhi, 23 October 1998.
78. *Times of India*, 28 May 1996.
79. Interview with senior government official, New Delhi, 22 October 1998.
80. *The Hindu*, 2 June 1996.
81. Sukumar Muralidharan, "A new equilibrium," *Frontline*, 28 June 1996, p. 7.
82. Bommai was the JD president until February 1996 when the media implicated him in the scandal. S.K. Pande, "New face," *Frontline*, 8 March 1996, pp. 25–7.
83. Interview with senior JD politician, New Delhi, 1 December 1998.
84. Muralidharan, "A new equilibrium," pp. 6–7.
85. Ramachandra Guha, *India after Gandhi: the history of the world's largest democracy* (New York: HarperCollins, 2007), p. 506.
86. Interview with senior government official, New Delhi, 7 February 2000.
87. Interview with senior government official, New Delhi, 16 November 1998.
88. Interview with senior JD politician, New Delhi, 23 October 1998.
89. Menon, "Road to the top," p. 14.
90. Francine Frankel and Sumantra Sen, *Andhra Pradesh's Long March towards 2020: electoral detours in a developmentalist state* (Philadelphia: Center for the Advanced Study of India, University of Pennsylvania, 2005), p. 4.
91. See Sunil Khilnani, *The Idea of India* (New York: Farrar Straus Giroux, 1999), pp. 103–6.
92. Interview with senior CPI politician, New Delhi, 23 October 1998.
93. Interview, New Delhi, 21 October 1998.
94. Interview with senior CPI official, New Delhi, 22 October 1998.
95. Interview with senior CPI official, New Delhi, 10 February 2000.
96. *Hindustan Times*, 4 June 1996.
97. *Times of India*, 5 June 1996.
98. *Janata Dal Election Manifesto 1996*.
99. *Janata Dal Election Manifesto 1996*.
100. Correspondent, "Less equal," *Economic and Political Weekly*, 31, 19 (11 May 1996): 1101–2.
101. The United Front, *A Common Approach to Major Policy Matters and a Minimum Programme* (hereafter *CMP*), p. 1.

102. *CMP*, p. 2.
103. *CMP*, p. 3.
104. *CMP*, p. 4.
105. *CMP*, pp. 11–2.
106. *CMP*, p. 6.
107. *CMP*, p. 9. The other significant fiscal measure concerned tax reform.
108. *CMP*, p. 6.
109. *CMP*, p. 6.
110. *CMP*, p. 6.
111. Editorial, "Glib promises," *Economic and Political Weekly*, 31, 24 (June 15, 1996): 1423.
112. Sukumar Muralidharan, "Getting down to work," *Frontline*, 28 June 1996, pp. 19–20.
113. Interviews with senior JD politician, New Delhi, 1 December 1998 and 23 October 1998, respectively.
114. Interview, New Delhi, 16 November 1998.
115. Interview with senior political journalist, New Delhi, 11 September 1998.
116. Interview, New Delhi, 16 November 1998.
117. Interview with senior regional politician, New Delhi, 12 November 1998.
118. "Interview with S. Jaipal Reddy", *Seminar*, 454 (June 1997), p. 55; quoted in Bidyut Chakrabarty, *Forging Power: coalition politics in India* (New Delhi: Oxford University Press, 2006), p. 144.
119. In November 1995, National Fertilizer Limited (NFL) awarded a contract to a Turkish company named Karsan to export two hundred thousand tons of urea, worth approximately Rs. 133 crore (US$?), by 4 May 1996. The chairman and vice-president of the company, respectively Tuncay Alankus and Cihan Karanci, disappeared after taking the entire payment in advance. Allegations of kickbacks, estimated at US$4 million, surfaced. The three main suspects were C. Ramakrishnan, managing director of NFL, D.S. Kanwar, former executive director of the NFL and Sambasiva Rao, Karsan's agent in Hyderabad. The latter told the CBI that Prabhakar Rao and Prakash Chandra, son of former Union minister Ram Lakhan Singh Yadav, had received kickbacks. *The Statesman*, 16 June 1996.
120. *Hindustan Times*, 13 June 1996.

8. ESTABLISHING POLITICAL AUTHORITY (JUNE–SEPTEMBER 1996)

1. See James Manor, "Understanding Deve Gowda," *Economic and Political Weekly*, 31, 39 (28 September 1996): 2675–8.
2. Lloyd I. Rudolph and Susanne Hoeber Rudolph, *In Pursuit of Lakshmi: the political economy of the Indian state* (Chicago: University of Chicago Press, 1987), pp. 96–8.
3. T.V. Sathyamurthy, "Epilogue," in T.V. Sathyamurthy (ed.), *Class Formation and Political Transformation in Post-Colonial India* (New Delhi: Oxford University Press, 1996), p. 473.

4. Achin Vanaik, *The Furies of Indian Communalism: religion, modernity and secularization* (London: Verso, 1997), pp. 349–50.
5. *Hindustan Times*, 14 June 1996.
6. Unless noted, the following analysis draws on S.K. Pande, "New face," *Frontline*, 8 March 1996, pp. 25–7; Sukumar Muralidharan, "Early turbulence" and Parvathi Menon, "Parting of ways," *Frontline*, 12 July 1996, pp. 4–8 and 8–12, respectively; and Rasheed Kidwai, *Sonia: a biography* (New Delhi: Penguin, 2011), p. 53.
7. Interview with senior JD politician, New Delhi, 10 February 2000.
8. Pande, "New face," pp. 25–7.
9. Interview with senior JD politician, New Delhi, 10 February 2000.
10. E. Raghavan and James Manor, *Broadening and Deepening Democracy: political innovation in Karnataka* (London: Routledge, 2009), p. 214.
11. Lewis P. Fickett Jr., "The rise and fall of the Janata Dal," *Asian Survey* 33, 12 (December 1993): 1153; Parvathi Menon, "Road to the top," *Frontline*, 14 June 1996, pp. 13–4.
12. *The Hindu*, 4 September 1996.
13. Environmental concerns led to calls for greater scrutiny in September 1994. But the project received environmental clearance on 11 June 1996, one day after Deve Gowda was sworn in as prime minister. Parvathi Menon, "A signal not so green," *Frontline*, 9 August 1996, pp. 52–4.
14. Sandeep Shastri, "Lok Shakti in Karnataka: regional party in bipolar alliance system," *Economic and Political Weekly*, 39, 14/15 (3 April 2004): 1492.
15. Interview with senior JD official, New Delhi, 10 February 2000.
16. Sukumar Muralidharan, "Pulls and pressures," *Frontline*, 26 July 1996, pp. 27–9.
17. *The Hindu*, 29 June 1996.
18. See Vamsicharan Vakulabharanam and Sripad Motiram, "Political economy of agrarian distress in India since the 1990s," in Sanjay Ruparelia, Sanjay Reddy, John Harriss and Stuart Corbridge (eds), *Understanding India's New Political Economy: a great transformation?* (New York and Oxford: Routledge, 2011), pp. 101–26.
19. *The Hindu*, 1 July 1996.
20. *Hindustan Times*, 20 June 1996.
21. Interview, New Delhi, 2 September 1998.
22. Interview, New Delhi, 15 February 2000.
23. Venkitesh Ramakrishnan, "Unequal allies," *Frontline*, 26 July 1996, pp. 33–6.
24. Kanchan Chandra, *Why Ethnic Parties Succeed: patronage and ethnic head counts in India* (Cambridge: Cambridge University Press, 2004), pp. 150–8.
25. Christophe Jaffrelot, "The Bahujan Samaj Party in north India: no longer just a dalit party?" *Comparative Studies of South Asia, Africa and the Middle East*, 18, 1 (Spring 1998): 35–52.
26. *The Hindu*, 26 July 1996.
27. *The Hindu*, 3 July 1996.
28. The BSP administration had appointed Ramesh Chandra, a senior civil servant, to investigate the episode and submit a report. *Hindustan Times*, 26 July 1996.

29. Paul R. Brass, *The Politics of India since Independence*, 2nd edition (Cambridge: Cambridge University Press, 1994), pp. 343–4.
30. Three unresolved issues required further specification: the "entry into force" mechanism of the treaty, guidelines for on-site inspections and definition of what constituted "national technical rules". *The Hindu*, 20 June 1996.
31. *The Hindu*, 20 June 1996.
32. Stephen P. Cohen, *India: emerging power* (New Delhi: Oxford University Press, 2003), p. 173.
33. The most powerful critique remains Praful Bidwai and Achin Vanaik, *South Asia on a Short Fuse: nuclear politics and the future of global disarmament* (New Delhi: Oxford University Press, 2001).
34. Cohen, *India: emerging power*, p. 174.
35. Cohen, *India: emerging power*, p. 175.
36. Interview with senior government official, New Delhi, 7 February 2000.
37. *Times of India*, 11 and 21 May 1996.
38. Interview with senior regional politician, New Delhi, 12 November 1998.
39. George Perkovich, *India's Nuclear Bomb: the impact of global proliferation* (Berkeley: University of California Press, 1999), pp. 374–5.
40. Ministry of External Affairs, Government of India, *Annual Report 1995–96*, pp. 7–8.
41. K. Shankar Bajpai, "India in 1991: new beginnings," *Asian Survey* 32, 2 (February 1992): 215–6.
42. P.S. Suryanarana, "Hope and gamble," *Frontline*, 12 July 1996, p. 48.
43. *The Hindu*, 3 July 1996.
44. *The Hindu*, 19 June 1996.
45. See Jayati Ghosh, "Tightening the wrong belts," *Frontline*, 12 July 1996, pp. 99–100.
46. *The Hindu*, 19 June 1996.
47. See Rob Jenkins, *Democratic Politics and Economic Reform in India* (Cambridge: Cambridge University Press, 1999), pp. 172–207.
48. Muralidharan, "Early turbulence," p. 8.
49. Interview with CPI(M) politburo official, New Delhi, 12 November 1998.
50. Interview with senior JD politician, New Delhi, 1 December 1998.
51. Interview, New Delhi, 23 October 1998.
52. *Hindustan Times*, 3 July 1996.
53. *The Hindu*, 30 July 1996.
54. Interview with CPI(M) Politburo member, New Delhi, 12 November 1998.
55. Interview with senior government official, New Delhi, 16 November 1998.
56. P. Sampathkumar, "Oil pricing and the consumer," *Frontline*, 26 July 1996, pp. 8–10.
57. Interviews with senior regional politician, New Delhi, 12 November 1998, and senior JD politician, New Delhi, 1 December 1998.
58. *Hindustan Times*, 7 July 1996.
59. Interview with senior regional politician, New Delhi, 1 December 1998.
60. Ministry of Finance, Government of India, *1996–97 Union Budget Speech*. The

following analysis draws on articles in the 9 August 1996 issue of *Frontline*: Sukumar Muralidharan, "Continuity with a difference," pp. 4–10; C.P. Chandrasekhar, "Beneath budget appearances," pp. 10–2; and Jayati Ghosh, "Leaving out the essentials," pp. 14–5.

61. The budget increased central assistance for state and UT plans for the provision of basic services by Rs. 2,466 crore. Sukumar Muralidharan, "More continuity than change," *Frontline*, 26 July 1996, pp. 4–7.
62. Lawrence Sáez, *Federalism without a Centre: the impact of political and economic reform on India's federal system* (New Delhi: Sage, 2002), p. 93.
63. Sumantra Bose, *The Challenge in Kashmir: democracy, self-determination and a just peace* (New Delhi: Sage, 1997), p. 59.
64. Jammu and Kashmir National Conference, "Political resolution adopted by the Central Working Committee on 2 November 1994 at Jammu."
65. Bose, *The Challenge in Kashmir*, pp. 152–4.
66. The national average was 58 per cent. See Election Commission of India, New Delhi, *Statistical Report on General Elections, 1996, to the Eleventh Lok Sabha, Volume I*, p. 9.
67. Praveen Swami, "The assembly round," *Frontline*, 12 July 1996, pp. 49–51.
68. The package included construction of a 290 km broad gauge railway line from Udhampur to Baramulla, an alternative national highway and the resumption of work on the Dulhasti power project. *Hindustan Times*, 24 July 1996.
69. The second bundle contained a waiver of interest-bearing loans below Rs. 50,000, and various measures to boost the tourism industry, improve conditions in migrant camps and upgrade public facilities in Jammu city. *The Statesman*, 3 August 1996.
70. See Stuart Corbridge, "Federalism, Hindu nationalism and mythologies of governance in India," in Graham Smith (ed.), *Federalism: the multiethnic challenge* (London: Longman, 1995), p. 119.
71. Interview, New Delhi, 11 February 2000.
72. *The Hindu*, 6 August 1996.
73. Interview, New Delhi, 11 February 2000.
74. Interview, New Delhi, 11 February 2000.
75. Praveen Swami, "For democratic change," *Frontline*, 26 July 1996, pp. 44–5.
76. *Hindustan Times*, 30 July 1996.
77. Praveen Swami, "Playing for power," *Frontline*, 9 August 1996, pp. 39–46.
78. *The Hindu*, 5 August 1996.
79. *The Hindu*, 5 August 1996.
80. *The Hindu*, 26 July 1996.
81. Praveen Swami, "Divided camps," *Frontline*, 20 September 1996, pp. 32–3.
82. See the *Hindustan Times*, 31 July and 12 August 1996.
83. *The Hindu*, 31 July 1996.
84. Interview with senior regional politician, New Delhi, 12 November 1998.
85. Praveen Swami, "The assembly round," *Frontline*, 12 July 1996, pp. 49–51.
86. *The Statesman*, 11 August 1996.
87. *The Hindu*, 10 August 1996.

88. See Parvathi Menon, "In favor of talks," *Frontline*, 9 August 1996, pp. 110–1.
89. Interviews with senior government officials, New Delhi, 24 September 1998 and 8 October 1998.
90. *The Hindu*, 2 August 1996.
91. *The Hindu*, 3 July 1996.
92. *Hindustan Times*, 10 July 1996.
93. *The Hindu*, 4 and 5 August 1996.
94. *The Hindu*, 6 August 1996.
95. Interviews, New Delhi, 8 October 1998 and 4 February 2000.
96. Interview with senior government official, New Delhi, 16 November 1998.
97. Parvathi Menon, "In full flow," *Frontline*, 3 October 1996, pp. 56–8.
98. *The Hindu*, 10 August 1996.
99. *The Hindu*, 9 August 1996.
100. M. Govinda Rao and Nirvikar Singh, *Political Economy of Federalism in India* (New Delhi: Oxford University Press, 2005), pp. 231. The following draws on reporting by T. Lakshmipathi, "Problems of sharing" and Ravi Sharma, "Stand-off continues," *Frontline*, 26 July 1996, pp. 122–4 and 125–7, respectively; R.J. Rajendra Prasad, "The question of height," *Frontline*, 6 September 1996, pp. 23–4.
101. *The Hindu*, 4 August 1996.
102. *The Hindu*, 10 August 1996.
103. Raghavan and Manor, *Broadening and Deepening Democracy*, p. 213.
104. The nine-member team would consist of two experts each from Bihar, Tamil Nadu and West Bengal, and one each from Assam, the Planning Commission and the Central Water Commission. *The Hindu*, 12 August 1996.
105. *The Hindu*, 4 August 1996.
106. V. Venkatesan, "A dam of contention," *Frontline*, 6 September 1996, pp. 21–2.
107. *The Hindu*, 15 August 1996.
108. Sukumar Muralidharan, "Treading softly," *Frontline*, 9 August 1996, pp. 19–20.
109. *Hindustan Times*, 10 and 12 July 1996.
110. *Hindustan Times*, 16 July 1996.
111. Muralidharan, "Treading softly," p. 20.
112. Interview with senior JD politician, New Delhi, 16 February 2000.
113. Interview with senior government official, New Delhi, 16 November 1998.
114. Granville Austin, *Working a Democratic Constitution: a history of the Indian experience* (New Delhi: Oxford University Press, 2000), pp. 582–90.
115. Interview with senior CPI politician, New Delhi, 23 October 1998.
116. *Times of India*, 6 August 1996.
117. See Praveen Swami, "Staying afloat," *Frontline*, 6 September 1996, pp. 18–20.
118. Charu Lata Joshi, "The yes man," *India Today*, October 31, 1996.
119. *Janata Dal Election Manifesto 1996*.
120. Interview, New Delhi, 14 February 2000.

121. Interview, New Delhi, 1 December 1998.
122. Interview, New Delhi, 4 February 2000.
123. See Ramesh Thakur, *The Government and Politics of India* (London: MacMillan Press, 1995), Chapter 8.
124. Interview with senior political journalist, New Delhi, 4 February 2000.
125. Interview with senior government official, New Delhi, 11 February 2000.
126. Interview, New Delhi, 26 October 1998.
127. Interview, New Delhi, 2 September 1998.
128. Interview with senior JD politician, New Delhi, 26 October 1998.
129. Kidwai, *Sonia*, p. 54.
130. Venkitesh Ramakrishnan, "The ascent of Sitaram Kesri," *Frontline*, 18 October 1996, p. 33.
131. Interview with senior JD politician, New Delhi, 16 February 2000.

9. EXERCISING NATIONAL POWER (SEPTEMBER–DECEMBER 1996)

1. Interview, New Delhi, 16 November 1998.
2. Interview, Calcutta, 19 November 1998.
3. See Sukumar Muralidharan, "An eventful budget session," *Frontline*, 4 October 1996, pp. 24–7.
4. *Hindustan Times*, 12 September 1996.
5. Aseema Sinha, "The changing political economy of federalism in India: a historical institutionalist approach," *India Review*, 3, 1 (January 2004): 44; Sudha Mahalingham, "Showcasing India," *Frontline*, 4 October 1996, pp. 100–2.
6. Zoya Hasan, "The 'politics of presence' and legislative reservations for women," in Zoya Hasan, E. Sridharan and R. Sudarshan (eds), *India's Living Constitution: ideas, practices, controversies* (New Delhi: Permanent Black, 2006), p. 415.
7. Hasan, "The 'politics of presence' and legislative reservations for women," p. 415.
8. *Hindustan Times*, 5 September 1996 and 18 September 1996.
9. I.K. Gujral, "Foreign policy objectives of India's United Front Government," Royal Institute of International Affairs, Chatham House, London, 23 September 1996, p. 5.
10. Interview with senior government official, New Delhi, 25 September 1998.
11. Interview with senior government official, New Delhi, 12 October 1998.
12. Interview with senior governmental official, New Delhi, 25 September 2000.
13. Stephen P. Cohen, *India: emerging power* (New Delhi: Oxford University Press, 2003), p. 40. Gujral encouraged this view, hanging a single portrait in his office, of Nehru. See Nalini Kant Jha, "Coalition governments and India's foreign policy," in Mahendra Prasad Singh and Anil Mishra (eds), *Coalition Politics in India: problems and prospects* (New Delhi: Manohar, 2004), p. 306.
14. C. Raja Mohan, *Crossing the Rubicon: the shaping of India's new foreign policy* (New York: Palgrave, 2004), pp. 241–6.
15. See Karuna Mantena, "Another realism: the politics of Gandhian nonviolence," *American Political Science Review*, 106, 2 (May 2012): 466.

16. Interview with senior governmental official, New Delhi, 6 September 1998.
17. See Devesh Kapur, "Public opinion and Indian foreign policy," *India Review*, 8, 3 (July–September 2009): 286–305.
18. See V. Venkatesan, "Breaking with the parivar" and Manini Chatterjee, "Vaghela's challenge," *Frontline*, 20 September 1996, pp. 10–11 and 12–13, respectively.
19. See Ghanshyam Shah, "The BJP's riddle in Gujarat: caste, factionalism and *Hindutva*." In Thomas Blom Hansen and Christophe Jaffrelot (eds), *The BJP and the Compulsions of Politics in India*, 2nd edition (New Delhi: Oxford University Press, 2001), pp. 243–266.
20. *Hindustan Times*, 19, 25 and 30 August 1996.
21. *Hindustan Times*, 10 September 1996.
22. *Hindustan Times*, 19 September 1996.
23. Granville Austin, *Working a Democratic Constitution: a history of the Indian experience* (New Delhi: Oxford University Press, 2000), pp. 611–2.
24. V. Venkatesan, "End of a government," *Frontline*, 18 October 1996, pp. 43–6.
25. Interview, New Delhi, 16 October 1998.
26. Interview, New Delhi, 11 February 2000.
27. Interview, New Delhi, 12 October 1998.
28. Interviews with senior CPI(M) politicians, New Delhi, 14 February 2000 and 8 February 2000; and senior CPI politician, New Delhi, 21 October 1998.
29. Interview, New Delhi, 18 February 2000.
30. Interviews, New Delhi, 16 November 1998 and 22 October 1998.
31. Interview with senior NF official, New Delhi, 16 February 2000.
32. Interview with senior government official, New Delhi, 16 November 1998.
33. Rekha Chowdhury and Nagendra Rao, "Elections 2002: implications for politics of separatism," *Economic and Political Weekly*, 38, 1 (4 January 2003): 16–7.
34. See the reports of Praveen Swami: "Hurriyat's prospects," *Frontline*, 18 October 1996, p. 41, and "Getting down to business," *Frontline*, 15 November 1996, pp. 44–7. For wider analysis, see Sumantra Bose, *The Challenge in Kashmir: democracy, self-determination and a just peace* (New Delhi: Sage, 1997), pp. 155–70.
35. Praveen Swami, "A vote for peace," *Frontline*, 1 November 1996, pp. 17–24.
36. See the *Hindustan Times*, 2 October 1996, 27 October 1996, and 1 November 1996.
37. *Hindustan Times*, 10 October 1996.
38. P.S. Suryanarayana, "The sacking of Benazir," *Frontline*, 29 November 1996, pp. 25–28.
39. The United Front announced the number of constituencies allotted to each party in mid-September: SP (281), JD (41), CPI (11), CPM (11), AIIC(T) (31), Bharatiya Kisan Kamgar Party (BKKP) (41) and BSP (Raj Bahadur) (10). Venkitesh Ramakrishnan and Sukumar Muralidharan, "Alliances and strategies," *Frontline*, 4 October 1996, pp. 40–43.
40. Interview with senior JD politician, New Delhi, 26 October 1998.

41. Interview, New Delhi, 26 October 1998.
42. Ramakrishnan and Muralidharan, "Alliances and strategies," p. 42.
43. The BJP+ won 177 seats (BJP 174, SAP 3); the United Front 134 (SP 110, JD 7, CPI 1, CPM 4, AIIC(T) 4, and BKKP 8) and Independents 14.
44. Venkitesh Ramakrishnan, "U.P. eludes the BJP," *Frontline*, 1 November 1996, pp. 4–10.
45. Venkitesh Ramakrishnan, "Exploring coalition possibilities," *Frontline*, 1 November 1996, pp. 10–15.
46. Francine R. Frankel, *India's Political Economy 1947–2004: the gradual revolution*, 2nd edition (New Delhi: Oxford University Press, 2006), pp. 726–7.
47. Interview, New Delhi, 14 February 2000.
48. Venkitesh Ramakrishnan, "'Shifting alignments," *Frontline*, 15 November 1996, pp. 16–21.
49. *Hindustan Times*, 9 November 1996.
50. Interview, New Delhi, 1 December 1998.
51. Interview, New Delhi, 21 October 1998.
52. Interview, New Delhi, 23 October 1998.
53. Interviews with senior NF official, New Delhi, 16 February 2000, and senior regional politician, New Delhi, 12 November 1998.
54. Interview with senior CPI politician, New Delhi, 21 October 1998.
55. Interview with senior CPI politician, New Delhi, 23 October 1998.
56. Rekha Saxena, *Situating Federalism: mechanisms of intergovernmental relations in Canada and India* (New Delhi: Manohar, 2006), pp. 294–5.
57. Ramakrishnan, "Shifting alignments," p. 18.
58. *Times of India*, 18 October 1996.
59. Interview, New Delhi, 15 February 2000.
60. See Venkitesh Ramakrishnan, "The judicial verdict," *Frontline*, 10 January 1997, pp. 17–9.
61. See Venkitesh Ramakrishnan, "A significant turn," *Frontline*, 13 December 1996, pp. 22–4, and "Stalemate continues," *Frontline*, 27 December 1996, pp. 28–30.
62. *The Hindu*, 28 October 1996; Tapas Ray, "Building bridges," *Frontline*, 29 November 1996, p. 49.
63. Tapas Ray, "Dilemmas over the Act," *Frontline*, 13 December 1996, pp. 50–52; Subir Bhaumik, "North-east India: the evolution of a post-colonial region," in Partha Chatterjee (ed.), *Wages of Freedom: fifty years of the Indian nation-state* (New Delhi: Oxford University Press, 1998), p. 327.
64. Interview, New Delhi, 7 February 2000.
65. Interview, New Delhi, 11 February 2000.
66. Interview with senior government official, New Delhi, 11 February 2000.
67. *The Hindu*, 30 October 1996.
68. Sanjib Baruah, *India against Itself: Assam and the politics of nationality* (Philadelphia: University of Pennsylvania Press, 1999), p. 194.
69. Ray, "Dilemmas over the Act," pp. 50–2.
70. Baruah, *India against Itself*, pp. 153–69.

71. Reportedly, government officers in six districts withdrew Rs. 253 crore for animal food and fodder between 1993–94 and 1995–96 as opposed to the Rs. 10.5 crore budgeted by the state treasury. *Hindustan Times*, 5 October 1996.
72. Praveen Swami, "Battered image," *Frontline*, 1 November 1996, pp. 26–30.
73. Interview with senior government official, New Delhi, 21 February 2000.
74. *Hindustan Times*, 8 October 1996.
75. *Hindustan Times*, 25 October 1996.
76. *Hindustan Times*, 6 November 1996.
77. Swami, "Battered image," p. 28.
78. Interview, New Delhi, 10 February 2000.
79. *The Statesman*, 14 November 1996.
80. Kesri removed three AICC general secretaries close to Rao: B.P. Maurya, Devendra Dwivedi and Janardhana Poojary. He also appointed Tariq Anwar, chairperson of the minority cell, as his political secretary, Ahmed Patel as treasurer and Ghulum Nabi Azad as general secretary. *Hindustan Times*, 24 and 26 October 1996.
81. *Times of India*, 2 November 1996.
82. *Hindustan Times*, 6 November 1996.
83. Venkitesh Ramakrishnan, "Little headway," *Frontline*, 29 November 1996, pp. 37–9.
84. *Hindustan Times*, 12 November 1996.
85. Zafar Agha and Javed Ansari, "Assertive moves," *India Today*, 30 November 1996, pp. 33–4.
86. Venkitesh Ramakrishnan, "Congress-U.F. relations," *Frontline*, 13 December 1996, p. 31.
87. *Hindustan Times*, 18 November 1996.
88. *Hindustan Times*, 12 December 1996.
89. They also signed three other agreements regarding crime prevention and drug trafficking, Indian consular services in Hong Kong and the status of vessels at their respective ports. *The Hindu*, 30 November 1996.
90. John Cherian, "Strengthening relations," *Frontline*, 27 December 1996, pp. 38–42.
91. *The Hindu*, 2 December 1996.
92. See Rajen Harshe, "South Asian regional co-operation," *Economic and Political Weekly*, 34, 19 (8 May 1999): 1105; Padmaja Murthy, "The Gujral doctrine and beyond," Mimeo, Institute for Defence Studies and Analyses, p. 6: http://www.idsa-india.org/an-jul9–8.html.
93. See John Cherian, "A historic accord," *Frontline*, 10 January 1997, pp. 47–9; and "Great expectations," *Frontline*, 13 December 1996, p. 52.
94. Interview, New Delhi, 26 October 1998.
95. Interview, New Delhi, 10 September 1998.
96. Interview, New Delhi, 15 February 2000.
97. Interview, New Delhi, 15 February 2000.
98. John Cherian, "Mending fences," *Frontline*, 20 September 1996, p. 54.
99. Interview, New Delhi, 10 September 1998.

100. Interview with senior government official, New Delhi, 16 November 1998.
101. Interview with senior government official, New Delhi, 12 October 1998.
102. See Haroon Habib, "Mission to Dhaka," *Frontline*, 27 December 1996, pp. 47–9; Punam Pandey, "Revisiting the politics of the Ganga water dispute between India and Bangladesh," *India Quarterly*, 68, 3 (2012): 267–81.
103. In later years, these provisions contributed to the Ganga flooding Bihar and forcing the Farraka power plant to close. *The Hindu*, September 13, 2011.
104. Interview with senior government official, New Delhi, 9 September 1998.
105. Cherian, "A historic Accord," p. 48.
106. Interview, New Delhi, 26 October 1998.
107. Interview with senior regional politician, New Delhi, 12 November 1998.
108. I.K. Gujral, "Aspects of India's foreign policy," Bandaranaike Center for International Studies, Colombo, 20 January 1997; Harshe, "South Asian regional co-operation," p. 1105.
109. Dasu Kesava Rao, "Some differences," *Frontline*, 27 December 1996, p. 30.
110. *India Today*, 31 January 1997.
111. *Hindustan Times*, 12 December 1996.
112. Sukumar Muralidharan, "Sidestepping core issues," *Frontline*, 27 December 1996, pp. 23–5.
113. *The Hindu*, 30 October 1996.
114. See Praful Bidwai, "An indifferent record," *Frontline*, 10 January 1997, pp. 101–3.
115. See Sukumar Muralidharan, "Points of divergence," *Frontline*, 29 November 1996, p. 32, and "Sidestepping core issues," *Frontline*, 27 December 1996, pp. 23–5.
116. Venkitesh Ramakrishnan, "Emerging stronger," *Frontline*, 27 December 1996, pp. 26–8.
117. Venkitesh Ramakrishnan, "Exit in disgrace," *Frontline*, 10 January 1997, pp. 4–11.

10. REFORM AMID CRISIS (JANUARY–APRIL 1997)

1. See Sukumar Muralidharan, "A bill in vain," *Frontline*, 10 January 1997.
2. *Hindustan Times*, 6 January 1997.
3. *Hindustan Times*, 6 January 1997.
4. *Hindustan Times*, 17 January 1997.
5. Venkitesh Ramakrishnan, "Kesri's concerns," *Frontline*, 21 February 1997, pp. 25–7.
6. Venkitesh Ramakrishnan, "Conflict in the Congress," *Frontline*, 13 December 1996, pp. 36–41.
7. *Hindustan Times*, 20 January 1997.
8. *The Hindu*, 20 January 1997.
9. Avirook Sen and Sayantan Chakravarty, "Political ammunition," *India Today*, 15 February 1997, pp. 26–7.
10. Prabu Chawla, "Quattrocchi connection," *India Today*, 28 February 1997, pp. 20–3.

11. *Hindustan Times*, 4 February 1997.
12. See Praveen Swami, "Emphatic victory," *Frontline*, 7 March 1997, pp. 29–34.
13. Venkitesh Ramakrishnan, "Rhetoric and reality," *Frontline*, 21 March 1997, pp. 114–6.
14. K.K. Kailash, "Middle game in coalition politics," *Economic and Political Weekly*, 42, 4 (27 January 2007): 310.
15. *Hindustan Times*, 17 February 1997.
16. Sukumar Muralidharan, "Hard choices to make," *Frontline*, 7 March 1997, pp. 97–100.
17. Editorial, "All for foreign capital," *Economic and Political Weekly* (11 January 1997): 3–6.
18. Francine Frankel and Sumantra Sen, *Andhra Pradesh's Long March towards 2020: electoral detours in a developmentalist state* (Philadelphia: Center for the Advanced Study of India, University of Pennsylvania, 2005), pp. 6, 12.
19. V. Venkatesan, "Concerns beyond the budget," *Frontline*, 7 March 1997, pp. 24–6.
20. Sukumar Muralidharan, "Economic questions" and Sudha Mahalingam, "NDC and growth," *Frontline*, 7 February 1997, pp. 22 and 107–9, respectively; C.P. Chandrasekhar, "On the beaten track," *Frontline*, 7 March 1997, pp. 107–9.
21. M.J. Akbar, "Interview with Jyoti Basu," *The Asian Age*, 2 January 1997.
22. "A settled matter: The CPI(M)'s position on participation in the Deve Gowda government," in Namboodiripad, *The Frontline Years: selected articles* (New Delhi: LeftWord Books, 2010), pp. 203–8.
23. Editorial, "Down the slippery slope," *Economic and Political Weekly* (1 February 1997): 183–186.
24. *Times of India*, 25 February 1997.
25. Interview with senior government official, New Delhi, 30 January 2000.
26. See Madhura Swaminathan, *Weakening Welfare: the public distribution of food in India* (New Delhi: LeftWord, 2000), pp. 94–100.
27. Ministry of Finance, Government of India, *Economic Survey 1997–98*.
28. The Laffer curve, named after the American economist Arthur Laffer, presumes an optimal value for tax rates. Increasing the rate above this point discourages productive activity, leading to lower tax revenues, whereas lower taxes might generate higher revenues through lower prices and higher output.
29. Unless otherwise noted, the preceding and following details are drawn from the *1997–1998 Union Budget*: http://indiabudget.nic.in/ub1997–98/welcome.html.
30. See Table 9.1 in M. Govinda Rao and Nirvikar Singh, *Political Economy of Federalism in India* (New Delhi: Oxford University Press, 2005), p. 192. The NDA passed the eightieth constitutional amendment in 2000, accepting the Tenth Finance Commission's recommendation, but reduced the states' allocation by specifying their net value as opposed to gross. See Katherine Adeney, "Hindu nationalists and federal structures in an era of regionalism," in Katharine Adeney and Lawrence Sáez (eds), *Coalition Politics and Hindu nationalism* (London: Routledge, 2005), p. 105.

31. *The Economic Times*, 1 March 1997.
32. See Ashutosh Varshney, "Mass politics or elite politics? India's economic reforms in comparative perspective," in Jeffrey Sachs, Ashutosh Varshney and Nirupam Bajpai (eds), *India in the Era of Economic Reforms* (New Delhi: Oxford University Press, 1999), pp. 222–61.
33. Interview with senior government official, New Delhi, 9 November 1998.
34. Interview, New Delhi, 9 November 1998.
35. See James Manor, "The Congress party and the great transformation," in Sanjay Ruparelia, Sanjay Reddy, John Harriss and Stuart Corbridge (eds), *Understanding India's New Political Economy: a great transformation?* (London: Routledge, 2011), p. 207.
36. Interview, New Delhi, 12 November 1998.
37. Interview, New Delhi, 7 February 2000.
38. Interview with senior JD politician, New Delhi, 3 January 2000.
39. Interview with senior JD politician, New Delhi, 3 January 2000.
40. *The Hindu*, 4 March 1997.
41. *The Hindu*, 4 March 1997.
42. Interview with CPI(M) politburo member, New Delhi, 14 February 2000.
43. Namboodiripad, *The Frontline Years*, pp. 216–20.
44. *The Hindu*, 4 March 1997.
45. Interview, New Delhi, 12 October 1998.
46. *The Hindu*, 2 March 1997.
47. See Sukumar Muralidharan, "Substance and artifice" and C.P. Chandrasekhar, "Hope as strategy," *Frontline*, 21 March 1997, pp. 4–10 and 10–14 respectively; Jayati Ghosh, "An analysis of the budget 1997–98: largesse for the rich, phrases for the poor," http://www.ieo.org/97anal.html; and Kamal Mitra Chenoy, "The budget 1997–98: an analysis," http://www.ieo.org/97anal2.html.
48. Chenoy, "The budget 1997–98."
49. *The Hindu*, 6 March 1997.
50. Interview with senior government official, New Delhi, 7 October 1998.
51. Interview with senior government official, New Delhi, 30 January 2000.
52. Interview with senior regional politician, New Delhi, 1 December 1998.
53. *Hindustan Times*, 12 March 1997.
54. *The Hindu*, 18 March 1997.
55. *Hindustan Times*, 16 March 1997.
56. *Times of India*, 25 February 1997.
57. *The Hindu*, 6 March 1997.
58. *The Hindu*, 5 March 1997.
59. Reportedly, Bhandari refused to permit the CBI to chargesheet SP advisors involved in an ayurveda scam. Dilip Aswathi and Subhash Mishra, "Governor reined," *India Today*, 31 March 1997, pp. 24–28.
60. *The Hindu*, 7 March 1997.
61. *The Hindu*, 11 March 1997.
62. *The Hindu*, 14 March 1997.

63. *The Hindu*, 20 March 1997.
64. K.K. Kailash, "Federal calculations in state level coalition governments," *India Review*, 10, 3 (July–September 2011): 264.
65. See Sudha Pai, "The state, social justice, and the Dalit movement: the BSP in Uttar Pradesh," in Niraja Gopal Jayal and Sudha Pai (eds), *Democratic Governance in India: challenges of poverty, development and identity* (New Delhi: Sage, 2001), pp. 212–16.
66. *The Hindu*, 28 February 1997.
67. *The Hindu*, 3 March 1997.
68. *The Hindu*, 21 March 1997.
69. *The Hindu*, 22 March 1997.
70. Interview, New Delhi, 15 February 2000.
71. *The Hindu*, 30 March 1997.
72. Sumit Mitra, "Gowda's gaffes ... and Kesri's curse," *India Today*, 30 April 1997, pp. 14–21.
73. *The Hindu*, 31 March 1997.
74. Transcript of interview with Shri Jyoti Basu, Acc. No. 781, Oral History Division, Nehru Memorial Museum and Library, New Delhi, pp. 156–7.
75. Interview with senior CPI politician, New Delhi, 23 October 1998.
76. Interview, New Delhi, 8 February 2000.
77. Interview with senior government official, New Delhi, 2 September 1998.
78. Interview with senior regional politician, New Delhi, 1 December 1998.
79. Interview with senior regional politician, New Delhi, 12 November 1998.
80. Interview with senior CPI(M) politician, Calcutta, 19 November 1998.
81. Interview with senior JD politician, New Delhi, 9 November 1998.
82. *The Hindu*, 31 March 1997.
83. Interview with senior political journalist, 4 February 2000, New Delhi.
84. Kailash, "Middle game in coalition politics," p. 311.
85. *The Hindu*, 31 March 1997.
86. Interviews with senior political journalist, New Delhi, 7 February 2000; senior JD politician, New Delhi, 1 December 1998; and senior political journalist, New Delhi, 4 February 2000.
87. Interview with senior political journalist, New Delhi, 4 February 2000.
88. Interview with senior Congress(I) official, New Delhi, 30 January 2000.
89. Interview with senior JD official, New Delhi, 10 February 2000.
90. Csaba Nikolenyi, *Minority Governments in India: the puzzle of elusive majorities* (London: Routledge, 2010), pp. 109–10.
91. Mitra, "Gowda's gaffes ... and Kesri's curse," pp. 14–21.
92. Quoted in "Voices," *India Today*, 30 April 1997.
93. Sumit Mitra and Javed M. Ansari, "Challenges of consensus," *India Today*, 15 May 1997, pp. 12–9.
94. N.K. Singh, "Once bitten twice shy," *India Today*, 30 April 1997, pp. 22–3.
95. Interview with senior CPI politician, New Delhi, 21 October 1998.
96. *The Hindu*, 31 March 1997.
97. Interview with senior JD politician, New Delhi, 26 October 1998.

98. Interviews with senior JD politician, New Delhi, 6 November 1998, and senior regional politician, New Delhi, 12 November 1998.
99. Interviews with senior government officials, New Delhi, 22 October 1998, 9 November 1998 and 11 February 2000.
100. Interview with senior JD politician, New Delhi, 26 October 1998.
101. Interview, New Delhi, 9 November 1998.
102. Interview with senior JD politician, New Delhi, 26 October 1998.
103. Interview with senior CPI(M) official, New Delhi, 8 February 2000.
104. Interviews with senior JD politicians, New Delhi, 13 October 1998 and 26 October 1998.
105. Interview with senior JD politician, New Delhi, 26 October 1998.

11. THE DECLINE OF THE UNITED FRONT (MAY 1997–MARCH 1998)

1. Interview with senior JD politician, New Delhi, 9 November 1998.
2. Interview with senior regional politician, New Delhi, 18 February 2000.
3. *The Statesman*, 9 May 1997.
4. *Times of India*, 22 May 1997.
5. Prabu Chawla and Sumit Mitra, "King Kesri," *India Today*, 9 June 1997, pp. 18–23.
6. *The Hindu*, 21 May 1997.
7. *The Hindu*, 29 May 1997.
8. Interview, New Delhi, 9 November 1998.
9. Interview, New Delhi, 26 October 1998.
10. Interviews with senior government official, New Delhi, 16 November 1998; and senior JD politician, New Delhi, 26 October 1998.
11. Swapan Dasgupta, "Betraying the doctrine," *India Today*, 31 May 1997, p. 23.
12. *Hindustan Times*, 28 April 1997.
13. *Times of India*, 29 April 1997.
14. *Hindustan Times*, 5 May 1997.
15. Pranab Bardhan, *The Political Economy of Development in India, expanded edition* (New Delhi: Oxford University Press, 2012), p. 133.
16. See Sudipta Kaviraj, "Democracy and social inequality," in Francine R. Frankel, Zoya Hasan, Rajeev Bhargava and Balveer Arora (eds), *Transforming India: social and political dynamics of democracy* (New Delhi: Oxford University Press, 2000), pp. 112–13; Guillermo O'Donnell, "Delegative democracy," in Larry Diamond and Marc F. Plattner (eds), *The Global Resurgence of Democracy*, 2nd edition (Baltimore: Johns Hopkins University Press, 1996), pp. 94–111.
17. *Times of India*, 6 May 1997.
18. *Hindustan Times*, 2 June 1997; *Times of India*, 17 June 1997.
19. *Hindustan Times*, 15 June 1997.
20. *Hindustan Times*, 8 June 1997.
21. *The Statesman*, 10 June 1997.

22. *The Hindu*, 29 June 1997.
23. Sukumar Muralidharan, "Marching orders," *Frontline*, 25 July 1997, p. 22.
24. Interview with senior government official, New Delhi, 21 February 2000.
25. *The Statesman*, 3 July 1997.
26. R.K. Raghavan, "The Indian police: expectations of a democratic polity," in Francine R. Frankel, Zoya Hasan, Rajeev Bhargava and Balveer Arora (eds), *Transforming India: social and political dynamics of democracy* (New Delhi: Oxford University Press, 2000), p. 302.
27. *Hindustan Times*, 6 July 1997.
28. *Hindustan Times*, 9 July 1997.
29. *Hindustan Times*, 10 July 1997.
30. *Hindustan Times*, 14 July 1997.
31. *Hindustan Times*, 18 July 1997.
32. *Hindustan Times*, 29 July 1997.
33. *The Hindu*, 21 May 1997.
34. *Hindustan Times*, 19 May 1997.
35. Francine R. Frankel, *India's Political Economy 1947–2004: the gradual revolution*, 2nd edition (New Delhi: Oxford University Press, 2006), p. 718.
36. *Times of India*, 13 June 1997.
37. *Hindustan Times*, 13 May 1997 and 4 June 1997.
38. N. Ram, "Dealing with Pakistan," *Frontline*, 17 October 1997, pp. 12–5.
39. *The Hindu*, 14 May 1997.
40. Subrata K. Mitra, "War and peace in South Asia: a revisionist view of India-Pakistan relations," in Subrata K. Mitra (ed.), *Politics of Modern South Asia, Volume V* (London: Routledge, 2009), pp. 171–2.
41. Reportedly, demands to restore RAW's covert operations wing grew louder following the Mumbai bomb blasts in November 2008, conducted by the Lashkar-e-Taiba with support from Pakistan's ISI. See Saikat Datta, "Shackled by a doctrine," *Outlook*, 16 March 2009. I thank Vipin Narang for pointing this out to me.
42. See M. Joshi, "Doctrine of grand gestures," *India Today*, 23 June 1997, p. 68.
43. Sangeeta Thapliyal, "India and Nepal treaty of 1950: the continuing discourse," *India Quarterly*, 68, 2 (2012), pp. 128–129.
44. Interviews, New Delhi, 9 and 10 September 1998.
45. Interview, New Delhi, 15 February 2000.
46. *The Hindu*, 24 June 1997.
47. *The Statesman*, 16 July 1997.
48. *Times of India*, 26 June 1997.
49. Interview with senior government official, New Delhi, 15 February 2000.
50. Interview, New Delhi, 15 February 2000.
51. *Times of India*, 24 September 1997.
52. See George Perkovich, *India's Nuclear Bomb: the impact of global proliferation* (Berkeley: University of California Press, 1999), p. 399.
53. The prime minister assigned several portfolios at the level of Minister of State: Jayanti Natarajan to Civil Aviation and Parliamentary Affairs; Kamala Sinha

and S.I. Shervani to External Affairs; Renuka Chowdhury to Health and Family Welfare; Ratnamala Savanoor to Planning and Programme Implementation. Gujral also shifted cabinet members of various rank to different portfolios: Y.K. Alagh from Planning to Power; J. Mishra from Water Resources to Petroleum and Natural Gas; M. Arunachalam from Labor to Chemicals and Fertilizers; S. Maharaj from Railways to Finance; S.R. Ola from Chemicals and Fertilizers to Water Resources (with independent charge); and S. Venugopalachari from Power to Agriculture. *The Hindu*, 10 June 1997.

54. Shefali Rekhi, "Time to deliver," India Today, June 9, 1997, pp. 56–8.
55. Shefali Rekhi, "Cheap laundering," *India Today*, 23 June 1997, p. 66.
56. *Times of India*, 27 June 1997.
57. *The Hindu*, 5 and 8 July 1997.
58. Interview, New Delhi, 26 October 1998.
59. Interview, New Delhi, 10 February 2000.
60. Interview, New Delhi, 23 October 1998.
61. Interview with CPI(M) politburo member, New Delhi, 14 February 2000.
62. Interview with senior CPI politician, New Delhi, 21 October 1998.
63. Interview with senior regional politician, New Delhi, 12 November 1998.
64. *Hindustan Times*, 11 July 1997.
65. *Times of India*, 7 August 1997.
66. Sukumar Muralidharan and T.K. Rajalakshmi, "Uneasy progress," *Frontline*, 5 September 1997, pp. 24–5.
67. Interview, New Delhi, 2 December 1998.
68. *Hindustan Times*, 26 August 1997.
69. Shefali Rekhi, "Slow combustion," *India Today*, 15 September 1997, p. 54.
70. Interview with senior government official, New Delhi, 7 February 2000.
71. Interview with senior JD politician, New Delhi, 3 January 2000.
72. *Hindustan Times*, 3 September 1997.
73. However, the committee recommended the elimination of vacant posts and overtime pay scales, which primarily affected lower-tier government workers. N.K. Singh, "More pay for less," *India Today*, 16 June 1997, pp. 28–9; K.P. Joseph, "Piped music and telephone attendants: report of the Fifth Pay Commission," *Economic and Political Weekly*, 32, 12 (22 March 1997): 563–5.
74. Interview, 26 September 1998, New Delhi.
75. Paul R. Brass, "Political parties and the radical left in South Asia," in Paul R. Brass and Marcus F. Franda (eds), *Radical Politics in South Asia* (Cambridge: MIT Press, 1973), p. 59.
76. Editorial, "Succumbing to blackmail," *India Today*, 22 September 1997, p. 6.
77. Interviews, New Delhi, 7 September and 26 October 1998.
78. Interview, New Delhi, 30 January 2000.
79. Interview with senior government official, New Delhi, 26 October 1998.
80. Interview with senior trade union official, New Delhi, 26 September 1998.
81. Interview with senior government official, New Delhi, 2 December 1998.
82. Interview with senior government official, New Delhi, 2 December 1998.

83. Interview with senior government official, New Delhi, 7 September 1998.
84. Interview with senior government official, New Delhi, 10 August 1998.
85. Prem Shankar Jha, "Coalition politics and economic decision-making," in Mahendra Prasad Singh and Anil Mishra (eds), *Coalition Politics in India: problems and prospects* (New Delhi: Manohar, 2004), p. 291.
86. Editorial, "Wrong medicine," *Economic and Political Weekly*, 32, 41 (11 October 1997): 2567; N.J. Kurian, "State government finances: a survey of recent trends," *Economic and Political Weekly*, 34, 19 (8 May 1999): 1115–25.
87. *Hindustan Times*, 18 June 1997.
88. Interview with senior government official, New Delhi, 11 February 2000.
89. Interview with senior government official, New Delhi, 11 February 2000.
90. *The Hindu*, 4 July 1997.
91. Interview with senior government official, New Delhi, 11 February 2000.
92. *The Hindu*, 18 July 1997. The final ISC meeting occurred on November 28, the day the Congress(I) withdrew its support to the Government.
93. *Hindustan Times*, 15 September 1997.
94. Sudha Pai, *Dalit Assertion and the Unfinished Democratic Revolution: the Bahujan Samaj Party in Uttar Pradesh* (New Delhi: Sage Publications, 2002), p. 17.
95. *Hindustan Times*, 19 October 1997.
96. *Hindustan Times*, 21 October 1997.
97. *Times of India*, 22 October 1997.
98. Farzand Ahmed, "Maya costs a lot," *India Today*, 22 September 1997, p. 16; Christophe Jaffrelot, *India's Silent Revolution: the rise of the low castes in north Indian politics* (London: Hurst/New Delhi: Permanent Black, 2003), p. 419.
99. Interview with senior government official, New Delhi, 26 October 1998.
100. Interview with senior JD politician, New Delhi, 9 November 1998.
101. Interview with senior government official, New Delhi, 2 February 2000.
102. *The Hindu*, 18 July 1997.
103. Parliament added section 13 of the Constitution (Forty-Second Amendment) Act, 1976. S.P. Sathe, *Judicial Activism in India: transgressing borders and enforcing limits*, 2nd edition, with a foreword by Upendra Baxi (New Delhi: Oxford University Press, 2004), p. 5.
104. *Times of India*, 23 October 1997.
105. *Hindustan Times*, 23 October 1997.
106. *Hindustan Times*, 30 October 1997.
107. *Times of India*, 17 and 18 October 1997.
108. Pravin Swami, "Public service broadcasting?" *Frontline*, November 15, 1997.
109. Nalini Kant Jha, "Coalition governments and India's foreign policy," in Mahendra Prasad Singh and Anil Mishra (eds), *Coalition Politics in India: problems and prospects* (New Delhi: Manohar, 2004), p. 310.
110. *Hindustan Times*, 9 November 1997.
111. *The Indian Express*, 29 November 2012. The subsequent BJP-led government

established the Cauvery River Authority in 1998 to ensure implementation of the interim order, while the Cauvery Water Dispute Tribunal finalized its award in 2007. However, the former did not meet between 2002 and 2012, and disputes keep flaring in lean seasons. See M.S. Menon, "Drawing lines in water," *The Indian Express*, 26 September 2012.

112. The following paragraph relies on Praveen Swami and Venkitesh Ramakrishnan, "The politics of blackmail," *Frontline*, 12 December 1997, pp. 4–12.
113. *Hindustan Times*, 10 November 1997.
114. Balveer Arora, "Negotiating differences," in Francine R. Frankel, Zoya Hasan, Rajeev Bhargava and Balveer Arora (eds), *Transforming India: social and political dynamics of democracy* (New Delhi: Oxford University Press, 2000), p. 187.
115. Zoya Hasan, "Introduction," in Zoya Hasan (ed.), *Parties and Party Politics in India* (New Delhi: Oxford University Press, 2004), p. 30.
116. Rasheed Kidwai, *Sonia: a biography* (New Delhi: Penguin, 2011), pp. 90–1.
117. *Hindustan Times*, 20 November 1997.
118. Interview with senior government official, New Delhi, 16 November 1998.
119. Interview with senior Congress(I) official, New Delhi, 30 January 2000.
120. *Hindustan Times*, 21 November 1997.
121. Sukumar Muralidharan and Venkitesh Ramakrishnan, "Out of the gridlock," *Frontline*, 26 December 1998, pp. 4–16.
122. *The Statesman*, 24 November 1997.
123. *Hindustan Times*, 25 November 1997.
124. *Hindustan Times*, 27 November 1997.
125. Interview, New Delhi, 24 October 1998.
126. Interview with Chandrababu Naidu, "We will get a majority on our own," *Frontline*, 26 December 1997, p. 9.
127. Editorial, "Dead men walking," *Economic and Political Weekly*, 32, 49 (6 December 1997): 3104.
128. See Sukumar Muralidharan, "Against the odds," *Frontline*, 9 January 1998, pp. 19–21; the articles of Praveen Swami, "For a coherent agenda," *Frontline*, 23 January 1998, pp. 17–9, and "Ideologies at odds," *Frontline*, 6 February 1998, pp. 20–4; and the articles of Sukumar Muralidharan, "Constitutional frailty," *Frontline*, 20 February 1998, pp. 21–3, and "Promises and policies," *Frontline*, 6 March 1998, pp. 27–9. On electoral alliances, see the general reporting in "The phase of alignments," *Frontline*, 6 February 1998, pp. 30–48; and "The line-up," *Frontline*, 20 February 1998, pp. 29–48.
129. The following summarizes Jayanta Sengupta, "Liberalization and the politics of Oriya identity," in Niraja Gopal Jayal and Sudha Pai (eds), *Democratic Governance in India: challenges of poverty, development and identity* (New Delhi: Sage, 2001), pp. 183–6; and Sunila Kale and Nimah Mazaheri, "Natural resources, development strategies and identity politics along India's mineral belt: Bihar and Odisha in the 1990s," (Mimeo, October 2012), pp. 26–36.

130. Bidyut Chakrabarty, *Forging Power: coalition politics in India* (New Delhi: Oxford University Press, 2006), p. 144, fn. 34.
131. Hasan, "Introduction," p. 28.
132. See the reporting of Venkitesh Ramakrishnan, "A party in disarray," *Frontline*, 26 December 1997, pp. 20–1; "Spreading turmoil," *Frontline*, 9 January 1998, pp. 21–2; "Counting on Sonia," *Frontline*, 23 January 1998, pp. 20–1; and "The Sonia effect," *Frontline*, 6 February 1998, pp. 4–10. On electoral alliances, see the general reporting in "The phase of alignments" and "The line-up".
133. See Christophe Jaffrelot, *The Hindu Nationalist Movement and Indian Politics: 1925 to the 1990s* (New Delhi: Penguin Books, 1999), p. 548.
134. See V. Venkatesan, "A nation-wide hunt for allies," *Frontline*, 9 January 1998, pp. 23–4; and Sukumar Muralidharan, "BJP and friends," *Frontline*, 23 January 1998, pp. 4–11. On electoral alliances, see the general reporting in "The phase of alignments," *Frontline*, 6 February 1998, pp. 30–48; and "The line-up," *Frontline*, 20 February 1998, pp. 29–48.
135. Narendra Subramanian, *Ethnicity and Populist Mobilization: political parties, citizens and democracy in south India* (New Delhi: Oxford University Press, 1999), pp. 307–9.
136. The following information is from Election Commission of India, New Delhi, *Statistical Report on General Elections, 1998, to the Twelfth Lok Sabha, Volume I* (hereafter *Election Commission Report 1998*), pp. 39 and 57–79. For general analysis of the election, see Atul Kohli, "Enduring another election," *Journal of Democracy*, 9, 3 (July 1998): 7–21.
137. The following analysis of the three electoral fronts draws on reporting from "Messages from the states," *Frontline*, 3 April 1998.
138. See *Rashtriya Janata Dal Election Manifesto—1998*; Sanjay Kumar, "New phase in backward caste politics in Bihar," in Ghanshyam Shah (ed.), *Caste and Democratic Politics in India* (London: Anthem Press, 2004), pp. 244–5.
139. Sanjib Baruah, *India against Itself: Assam and the politics of nationality* (Philadelphia: University of Pennsylvania Press, 1999), p. 159.
140. See the following articles by Sukumar Muralidharan: "Post-poll scenarios," *Frontline*, 20 March 1998, pp. 23–26; and "Retreat of a front," *Frontline*, 3 April 1998, pp. 21–24.
141. Alistair McMillan, 'The BJP coalition: partisanship and power-sharing in government', in Katharine Adeney and Lawrence Sáez (eds), *Coalition Politics and Hindu nationalism* (London: Routledge, 2005), pp. 14–5.
142. Sukumar Muralidharan, "BJP's turn" and V. Venkatesan, "Pulls and pressures," *Frontline*, 17 April 1998, pp. 114–8 and 118–20, respectively.
143. McMillan, 'The BJP coalition," p. 16.
144. S. Nagesh Kumar, "TDP changes tack," *Frontline*, 17 April 1998, p. 120.
145. Rob Jenkins, "India's electoral result: an unholy alliance between nationalism and regionalism," (Mimeo: 1998), p. 4.
146. Jenkins, "India's electoral result," pp. 5–6.
147. Interview with senior regional leader, New Delhi, 18 February 2000.

12. THE DISSOLUTION OF THE THIRD FORCE (1998–2012)

1. Apart from other works cited, this chapter draws on my two previous essays: "Rethinking institutional theories of political moderations: the case of Hindu nationalism in India, 1996–2004," *Comparative Politics*, 38, 3 (April 2006): 317–37; and "Expanding Indian democracy: the paradoxes of the third force," in Sanjay Ruparelia, Sanjay Reddy, John Harriss and Stuart Corbridge (eds), *Understanding India's New Political Economy: a great transformation?* (London: Routledge, 2011), pp. 186–203.
2. Nalini Kant Jha, "Coalition governments and India's foreign policy," in Mahendra Prasad Singh and Anil Mishra (eds), *Coalition Politics in India: problems and prospects* (New Delhi: Manohar, 2004), p. 312.
3. *Deccan Herald*, 4 May 1998.
4. See Sukumar Muralidharan, "BJP and friends," *Frontline*, 23 January 1998; and Thomas Blom Hansen and Christophe Jaffrelot, "Introduction: the rise to power of the BJP," in Thomas Blom Hansen and Christophe Jaffrelot (eds), *The BJP and the Compulsions of Politics in India*, 2nd edition (New Delhi: Oxford University Press, 2001), pp. 1–22.
5. V. Ramakrishnan, "All for survival," *Frontline*, 24 April 1998.
6. Christophe Jaffrelot, *The Hindu Nationalist Movement and Indian Politics: 1925 to the 1990s* (New Delhi: Penguin Books, 1999), p. 548.
7. Adam Przeworksi, *Capitalism and Social Democracy* (Cambridge: Cambridge University Press, 1985), p. 112.
8. T.V. Paul, *Power versus Prudence: why nations forgo nuclear weapons* (Montreal and Kingston: McGill-Queens University Press, 2000), pp. 129–30.
9. See C. Raja Mohan, *Crossing the Rubicon: the shaping of India's new foreign policy* (New York: Palgrave MacMillan, 2004), pp. 1–29.
10. Communist Party of India (Marxist), *Election Manifesto for the 1999 Lok Sabha Elections*, p. 4.
11. The South Asian anti-nuclear movement was the major exception. See Praful Bidwai and Achin Vanaik, *South Asia on a Short Fuse: nuclear politics and the future of global disarmament* (New Delhi: Oxford University Press, 2001).
12. Padmaja Murthy, "The Gujral doctrine and beyond," Mimeo, Institute for Defence Studies and Analyses, p. 9: http://www.idsa-india.org/an-jul9–8.html.
13. Mohan, *Crossing the Rubicon*, p. 245.
14. K.K. Kailash, "Middle game in coalition politics," *Economic and Political Weekly*, 42, 4 (27 January 2007): 311 and 313.
15. Sukumar Muralidharan and S.K. Pande, "Taking Hindutva to school," *Frontline*, 20 November 1998.
16. M. Venkatesan, "Communal outrages in M.P.," *Frontline*, 23 October 1998.
17. Philip Oldenburg, *The Thirteenth General Election of India's Lok Sabha* (New York: The Asia Society, 1999), p. 1.
18. Venkitesh Ramakrishnan, "The numbers game," *Frontline*, 24 April 1999.
19. Zoya Hasan, "Introduction," in Zoya Hasan (ed.), *Parties and Party Politics in India* (New Delhi: Oxford University Press, 2004), pp. 29–30.

20. Sukumar Muralidharan, "The end of an ordeal," *Frontline*, 24 April 1999.
21. See Arvind Das, "The future postponed," *Economic and Political Weekly*, 34, 20 (15 May 1999): 1167.
22. "Questionable shift," *Economic and Political Weekly*, 34, 26 (26 June 1999): 1647–8.
23. See E. Sridharan, "Electoral coalitions in 2004 general elections: theory and evidence," *Economic and Political Weekly*, 39, 51 (December 18, 2004): 5418–25.
24. Transcript of interview with Shri Jyoti Basu, Acc. No. 781, Oral History Division, Nehru Memorial Museum and Library, New Delhi, pp. 155–7.
25. Devesh Kapur, "India in 1999," *Asian Survey*, 40, 1 (January/February 2000): 196.
26. Venkitesh Ramakrishnan, "Alliance qualms," *Frontline*, 10 September 1999.
27. See Christophe Jaffrelot, "The BJP at the Centre," in Thomas Blom Hansen and Christophe Jaffrelot (eds), *The BJP and the Compulsions of Politics in India*, 2nd edition (New Delhi: Oxford University Press, 2001), p. 344.
28. See Yogendra Yadav, "Open contest, closed options," *Seminar*, 534 (February 2004): 62–6.
29. Sudha Pai, *Dalit Assertion and the Unfinished Democratic Revolution: the Bahujan Samaj Party in Uttar Pradesh* (New Delhi: Sage Publications, 2002), pp. 181–6. For a general sociological analysis, see D.L. Sheth, "Secularisation of caste and making of new middle class," *Economic and Political Weekly*, 34, 34/35 (21 August 1999): 2502–10.
30. See Venkitesh Ramakrishnan and S.K. Pande, "The split and the wait," *Frontline*, 31 July 1999.
31. See Sukumar Muralidharan, "An incipient third force," *Frontline*, 3 July 1999.
32. James Manor, "In part, a myth: the BJP's organisational strength," in Katharine Adeney and Lawrence Sáez (eds), *Coalition Politics and Hindu nationalism* (London: Routledge, 2005), p. 59.
33. Formally, the second NDA ended its tenure in February 2004, less than five years after it took office. However, it called for early polls.
34. Rob Jenkins, "Appearance and reality in Indian politics: making sense of the 1999 general election," *Government and Opposition*, 35, 1 (January 2000): 66.
35. See Table 6 in Eswaran Sridharan, "Why are multi-party minority governments viable in India? theory and comparison," *Commonwealth & Comparative Politics*, 50, 3 (July 2012): 323–5.
36. Jana Krishnamurthy replaced Laxman after the latter was caught accepting bribes in exchange for defense contracts in Operation West End, a sting by the Tehelka media group. *The Hindu*, 20 March 2001.
37. See Subrata K. Mitra, ""The NDA and the politics of 'minorities' in India," in Katharine Adeney and Lawrence Sáez (eds), *Coalition Politics and Hindu nationalism* (London: Routledge, 2005), pp. 77–96.
38. Unless noted, the following summarizes Katherine Adeney, "Hindu nationalists and federal structures in an era of regionalism," in Katharine Adeney and

Lawrence Sáez (eds), *Coalition Politics and Hindu nationalism* (London: Routledge, 2005), pp. 97–115.
39. Kanti Bajpai, "Foreign policy in 2001," *Seminar*, 509 (January 2002): 40–4.
40. Yogendra Yadav and Sanjay Kumar, "Interpreting the mandate," *Frontline*, 5 November 1999.
41. Lloyd I. Rudolph and Susanne Hoeber Rudolph, "Iconisation of Chandrababu: sharing sovereignty in India's federal market economy," *Economic and Political Weekly*, 36, 18 (5 May 2001): 1547; Sukumar Muralidharan, "A conclave of eight," *Frontline*, 2 September 2000.
42. Adeney, "Hindu nationalists and federal structures in an era of regionalism," p. 105.
43. Francine Frankel and Sumantra Sen, *Andhra Pradesh's Long March towards 2020: electoral detours in a developmentalist state* (Philadelphia: Center for the Advanced Study of India, University of Pennsylvania, 2005), pp. 6–9.
44. Jean Dréze and Amartya Sen, *India: economic development and social opportunity* (Oxford: Clarendon Press, 1998): 319.
45. Yogendra Yadav and Suhas Palshikar, "From hegemony to convergence: party system and electoral politics in the Indian states, 1952–2002," *Journal of Indian School of Political Economy*, XV, 1&2 (January–June 2003): p. 37; Francine R. Frankel, *India's Political Economy 1947–2006: the gradual revolution*, 2nd edition (New Delhi: Oxford University Press, 2006), pp. 616–25.
46. James Manor, "Explaining political trajectories in Andhra Pradesh and Karnataka," in Rob Jenkins (ed.), *Regional Reflections: comparing politics across India's states* (New Delhi: Oxford University Press, 2004), pp. 264–5; and Sen and Frankel, *Andhra Pradesh's long march towards 2020*, pp. 28–9.
47. See Sudha Pai, "The problem," *Seminar*, 571 (March 2007).
48. Rudolph and Rudolph, "Iconisation of Chandrababu," p. 1545.
49. Devesh Kapur, "Explaining democratic durability and economic performance: the role of India's institutions," in Devesh Kapur and Pratap Bhanu Mehta (eds), *Public Institutions in India: performance and design* (New Delhi: Oxford University Press, 2008), pp. 49–50; Rudolph and Rudolph, "Iconisation of Chandrababu," p. 1546.
50. See Human Rights Watch, *'We Have No Orders to Save You': state participation and complicity in communal violence in Gujarat* (April 2002).
51. *Financial Times*, 16 April 2002.
52. See Dionne Bunsha, "Narendra Modi's long haul," *Frontline*, 27 September 2002; Achin Vanaik, "Indian foreign policy since the end of the Cold War: domestic determinants," in Sanjay Ruparelia, Sanjay Reddy, John Harriss and Stuart Corbridge (eds), *Understanding India's New Political Economy: a great transformation?* (London: Routledge, 2011), p. 227.
53. Alistair McMillan, "The BJP coalition: partisanship and power-sharing in government," in Katharine Adeney and Lawrence Sáez (eds), *Coalition Politics and Hindu nationalism* (London: Routledge, 2005), p. 34, fn. 14.
54. Frankel, *India's Political Economy*, p. 754, and McMillan, "The BJP coalition," p. 32.

55. Frankel and Sen, *Andhra Pradesh's Long March towards 2020*, pp. 17–8.
56. Max Weber, "The profession and vocation of politics," in Peter Lassman and Ronald Speirs (eds), *Weber: political writings* (Cambridge: Cambridge University Press, 1994), pp. 366–7.
57. Baldev Raj Nayar, "India in 2004: regime change in a divided democracy," *Asian Survey*, 45, 1 (2005): 72.
58. See Pankaj Mishra, "India: the neglected majority wins!" *New York Review of Books*, 12 August 2004.
59. Alfred Stepan, Juan Linz and Yogendra Yadav, *Crafting State-Nations: India and other multinational democracies* (Baltimore: Johns Hopkins University Press, 2011), pp. 81–8.
60. Christophe Jaffrelot, "The BJP and the 2004 general election: dimensions, causes and implications of an unexpected defeat," in Katharine Adeney and Lawrence Sáez (eds), *Coalition Politics and Hindu nationalism* (London: Routledge, 2005), pp. 237–53.
61. See Yogendra Yadav, "The elusive mandate of 2004," *Economic and Political Weekly*, 39, 51 (18 December 2004): 5383–5395.
62. See Sridharan, "Electoral coalitions in 2004 general elections."
63. K.K. Kailash, "Federal calculations in state level coalition governments," *India Review*, 10, 3 (July–September 2011): 267–8.
64. Dwaipayan Bhattacharya, "West Bengal: permanent incumbency and political stability," *Economic and Political Weekly*, 39, 51 (18 December 2004): 5478.
65. Communist Party of India (Marxist), *Election Manifesto 2004*, p. 8.
66. "An Appeal to Comrade Harkishan Surjeet/Comrade A.B. Bardhan and the Central Committee and Polit Bureau members of the CPM and CPI." Signatories included Achin Vanaik, Bipan Chandra, Irfan Habib, K.N. Pannikar, Praful Bidwai, Sumit and Tanika Sarkar and Zoya Hasan.
67. Bidyut Chakrabarty, *Forging Power: coalition politics in India* (New Delhi: Oxford University Press, 2006), p. 154, fn. 59.
68. Chakrabarty, *Forging Power*, p. 208.
69. Communist Party of India (Marxist), "Review Report of the Lok Sabha elections, 2004," p. 19.
70. See Samuel Paul and M. Vivekananda, "Holding a mirror to the new Lok Sabha," *Economic and Political Weekly*, 39, 45 (6 November 2004): 4928–31.
71. *The Hindu*, 26 March 2004.
72. See Aditya Nigam, "The market is not God," *Tehelka*, June 12, 2004; Ramachandra Guha, "After the fall," *The Caravan* (June 2011): http://www.caravanmagazine.in/essay/after-fall.
73. Yadav and Palshikar, "From hegemony to convergence," pp. 35 and 39.
74. D.L. Sheth, "The change of 2004," *Seminar*, 545 (January 2005). For data comparing state-based parties as regionalist and non-regionalist, see the appendix to Adam Ziegfeld, "Coalition government and party system change: explaining the rise of regional political parties in India," *Comparative Politics* 45, 1 (October 2012): 69–87.

75. Mahesh Rangarajan, "Polity in transition: India after the 2004 general elections," *Economic and Political Weekly*, 40, 32 (6 August 2005): 3598–605.
76. The UPA used other vehicles towards the same end, such as the Group of Ministers. It employed all party meetings and chief ministers' conferences relatively less, however. See Kailash, "Middle game in coalition politics," p. 312.
77. James Manor, "Prologue: caste and politics in recent times," In Rajni Kothari (ed.), *Caste in Indian Politics*, 2nd edition (New Delhi: Orient Blackswan, 2010), xviii.
78. Francine Frankel and Sumantra Sen, *Ideology, Politics and Economic Reforms: the national democratic alliance, 1999–2004* (Philadelphia: The Center for the Advanced Study of India, University of Pennsylvania, 2004), p. 40.
79. *The Hindu*, 28 May 2004.
80. Nayar, "India in 2004," p. 79.
81. The following quotations are drawn from Prakash Karat, "18th Congress of the CPI(M): favorable situation for party advance," *The Marxist*, 21, 2/3 (April–September 2005): 1–7.
82. Editorial, "CPI(M): change of guard, change of focus," *Economic and Political Weekly*, 40, 17 (23 April 2005): 1660.
83. The following discussion relies on Richard Sandbrook, Marc Edelman, Patrick Heller and Judith Teichman, *Social Democracy in the Global Periphery: origins, challenges, prospects* (Cambridge: Cambridge University Press, 2007), pp. 65–92.
84. René Véron, "The 'new' Kerala model: lessons for sustainable development," *World Development*, 29, 4 (April 2001): 607.
85. Swagato Sarkar, "Between egalitarianism and domination: governing differences in a transitional society," *Third World Quarterly*, 33, 4 (2012), p. 681.
86. See Dwaipayan Bhattacharyya, "Election 1999: ominous outcome for Left in West Bengal," *Economic and Political Weekly*, 34, 46/47 (20 November 1999): 3267–9.
87. The CPI(Maoist) formed in 2004, an amalgamation of the People's War Group, CPI(ML) Party Unity and Maoist Communist Centre. See Nivedita Menon and Aditya Nigam, *Power and Contestation: India since 1989* (Hyderabad: Orient Longman, 2008), p. 123.
88. Menon and Nigam, *Power and Contestation*, pp. 105–7.
89. Sumit Ganguly, "India in 2008: domestic turmoil and external hopes," *Asian Survey*, 49, 1 (2008): 41–2.
90. See Venkatesh Athreya, "Progress and challenges," *Frontline*, 27 August 2004; Sumanta Banerjee, "Hobson's choice for Indian communists," *Economic and Political Weekly* 40, 19 (7 May 2005): 1935–7.
91. See T.K. Oommen, "Development policy and the nature of society: understanding the Kerala model," *Economic and Political Weekly*, 44, 13 (28 March 2009): 25–31; John Harriss, "Do political regimes matter? poverty reduction and regime differences across India," in Peter R. Houtzager and Mick Moore (eds), *Changing Paths: international development and the new politics of inclusion* (Ann Arbor: University of Michigan Press, 2005), pp. 204–32.

92. Bidyut Chakrabarty, "Left Front's 2006 victory in West Bengal: continuity or a trendsetter?" *Economic and Political Weekly*, 41, 32 (12 August 2006): 3524; Partha Sarathi Banerjee and Dayabati Roy, "Behind the present peasant unrest in West Bengal," *Economic and Political Weekly*, 42, 22 (2 June 2007): 2048–50.
93. Prabhat Patnaik, "In the aftermath of Nandigram," *Economic and Political Weekly*, 42, 21 (26 May 2007): 1893–5.
94. Mritunjoy Mohanty, "Political economic of agrarian transformation: another view of Singur," *Economic and Political Weekly*, 42, 9 (3 March 2007): 737–41.
95. Sumanta Banerjee, "Moral betrayal of a leftist dream," *Economic and Political Weekly* 42, 40 (7 April 2007): 1241.
96. *The Hindu*, August 10, 2006.
97. See Sohini Guha, "From ethnic to multiethnic: the transformation of the Bahujan Samaj Party in north India," *Ethnopolitics*, 12, 1 (2013): 1–29; and *Indian Express*, 24 July 2009.
98. Prakash Karat, *Subordinate Ally: the nuclear deal and India-US strategic relations* (New Delhi: LeftWord Books, 2007), p. 13. On the United States' perspective, see Strobe Talbott, *Engaging India: diplomacy, democracy, and the bomb* (Washington, D.C.: The Brookings Institution, 2006).
99. Baldev Raj Nayar, "India in 2005: India rising, but uphill road ahead," *Asian Survey*, 46, 1 (2005): 96.
100. Karat, *Subordinate Ally*, p. 15.
101. See Peter R. Lavoy, "India in 2006: a new emphasis on engagement," *Asian Survey*, 47, 1 (2007): 116–7.
102. Harsh V. Pant, *The US-India Nuclear Pact: policy, process, and great power politics* (New Delhi: Oxford University Press, 2011), pp. 84–5.
103. *The Hindu*, 13 January 2006.
104. Achin Vanaik, "Indian foreign policy since the end of the Cold War," pp. 229–31.
105. Pant, *The US-India Nuclear Pact*, pp. 80–91.
106. See Sumit Ganguly, "India in 2007: a year of opportunities and disappointments," *Asian Survey*, 48, 1 (2008): 172–4; Vipin Narang and Paul Staniland, "Institutions and worldviews in Indian foreign security policy," *India Review*, 11, 2 (2012): 89.
107. Vanaik, "Indian foreign policy since the end of the Cold War," pp. 228–9.
108. Jo Johnson and Edward Luce, "Welcome to the club," *Financial Times*, 3 August 2007.
109. *Indian Express*, 9 July 2008.
110. T.J. Rajalakshmi, "Feeling betrayed," *Frontline*, 19 July 2008.
111. *The Indian Express*, 9 July 2008.
112. See V. Ramakrishnan, "What price victory?" *Frontline*, August 2, 2008. Subsequent disclosures by Wikileaks seemed to corroborate the accusation. See *The Hindu*, 17 November 2011.
113. V. Ramakrishnan, "No deals here," *Frontline*, 19 July 2008.

114. *The Hindu*, 23 July 2008.
115. See Pratap Bhanu Mehta, "18 Karat light," *Outlook*, 27 October 2007; and Vanaik, "Indian foreign policy since the end of the Cold War," p. 235.
116. *The Hindu*, 5 April 2012.
117. Pant, *The US-India Nuclear Pact*, pp. 3–12.
118. Siddharth Varadarajan, "Turn the nuclear bill from liability to asset," *The Hindu*, 16 June 2010.
119. The most notable was an $11 billion deal to supply 126 jet fighters, which the US lost on technical grounds. *Financial Times*, 29 April 2011.
120. Ramashray Roy, "Regional base and national dream: alliance formation, 2009 national elections," in Paul Wallace and Ramashray Roy (eds), *India's 2009 Elections: coalition politics, party competition, and Congress continuity* (New Delhi: Sage, 2011), p. 36.
121. See K.K. Kailash, "Alliances and lessons of election 2009," *Economic and Political Weekly*, 44, 39 (26 September 2009): 52–7.
122. *Times of India*, 10 May 2009.
123. *Indian Express*, 19 March 2009.
124. Yogendra Yadav, "Suboptimal alliances, wide open race," *The Hindu*, 2 April 2009.
125. Yogendra Yadav and Suhas Palshikar, "Between fortuna and virtu: explaining the Congress' ambiguous victory in 2009," *Economic and Political Weekly*, 44, 39 (26 September 2009): 33–51.
126. Jeffrey Witsoe and Francine Frankel, *Social Justice and Stalled Development: caste empowerment and the breakdown of governance in Bihar* (Philadelphia: Center for the Advanced Study of India, University of Pennsylvania, 2006), pp. 9–27; Mohd Sanjeer Alam, "Bihar: can Lalu Prasad reclaim lost ground?" *Economic and Political Weekly*, 44, 17 (25 April 2009): 13–4; Manor, "Prologue," xxxvii.
127. Saba Naqvi, "Without a one-liner," *Outlook*, 11 May 2009.
128. Christophe Jaffrelot, "The BSP in 2009: still making progress, but only as a Dalit party," in Paul Wallace and Ramashray Roy (eds), *India's 2009 Elections: coalition politics, party competition, and Congress continuity* (New Delhi: Sage, 2011), pp. 140–62. Jaffrelot offers a more positive assessment, however.
129. Raghavendra Keshavarao Hebsur, "The surge of saffron: some genuine and some imitation?" in Paul Wallace and Ramashray Roy (eds), *India's 2009 Elections: coalition politics, party competition, and Congress continuity* (New Delhi: Sage, 2011), pp. 270–85.
130. Karli Srinivasulu, "Political mobilization, competitive populism, and changing party dynamics in Andhra Pradesh," in Paul Wallace and Ramashray Roy (eds), *India's 2009 Elections: coalition politics, party competition, and Congress continuity* (New Delhi: Sage, 2011), pp. 286–308.
131. See G. Gopa Kumar, "The LDF's debacle: Kerala votes for national stability," in Paul Wallace and Ramashray Roy (eds), *India's 2009 Elections: coalition politics, party competition, and Congress continuity* (New Delhi: Sage, 2011), pp. 234–51.

132. Ashok Mitra, "The state of the CPI(M) in West Bengal," *Economic and Political Weekly*, 44, 30 (25 July 2009): 8–12; Dwaipayan Bhattacharya, "Left in the lurch: the demise of the world's longest elected regime?" *Economic and Political Weekly*, 45, 3 (16 January 2010): 51–9.
133. Yadav and Palshikar, "Between *fortuna* and *virtu*," p. 36.
134. Unless noted otherwise, the following summarizes Yadav and Palshikar, "Between *fortuna* and *virtu*."
135. K.C. Suri, "The economy and voting in the 15th Lok Sabha elections," *Economic and Political Weekly*, 44, 39 (26 September 2009): 64–70.
136. Yogendra Yadav and Suhas Palshikar, "Principal state level contests and derivative national choices: electoral trends in 2004–9," *Economic and Political Weekly*, 44, 6 (7 February 2009): 59.
137. Balveer Arora and Stéphanie Tawa Lama-Rewal, "Contextualizing and interpreting the 15th Lok Sabha elections," *South Asia Multidisciplinary Academic Journal*, 3 (2009): 4.
138. *The Hindu*, 26 May 2009.
139. See Prakash Karat, "On the Lok Sabha election results: reviewing the party's performance," *The Marxist*, 25, 1–2 (January–June 2009): 3–15; and remarks made by Sitaram Yechury in *The Hindu*, 5 May 2009
140. Prabhat Patnaik, "Reflections on the Left," *Economic and Political Weekly*, 44, 28 (11 July 2009): 8–10.
141. Prasenjit Bose, "Verdict 2009: an appraisal of critiques of the Left," *Economic and Political Weekly*, 44, 40 (3 October 2009): 38.
142. *The Hindu*, 14 May 2011.
143. Election Commission of India, *Statistical Report on General Election, 2011, to Legislative Assembly of West Bengal*, New Delhi, pp. 13–15.
144. Suhit Sen, "The Left rout: patterns and prospects," *Economic and Political Weekly*, 46, 24 (11 June 2011): 14–16.
145. Sumanta Banerjee, "West Bengal's next quinquennium, and the future of the Indian left," *Economic and Political Weekly*, 46, 23 (4 June 2011): 14–9.
146. Prabhat Patnaik, "The CPI(M) and the building of capitalism," International Development Economic Associates (30 January 2008): http://www.networkideas.org/news/jan2008/news30_Capitalism.htm.
147. Prabhat Patnaik, "The left in decline," *Economic and Political Weekly*, 46, 29 (16 July 2011): 16.
148. *The Hindu*, 4 April 2012.
149. *The Hindu*, 11 April 2012.
150. *The Hindu*, 3 April 2012 and 6 April 2012.
151. See *The Indian Express*, 28 March 2012.

CONCLUSION

1. Atul Kohli, *The State and Poverty in India: the politics of reform* (Cambridge: Cambridge University Press, 1987), p. 91.
2. Lloyd I. Rudolph and Susanne Hoeber Rudolph, *In Pursuit of Lakshmi: the*

political economy of the Indian state (Chicago: University of Chicago Press, 1987), p. 177.
3. Seema Mustafa, *The Lonely Prophet: V.P. Singh* (New Delhi: New Age International, 1995), p. 101.
4. K.K. Kailash, "Middle game in coalition politics," *Economic and Political Weekly*, 42, 4 (January 27, 2007): 310 and 313.
5. Editorial, "Virtue of abstinence," *Economic and Political Weekly*, 31, 19 (11 May 1996): 1099, and Ramesh Thakur, "A changing of the guard in India," *Asian Survey*, 38, 6 (June 1998): 618. Thakur concedes that some political learning occurred, however.
6. Sunil Khilnani, "Branding India," *Seminar 533* (2004): http://www.india-seminar.com/2004/533/533%20sunil%20khilnani.htm.
7. To invoke Gabriel García Márquez, *Chronicle of a Death Foretold* (New York: Vintage, 2003).
8. Jyotirindra Das Gupta, "The Janata phase: reorganization and redirection in Indian politics," in Zoya Hasan (ed.), *Parties and Party Politics in India* (New Delhi: Oxford University Press, 2004), p. 369.
9. See Bidyut Chakrabarty, *Forging Power: coalition politics in India* (New Delhi: Oxford University Press, 2006), pp. 133–50; Mahendra Prasad Singh, "India's National Front and United Front coalition governments: a phase in federalized governance," in Mahendra Prasad Singh and Anil Mishra (eds), *Coalition Politics in India: problems and prospects* (New Delhi: Manohar, 2004), p. 95.
10. Stuart Corbridge and John Harriss, *Reinventing India: liberalization, Hindu nationalism and popular democracy* (Cambridge: Polity Press, 2000), pp. 88–9.
11. See Christophe Jaffrelot, *India's Silent Revolution: the rise of the low castes in north Indian politics* (London: Hurst/New Delhi: Permanent Black, 2003).
12. E. Sridharan, "Unstable parties and unstable alliances: births, splits, mergers and deaths of parties, 1952–2000," in *Coalition Politics in India*, pp. 65–6.
13. See Table 7.3 in Rekha Saxena, *Situating Federalism: mechanisms of intergovernmental relations in Canada and India* (New Delhi: Manohar, 2006), p. 299. For greater analysis, see Lawrence Sáez, *Federalism without a Centre: the impact of political and economic reform on India's federal system* (New Delhi: Sage, 2002), pp. 101–34.
14. For a similar assessment, see James Chiriyankandath and Andrew Wyatt, "The NDA and Indian foreign policy," in Katharine Adeney and Lawrence Sáez (eds), *Coalition Politics and Hindu nationalism* (London: Routledge, 2005), p. 210.
15. Subrata K. Mitra, "War and peace in South Asia: a revisionist view of India-Pakistan relations," in Subrata K. Mitra (ed.), *Politics of Modern South Asia, volume V* (London: Routledge, 2009), pp. 171–2.
16. Sumit Ganguly, "The prime minister and foreign and defence policies," in James Manor (ed.), *Nehru to the Nineties: the changing office of the prime minister of India* (London: Hurst, 1994), p. 160.
17. See Balveer Arora, "Political parties and the party system: the emergence of new coalitions," in *Parties and Party Politics in India*, pp. 504–33; and Alfred

Stepan, Juan Linz and Yogendra Yadav, *Crafting State-Nations: India and other multinational democracies* (Baltimore: Johns Hopkins University Press, 2011), pp. 17–22.
18. Yogendra Yadav, "The third electoral system," *Seminar*, 480 (August 1999): 14–20.
19. Atul Kohli, *Democracy and Discontent: India's growing crisis of ungovernability* (Cambridge: Cambridge University Press, 1990).
20. E. Sridharan, "Principles, power and coalition politics in India: lessons from theory, comparison and recent history," in D.D. Khanna and Gert W. Kueck (eds), *Principles, Power and Politics* (New Delhi: Macmillan India, 1999), pp. 270–91.
21. E. Sridharan, "Why are multi-party minority governments viable in India?" *Commonwealth & Comparative Politics*, 50, 3 (July 2012): 328–32.
22. Sridharan, "Principles, power and coalition politics in India."
23. For general analysis, see Herbert Kitschelt, "Party systems," in Carles Boix and Susan Stokes (eds), *The Oxford Handbook of Comparative Politics* (New York: Oxford University Press, 2009), pp. 530–1.
24. David Runciman, *The Politics of Good Intentions: history, fear and hypocrisy in the new world order* (Princeton: Princeton University Press, 2006), p. 120.
25. See Kaare Strøm and Benjamin Nyblade, "Coalition theory and government formation," in Carles Boix and Susan C. Stokes (eds), *The Oxford Handbook of Comparative Politics* (New York: Oxford University Press, 2009), pp. 784–5.
26. See Henry Hale, "Divided we stand: institutional sources of ethnofederal state survival and collapse," *World Politics*, 56, 2 (January 2004): 165–93.
27. See Sara B. Hobolt and Jeffrey A. Karp, "Voters and coalition governments," *Electoral Studies*, 29, 3 (September 2010): 299–307.
28. Unless otherwise stated, the following draws on Akash Paun, "After the age of majority? multi-party governance and the Westminster model," *Commonwealth & Comparative Politics*, 49, 4 (November 2011): 440–456.
29. See Vernon Bogdanor, "This general election will be fought along regional lines," *Financial Times*, 4 January 2015.
30. In addition to Paun, "After the age of majority?", and Bogdanor, "This general election will be fought along regional lines", see Felicity Matthews, "Constitutional stretching: coalition governance and the Westminster model," *Commonwealth & Comparative Politics*, 49, 4 (November 2011): 486–509.
31. For an overview, see Mark P. Jones and Scott Mainwaring, "The nationalization of parties and party systems: an empirical measure and an application to the Americas," *Party Politics*, 9, 2 (March 2003): 139-166; Dawn Brancati, "The origins and strengths of regional parties," *British Journal of Political Science*, 38, 1 (January 2008): 135-159; Imke Harbers, "Decentralization and the development of nationalized party systems in new democracies: evidence from Latin America," *Comparative Political Studies*, 43, 5 (May 2010): 606–627.
32. See Gosta Esping-Andersen, *The Three Worlds of Welfare Capitalism* (Princeton: Princeton University Press, 1990), p. 30.

33. Ronald J. Herring, "Class politics in India: euphemization, identity and power," in Atul Kohli and Prerna Singh (eds), *Routledge Handbook of Indian Politics* (New York: Routledge, 2013), p. 133.
34. Gregory M. Luebbert, *Liberalism, Fascism or Social Democracy: social classes and the political origins of regimes in interwar Europe* (Oxford: Oxford University Press, 1991), pp. 267–305.
35. John D. Stephens, "Democratization and social policy development in the advanced capitalist societies," in Yusuf Bangura (ed.), *Democracy and Social Policy* (New York: Palgrave Macmillan, 2007), pp. 33–61.
36. I develop these claims in "How the politics of recognition enabled India's democratic exceptionalism." *International Journal for Politics, Culture and Society—Special Issue on the Work of Charles Taylor*, 21, 4 (December 2008): 39–56.
37. Yogendra Yadav, "Reconfiguration in Indian politics: state assembly elections, 1993-1995," in Partha Chatterjee (ed.), *State and Politics in India* (New Delhi: Oxford University Press, 1997), pp. 177–208. The original phrase comes from Ernesto Laclau and Chantal Mouffe, *Hegemony and Socialist Strategy: towards a radical democratic politics* (London: Verso, 2001). For an analysis that seeks to reconcile both logics, see Nancy Fraser and Axel Honneth, *Redistribution or Recognition? a political-philosophical exchange* (London: Verso, 2004).
38. For recent evidence, see Stéphanie Tawa Lama-Rewal, "The resilient *bhadralok*: a profile of the West Bengal MLAs," and G. Gopa Kumar, "Socioeconomic background of legislators in Kerala," in Christophe Jaffrelot and Sanjay Kumar (eds), *Rise of the Plebeians? The changing face of Indian legislative assemblies* (New Delhi: Routledge, 2009), pp. 361–92 and 393–406, respectively.
39. Sudipta Kaviraj, "Democracy and social inequality," in Francine R. Frankel, Zoya Hasan, Rajeev Bhargava and Balveer Arora (eds), *Transforming India: social and political dynamics of democracy* (New Delhi: Oxford University Press, 2000), pp. 113–4.
40. See Perry Anderson, "The antinomies of Antonio Gramsci," *New Left Review*, I/100 (November–December 1976): 5–78.
41. Kaviraj, "Democracy and social inequality," p. 110.
42. Reproduced in D.L. Sheth, "Ram Manohar Lohia on caste in Indian politics," in Gyanshyam Shah (ed.), *Caste and Democratic Politics in India* (London: Anthem Press, 2004), pp. 84–5.
43. See Adam Ziegfeld, "Coalition government and party system change: explaining the rise of regional political parties in India," *Comparative Politics*, 45, 1 (October 2012): 69–87; Csaba Nikolenyi, *Minority Governments in India: the puzzle of elusive majorities* (London: Routledge, 2010).
44. See Rajni Kothari, "Rise of the dalits and the renewed debate on caste," in *State and Politics in India*, p. 444.
45. See Anthony Heath and Yogendra Yadav, "The united colours of the Congress: social profile of Congress voters, 1996 and 1998," in *Parties and Party Politics in India*, pp. 107–51.

46. Interview with senior JD official, New Delhi, 10 February 2000.
47. Interview with senior JD politician, New Delhi, 24 October 1998.
48. Interview with senior JD politician, New Delhi, 23 September 1998.
49. Madhu Limaye, *Janata Party Experiment: an insider's account of opposition politics: 1977–1980, volume two* (New Delhi: B.R. Publishing, 1994), p. 552.
50. Limaye, *Janata Party Experiment*, p. 553.
51. See Lucia Michelutti, *The Vernacularisation of Democracy: politics, caste and religion in India* (New Delhi: Routledge, 2008), especially pp. 161–216.
52. For a broader development of this argument, see Pratap Bhanu Mehta, *The Burden of Democracy* (New Delhi: Penguin Books, 2003), p. 46.
53. See Sidney Tarrow, "Transforming enemies into allies: non-ruling communist parties in multiparty coalitions," *The Journal of Politics*, 44, 4 (November 1982): 924–54.
54. Roberto Sarti, "The dissolution of the Italian Communist Party," *The Marxist*, http://www.marxist.com/the-dissolution-of-the-italian-communist-party-1991.htm.
55. Giuseppe Di Palma, "Eurocommunism?" *Comparative Politics*, 9, 3 (April 1977): 372 and 365.
56. The political reasons seem different, however. See Stéphane Courtois and Denis Peschanski, "From decline to marginalization: the PCF breaks with French society," in Michael Waller and Meindert Fennema (eds.), *Communist Parties in Western Europe: decline or adaptation?* (Oxford: Basil Blackwell, 1988), pp. 47–68.
57. Amrita Basu, "Parliamentary communism as a historical phenomenon: the CPI(M) in West Bengal," in Zoya Hasan (ed.), *Parties and Party Politics in India* (New Delhi: Oxford University Press, 2004), p. 318.
58. See Perry Anderson, "An invertebrate left," *London Review of Books*, 31, 5 (12 March 2009); and Meindert Fennema, "Conclusions," in *Communist Parties in Western Europe*, pp. 244–60.
59. Maurizio Vannicelli, "Introduction to part two: alliance policy," in Peter Lange and Maurizio Vannicelli (eds), *The Communist Parties of Italy, France and Spain: postwar change and continuity—a casebook* (Cambridge: George Allen & Unwin, 1991), p. 106.
60. Aijaz Ahmad, *Lineages of the Present: ideology and politics in contemporary South Asia* (London: Verso, 2000), p. 234.
61. David Samuels, "Separation of powers," in Carles Boix and Susan C. Stokes (eds), *The Oxford Companion to Comparative Politics* (Oxford: Oxford University Press, 2009), pp. 709–10.
62. See Michael Laver, "Legislatures and parliaments in comparative context," and Gary W. Cox, "The organization of democratic legislatures," in Barry M. Weingast and Donald A. Wittman (eds), *The Oxford Handbook of Political Economy* (New York: Oxford University Press, 2008), pp. 121–40 and 141–61, respectively; and Akhtar Majeed, "Coalitions as power-sharing arrangements," in Akhtar Majeed (ed.), *Coalition Politics and Power Sharing* (New Delhi: Manak Publications PVT Ltd., 2000), p. 26.

63. Nikolenyi, *Minority Governments in India*, p. 91.
64. Wolfgang C. Müller and Kaare Strøm, "Political parties and hard choices," in Wolfgang C. Müller and Kaare Strøm (eds), *Policy, Office or Votes? How political parties in Western Europe make hard decisions* (Cambridge: Cambridge University Press, 1999), p. 23.
65. See Praful Bidwai, "Jyoti Basu for prime minister," *Frontline*, 26 December 1997.
66. See Adam Przeworski, *Capitalism and Social Democracy* (Cambridge: Cambridge University Press, 1991) and Sheri Berman, *The Primacy of Politics: social democracy and the making of Eruope's twentieth century* (New York: Cambridge University Press, 2007).
67. See Anna Gryzmala-Busse, "Time will tell? temporality and the analysis of causal mechanisms and processes," *Comparative Political Studies*, 44, 9 (December 2010): 1267–97.
68. Aditya Nigam, "Logic of failed revolution: federalization of CPM," *Economic and Political Weekly*, 35, 5 (29 January 2000): 263–6.
69. Achin Vanaik, "Subcontinental strategies," *New Left Review*, II/70 (July–August 2011), p. 107.
70. For example, see Wendy Hunter, *The Transformation of the Workers' Party in Brazil, 1989–2009* (New York: Cambridge University Press, 2010).
71. That said, its institutional projects of political decentralization and land reform suggest a more radical outlook. Zoya Hasan, "Introduction," in *Parties and Party Politics in India*, p. 22.
72. Vanaik, "Subcontinental strategies," pp. 112–13.
73. Steven Lukes, *Marxism and Morality* (Oxford: Clarendon Press, 1985), p. 146.
74. Alexander L. George and Andrew Bennett, *Case Studies and Theory Development in the Social Sciences* (Cambridge: MIT Press, 2005), pp. 205–32.
75. Sudipta Kaviraj, "The politics of subalternity," Plenary lecture, 33rd Annual Conference on South Asia, University of Wisconsin-Madison, 16 October 2004 (unpublished).
76. *The Hindu*, 28 February 2014.
77. *The Hindu*, 5 February 2014.
78. See E. Sridharan, "Pre-electoral coalitions in 2014," available at http://nottspolitics.org/2014/04/07/pre-electoral-coalitions-in-2014/.
79. *The Hindu*, 28 April 2014 and *Times of India*, 11 May 2014.
80. See Milan Vaishnav and Danielle Smogard, "A new era in Indian politics?" available at http://carnegieendowment.org/2014/06/10/new-era-in-indian-politics/hdc6.
81. A.K. Verma, Mirza Asmer beg and Sudhir Kumar, "A saffron sweep in Uttar Pradesh," *The Hindu*, 23 May 2014.
82. See Amitabh Dubey, "The Rajya Sabha will remain a headache for the NDA," http://chunauti.org/2014/05/31/rajya-sabha-remain-headache/.
83. See Editorial, "Anger, aspiration, apprehension," *Economic and Political Weekly*, 49, 21 (24 May 2014): 7–8.
84. See Praful Bidwai, "Resisting Modi through mass struggle," *The News*, 7 June 2014.

BIBLIOGRAPHY

Interviews, Private Papers, and Political Speeches and Writings
Advani, Lal Krishna. "Opening speech," BJP National Executive, Udaipur, 3–5 March 1989. In A.M. Zaidi (ed.), *Annual Register of Indian Political Parties: proceedings and fundamental texts 1989: part one.* New Delhi: Indian Institute of Applied Political Research, 1992.
Basu, Jyoti. "Interview." Acc. No. 781, Oral History Division, Nehru Memorial Museum and Library, New Delhi.
Bommai, S.R. "Presidential Address." Janata Dal Annual Convention, 8–10 February 1991, Puri, Orissa. In O.P. Rahlan (ed.), *Encyclopedia of Political Parties Post-Independence India, Volume 96: Janata Dal proceedings.* New Delhi: Anmol Publications, 2001, pp. 76–87.
Chandrashekhar. "Betrayal of Janata Party." In *Ideals or Opportunism?* A Janata Party Publication, 1979, pp. 1–9.
Dandavate, Madhu. "Crisis rooted in summit power politics." In *Ideals or Opportunism?* A Janata Party Publication, 1979, pp. 18–27.
———. "Interview." Acc. No. 788, Oral History Division, Nehru Memorial Museum and Library, New Delhi.
Gujral, Inder Kumar. "Interview." Acc. No. 797, Oral History Division, Nehru Memorial Museum and Library, New Delhi.
———. "Foreign policy objectives of India's United Front government." Royal Institute of International Affairs, Chatham House, London, 23 September 1996.
———. "Aspects of India's foreign policy." Bandaranaike Center for International Studies, Colombo, 20 January 1997.
Karat, Prakash. "18th Congress of the CPI(M): favorable situation for party advance." *The Marxist*, 21, 2/3 (April–September 2005): www.cpim.org/marxist/200502_18%20Congress.doc
———. *Subordinate Ally: the nuclear deal and India-US strategic relations.* New Delhi: LeftWord Books, 2007.
———. "On the Lok Sabha election results: reviewing the party's performance." *The Marxist*, 25, 1–2 (January–June 2009): 3–15.

Naidu, N. Chandrababu. "We will get a majority on our own." *Frontline*, 26 December 1997.
Namboodiripad, E.M.S. "The Left in India's freedom movement and in free India." *Social Scientist*, 14, 8/9 (August–September 1986): 3–17.
———. "Opening speech." 13th Congress of CPI(M), Calcutta, 25–30 December 1988. In A.M. Zaidi (ed.), *Annual Register of Indian Political Parties: proceedings and fundamental texts 1989: part one*. New Delhi: Indian Institute of Applied Political Research, 1992.
———. *The Frontline Years: selected articles*. New Delhi: LeftWord Books, 2010.
Narayan, Jayaprakash. "Free India from the dictators." In Arun Shourie (ed.), *Institutions in the Janata Phase*. Bombay: Popular Prakash Private Ltd, 1980, pp. 5–8.
———. Letter to Chandrasekhar, 9 May 1978. Janata Party Papers, Subject File 311, Nehru Memorial Museum and Library, New Delhi.
Rama Rao, N.T. "Outlines for the proposed confederation of like-minded parties, 4 January 1986."
———. "Speech to meeting of like minded non-Congress political parties, May 1987."
Reddy, S. Japial. "Interview." *Seminar*, 454 (June 1997).
Shekhar, Chandra. "Betrayal of Janata Party". In *Ideals or Opportunism? A Janata Party Publication*, 1979, pp. 1–9.
Singh, Charan. Letter to Jayaprakash Narayan, 1 June 1976. In Dhirendra Sharma (ed.), *The Janata (People's) Struggle: the finest hour of the Indian people—with underground documents, resistance literature and correspondence relating to the advent of the Janata Party*. New Delhi: Sanjivan Press, 1977, pp. 300–304.
———. Letter to Ashok Mehta, 27 September 1976. In Dhirendra Sharma (ed.), *The Janata (People's) Struggle: the finest hour of the Indian people—with underground documents, resistance literature and correspondence relating to the advent of the Janata Party*. New Delhi: Sanjivan Press, 1977, p. 312.
Surjeet, Harkishan Singh. "The present political situation and the evolution of our tactics in perspective." *The Marxist*, 8, 4 (October–December 1990): 1–21.
Thakre, Kushabhau. "Presidential Address." National Council Meeting, Gandhinagar, 3–4 May 1998. In O.P. Rahlan (ed.), *Encyclopedia of Political Parties Post-Independence India, Volume 82: BJP National Council Meetings (1981–1998)*. New Delhi: Anmol Publications, 2000, pp. 181–217.
Vajpayee, Atal Bihari. *New Dimensions of India's Foreign Policy*, with a foreword by M.C. Chagla. New Delhi: Vision Books, 1979.
———. "All responsible for the Janata crisis." In *Ideals or Opportunism? A Janata Party Publication*, 1979, pp. 9–18.
Yechuri, Sitaram. "Caste and class in Indian politics today." *P. Sundarayya Memorial Lecture*, 1997.

BIBLIOGRAPHY

Political Party Documents

"An Appeal to Comrade Harkishan Surjeet/Comrade A.B. Bardhan and the Central Committee and Polit Bureau members of the CPM and CPI."
Bharatiya Janata Party. *Election Manifesto 1996*.
———. *Election Manifesto 1998*.
Communist Party of India (CPI). *Election Manifesto 1996*.
———. "Platform for secular democratic alternative." *New Age Weekly*, 21–27 April 1996.
———. *Election Manifesto 1998*.
Communist Party of India (Marxist). *Election Manifesto 1996*.
———. "Review Report on 1996 General Elections."
———. *Election Manifesto 1998*.
———. *Election Manifesto 1999*.
———. "Review of the 1999 Lok Sabha elections."
———. *Election Manifesto 2004*.
———. "Review Report of the Lok Sabha elections, 2004."
———. "Left Parties Appeal." New Delhi, 1 October 1989. In O.P. Rahlan (ed.), *Encyclopedia of Political Parties Post-Independence India, Volume 97: Samajwadi Janata Party and other smaller groups*. New Delhi: Anmol Publications, 2001, pp. 137–139.
Jammu and Kashmir National Conference. "Political resolution adopted by the Central Working Committee on 2 November 1994 at Jammu."
Janata Dal. *The Peoples Charter—The Party Manifesto*. In O.P. Rahlan (ed.), *Encyclopedia of Political Parties Post-Independence India, Volume 96*. New Delhi: Anmol Publications, 2001, pp. 31–58.
———. *Election Manifesto 1996*.
———. *Election Manifesto 1998*.
Janata Party. Minutes of meeting of four Opposition parties, 8 July 1976. In Dhirendra Sharma (ed.), *The Janata (People's) Struggle: the finest hour of the Indian people—with underground documents, resistance literature and correspondence relating to the advent of the Janata Party*. New Delhi: Sanjivan Press, 1977.
———. *Election Manifesto 1977*.
———. *Election Manifesto 1980*.
Indian National Congress(I). *Election Manifesto 1980*.
———. *Election Manifesto 1984*.
———. *Election Manifesto 1996*.
———. *Election Manifesto 1998*.
National Front. "Constitution." In A.M. Zaidi (ed.), *Annual Register of Indian Political Parties: proceedings and fundamental texts 1990: part one*. New Delhi: Indian Institute of Applied Political Research, 1992.
———. *Manifesto: Lok Sabha Elections, 1989*.
———. *Manifesto: Lok Sabha Elections, May 1991*.
Rashtriya Janata Dal. *Election Manifesto 1998*.

BIBLIOGRAPHY

Samata Party. "Statement of Policy and Programmes." National Convention, New Delhi, 19 August 1994. In O.P. Rahlan (ed.), *Encyclopedia of Political Parties Post-Independence India, Volume 97: Samajwadi Janata Party and other smaller groups.* New Delhi: Anmol Publications, 2001, pp. 65–92.

Telugu Desam Party (TDP). *Manifesto 1996.*

———. *Manifesto 1998.*

United Front. *A Common Approach to Major Policy Matters and a Minimum Programme.*

Government Documents

Disinvestment Commission, Government of India. "Strategy and Issues, December 1996."

Election Commission of India. *Statistical Report on General Elections, 1971, to the Sixth Lok Sabha: Volume I.* New Delhi.

———. *Statistical Report on General Elections, 1977, to the Sixth Lok Sabha: Volume I.* New Delhi.

———. *Statistical Report on General Elections, 1980, to the Seventh Lok Sabha: Volume I.* New Delhi.

———. *Statistical Report on General Elections, 1989, to the Ninth Lok Sabha: Volume I.* New Delhi.

———. *Statistical Report on General Elections, 1996, to the Eleventh Lok Sabha: Volume I.* New Delhi.

———. *Statistical Report on General Elections, 1998, to the Twelfth Lok Sabha: Volume I.* New Delhi.

———. *Statistical Report on General Elections, 1999, to the Twelfth Lok Sabha: Volume I.* New Delhi.

———. *Statistical Report on General Elections, 2004, to the Twelfth Lok Sabha: Volume I.* New Delhi.

———. *Statistical Report on General Elections, 2009, to the Twelfth Lok Sabha: Volume I.* New Delhi.

———. *Statistical Report on General Election, 2011, to Legislative Assembly of West Bengal.* New Delhi.

Home Ministry. *Report of the Fifth Central Pay Commission, Volumes I-III.*

Ministry of External Affairs, Government of India. *Annual Report 1995–96.*

Ministry of Finance, Government of India. *1979–80 Union Budget Speech.*

———. *1990–91 Union Budget Speech.*

———. *Economic Survey 1997–98.*

———. *1996–97 Union Budget Speech.*

———. *1997–98 Union Budget Speech.*

Planning Commission. *Towards Social Transformation: Approach to the Eighth Five-Year Plan, 1990–95.*

———. *Ninth Five-Year Plan, 1997–2002.*

Srikrishna Commission. *Inquiry into the Riots at Mumbai during December 1992—January 1993 and the March 12, 1993 Bomb Blasts. Report: Volume I*, by Hon. Justice B.N. Srikrishna, Mumbai High Court.

BIBLIOGRAPHY

Indian Newspapers and Periodicals

Asian Age
Frontline
India Today
Indian Express
Outlook
The Hindu
The Hindustan Times
The Statesman
Times of India

Editorials

"What is different?" *Economic and Political Weekly* 25, 12 (24 March 1990): 579–580.
"Less equal." *Economic and Political Weekly*, 31, 19 (11 May 1996): 1101–1102.
"Virtue of abstinence." *Economic and Political Weekly*, 31, 19 (11 May 1996): 1099.
"Glib promises." *Economic and Political Weekly*, 31, 24 (15 June 1996): 1423.
"All for foreign capital." *Economic and Political Weekly*, 11 January 1997: 3–6.
"Down the slippery slope." *Economic and Political Weekly*, 32, 5 (1 February 1997): 183–186.
"Perils: real and fanciful." *Economic and Political Weekly*, 33, 10 (7 March 1998): 491.
"Minus the plumage." *Economic and Political Weekly*, 33, 13 (26 March 1998): 679.
"Questionable shift." *Economic and Political Weekly*, 34, 26 (26 June 1999): 1647–1648.
"CPI(M): change of guard, change of focus." *Economic and Political Weekly*, 40, 17 (23 April 2005): 1660.
"Wrong medicine." *Economic and Political Weekly*, 32, 41 (11 October 1997): 2567.
"Anger, aspiration, apprehension." *Economic and Political Weekly*, 49, 21 (24 May 2014): 7–8.
"Dead men walking," *Economic and Political Weekly*, 32, 49 (6 December 1997): 3104.

Articles, Chapters, Books

Abbott, Andrew. "From causes to events." In *Time Matters*. Chicago: University of Chicago Press, 2001, pp. 183–209.
Adams, John. "Breaking away: India's economy vaults into the 1990s." In

Marshall M. Bouton and Philip Oldenburg (eds), *India Briefing, 1990*. Boulder: Westview Press, 1990, pp. 77–100.

Adeney, Katherine. "Hindu nationalists and federal structures in an era of regionalism." In Katharine Adeney and Lawrence Sáez (eds), *Coalition Politics and Hindu nationalism*. London: Routledge, 2005, pp. 97–115.

Alam, Javeed. "Communist politics in search of hegemony." In Partha Chatterjee (ed.), *Wages of Freedom: fifty years of the Indian nation-state*. Delhi: Oxford University Press, 1998, pp. 179–207.

Alam, Mohd Sanjeer. "Bihar: can Lalu Prasad reclaim lost ground?" *Economic and Political Weekly*, 44, 17 (25 April 2009): 11–4.

Ahmad, Aijaz. *Lineages of the Present: ideology and politics in contemporary South Asia*. London: Verso, 2000.

Andersen, Walter K. "India in 1981: stronger political authority and social tension." *Asian Survey*, 22, 2 (February 1982): 119–35.

———. "Election 1989 in India: the dawn of coalition politics?" *Asian Survey*, 30, 6 (June 1990): 527–40.

———. "India's 1991 election: the uncertain verdict." *Asian Survey*, 31, 10 (October 1991): 976–89.

Anderson, Perry. "The antinomies of Antonio Gramsci." *New Left Review*, I/100 (November–December 1976): 5–78.

———. "An invertebrate left." *London Review of Books*, 31, 5 (12 March 2009).

Appadorai, A. and M.S. Rajan. *India's Foreign Policy and Relations*. New Delhi: South Asian Publishers, 1985.

Arora, Balveer. "India's federal system and the demands of pluralism: crisis and reform in the '80s." In Joyotpaul Chaudhari (ed.), *India's Beleaguered Federalism: the pluralist challenge*. Tempe: Arizona State University, 1992, pp. 5–26.

———. "Negotiating differences." In Francine R. Frankel, Zoya Hasan, Rajeev Bhargava and Balveer Arora (eds), *Transforming India: social and political dynamics of democracy*. New Delhi: Oxford University Press, 2000, pp. 176–207.

———. "Political parties and the party system: the emergence of new coalitions." In Zoya Hasan (ed.), *Parties and Party Politics in India*. Delhi: Oxford University Press, 2004, pp. 504–33.

Arora, Balveer and Stéphanie Tawa Lama-Rewal. "Contextualizing and interpreting the 15th Lok Sabha elections." *South Asia Multidisciplinary Academic Journal*, 3 (2009): 2–11.

Arora, Balveer and Douglas Verney (eds). *Multiple Identities in a Single State: Indian federalism in comparative perspective*. Delhi: Kanark Publications, 1995.

Athreya, Venkatesh. "Progress and challenges." *Frontline*, 27 August 2004.

Austin, Granville. *Working a Democratic Constitution: a history of the Indian experience*. Delhi: Oxford University Press, 2000.

Axelrod, Robert. *Conflict of Interest: a theory of divergent goals with applications to politics*. Chicago: Markham Publishing Company, 1970.

BIBLIOGRAPHY

Bajpai, Kanti. "Foreign policy in 2001." *Seminar*, 509 (January 2002): 40–44.

Bajpai, K. Shankar. "India in 1991: new beginnings." *Asian Survey*, 32, 2 (February 1992): 207–16.

Balakrishnan, Pulapre. "Union budget for 1990-91." *Economic and Political Weekly*, 25, 16 (21 April 1990): 893–6.

Ball, Terence. "Party." In Terence Ball, James Farr and Russell Hanson (eds), *Political Innovation and Conceptual Change*. Cambridge: Cambridge University Press, 1989, pp. 155–76.

Banerjee, Partha Sarathi and Dayabati Roy. "Behind the present peasant unrest in West Bengal." *Economic and Political Weekly*, 42, 22 (2 June 2007): 2048–50.

Banerjee, Abhijit, Pranab Bardhan, Kaushik Basu, Mrinal Datta Chaudhuri, Maitresh Ghatak, Ashok Sanjay Guha, Mukul Majumdar, Dilip Mookherjee and Debraj Ray. "Strategy for economic reform in West Bengal." *Economic and Political Weekly*, 37, 41 (12 October 2002): 4203–18.

Banerjee, Sumanta. "Strategy, tactics, and forms of political participation among left parties." In T.V. Sathyamurthy (ed.), *Class Formation and Political Transformation in Post-Colonial India*. New Delhi: Oxford University Press, 1999, pp. 202–38.

———. "Hobson's choice for Indian communists." *Economic and Political Weekly* 40, 19 (7 May 2005): 1935–7.

———. "Moral betrayal of a leftist dream." *Economic and Political Weekly* 42, 40 (7 April 2007): 1240–2.

———. "West Bengal's next quinquennium, and the future of the Indian left." *Economic and Political Weekly*, 46, 23 (4 June 2011): 14–9.

Bardhan, Pranab. "The political economy of reform in India." In Zoya Hasan (ed.), *Politics and the State in India*. New Delhi: Sage Publications, 2000, pp. 158–77.

———. "Sharing the spoils: group equity, development and democracy." In Atul Kohli (ed.), *The Success of India's Democracy*. Cambridge: Cambridge University Press, 2001, pp. 226–42.

———. *The Political Economy of Development in India, expanded edition with an epilogue on the political economy of reform in India*. Delhi: Oxford University Press, 2012.

Baru, Sanjaya. "A quiet coup in the fisc." *Economic and Political Weekly* 25, 16 (21 April 1990): 887–92.

———. "Economic policy and the development of capitalism in India: the role of regional capitalists and political parties." In Francine R. Frankel, Zoya Hasan, Rajeev Bhargava and Balveer Arora (eds), *Transforming India: social and political dynamics of democracy*. New Delhi: Oxford University Press, pp. 207–31.

Baruah, Sanjib. *India against Itself: Assam and the politics of nationality*. Philadelphia: University of Pennsylvania Press, 1999.

Basu, Amrita. "The dialectics of Hindu nationalism." In Atul Kohli (ed.), *The Success of India's Democracy*. Cambridge: Cambridge University Press, 2001, pp. 163–91.

———. "Parliamentary communism as a historical phenomenon: the CPI(M) in West Bengal." In Zoya Hasan (ed.), *Parties and Party Politics in India*. New Delhi: Oxford University Press, 2004, pp. 317–51.

Bates, Robert H., Avner Grief, Margaret Levi, Jean-Laurent Rosenthal and Barry R. Weingast. *Analytic Narratives*. Princeton: Princeton University Press, 1998.

Béteille, André. "Distributive justice and institutional well-being." In André Béteille, *The Backward Classes in Contemporary India*. New Delhi: Oxford University Press, 1992, pp. 45–72.

———. "The future of the Backward Classes: the competing demands of status and power." In André Béteille, *Society and Politics in India: essays in comparative perspective*. New Delhi: Oxford University Press, 1997, pp. 150–92.

Berk, Gerald and Dennis Galvan. "How people experience and change institutions: a field guide to creative syncretism." *Theory and Society*, 38, 6 (2009): 543–80.

Berlin, Isaiah. "Political judgment." In Henry Hardy (ed.), *The Sense of Reality: studies in ideas and their history*, with an introduction by Patrick Gardiner. London: Pimlico, 1997, pp. 40–54.

Berman, Sheri. *The Primacy of Politics: social democracy and the making of Europe's twentieth century*. New York: Cambridge University Press, 2007.

Bhagwati, Jagdish. "The design of Indian development." In Isher Judge Ahluwalia and I.M.D. Little (eds), *India's Economic Reforms and Development: essays for Manmohan Singh*. New Delhi: Oxford University Press, 1998, pp. 23–40.

Bhaduri, Amit and Deepak Nayyar. *The Intelligent Person's Guide to Liberalization*. New Delhi: Penguin, 1996.

Bhattacharya, Dwaipayan. "Election 1999: ominous outcome for Left in West Bengal." *Economic and Political Weekly*, 34, 46/47 (20 November 1999): 3267–9.

———. "West Bengal: permanent incumbency and political stability." *Economic and Political Weekly*, 39, 51 (18 December 2004): 5477–83.

———. "Left in the lurch: the demise of the world's longest elected regime?" *Economic and Political Weekly*, 45, 3 (16 January 2010): 51–9.

Bhaumik, Subir. "North-east India: the evolution of a post-colonial region." In Partha Chatterjee (ed.), *Wages of Freedom: fifty years of the Indian nation-state*. New Delhi: Oxford University Press, 1998, pp. 310–28.

Bidwai, Praful. "Resisting Modi through mass struggle," *The News*, 7 June 2014.

Bidwai, Praful and Achin Vanaik. *South Asia on a Short Fuse: nuclear politics and the future of global disarmament*. New Delhi: Oxford University Press, 2001.

Blondel, Jean and Ferdinand Müller-Rommel (eds). *Governing Together: the extent and limits of joint decision-making in western European cabinets*. London: Macmillan, 1993.

Blyth, Mark. "Structures do not come with an instruction sheet: interests, ideas, and progress in political science." *Perspectives on Politics*, 1, 4 (December 2003): 695–706.

BM. "Liberalisation lobby's attempted coup." *Economic and Political Weekly*, 25, 23 (9 June 1990): 1237–8.

———. "Precarious balancing act." *Economic and Political Weekly*, 25, 13 (31 March 1990): 648–9.

Bogdanor, Vernon (ed.). *Coalition Government in Western Europe*. London: Heinemann Educational Books, 1983.

————. "This general election will be fought along regional lines." *Financial Times*, 4 January 2015.

Bose, Sumantra. *The Challenge in Kashmir: democracy, self-determination and a just peace*. New Delhi: Sage, 1997.

Bose, Prasenjit. "Verdict 2009: an appraisal of critiques of the Left." *Economic and Political Weekly*, 44, 40 (3 October 2009): 32–8.

Brancati, Dawn. "The origins and strengths of regional parties." *British Journal of Political Science*, 38, 1 (January 2008): 135–159.

Brass, Paul R. "Political parties and the radical left in South Asia." In Paul R. Brass and Marcus F. Franda (eds), *Radical Politics in South Asia*. Cambridge: MIT Press, 1973, pp. 3–116.

———. "National power and local politics in India: a twenty-year perspective." *Modern Asian Studies*, 18, 1 (1984): 89–118.

———. "Party systems and governmental instability in Indian states." In Paul R. Brass, *Caste, Faction and Party in Indian Politics, Volume One: faction and party*. New Delhi: Chanakya Publications, 1984, pp. 19–62.

———. "Coalition politics in north India." In Paul R. Brass, *Caste, Faction and Party in Indian Politics, Volume One: faction and party*. New Delhi: Chanakya Publications, 1984, pp. 97–135.

———. "Leadership conflict and the disintegration of the Indian socialist movement: personal ambition, power and policy." In Paul R. Brass, *Caste, Faction and Party in Indian Politics, Volume One: faction and party*. New Delhi: Chanakya Publications, 1984, pp. 155–89.

———. "Congress, the Lok Dal, and the middle-peasant castes: an analysis of the 1977 and 1980 parliamentary elections in Uttar Pradesh." In Paul R. Brass, *Caste, Faction and Party in Indian Politics, Volume Two: election studies*. New Delhi: Chanakya Publications, 1984, pp. 162–204.

———. "The Punjab crisis and the unity of India." In Atul Kohli (ed.), *India's Democracy: an analysis of changing state-society relations*. Princeton: Princeton University Press, 1988, pp. 169–213.

———. "Chaudhuri Charan Singh: an Indian political life." *Economic and Political Weekly*, 28, 39 (25 September 1993): 2087–90.

———. *The Politics of India since Independence*, 2nd edition. Cambridge: Cambridge University Press, 1994.

Bratersky, M.V. and S.I. Lunyov. "India at the end of the century: transformation into an Asian regional power." *Asian Survey*, 30, 10 (October 1990): 927–42.

Browne, Eric C. and John Dreijmanis (eds). *Coalition Governments in Western Democracies*. London: Longman, 1982.

Browne, Eric C. and D.W. Gleiber. "An 'events' approach to the problem of cabinet stability." *Comparative Political Studies*, 17, 2 (July 1984): 167–97.

Bueno de Mesquita, Bruce. *Strategy, Risk and Personality in Coalition Politics: the case of India*. Cambridge: Cambridge University Press, 1975.

Büthe, Tim. "Taking temporality seriously: modeling history and the use of narratives as evidence." *American Political Science Review*, 96, 3 (September 2002): 481–94.

Butler, David, Ashok Lahiri and Prannoy Roy. *India Decides: elections 1952–1995*, 3rd edition. New Delhi: Books and Things, 1995.

———. *India Decides: update 1996*. New Delhi: Books and Things, 1996.

Byres, Terence J. "Charan Singh, 1902–87: an assessment." *The Journal of Peasant Studies*, 15, 2 (1988): 139–89.

Carpenter, Daniel. "Commentary: what is the marginal value of 'analytic narratives'?" *Social Science History*, 24, 4 (Winter 2000): 653–67.

Chakrabarty, Bidyut. *Forging Power: coalition politics in India*. New Delhi: Oxford University Press, 2006.

———. "Left Front's 2006 victory in West Bengal: continuity or a trendsetter?" *Economic and Political Weekly*, 41, 32 (12 August 2006): 3521–7.

Chandra, Bipan. *In the Name of Democracy: JP movement and the Emergency*. New Delhi: Penguin Books, 2003.

Chandra, Kanchan. *Why Ethnic Parties Succeed: patronage and ethnic head counts in India*. Cambridge: Cambridge University Press, 2004.

Chatterjee, Partha. *A Possible India: essays in political criticism*. New Delhi: Oxford University Press, 1997.

Chatterji, Saral K. (ed.). *The Coalitional Government: a critical examination of the concept of coalition, the performance of some coalitional governments and the future prospects of coalition in India*. Madras: The Christian Literature Society, 1974.

Chenoy, Kamal Mitra. "The budget 1997-98: an analysis." http://www.ieo.org/97anal2.html.

Chibber, Pradeep K. and John R Petrocik. "The puzzle of Indian politics: social cleavages and the Indian party system." *British Journal of Political Science*, 19, 2 (April 1989): 191–210.

Chibber, Vivek. *Locked in Place: state-building and late industrialization in India*. Princeton: Princeton University Press, 2003.

Chiriyankandath, James. "'Unity in diversity?' coalition politics in India (with special reference to Kerala)." *Democratization*, 4, 4 (Winter 1997): 16–39.

Chiriyankandath, James, and Andrew Wyatt. "The NDA and Indian foreign policy." In Katharine Adeney and Lawrence Sáez (eds), *Coalition Politics and Hindu nationalism*. London: Routledge, 2005, pp. 193–211.

Chowdhury, Rekha and Nagendra Rao. "Elections 2002: implications for politics of separatism." *Economic and Political Weekly*, 38, 1 (4 January 2003): 15–21.

Cohen, Stephen P. *India: emerging power*. New Delhi: Oxford University Press, 2003.
Collier, David, James Mahoney and Jason Seawright. "Claiming too much: warnings about selection bias." In Henry E. Brady and David Collier (eds), *Rethinking Social Inquiry: diverse tools, shared standards*. Lanham, MD: Rowman and Littlefield, 2004, pp. 85–102.
Collier, Ruth Berins and David Collier. *Shaping the Political Arena: critical junctures, the labor movement, and regime dynamics in Latin America*. Princeton: Princeton University Press, 1991.
Corbridge, Stuart. "Federalism, Hindu nationalism and mythologies of governance in India." In Graham Smith (ed.), *Federalism: the multiethnic challenge*. London: Longman, 1995, pp. 101–28.
Corbridge, Stuart and John Harriss. *Reinventing India: liberalization, Hindu nationalism and popular democracy*. Cambridge: Polity Press, 2000.
Courtois, Stéphane and Denis Peschanski. "From decline to marginalization: the PCF breaks with French society." In Michael Waller and Meindert Fennema (eds), *Communist Parties in Western Europe: decline or adaptation?* Oxford: Basil Blackwell, 1988, pp. 47–68.
Cox, Gary W. "The organization of democratic legislatures." In Barry M. Weingast and Donald A. Wittman (eds), *The Oxford Handbook of Political Economy*. New York: Oxford University Press, 2008, pp. 141–61.
Das, Arvind. "The future postponed." *Economic and Political Weekly*, 34, 20 (15 May 1999): 1166–7.
Das Gupta, Jyotirindra. "Ethnicity, democracy and development in India: Assam in a general perspective." In Atul Kohli (ed.), *India's Democracy: an analysis of changing state-society relations*. Princeton: Princeton University Press, 1988, pp. 144–169.
———. "The Janata phase: reorganization and redirection in Indian politics." In Zoya Hasan (ed.), *Parties and Party Politics in India*. New Delhi: Oxford University Press, 2004, pp. 353–70.
Datta, Polly. "The issue of discrimination in Indian federalism in the post-1977 politics of West Bengal." *Comparative Studies of South Asia, Africa and the Middle East*, 1, 2 (2005): 449–64.
Desai, Radhika. "Forward march of Hindutva halted?" *New Left Review*, 30 (November–December 2004): 49–67.
Di Palma, Giuseppe. "Eurocommunism?" *Comparative Politics*, 9, 3 (April 1977):357–75.
Diermeier, Daniel. "Coalition government." In Barry M. Weingast and Donald A. Wittman (eds), *The Oxford Handbook of Political Economy*. New York: Oxford University Press, 2008, p. 162–79.
Diermeier, D., and R. Stevenson. "Cabinet survival and competing risks." *American Journal of Political Science*, 43, 4 (October 1999): 1051–68.
Diermeier, Daniel, Hulya Eraslan and Antonio Merlo. "Coalition government and comparative constitutional design." *European Economic Review*, 46, 4 (2002): 893–907.

BIBLIOGRAPHY

Dodd, Lawrence C. *Coalitions in Parliamentary Government.* Princeton: Princeton University Press, 1976.

Downs, Anthony. *An Economic Theory of Democracy.* New York: Harper, 1957.

Dréze, Jean and Amartya Sen. *India: economic development and social opportunity.* Oxford: Clarendon Press, 1998.

Dua, B.D. "The prime minister and the federal system." In James Manor (ed.), *Nehru to the Nineties: the changing office of the prime minister of India.* London: Hurst, 1994, pp. 20–47.

Dubey, Amitabh. "The Rajya Sabha will remain a headache for the NDA." http://chunauti.org/2014/05/31/rajya-sabha-remain-headache/.

Dunn, John. "Unger's Politics and the appraisal of political possibility." In John Dunn, *Interpreting Political Responsibility: essays 1981–1989.* Cambridge: Cambridge University Press, 1990.

———. *The Cunning of Unreason: making sense of politics.* London: HarperCollins, 2001.

Duverger, Maurice. *Political Parties: their organization and activity in the modern state.* New York: Wiley, 1963.

Elster, Jon. "Rational choice history." *American Political Science Review*, 94, 3 (September 2000): 685–95.

———. *Explaining Social Behavior: more nuts and bolts for the social sciences.* Cambridge: Cambridge University Press, 2007.

Esping-Andersen, Gosta. *The Three Worlds of Welfare Capitalism.* Princeton: Princeton University Press, 1990.

Fearon, James D. "Counterfactuals and hypothesis testing in political science." *World Politics*, 43, 2 (January 1991): 169–95.

Fennema, Meindert. "Conclusions." In Michael Waller and Meindert Fennema (eds), *Communist Parties in Western Europe: decline or adaptation?* Oxford: Basil Blackwell, 1988, pp. 244–60.

Ferejohn, John. "External and internal explanation." In Ian Shapiro, Rogers M. Smith and Tarek E. Masoud (eds), *Problems and Methods in the Study of Politics.* New York: Cambridge University Press, 2004, pp. 144–66.

Fickett, Lewis P., Jr. "The major socialist parties of India in the 1967 election." *Asian Survey*, 8, 6 (June 1968): 489–98.

———. "The Praja Socialist Party of India–1952–1972: a final assessment." *Asian Survey*, 13, 9 (September 1973): 826–32.

———. "The rise and fall of the Janata Dal." *Asian Survey* 33, 12 (December 1993): 1151–62.

Flyvbjerg, Bent. *Rationality and Power: democracy in practice.* Chicago: University of Chicago Press, 1998.

———. *Making Social Science Matter: why social inquiry fails and how it can succeed again.* Cambridge: Cambridge University Press, 2001.

Franda, Marcus F. *Radical Politics in West Bengal.* Cambridge: MIT Press, 1971.

Frankel, Francine R. "Middle classes and castes in Indian politics: prospects for political accommodation." In Atul Kohli (ed.), *India's Democracy: an analysis of changing state-society relations*. Princeton: Princeton University Press, 1988, pp. 225–62.

———. "Conclusion." In Francine Frankel and M.S.A. Rao (eds), *Dominance and State Power in Modern India: decline of a social order, Volume II*. New Delhi: Oxford University Press, 1990, pp. 482–518.

———. *India's Political Economy 1947–2004: the gradual revolution*, 2nd edition. New Delhi: Oxford University Press, 2006.

Frankel, Francine and Sumantra Sen. *Ideology, Politics and Economic Reforms: the national democratic alliance, 1999–2004*. Philadelphia: The Center for the Advanced Study of India, University of Pennsylvania, 2004.

———. *Andhra Pradesh's Long March towards 2020: electoral detours in a developmentalist state*. Philadelphia: Center for the Advanced Study of India, University of Pennsylvania, 2005.

Franklin, M. and T. Mackie. "Reassessing the importance of size and ideology for the formation of governing coalitions in parliamentary democracies." *American Journal of Political Science*, 28 (1984): 94–103.

Fraser, Nancy and Axel Honneth. *Redistribution or Recognition? a political-philosophical exchange*. London: Verso, 2004.

Galanter, Marc. "Pursuing equality in the land of hierarchy: an assessment of India's policies of compensatory discrimination for historically disadvantaged groups." In Rajeev Dhavan (ed.), *Law and Society in Modern India*. New Delhi: Oxford University Press, 1997, pp. 185–208.

Gamson, William. "A theory of coalition formation." *American Sociological Review*, 26 (1961): 373–82.

Ganguly, Sumit. "The prime minister and foreign and defence policies." In James Manor (ed.), *Nehru to the Nineties: the changing office of the prime minister of India*. London: Hurst, 1994, pp. 138–60.

———. "India in 2007: a year of opportunities and disappointments." *Asian Survey*, 48, 1 (2008): 164–76.

———. "India in 2008: domestic turmoil and external hopes." *Asian Survey*, 49, 1 (2008): 39–52.

Garton Ash, Timothy. *History of the Present: essays, sketches and dispatches from Europe in the 1990s*. London: Penguin, 1999.

George, Alexander L. and Andrew Bennett. *Case Studies and Theory Development in the Social Sciences*. Cambridge: MIT Press, 2005.

Gerring, John. *Case Study Research: principles and practices*. Cambridge: Cambridge University Press, 2007.

Geuss, Raymond. *Philosophy and Real Politics*. Princeton: Princeton University Press, 2008.

———. *Politics and the Imagination*. Princeton: Princeton University Press, 2010.

Ghosh, Jayati. "Tightening the wrong belts." *Frontline*, 12 July 1996: 99–100.

———. "Leaving out the essentials," *Frontline*. 9 August 1996.

———. "An analysis of the Budget 1997–98: largesse for the rich, phrases for the poor." http://www.ieo.org/97anal.html.

Gould, Harold A. "The second coming: the 1980 elections in India's Hindi belt." *Asian Survey*, 20, 6 (June 1980): 595–616.

Grofman, B. "The comparative analysis of coalition formation and duration: distinguishing between-country and within-country effects." *British Journal of Political Science*, 19, 2 (April 1989): 291–302.

Gryzmala-Busse, Anna. "Time will tell? temporality and the analysis of causal mechanisms and processes." *Comparative Political Studies*, 44, 9 (December 2010): 1267–97.

Guha, Ramachandra. *India after Gandhi: the history of the world's largest democracy*. New York: HarperCollins, 2007.

———(ed.). *The Makers of Modern India*. Cambridge: Harvard University Press, 2011.

———. "After the fall." *The Caravan*. June 2011: http://www.caravanmagazine.in/essay/after-fall

Guha, Sohini. "From ethnic to multiethnic: the transformation of the Bahujan Samaj Party in north India." *Ethnopolitics*, 12, 1 (2013): 1–29.

Hale, Henry. "Divided we stand: institutional sources of ethnofederal state survival and collapse." *World Politics*, 56, 2 (January 2004): 165–93.

Hall, Peter A. and Rosemary Taylor. "Political science and the three new institutionalisms." *Political Studies*, 44, 5 (December 1996): 936–57.

Hansen, Thomas Blom. "The ethics of Hindutva and the spirit of capitalism." In Thomas Blom Hansen and Christophe Jaffrelot (eds), *The BJP and the Compulsions of Politics in India*, 2nd edition. New Delhi: Oxford University Press, 2001, pp. 291–315.

Hansen, Thomas Blom and Christophe Jaffrelot. "Introduction: the rise to power of the BJP." In Thomas Blom Hansen and Christophe Jaffrelot (eds), *The BJP and the Compulsions of Politics in India*, 2nd edition. New Delhi: Oxford University Press, 2001, pp. 1–22.

Harbers, Imke. "Decentralization and the development of nationalized party systems in new democracies: evidence from Latin America." *Comparative Political Studies*, 43, 5 (May 2010): 606–627.

Hardgrave, Robert L., Jr. "India on the eve of elections: Congress and the opposition." *Pacific Affairs*, 57, 3 (Autumn 1984): 404–28.

Harriss, John. "Do political regimes matter? poverty reduction and regime differences across India." In Peter R. Houtzager and Mick Moore (eds), *Changing Paths: international development and the new politics of inclusion*. Ann Arbor: University of Michigan Press, 2005, pp. 204–32.

Harshe, Rajen. "South Asian regional co-operation." *Economic and Political Weekly*, 34, 19 (8 May 1999): 1100–5.

Hart, Henry C. "Political leadership in India: dimensions and limits." In Atul Kohli (ed.), *India's Democracy: an analysis of changing state-society relations*. Princeton: Princeton University Press, 1988, pp. 18–62.

Hasan, Zoya. "The prime minister and the left." In James Manor (ed.), *Nehru to the Nineties: the changing office of the prime minister of India*. London: Hurst, 1994, pp. 207–29.

BIBLIOGRAPHY

———. *Quest for Power: oppositional movements and post-Congress politics in Uttar Pradesh*. New Delhi: Oxford University Press, 1999.

———. "Representation and redistribution: the new lower caste politics of north India." In Francine R. Frankel, Zoya Hasan, Rajeev Bhargava and Balveer Arora (eds), *Transforming India: social and political dynamics of democracy*. New Delhi: Oxford University Press, 2000, pp. 146–76.

———. "Introduction." In Zoya Hasan (ed.), *Parties and Party Politics in India*. New Delhi: Oxford University Press, 2004, pp. 1–37.

———. "The 'politics of presence' and legislative reservations for women." In Zoya Hasan, E. Sridharan and R. Sudarshan (eds), *India's Living Constitution: ideas, practices, controversies*. Delhi: Permanent Black, 2006, pp. 405–27.

Hawthorn, Geoffrey. *Plausible Worlds: possibility and understanding in history and the social sciences*. Cambridge: Cambridge University Press, 1991.

———. "Pericles' unreason." In Richard Bourke and Raymond Geuss (eds), *Political Judgment: essays for John Dunn*. Cambridge: Cambridge University Press, 2009, pp. 203–28.

Hazarika, Sanjoy. *Strangers of the Mist: tales of war and peace from India's northeast*. New Delhi: Penguin, 1995.

Heath, Anthony and Yogendra Yadav. "The united colours of the Congress: social profile of Congress voters, 1996 and 1998." In Zoya Hasan (ed.), *Parties and Party Politics in India*. New Delhi: Oxford University Press, 2004, pp. 107–51.

Heath, Oliver. "Anatomy of BJP's rise to power: social, regional, and political expansion in the 1990s." In Zoya Hasan (ed.), *Parties and Party Politics in India*. New Delhi: Oxford University Press, 2002, pp. 232–57.

Hebsur, Raghavendra Keshavarao. "The surge of saffron: some genuine and some imitation?" In Paul Wallace and Ramashray Roy (eds), *India's 2009 Elections: coalition politics, party competition, and Congress continuity*. New Delhi: Sage, 2011, pp. 270–85.

Heller, Patrick. *The Labor of Development: workers and the transformation of capitalism in Kerala, India*. Ithaca: Cornell University Press, 1999.

———. "Degrees of democracy: some comparative lessons from India." *World Politics*, 52, 4 (2000): 484–519.

Helmke, Gretchen and Steven Levitsky. "Informal institutions and comparative politics: A research agenda." *Perspectives on Politics*, 2, 4 (December 2004): 725–40.

Herring, Ronald J. *Land to the Tiller: the political economy of agrarian reform in South Asia*. New Haven: Yale University Press, 1983.

———. "Class politics in India: euphemization, identity and power." In Atul Kohli and Prerna Singh (eds), *Routledge Handbook of Indian Politics*. New York: Routledge, 2013, pp. 129–43.

Hobolt, Sara B. and Jeffrey A. Karp. "Voters and coalition governments." *Electoral Studies*, 29, 3 (September 2010): 299–307.

Human Rights Watch. *'We Have No Orders to Save You': state participation and complicity in communal violence in Gujarat*. April 2002.

Hunter, Wendy. *The Transformation of the Workers' Party in Brazil, 1989–2009*. New York: Cambridge University Press, 2010.
Ilaiah, Kancha. "Towards the dalitization of the nation." In Partha Chatterjee (ed.), *Wages of Freedom: fifty years of the Indian nation-state*. Delhi: Oxford University Press, 1998, pp. 267–92.
Jaffrelot, Christophe. "The Bahujan Samaj Party in north India: no longer just a dalit party?" *Comparative Studies of South Asia, Africa and the Middle East* 18, 1 (Spring 1998): 35–52.
———. *The Hindu Nationalist Movement and Indian Politics: 1925 to the 1990s*. New Delhi: Penguin Books, 1999.
———. "The BJP at the Centre." In Thomas Blom Hansen and Christophe Jaffrelot (eds), *The BJP and the Compulsions of Politics in India*, 2nd edition. New Delhi: Oxford University Press, 2001, pp. 315–369
———. *India's Silent Revolution: the rise of the low castes in north Indian politics*. London: Hurst/Delhi: Permanent Black, 2003.
———. "The BJP and the 2004 general election: dimensions, causes and implications of an unexpected defeat." In Katharine Adeney and Lawrence Sáez (eds), *Coalition Politics and Hindu nationalism*. London: Routledge, 2005, pp. 237–53.
———. "The BSP in 2009: still making progress, but only as a Dalit party." In Paul Wallace and Ramashray Roy (eds), *India's 2009 Elections: coalition politics, party competition, and Congress continuity*. New Delhi: Sage, 2011, pp. 140–62.
Jaffrelot, Christophe and Sanjay Kumar (eds.). *Rise of the Plebeians? the changing face of Indian legislative assemblies*. New Delhi: Routledge, 2009.
Jayal, Niraja Gopal. *Democracy and the State: welfare, secularism and development in contemporary India*. New Delhi: Oxford University Press, 2002.
Jenkins, Rob. "India's electoral result: an unholy alliance between nationalism and regionalism." Mimeo: 1998.
———. *Democratic Politics and Economic Reform in India*. Cambridge: Cambridge University Press, 1999.
———. "Appearance and reality in Indian politics: making sense of the 1999 general election." *Government and Opposition*, 35, 1 (January 2000): 49–66.
Jha, Nalini Kant. "Coalition governments and India's foreign policy." In Mahendra Prasad Singh and Anil Mishra (eds), *Coalition Politics in India: problems and prospects*. New Delhi: Manohar, 2004, pp. 293–326.
Jha, Prem Shankar. "Coalition politics and economic decision-making." In Mahendra Prasad Singh and Anil Mishra (eds), *Coalition Politics in India: problems and prospects*. New Delhi: Manohar, 2004, pp. 285–92.
Jones, Mark P. and Scott Mainwaring. "The nationalization of parties and party systems: an empirical measure and an application to the Americas." *Party Politics*, 9, 2 (March 2003): 139–166.
Joseph, K.P. "Piped music and telephone attendants: report of the Fifth Pay Commission." *Economic and Political Weekly*, 32, 12 (22 March 1997): 563–5.

BIBLIOGRAPHY

Joshi, Vijay and I.M.D. Little. *India: macroeconomics and political economy, 1964–1991*. New Delhi: Oxford University Press, 1994.

Kailash, K.K. "Middle game in coalition politics." *Economic and Political Weekly*, 42, 4 (27 January 2007): 307–17.

———. "Alliances and lessons of election 2009." *Economic and Political Weekly*, 44, 39 (26 September 2009): 52–57.

———. "Federal calculations in state level coalition governments." *India Review*, 10, 3 (July–September 2011): 246–82.

Kale, Sunila and Nimah Mazaheri. "Natural resources, development strategies and identity politics along India's mineral belt: Bihar and Odisha in the 1990s." Mimeo, October 2012.

Kahneman, Daniel. "Nobel lecture: maps of bounded rationality: psychology for behavioral economics." *American Economic Review*, 93, 5 (2003): 1449–75.

Kapur, Devesh. "India in 1999." *Asian Survey*, 40, 1 (January/February 2000): 195–207.

———. "Explaining democratic durability and economic performance: the role of India's institutions." In Devesh Kapur and Pratap Bhanu Mehta (eds), *Public Institutions in India: performance and design*. New Delhi: Oxford University Press, 2008, pp. 28–76.

———. "Public opinion and Indian foreign policy." *India Review*, 8, 3 (July–September 2009): 286–305.

Kapur, Devesh and Pratap Bhanu Mehta (eds). *Public Institutions in India: performance and design*. New Delhi: Oxford University Press, 2008.

Katznelson, Ira. "Periodization and preferences: reflections on purposive action in comparative historical social science." In James Mahoney and Dietrich Rueschemeyer (eds), *Comparative Historical Analysis in the Social Sciences*. Cambridge: Cambridge University Press, 2003, pp. 270–301.

Kaviraj, Sudipta. "The general elections in India." *Government and Opposition*, 32, 1 (Winter 1997): 3–25.

———. "Democracy and social inequality." In Francine R. Frankel, Zoya Hasan, Rajeev Bhargava and Balveer Arora (eds), *Transforming India: social and political dynamics of democracy*. New Delhi: Oxford University Press, 2000, pp. 89–120.

———. "The politics of subalternity." Plenary lecture, 33rd Annual Conference on South Asia, University of Wisconsin-Madison, 16 October 2004 (unpublished).

———. "Marxism in translation: critical reflections on Indian radical thought." In Richard Bourke and Raymond Geuss (eds), *Political Judgment: essays in honor of John Dunn*. Cambridge: Cambridge University Press, 2009, pp. 172–200.

Khilnani, Sunil. *The Idea of India*. New York: Farrar Straus Giroux, 1999.

———. "Branding India." *Seminar 533* (2004): http://www.india-seminar.com/2004/533/533%20sunil%20khilnani.htm

Kidwai, Rasheed. *Sonia: a biography*. New Delhi: Penguin, 2011.

King, Gary, James Alt, Nancy Burns and Michael Laver. "A unified model of cabinet dissolution in parliamentary democracies." *American Journal of Political Science*, 34, 3 (August 1990): 846–71.

King, Gary, Robert O. Keohane and Sidney Verba. *Designing Social Inquiry: scientific inference in qualitative research*. Princeton: Princeton University Press, 1994.

Kiser, Edgar and Howard T. Welser. "The microfoundations of analytic narratives." *Sociologica*, 3 (November–December 2007): 1–19.

Kitschelt, Herbert. "Party systems." In Carles Boix and Susan Stokes (eds), *The Oxford Handbook of Comparative Politics*. New York: Oxford University Press, 2009, pp. 522–54.

Kohli, Atul. *The State and Poverty in India: the politics of reform*. Cambridge: Cambridge University Press, 1987.

———. *Democracy and Discontent: India's growing crisis of ungovernability*. Cambridge: Cambridge University Press, 1990.

———. "From majority to minority rule: making sense of the 'new' Indian politics." In Marshall M. Bouton and Philip Oldenburg (eds), *India Briefing, 1990*. Boulder: Westview Press, 1990, pp. 1–24.

———. "Enduring another election." *Journal of Democracy*, 9, 3 (July 1998): 7–21.

———. "Politics of economic growth in India, 1980–2005—Part I: the 1980s." *Economic and Political Weekly*, 41, 13 (1 April 2006): 1251–9.

Kohli, Atul, Peter Evans, Peter J. Katzenstein, Adam Przeworski, Susanne Hoeber Rudolph, James C. Scott and Theda Skocpol. "The role of theory in comparative politics: a symposium." *World Politics*, 48, 1 (October 1995): 1–49.

Kothari, Rajni. "The Congress 'system' in India." *Asian Survey*, 4, 12 (December 1964): 1161–73.

———. *Politics and the People: in search of a humane India, volume I*. New Delhi: Ajanta, 1989.

———. "Rise of the *dalits* and the renewed debate on caste." In Partha Chatterjee (ed.), *State and Politics in India*. New Delhi: Oxford University Press, 1997, pp. 439–59.

Krishnaswamy, K.S., I.S. Gulati and A. Vaidyanathan. "Economic aspects of federalism in India." In Nirmal Mukarji and Balveer Arora (eds), *Federalism in India: origins and development*. New Delhi: Vikas Publishing House, 1992, pp. 180–212.

Kumar, Dhruba. "Managing Nepal's India policy?" *Asian Survey*, 30, 7 (July 1990): 697–710.

Kumar, G. Gopa. "Socio-economic background of legislators in Kerala." In Christophe Jaffrelot and Sanjay Kumar (eds), *Rise of the Plebeians? The changing face of Indian legislative assemblies*. New Delhi: Routledge, 2009, pp. 393–406.

———. "The LDF's debacle: Kerala votes for national stability." In Paul Wallace and Ramashray Roy (eds), *India's 2009 Elections: coalition poli-*

tics, party competition, and Congress continuity. New Delhi: Sage, 2011, pp. 234–51.

Kumar, Sanjay. "New phase in backward caste politics in Bihar." In Ghanshyam Shah (ed.), *Caste and Democratic Politics in India*. London: Anthem Press, 2004, pp. 235–268.

———. "Regional parties, coalition government, and functioning of Indian parliament: the changing patterns." *Journal of Parliamentary Studies*, 1, 1 (2010): 75–91.

Kurian, N.J. "State government finances: a survey of recent trends." *Economic and Political Weekly*, 34, 19 (8 May 1999): 1115–25.

Laclau, Ernesto and Chantal Mouffe. *Hegemony and Socialist Strategy: towards a radical democratic politics*. London: Verso, 2001.

Laver, Michael. "Legislatures and parliaments in comparative context." In Barry Weingast and Donald A. Wittman (eds), *The Oxford Handbook of Political Economy*. New York: Oxford University Press, 2008, pp. 121–40.

Laver, Michael and Norman Schofield. *Multiparty Government: the politics of coalition in Europe*. Oxford: Oxford University Press, 1990.

Laver, Michael and Kenneth A. Shepsle (eds). *Cabinet Ministers and Parliamentary Government*. Cambridge: Cambridge University Press, 1994.

Lavoy, Peter R. "India in 2006: a new emphasis on engagement." *Asian Survey*, 47, 1 (2007): 113–24.

Levi, Margaret. "A model, a method and a map: rational choice in comparative and historical analysis." In Mark Irving Lichbach and Alan S. Zuckerman (eds), *Comparative Politics: rationality, culture and structure*. New York: Cambridge University Press, 1999, pp. 19–41.

———. "An analytic narrative approach to puzzles and problems." In Ian Shapiro, Rogers M. Smith and Tarek E. Masoud (eds), *Problems and Methods in the Study of Politics*. New York: Cambridge University Press, 2004.

Limaye, Madhu. *Janata Party Experiment: an insider's account of opposition politics: 1977–1980, volume two*. New Delhi: B.R. Publishing, 1994.

Linz, Juan J. "The perils of presidentialism." In Larry Diamond and Marc F. Plattner (eds), *The Global Resurgence of Democracy*, 2nd edition. Baltimore: The John Hopkins University Press, 1996, pp. 124–43.

———. "The virtues of parliamentarism." In Larry Diamond and Marc F. Plattner (eds), *The Global Resurgence of Democracy*, 2nd edition. Baltimore: The John Hopkins University Press, 1996, pp. 154–62.

Lijphart, Arend. *Patterns of Democracies: government forms and performance in thirty-six countries*. New Haven: Yale University Press, 1999.

Luebbert, Gregory M. "Coalition theory and government formation in multiparty democracies," *Comparative Politics*, 15, 2 (January 1983): 235–49.

———. "A theory of government formation." *Comparative Political Studies*, 17, 2 (July 1984): 229–64.

———. *Comparative Democracy: policymaking and governing coalitions in Europe and Israel*. New York: Columbia University Press, 1986.

———. *Liberalism, Fascism or Social Democracy: social classes and the political origins of regimes in interwar Europe*. Oxford: Oxford University Press, 1991.

Lukes, Steven. *Marxism and Morality*. Oxford: Clarendon Press, 1985.

Lupia, A., and K. Strøm. "Coalition termination and the strategic timing of parliamentary elections." *American Political Science Review*, 89, 3 (September 1995): 33–50.

Machiavelli, Niccolò. *The Prince*, edited by Quentin Skinner and Russell Price. Cambridge: Cambridge University Press, 2008.

Mahoney, James. "After KKV: the new methodology of qualitative research." *World Politics*, 62, 1 (January 2010): 120–47.

Mainwaring, Scott and Matthew S. Shugart. "Juan Linz, presidentialism, and democracy: a critical appraisal." *Comparative Politics*, 29, 4 (July 1997): 449–71.

Mair, Peter. "Introduction." In Peter Mair (ed.), *The West European Party System*. Oxford: Oxford University Press, 1990, pp. 1–25.

Majeed, Akhtar. "Coalitions as power-sharing arrangements." In Akhtar Majeed (ed.), *Coalition Politics and Power Sharing*. New Delhi: Manak Publications PVT Ltd., 2000, pp. 6–35.

Manor, James. "The electoral process amid awakening and decay: reflections on the Indian general election of 1980." In Peter Lyon and James Manor (eds), *Transfer and Transformation: political institutions in the new commonwealth*. Leicester: Leicester University Press, 1983, pp. 87–116.

———. "The Indian general election of 1984." *Electoral Studies*, 4, 2 (1985): 149–58.

———. "Parties and the party system." In Atul Kohli (ed.), *India's Democracy: an analysis of changing state-society relations*. Princeton: Princeton University Press, 1988, pp. 62–99.

———. "Introduction." In James Manor (ed.), *Nehru to the Nineties: the changing office of the prime minister of India*. London: Hurst, 1994.

———. "The prime minister and the president." In James Manor (ed.), *Nehru to the Nineties: the changing office of the prime minister of India*. London: Hurst, 1994, pp. 115–37.

———. "India's chief ministers and the problem of governability." In Philip Oldenburg (ed.), *India Briefing: staying the course*. New York: M.E. Sharpe, 1995, 47–75.

———. "'Ethnicity' and politics in India." *International Affairs*, 72, 3 (July 1996): 459–75.

———. "Regional parties in federal systems." In Balveer Arora and Douglas Verney (eds), *Multiple Identities in a Single State: Indian federalism in comparative perspective*. New Delhi: Konark, 1995, 105–35.

———. "Understanding Deve Gowda." *Economic and Political Weekly*, 31, 39 (28 September 1996): 2675–8.

BIBLIOGRAPHY

———. "Explaining political trajectories in Andhra Pradesh and Karnataka." In Rob Jenkins (ed.), *Regional Reflections: comparing politics across India's states*. New Delhi: Oxford University Press, 2004.

———. "In part, a myth: the BJP's organisational strength." In Katharine Adeney and Lawrence Sáez (eds), *Coalition Politics and Hindu nationalism*. London: Routledge, 2005, pp. 55–74.

———. "Prologue: caste and politics in recent times." In Rajni Kothari (ed.), *Caste in Indian Politics*, 2nd edition. New Delhi: Orient Blackswan, 2010, xi-lxi.

———. "The Congress party and the great transformation." In Sanjay Ruparelia, Sanjay Reddy, John Harriss and Stuart Corbridge (eds), *Understanding India's New Political Economy: a great transformation?* London: Routledge, 2011, pp. 204–20.

Mantena, Karuna. "Another realism: the politics of Gandhian nonviolence." *American Political Science Review*, 106, 2 (May 2012): 455–70.

Maor, Moshe. *Parties, Conflicts and Coalitions in Western Europe: organisational determinants of coalition bargaining*. London: Routledge/LSE, 1998.

Márquez, Gabriel García. *Chronicle of a Death Foretold*. New York: Vintage, 2003.

Matthews, Felicity. "Constitutional stretching: coalition governance and the Westminster model." *Commonwealth & Comparative Politics*, 49, 4 (November 2011): 486–509.

McAdam, Doug. *Political Process and the Development of Black Insurgency, 1930–1970*. Chicago: University of Chicago press, 1982.

McMillan, Alistair. "The BJP coalition: partisanship and power-sharing in government." In Katharine Adeney and Lawrence Sáez (eds), *Coalition Politics and Hindu nationalism*. London: Routledge, 2005, pp. 13–35.

Mehta, Pratap B. "Reform political parties first." *Seminar*, 497 (January 2001): 38–41.

———. *The Burden of Democracy*. New Delhi: Penguin Books, 2003.

Mendelsohn, Oliver. "The collapse of the Indian National Congress." *Pacific Affairs*, 51, 1 (Spring 1978): 41–66.

Menon, M.S. "Drawing lines in water." *The Indian Express*, 26 September 2012.

Menon, Nivedita and Aditya Nigam. *Power and Contestation: India since 1989*. Hyderabad: Orient Longman, 2008.

Michelutti, Lucia. *The Vernacularisation of Democracy: politics, caste and religion in India*. New Delhi: Routledge, 2008.

Mishra, Pankaj. "India: the neglected majority wins!" *New York Review of Books*, 12 August 2004.

Mitra, Ashok. "The state of the CPI(M) in West Bengal." *Economic and Political Weekly*, 44, 30 (25 July 2009): 8–12.

Mitra, Subrata K. *Governmental Instability in Indian States: West Bengal, Bihar, Uttar Pradesh and Punjab*. New Delhi: Ajanta Publications, 1978.

———. "The NDA and the politics of 'minorities' in India." In Katharine

Adeney and Lawrence Sáez (eds), *Coalition Politics and Hindu nationalism*. London: Routledge, 2005, pp. 77–96.

———. "War and peace in South Asia: a revisionist view of India-Pakistan relations." In Subrata K. Mitra (ed.), *Politics of Modern South Asia, volume V*. London: Routledge, 2009, pp. 157–77.

Mohanty, Mritunjoy. "Political economic of agrarian transformation: another view of Singur." *Economic and Political Weekly*, 42, 9 (3 March 2007): 737–41.

Morris-Jones, W.H. *Politics Mainly Indian*. Madras: Orient Longman, 1978.

Moury, Catherine. *Coalition Government and Party Mandate: how coalition agreements constrain ministerial action*. London: Routledge, 2013.

Müller, Wolfgang C. and Kaare Strøm. "Political parties and hard choices." In Wolfgang C. Müller and Kaare Strøm (eds), *Policy, Office or Votes? How political parties in Western Europe make hard decisions*. Cambridge: Cambridge University Press, 1999, pp. 1–35.

Murthy, Padmaja. "The Gujral doctrine and beyond." Mimeo, Institute for Defence Studies and Analyses, available at: http://www.idsa-india.org/anjul9-8.html.

Mustafa, Seema. *The Lonely Prophet: V.P. Singh*. New Delhi: New Age International, 1995.

Nandy, Ashis. "Federalism, the ideology of the state and cultural pluralism." In Nirmal Mukherji and Balveer Arora (eds), *Federalism in India: origins and development*. New Delhi: Vikas Publishing House, 1992, pp. 27–41.

Narang, Vipin and Paul Staniland. "Institutions and worldviews in Indian foreign security policy." *India Review*, 11, 2 (2012): 76–94.

Nayar, Baldev Raj. "India and the super powers: deviation or continuity in foreign policy?" *Economic and Political Weekly*, 12, 30 (23 July 1977): 1185–9.

———. "Policy and performance under democratic coalitions." *Journal of Commonwealth and Comparative Politics*, 37, 2 (July 1999): 22–56.

———. "India in 2004: regime change in a divided democracy." *Asian Survey*, 45, 1 (2005): 71–82.

———. "India in 2005: India rising, but uphill road ahead." *Asian Survey*, 46, 1 (2005): 95–106.

Nayar, Baldev Raj and T.V. Paul. *India in the World Order: searching for major-power status*. Cambridge: Cambridge University Press, 2003.

Nayyar, Deepak. "Economic development and political democracy: interaction of economics and politics in independent India." *Economic and Political Weekly*, 33, 49 (5 December 1998): 3121–31.

Nigam, Aditya. "India after the 1996 elections: nation, locality and representation." *Asian Survey*, 36, 12 (December 1996): 1157–70.

———. "Communist politics hegemonized." In Partha Chatterjee (ed.), *Wages of Freedom: fifty years of the Indian nation-state*. Delhi: Oxford University Press, 1998, 207–41.

———. "Logic of failed revolution: federalisation of CPI(M)." *Economic and Political Weekly*, 35, 5 (29 January 2000): 263–6.

―――. "The market is not God." *Tehelka*, June 12, 2004.
Nikolenyi, Csaba. *Minority Governments in India: the puzzle of elusive majorities*. London: Routledge, 2010.
Noorani, A.G. "The prime minister and the judiciary." In James Manor (ed.), *Nehru to the Nineties: the changing office of the prime minister of India*. London: Hurst, 1994, pp. 94–114.
―――. "A gross case," *Frontline*, 9 February 1996.
―――. "A grave lapse," *Frontline*, 14 June 1996.
―――. "Communist memories." *Frontline*, 3 December 2011.
―――. "Of Stalin, Telengana & Indian revolution." *Frontline*, 17 December 2011.
―――. "Of Quit India, Nehru and CPI split." *Frontline*, 31 December 2011.
Nossiter, T.J. *Marxist State Governments in India: politics, economics and society*. London: Pinter Publishers, 1988.
O'Donnell, Guillermo. "Delegative democracy." In Larry Diamond and Marc F. Plattner (eds), *The Global Resurgence of Democracy*, 2nd edition. Baltimore: Johns Hopkins University Press, 1996, pp. 94–111.
Oldenburg, Philip. "Pollsters, pundits and a mandate to rule: interpreting India's 1984 parliamentary elections." *Journal of Commonwealth and Comparative Politics* 26, 3 (1988): 296–317.
―――. *The Thirteenth General Election of India's Lok Sabha*. New York: The Asia Society, 1999.
Oommen, T.K. "Development policy and the nature of society: understanding the Kerala model." *Economic and Political Weekly*, 44, 13 (28 March 2009): 25–31.
Orren, Karen and Stephen Skowroneck. *The Search for American Political Development*. Cambridge: Cambridge University Press, 2004.
Overstreet, Gene D. and Marshall Windmiller. *Communism in India*. Berkeley: University of California Press, 1959.
Pai, Sudha. "The state, social justice, and the Dalit movement: the BSP in Uttar Pradesh." In Niraja Gopal Jayal and Sudha Pai (eds), *Democratic Governance in India: challenges of poverty, development and identity* (New Delhi: Sage, 2001), pp. 201–20.
―――. *Dalit Assertion and the Unfinished Democratic Revolution: the Bahujan Samaj Party in Uttar Pradesh*. New Delhi: Sage Publications, 2002.
―――. "The problem." *Seminar*, 571 (March 2007).
Pandey, Punam. "Revisiting the politics of the Ganga water dispute between India and Bangladesh." *India Quarterly*, 68, 3 (2012): 267–281.
Pandian, M.S.S. "Stepping outside history? new dalit writings from Tamil Nadu." In Partha Chatterjee (ed.), *Wages of Freedom: fifty years of the Indian nation-state*. New Delhi: Oxford University Press, 1998, 292–310.
Pant, Harsh V. *The US-India Nuclear Pact: policy, process, and great power politics*. New Delhi: Oxford University Press, 2011.
Patnaik, Prabhat. "In the aftermath of Nandigram." *Economic and Political Weekly*, 42, 21 (26 May 2007): 1893–5.

———. "The CPI(M) and the building of capitalism." International Development Economic Associates (30 January 2008). http://www.networkideas.org/news/jan2008/news30_Capitalism.htm

———. "Reflections on the Left." *Economic and Political Weekly*, 44, 28 (11 July 2009): 8–10.

———. "The left in decline." *Economic and Political Weekly*, 46, 29 (16 July 2011): 12–6.

Paul, Samuel and M. Vivekananda. "Holding a mirror to the new Lok Sabha." *Economic and Political Weekly*, 39, 45 (6 November 2004): 4927–34.

Paul, T.V. *Power versus Prudence: why nations forgo nuclear weapons*. Montreal and Kingston: McGill-Queens University Press, 2000.

Paun, Akash. "After the age of majority? multi-party governance and the Westminster model." *Commonwealth & Comparative Politics*, 49, 4 (November 2011): 440–456.

Pedersen, Jùrgen Dige. "India's industrial dilemmas in West Bengal." *Asian Survey*, 41, 4 (2001): 646–68.

Perkovich, George. *India's Nuclear Bomb: the impact of global proliferation*. Berkeley: University of California Press, 1999.

Pierson, Paul. *Politics in Time: history, institutions and social analysis*. Princeton: Princeton University Press, 2004.

Pridham, Geoffrey (ed.). *Coalitional Behaviour in Theory and Practice*. Cambridge: Cambridge University Press, 1986.

Przeworski, Adam. *Capitalism and Social Democracy*. Cambridge: Cambridge University Press, 1991.

Pulshikar, Suhas. "Regional and caste parties." In Atul Kohli and Prerna Singh (eds), *Routledge Handbook of Indian Politics*. New York: Routledge, 2013, pp. 91–104.

Puri, Balraj. *Kashmir: towards insurgency*. Hyderabad: Orient Longman, 1995.

Raghavan, E. and James Manor. *Broadening and Deepening Democracy: political innovation in Karnataka*. London: Routledge, 2009.

Raghavan, R.K. "The Indian police: expectations of a democratic polity." In Francine R. Frankel, Zoya Hasan, Rajeev Bhargava and Balveer Arora (eds), *Transforming India: social and political dynamics of democracy*. New Delhi: Oxford University Press, 2000, pp. 288–314.

Ragin, Charles C. "Turning the tables: how case-oriented research challenges variable-oriented research." In Henry E. Brady and David Collier (eds), *Rethinking Social Inquiry: diverse tools, shared standards*. New York: Rowman & Littlefield Publishers Ltd., 2004, p. 123–38.

Raja Mohan, C. *Crossing the Rubicon: the shaping of India's new foreign policy*. New York: Palgrave, 2004.

Rajeshekara, H. M. "The nature of Indian federalism: a critique." *Asian Survey*, 37, 3 (March 1997): 245–54.

Rangarajan, Mahesh. "Polity in transition: India after the 2004 general elections." *Economic and Political Weekly*, 40, 32 (6 August 2005): 3598–605.

BIBLIOGRAPHY

Rao, M. Govinda and Nirvikar Singh. *Political Economy of Federalism in India*. New Delhi: Oxford University Press, 2005.

Riker, William H. *The Theory of Political Coalitions*. New Haven: Yale University Press, 1962.

Rose, Leo E. "India's foreign relations: reassessing basic policies." In Marshall M. Bouton and Philip Oldenburg (eds), *India Briefing, 1990*. Boulder: Westview Press, 1990, pp. 51–76.

Roth, Paul A. "Narrative explanations: the case of history." In Michael Martin and Lee C. McIntyre (eds), *Readings in the Philosophy of Social Science*. Cambridge: MIT Press, 1995, pp. 701–13.

Roy, Ramashray. "Regional base and national dream: alliance formation, 2009 national elections." In Paul Wallace and Ramashray Roy (eds), *India's 2009 Elections: coalition politics, party competition, and Congress continuity*. New Delhi: Sage, 2011, pp. 21–41.

Rudolph, Lloyd I. and Susanne Hoeber Rudolph. *In Pursuit of Lakshmi: the political economy of the Indian state*. Chicago: University of Chicago Press, 1987.

———. "Iconisation of Chandrababu: sharing sovereignty in India's federal market economy." *Economic and Political Weekly*, 36, 18 (5 May 2001): 1541–1552.

———. "Redoing the constitutional design: from an interventionist to a regulatory state." In Atul Kohli (ed.), *The Success of India's Democracy*. Cambridge: Cambridge University Press, 2001, pp. 127–63.

Rudolph, Susanne Hoeber and Lloyd I. Rudolph. "The centrist future of Indian politics." *Asian Survey*, 20, 6 (June 1980): 575–94.

Runciman, David. *The Politics of Good Intentions: history, fear and hypocrisy in the new world order*. Princeton: Princeton University Press, 2006.

Ruparelia, Sanjay. "Rethinking institutional theories of political moderation: the case of Hindu nationalism in India, 1996–2004." *Comparative Politics*, 38, 3 (April 2006): 317–37.

———. "How the politics of recognition enabled India's democratic exceptionalism." *International Journal for Politics, Culture and Society—Special Issue on the Work of Charles Taylor*, 21, 4 (December 2008): 39–56.

———. "The role of judgment in explanations of politics." Political Methodology Working Paper 29 (May 2010), Committee on Concepts & Methods, International Political Science Association, www.concepts-methods.org.

———. "Expanding Indian democracy: the paradoxes of the third force," in Sanjay Ruparelia, Sanjay Reddy, John Harriss and Stuart Corbridge (eds), *Understanding India's New Political Economy: a great transformation?* London: Routledge, 2011, pp. 186–203.

Sáez, Lawrence. *Federalism without a Centre: the impact of political and economic reform on India's federal system*. New Delhi: Sage Publications, 2002.

Samuels, David. "Separation of powers." In Carles Boix and Susan C. Stokes (eds), *The Oxford Companion to Comparative Politics*. New York: Oxford University Press, 2009, pp. 703–26.

Sandbrook, Richard, Marc Edelman, Patrick Heller and Judith Teichman. *Social Democracy in the Global Periphery: origins, challenges, prospects*. Cambridge: Cambridge University Press, 2007.

Sanghvi, Vir. "The beginning of the end." *Seminar*, 461 (January 1998): 39–43.

Sarkar, Swagato. "Between egalitarianism and domination: governing differences in a transitional society." *Third World Quarterly*, 33, 4 (2012): 669–84.

Sarkar, Sumit. *Modern India 1885–1947*. Delhi: Macmillan Indian Limited, 1986.

Sarti, Roberto. "The dissolution of the Italian Communist Party." *The Marxist*, http://www.marxist.com/the-dissolution-of-the-italian-communist-party-1991.htm.

Sathe, S.P. *Judicial Activism in India: transgressing borders and enforcing limits, second edition, with a foreword by Upendra Baxi*. New Delhi: Oxford University Press, 2004.

Sathyamurthy, T.V. "Epilogue." In T.V. Sathyamurthy (ed.), *Class Formation and Political Transformation in Post-Colonial India*. New Delhi: Oxford University Press, 1996, pp. 465–475.

———. "Impact of Centre-state relations on Indian politics: an interpretative reckoning, 1947–84." In Partha Chatterjee (ed.), *State and Politics in India*. New Delhi: Oxford University Press, 1998, pp. 232–71.

Saxena, Rekha. *Situating Federalism: mechanisms of intergovernmental relations in Canada and India*. New Delhi: Manohar, 2006.

Schumpeter, Joseph A. *Capitalism, Socialism and Democracy*. London: Routledge, 1994.

Scott, James C. *Seeing like a State: how certain schemes to improve the human condition have failed*. New Haven: Yale University Press, 1998.

Sen, Amartya. *Rationality and Freedom*. Cambridge: Harvard University Press, 2004.

Sen, Suhit. "The Left rout: patterns and prospects." *Economic and Political Weekly*, 46, 24 (11 June 2011): 14–6.

Sengupta, Jayanta. "Liberalization and the politics of Oriya identity." In Niraja Gopal Jayal and Sudha Pai (eds), *Democratic Governance in India: challenges of poverty, development and identity*. New Delhi: Sage, 2001, pp. 179–200.

Shafiuzzaman, *The Samajwadi Party: a study of its social base, ideology and programme*. New Delhi: A.P.H. Publishing Corporation, 2003.

Shah, Ghanshyam. "The prime minister and the 'weaker sections of society'". In James Manor (ed.), *Nehru to the Nineties: the changing office of the prime minister of India*. London: Hurst, 1994, pp. 230–55.

———. "The BJP's riddle in Gujarat: caste, factionalism and *Hindutva*." In Thomas Blom Hansen and Christophe Jaffrelot (eds), *The BJP and the*

Compulsions of Politics in India, 2nd edition. New Delhi: Oxford University Press, 2001, pp. 243–266.

———. "Social backwardness and the politics of reservation." In Ghanshyam Shah (ed.), *Caste and Democratic Politics in India*. London: Anthem Press, 2004, 296–315.

Shapiro, Ian. "Problems, methods and theories in the study of politics, or: what's wrong with political science and what to do about it." *Political Theory*, 30, 4 (August 2002): 596–620.

Shastri, Sandeep. "Lok Shakti in Karnataka: regional party in bipolar alliance system." *Economic and Political Weekly*, 39, 14/15 (3 April 2004): 1491–6.

Sheingate, Adam D. "Political entrepreneurship, institutional change, and American political development." *Studies in American Political Development*, 17, 2 (Fall 2003): 185–203.

Sheth, D.L. "Secularisation of caste and making of new middle class." *Economic and Political Weekly*, 34, 34/35 (21 August 1999): 2502–10.

———. "Ram Manohar Lohia on caste in Indian politics." In Ghanshyam Sheth (ed.), *Caste and Democratic Politics in India*. London: Anthem Press, 2004, pp. 79–99.

———. "The change of 2004." *Seminar*, 545 (January 2005).

Singh, Mahendra Prasad. "India's National Front and United Front coalition governments: a phase in federalized governance." In Mahendra Prasad Singh and Anil Mishra (eds), *Coalition Politics in India: problems and prospects*. New Delhi: Manohar, 2004, pp. 75–99.

Singh, V.B. and Yogendra Yadav. *National Election Study, 1996, Post-Poll Survey*. Centre for the Study of Developing Societies, Delhi.

Sinha, Aseema. "The changing political economy of federalism in India: a historical institutionalist approach." *India Review*, 3, 1 (January 2004): 25–63.

Sisson, Richard. "India in 1989: a year of elections in a culture of change." *Asian Survey*, 30, 2 (February 1990): 111–25.

Sisson, Richard and Munira Majumdar. "India in 1990: political polarization." *Asian Survey*, 31, 2 (February 1991): 103–12.

Skocpol, Theda. *States and Social Revolutions: a comparative analysis of France, Russia and China*. Cambridge: Cambridge University Press, 1979.

Smith, Rogers M. "Science, non-science and politics." In Terrence J. McDonald (ed.), *The Historic Turn in the Human Sciences*. Ann Arbor: University of Michigan Press, 1996, pp. 119–48.

Sridharan, E. "Principles, power and coalition politics in India: lessons from theory, comparison and recent history." In D.D. Khanna and Gert W. Kueck (eds), *Principles, Power and Politics*. New Delhi: Macmillan India, 1999, pp. 270–91.

———. "Coalitions and party strategies in India's parliamentary federation." *Publius*, 33, 4 (Autumn 2003): 135–52.

———. "The fragmentation of the Indian party system, 1952–1999: seven competing explanations." In Zoya Hasan (ed.), *Parties and Party Politics in India*. New Delhi: Oxford University Press, 2004, pp. 478–504.

———. "Unstable parties and unstable alliances: births, splits, mergers and deaths of parties, 1952–2000." In Mahendra Prasad Singh and Anil Mishra (eds), *Coalition Politics in India: problems and prospects*. New Delhi: Manohar, 2004, pp. 43–74.

———. "Electoral coalitions in 2004 general elections: theory and evidence." *Economic and Political Weekly*, 39, 51 (18 December 2004): 5418–25.

———. "Why are multi-party minority governments viable in India? theory and comparison." *Commonwealth & Comparative Politics*, 50, 3 (July 2012): 314–43.

———. "Pre-electoral coalitions in 2014." Available at http://nottspolitics.org/2014/04/07/pre-electoral-coalitions-in-2014/.

Sridharan, E. and Ashutosh Varshney. "Toward moderate pluralism: political parties in India.' In Larry Diamond and Richard Gunther (eds), *Political Parties and Democracy*. Baltimore: Johns Hopkins University Press, 2001, pp. 206–38.

Srinivasulu, Karli. "Political mobilization, competitive populism, and changing party dynamics in Andhra Pradesh." In Paul Wallace and Ramashray Roy (eds), *India's 2009 Elections: coalition politics, party competition, and Congress continuity*. New Delhi: Sage, 2011, pp. 286–308.

Staniland, Paul. "Foreign policy making in India in the pre-liberalization and coalition era: unit level variables as determinants." In Amitabh Matoo and Happymon Jacob (eds), *Shaping India's Foreign Policy: people, politics and places*. New Delhi: Har-Anand Publications, 2010, pp. 255–74.

Stepan, Alfred, Juan Linz and Yogendra Yadav. *Crafting State-Nations: India and other multinational democracies*. Baltimore: Johns Hopkins University Press, 2011.

Stephens, John D. "Democratization and social policy development in the advanced capitalist societies." In Yusuf Bangura (ed.), *Democracy and Social Policy*. New York: Palgrave Macmillan, 2007, pp. 33–61.

Streeck, Wolfgang and Kathleen Thelen (eds). *Beyond Continuity*. Oxford: Oxford University Press, 2004.

Strøm, Kaare, Wolfgang C. Müller and Torbjörn Bergman (eds). *Cabinets and Coalition Bargaining: the democratic life cycle in western Europe*. Oxford: Oxford University Press, 2008.

Strøm, Kaare, and Benjamin Nyblade. "Coalition theory and government formation." In Carles Boix and Susan C. Stokes (eds), *The Oxford Handbook of Comparative Politics*. New York: Oxford University Press, 2009, pp. 782–802.

Subramanian, Narendra. *Ethnicity and Populist Mobilization: political parties, citizens and democracy in south India*. New Delhi: Oxford University Press, 1999.

Suri, K.C. "The economy and voting in the 15th Lok Sabha elections." *Economic and Political Weekly*, 44, 39 (September 26, 2009): 64–70.

Swaminathan, Madhura. *Weakening Welfare: the public distribution of food in India*. New Delhi: LeftWord, 2000.

BIBLIOGRAPHY

Swann, Abram de. *Coalition Theories and Cabinet Formations: a study of formal theories of coalition formation applied to nine European parliaments after 1918.* New York: Elsevier Scientific Publishing Company, 1973.

Talbott, Strobe. *Engaging India: diplomacy, democracy, and the bomb.* Washington, D.C.: The Brookings Institution, 2006.

Tarrow, Sidney. "Transforming enemies into allies: non-ruling communist parties in multiparty coalitions." *The Journal of Politics*, 44, 4 (November 1982): 924–954.

Tawa Lama-Rewal, Stéphanie. "The resilient *bhadralok*: a profile of the West Bengal MLAs." In Christophe Jaffrelot and Sanjay Kumar (eds), *Rise of the Plebeians? The changing face of Indian legislative assemblies.* New Delhi: Routledge, 2009, pp. 361–92.

Taylor, Charles. "Neutrality in political science." In Alan Ryan (ed.), *The Philosophy of Social Explanation.* Oxford: Oxford University Press, 1973, pp. 139–71.

Thachil, Tariq. *Elite Parties, Poor Voters: how social services win votes in India.* Cambridge: Cambridge University Press, forthcoming/October 2014.

Thakur, Ramesh. *The Government and Politics of India.* London: MacMillan Press, 1995.

———. "A changing of the guard in India." *Asian Survey*, 38, 6 (June 1998): 603–623.

Thapliyal, Sangeeta. "India and Nepal treaty of 1950: the continuing discourse." *India Quarterly*, 68, 2 (2012): 119–133.

Thomas, E.J. *Coalition Game Politics in Kerala.* New Delhi: Intellectual Publishing House, 1985.

Tilly, Charles and Robert E. Goodin. "It depends." In Robert E. Goodin and Charles Tilly (eds), *The Oxford Handbook of Contextual Political Analysis.* New York: Oxford University Press, 2008, pp. 3–34.

Vaishnav, Milan and Danielle Smogard. "A new era in Indian politics?" Available at http://carnegieendowment.org/2014/06/10/new-era-in-indian-politics/hdc6.

Vakulabharanam, Vamsicharan and Sripad Motiram. "Political economy of agrarian distress in India since the 1990s." In Sanjay Ruparelia, Sanjay Reddy, John Harriss and Stuart Corbridge (eds), *Understanding India's New Political Economy: a great transformation?* London: Routledge, 2011, pp. 101–26.

Vanaik, Achin. *The Painful Transition: bourgeois democracy in India.* London: Verso, 1990.

———. *India in a Changing World.* New Delhi: Orient Longman, 1995.

———. *The Furies of Indian Communalism: religion, modernity and secularization.* London: Verso, 1997.

———. "Indian foreign policy since the end of the Cold War: domestic determinants." In Sanjay Ruparelia, Sanjay Reddy, John Harriss and Stuart Corbridge (eds), *Understanding India's New Political Economy: a great transformation?* London: Routledge, 2001, pp. 221–36.

BIBLIOGRAPHY

———. "Subcontinental strategies." *New Left Review*, II/70 (July–August 2011): 101–14.

Vanderbok, William G. "The tiger triumphant: the mobilization and alignment of the Indian electorate." *British Journal of Political Science*, 20, 2 (April 1990): 237–61.

Vannicelli, Maurizio. "Introduction to part two: alliance policy." In Peter Lange and Maurizio Vannicelli (eds), *The Communist Parties of Italy, France and Spain: postwar change and continuity—a casebook*. Cambridge: George Allen & Unwin, 1991, pp. 103–06.

Varshney, Ashutosh. "Contested meanings: India's national identity, Hindu nationalism, and the politics of anxiety." *Daedalus*, 122, 3 (1993): 227–61.

———. *Democracy, Development and the Countryside: urban-rural struggles in India*. Cambridge: Cambridge University Press, 1995.

———. "'Mass politics' or 'elite politics?' India's economic reforms in comparative perspective." In Jeffrey Sachs, Ashutosh Varshney and Nirupam Bajpai (eds), *India in the Era of Economic Reforms*. New Delhi: Oxford University Press, 1999, pp. 222–261.

Verma, A.K., Mirza Asmer Beg and Sudhir Kumar. "A saffron sweep in Uttar Pradesh." *The Hindu*, 23 May 2014.

Verma, Manju. *The Coalition Ministries in Punjab*. Patiala: Shivalik Printing House, 1978.

Véron, René. "The 'new' Kerala model: lessons for sustainable development." *World Development*, 29, 4 (April 2001): 601–17.

Watkins, J. W. N. "Ideal types and historical explanation." In Alan Ryan (ed.), *The Philosophy of Social Explanation*. Oxford: Oxford University Press, 1973, pp. 82–105.

Weber, Max. 'The profession and vocation of politics.' In Peter Lassman and Ronald Speirs (eds), *Weber: political writings*. Cambridge: Cambridge University Press, 1994, pp. 309–70.

Weiner, Myron. *Party Politics in India: the development of a multi-party system*. Princeton: Princeton University Press, 1957.

———. "The 1971 elections and the Indian party system." *Asian Survey*, 11, 12 (December 1971): 1153–66.

———. "India's new political institutions." *Asian Survey*, 16, 9 (September 1976): 898–901.

———. "The 1977 parliamentary elections in India." *Asian Survey*, 17, 7 (July 1977): 619–26.

———. "Congress restored: continuities and discontinuities in Indian politics." *Asian Survey*, 22, 4 (April 1982): 339–55.

———. "The struggle for equality: caste in Indian politics." In Atul Kohli (ed.), *The Success of India's Democracy*. Cambridge: Cambridge University Press, 2001, pp. 193–226.

Weyland, Kurt. *Bounded Rationality and Policy Diffusion: social sector reform in Latin America*. Princeton: Princeton University Press, 2007.

Whelpton, John. *A History of Nepal*. Cambridge: Cambridge University Press, 2005.

Widmalm, Sten. "The rise and fall of democracy in Jammu and Kashmir: 1975–1989." In Amrita Basu and Atul Kohli (eds), *Community Conflicts and the State*. New Delhi: Oxford University Press, 2000, pp. 149–82.

Williams, Bernard. "Realism and moralism in political theory." In Geoffrey Hawthorn (ed.), *In the Beginning Was the Deed: realism and moralism in political argument*. Princeton: Princeton University Press, 2005.

Witsoe, Jeffrey and Francine Frankel. *Social Justice and Stalled Development: caste empowerment and the breakdown of governance in Bihar*. Philadelphia: Center for the Advanced Study of India, University of Pennsylvania, 2006.

Wood, John R. "On the periphery but in the thick of it: some recent Indian political crises viewed from Gujarat." In Philip Oldenburg (ed.), *India Briefing: Staying the Course*. Armonk: M.E. Sharpe, 1995, pp. 23–46.

Wyatt, Andrew. "The limitations on coalition politics in India: the case of electoral alliances in Uttar Pradesh." *Commonwealth & Comparative Politics*, 37, 2 (July 1999): 1–21.

Yadav, Yogendra. "Reconfiguration in Indian politics: state assembly elections, 1993–1995." In Partha Chatterjee (ed.), *State and Politics in India*. New Delhi: Oxford University Press, 1997, pp. 177–208.

———. "The third electoral system." *Seminar*, 480 (August 1999): 14–20.

———. "Open contest, closed options." *Seminar*, 534 (February 2004): 62–9.

———. "The elusive mandate of 2004." *Economic and Political Weekly*, 39, 51 (18 December 2004): 5383–95.

Yadav, Yogendra and Sanjay Kumar. "Interpreting the mandate." *Frontline*, 5 November 1999.

Yadav, Yogendra and Suhas Palshikar. "From hegemony to convergence: party system and electoral politics in the Indian states, 1952–2002." *Journal of Indian School of Political Economy*, XV, 1&2 (January–June 2003): 5–44.

———. "Principal state level contests and derivative national choices: electoral trends in 2004–9." *Economic and Political Weekly*, 44, 6 (7 February 2009): 55–62.

———. "Between fortuna and virtu: explaining the Congress' ambiguous victory in 2009." *Economic and Political Weekly*, 44, 39 (26 September 2009): 33–51.

Ziegfeld, Adam. "Coalition government and party system change: explaining the rise of regional political parties in India." *Comparative Politics*, 45, 1 (October 2012): 69–87.

INDEX

aam aadmi (common man) 300
Aam Aadmi Party (AAP) 341–2
Abdullah, Farooq 92, 94, 202, 203, 222
Abdullah, Sheikh Mohammed 79–80
Acharya, Shankar 176
Adivasis 46, 140, 154, 157, 177, 215, 270, 295, 316
Advani, Lal Krishna 74, 97, 100, 120, 126, 159, 309
Agra, Uttar Pradesh 295
Agribusiness Vision 2010 305
Agricultural and Rural Debt Scheme (1990) 120
agriculture 47, 50, 55, 59, 61, 64, 68, 69, 81, 90, 117, 127, 133, 135, 139, 174, 176, 187, 197, 237, 246, 247, 305, 331
Ahirs 59
Ahluwalia, Montek Singh 176
Ahmadi, Aziz Mushabber 228
Ahmed, Waseem 203
AJGAR (Ahirs, Jats, Gujars, Rajputs) 73, 104
Akali Dal 58, 97
Akhil Bharatiya Vidyarthi Parishad (ABVP) 63
alcohol taxes 174

All Assam Students Union (AASU) 90, 96, 227
All Bodo Students Union 226
All-India Agricultural Workers Union 165
All-India Congress (I) Committee (AICC) 130, 188, 211, 213, 233, 257
All-India Democratic Women's Association 165
All-India Forward Bloc (AIFB) 136, 144, 158, 188, 290
All-India Indira Congress-Tiwari (AIIC-T) 131, 161, 187, 188, 224, 229
All-India Kisan Sabha (AIKS) 52, 165
All-India Radio and Doordarshan 120, 272
All-India Railwaymen's Federation 64
All-India Trade Union Congress (AITUC) 47
All-India Trinamool Congress (AITC) 278, 280, 281, 285, 297, 298, 305, 315, 316, 342
All-Indian Anna Dravida Munnetra Kazhagam (AIADMK) 65, 95, 105, 132, 147, 162, 168, 273,

INDEX

274, 277, 279, 280, 281, 285, 290, 294, 297, 298, 307, 311, 315, 342
All-Party Hurriyat Conference (APHC) 203, 223
Allahabad, Uttar Pradesh 64, 99
Allende, Salvador 336
Almatti dam 206–8
Alva, Margaret 168
Ambedkar Udyan Complex 271
Ambedkar Village Scheme 249, 270
Ambedkar, Bhimrao Ramji 117, 270
Amritsar, Punjab 55, 94, 114
Andhra Pradesh 49, 56, 59, 62, 69, 70, 91–3, 95, 143–4, 149, 153, 158, 166, 172, 174, 206–8, 232, 238, 259, 277, 280, 284, 285, 295–6, 300, 306, 311–12, 315, 323, 342
animal husbandry 187
animal husbandry scam *see* fodder scam
Anjaiah, Tanguturi 91
anti-Americanism 21, 138, 304
anti-Brahmanism 57
anti-capitalism 31, 50, 108, 340
anti-caste 53
anti-colonial movement 9, 36, 49
anti-colonialism 129
anti-communism 55
Anti-Defection Law (1985) 17, 95, 97, 334
anti-Hindi 57
anti-imperialism 31, 50, 52–3, 108, 138, 190, 289, 301, 304, 308–9, 315, 340
anti-Muslim pogrom (2002) 3, 13, 296–7, 342
anti-Sikh pogrom (1984) 94, 96, 114
Antyodaya (food for work scheme) 81
Approach Paper to the Eighth Five-Year Plan (1990–95) 116

Approach Paper to the Fifth Five Year Plan (1997–2002) 238
Archimedes 41
Armed Forces Special Powers Act (AFSPA) 113, 227
Arunachal Pradesh 149
ASEAN (Association of Southeast Asian Nations) 272
Ashok Mehta Committee 79
Asom Gana Parishad (AGP) 96, 98, 99, 105, 107, 110, 143, 144, 158, 161, 162, 166, 167, 168, 171, 227, 277, 282, 284, 320
Assam 62, 90, 96, 97, 105, 144, 149, 153, 158, 172, 207, 227, 277, 282, 295
Atomic Energy Act (1954) 307
Australia 26, 308
Austria 307, 327
Awadi declaration 54, 55
Awami League 203
Ayodhya, Uttar Pradesh 2, 97, 99, 120–2, 125–6, 129, 133, 167, 169, 177, 248, 281, 287, 290, 291, 296, 321

Babbar, Raj 306
Babri masjid 2, 97, 99, 120–2, 125–6, 129, 133, 177, 248, 291
Bachawat Tribunal Award (1976) 206, 207
Backward Classes Commission 83
Badal faction 159
Bahuguna, Hemvati Nandan 89
Bahujan Samaj Party (BSP) 120, 140–1, 154, 188–9, 208, 224–5, 248–50, 269–71, 279, 280, 290, 302, 306, 309, 311–12, 314, 332, 334, 342
Baishya, Birendra Prasad 184
Bajrang Dal 97
balance of payments 116, 121, 126
Balayogi, Ganti Mohana Chandra 284, 297
bandh (strike) 233

INDEX

Banerjee, Mamata 278, 280, 305, 316
Bangalore, Karnataka 93, 98, 205
Bangarappa, Sarekoppa 229
Bangla Congress 59
Bangladesh 12, 82, 214, 218, 230–1, 324
Bangladesh Nationalist Party 231
Bardhan, Ardhendu Bhushan 302, 317
bargaining, bargaining complexity 7, 19, 22–3, 25
Basapunniah, Makineni 52
Basu, Jyoti v, 12, 36, 56, 71–2, 91, 99, 101, 107, 119, 138, 162–4, 167, 172, 179, 197, 231, 241–2, 246, 250, 254, 290, 291, 296, 305, 311, 317, 337, 338, 340
Belchi, Bihar 78
Belgium 26
benami 60
Bhagat, Hari Krishan Lal 129
Bhagirathi-Hooghly river 231
Bhandari, Romesh 209, 225, 226, 248, 270
Bharat 331
Bharat Ratna 117
Bharatiya Janata Party (BJP) 1–4, 11, 13, 18, 27, 28, 32, 315, 320, 321, 326, 331, 333, 337, 338, 342, 343
  1980–1989 45, 89–101 89, 95, 97, 99, 100, 101
  1989–1991 104–5, 107, 109, 110, 113, 120–2
  1991–1996 125, 133–5, 142, 144–5, 149, 153–7, 159, 160, 162, 165, 168–70, 178, 182, 184, 190–1, 203, 219, 222, 224–6
  1997–1998 237–8, 243, 246, 248–50, 252, 254, 259, 261, 265, 269, 270, 271, 272, 275, 277, 279–85, 287
  1998–2012 288, 290–303, 307–12
Bharatiya Kisan Kamgar Party (BKKP) 224, 279
Bharatiya Kranti Dal (BKD) 59, 73, 77
Bharatiya Lok Dal (BLD) 67, 70, 71, 73, 74, 83, 84, 85, 86, 337
Bharatvarsha 125
Bhattacharya, Buddhadeb 305, 317
Bhave, Gandhian Vinoba 53, 54
Bhindranwale, Sant Jarnail Singh 94
bhoodan movement 54
Bhutan 218
Bhutto, Benazir 113, 176, 191, 204, 223
Bihar 58, 64, 69, 73, 77–8, 84, 103, 105, 110, 113, 120, 134, 141–2, 144, 148–9, 153–4, 157–8, 167, 172, 182–3, 187, 207, 218, 228, 231–3, 253, 258–61, 269, 274, 276, 278–80, 290, 292, 295–6, 300, 311–12, 334, 342
Bihar State Struggle Committee of Workers and Employees Against Price Rise and Professional Tax 63
Biju Janata Dal (BJD) 276, 311, 334
Biswas, Upendranath 228
Bloc Quebecois 330
Bodo 97, 105, 226–7, 295
Bodo Territorial Council 295
Bodoland State Movement Council 226
Bofors scandal 98, 236–7, 248, 257
Bombay riots (1992–1993) 170
Bombay, Maharashtra 191
Bommai, Somappa Rayappa 167, 172, 174, 177, 221
Border Security Forces 66
Brahmin-Dalit alliance 63
Brahmins 57, 63, 77, 172, 276
Brazil 340
Broadcasting Corporation of India (Prasar Bharati) Bill 272
Bush, George Walker 304, 307

INDEX

Cabinet Mission Plan (1946) 49
Calcutta, West Bengal 98, 99, 108, 197, 231, 265
Camdessus, Michel 247
Canada 26, 327, 330
Canadian Alliance 330
capitalism 2, 3, 9, 14, 31, 50, 55, 88, 91, 108, 143, 169, 245, 316, 320, 336, 338–40
Capitalism and Social Democracy (Przeworksi) v
Cardoso, Fernando Henrique v
Carter, James Earl "Jimmy" 82
caste 1, 2, 4, 6, 9, 20–1, 27, 46, 53–4, 57, 61, 63–4, 70–1, 73, 77–8, 83–4, 86, 88, 106, 109, 117–20, 125, 134–5, 139–42, 154, 165, 167, 177, 189, 215, 270, 292, 294–5, 312, 322, 326, 331–2, 334
   Brahmins 57, 63, 77, 172, 276
   Dalits 46, 57, 63, 70–1, 77–8, 84, 86, 88, 89, 117, 120, 135, 140–1, 154, 165, 172, 177, 189, 224, 226, 270, 294, 316, 322, 334
   jatis 141, 334, 335
   Kshatriyas 63, 142
   shudra castes 57, 61, 83, 166
   varna classifications 141
Cauvery River Water Authority 272
Cauvery Waters Dispute Tribunal 114, 204–6, 323
Central Bureau of Investigation (CBI) 130–1, 174, 179, 187, 188, 209, 210, 211, 228, 233, 236, 248, 257, 259, 261
Central Industrial Security Forces 66
Central Intelligence Agency (CIA) 64
Central Pay Commission 137, 237, 266–7, 268, 338
Central Reserve Police 66
Central Water Commission 207, 208

Centre for the Study of Developing Societies (CSDS) 40, 149–52, 156, 192, 194–5, 199, 201, 216–17, 220, 239, 240, 314
Centre of Indian Trade Unions (CITU) 72
Centre-state relations 1, 5, 10, 11, 12, 13, 22, 113, 114, 144, 181, 198, 187, 214, 219, 235, 267–72, 321, 323
Chadha, Win 236
Chakma 231
Chamars 140, 334
Chand, Lokendra Bahadur 263
Chandra, Ramesh 189
Chandraswami 209
Chatra Sangharsh Samiti (CSS) 63
Chatterjee, Somnath 302
Chautala, Om Prakash 118, 141
Chavan, Yashwantrao Balwantrao 85
Chennai, Tamil Nadu 58, 110, 205
Chhattisgarh 294
Chidambaram, Palaniyappan 174, 176, 178, 193, 196, 197, 214, 233, 238, 242–7, 266, 268, 272
Chief Ministers' Conference 93
Chile 336
China 31, 50, 51, 53, 55, 56, 81, 82, 83, 114, 115, 190, 219, 230, 246, 288, 304, 306, 307–8, 337, 340
   1959 invasion of Tibet; Sino-Indian border clash 53, 55
   1962 Sino-Indian war 55
   1976 resumption of diplomatic links with India 82
   1978 Atal Bihari Vajpayee makes visit to Peking 82
   1979 Sino-Vietnamese war 82
   1989 Tiananmen Square massacre 105
   1993 Peace and Tranquility Agreement with India 230
   1995 sale of nuclear ring magnets to Pakistan 190

INDEX

1996 Jiang Zemin makes state visit to India 230
Chowdhury, Renuka 261
Christian Democracy (Italy) 336
Christianity 3, 135, 177, 290, 311
Churchill, Winston 65
Civil Liability for Nuclear Damages Bill (2010) 310
civil rights violations 62, 70, 71
civil society 58, 91
class 1, 2, 4, 9, 19, 20, 31, 53, 57, 64, 106, 109, 331; *see also* caste
class struggle 49, 53, 58–9, 63, 139–40, 332
Clinton, William Jefferson "Bill" 227
coalition attributes model 24
coalition theory 15, 22, 23
colonialism 9, 36, 49, 129
Comintern 50
Commission on Agricultural Costs and Prices (CACP) 115–16
Common Minimum Programme (CMP) 163, 176–9, 196, 215, 233, 301, 303
communal violence 2, 3, 11, 60, 83–4, 99, 104, 120, 290, 296–7, 311
communalism 64, 74, 85, 93–4, 125, 135, 141, 219, 248, 343
communism 2, 3, 4, 6, 7, 8, 10, 13, 14, 18, 21, 27, 31, 320, 321, 331–3, 335–42
 1934–1977 50–60, 65, 66
 1977–1980 71–2, 77, 85
 1980–1989 90, 91, 99–101
 1989–1991 108, 116, 119–20
 1991–1996 136–9, 144, 158, 162–7, 175–6, 178–9, 182, 187–8, 193, 196–7, 209, 233
 1997–1998 238, 241, 245–6, 250, 268, 278, 282, 285
 1998–2012 290, 301–2, 304, 308–9, 314–17
Communist Party of India (CPI) 11, 49–56, 58, 59–62, 65, 70, 86, 90, 100, 105, 136, 145, 158, 175–6, 184, 187–8, 193, 196, 209, 224, 238, 245, 259, 265, 272, 277, 278, 282, 290–1, 301–2, 321, 333, 336–8
Communist Party of India–Maoist 305
Communist Party of India–Marxist (CPI-M) 7, 11, 12, 13, 32, 33, 36, 320, 322, 333, 336–42
 1934–1977 56, 58, 59–61
 1977–1980 71–2, 77, 87
 1980–1989 90, 91, 99, 101
 1989–1991 105, 108, 119, 122
 1991–1996 136–9, 144, 145, 158, 162–7, 172, 175–6, 178–9, 182, 188, 193, 196–7, 200, 210, 213, 225, 231–3
 1997–1998 237, 241–2, 245–6, 250, 251, 254, 266, 267, 271, 272, 277, 278, 282, 284, 285, 287
 1998–2012 289, 290–1, 301–2, 304, 308–9, 311, 314–17
Communist Party of India–Marxist-Leninist (CPI-ML) 60, 278
Communist Party of the Soviet Union (CPSU) 53
Companies Act 244
comparativists 15, 18, 22–5, 31, 36, 39, 40, 329
Comprehensive Test Ban Treaty (CTBT) 133, 176, 190, 191, 288
Comptroller and Auditor General (CAG) 228, 233
Congentrix 184
Congress (I) 320, 321, 324, 326, 331, 338, 339, 341, 342
 1977–1980 83, 86–88
 1980–1989 89–101
 1989–1991 103, 105–8, 114, 116, 121–2
 1991–1996 125–31, 141, 143–5, 147–9, 153–4, 157, 159–62,

INDEX

164, 167–71, 177, 179–89, 193, 198, 200, 206, 208–11, 213, 219, 224, 226–9, 232
1997–1998 235–7, 247, 250–4, 257, 261, 267, 271–5, 277–9, 282, 284–5
1998–2012 289, 290, 296, 297, 298, 300, 302–3, 307, 308, 309, 311, 312, 315
Congress (I) Parliamentary Party (CPP) 130, 226, 233, 251, 273
Congress for Democracy (CFD) 71, 73, 74
Congress Parliamentary Party 160
Congress *see* Indian National Congress
Congress Socialist Party (CSP) 49, 51, 52, 333
Congress system 45–61
Congress Working Committee (CWC) 160, 188, 208, 211, 225, 226, 229, 233, 237, 251, 252, 257, 274
Conservative Party (UK) 330
Constitution
  38th amendment 66
  39th amendment 66
  42nd amendment 66, 67, 79
  43rd amendment 79
  44th amendment 79
  52nd amendment 95
  65th amendment 117
  67th amendment 120
  73rd amendment 215
  74th amendment 215
  81st amendment 214
  Article *180* 221
  Article *263* 114
  Article *352* 67, 79
  Article *356* 68, 79, 80, 90, 93, 106, 177, 222, 226, 229, 261, 269, 270
  Article *370* 80, 100, 113, 133, 169, 177, 225, 281, 287, 290
Constitution (65th Amendment) Bill (1990) 117
Constitution (67th Amendment) Bill (1990) 120
Constitution (81st Amendment) Bill (1996) 214–15, 235
Constitutional Amendment Act 95
constitutional reform 1
cooperative development 1
corruption 10, 57, 64, 65, 95, 98, 101, 163, 179, 208–11, 218, 223, 233, 236–7, 257–8, 341
Council of Ministers 27, 32, 77, 110, 119, 159, 169, 172, 181–2, 184, 188–9, 193, 215, 258, 264, 266, 270, 271, 274, 337–8
crorepatis 302
Cultural Nationalism 125, 136

dairy 187
dalitbahujan 154, 189, 249, 335
Dalits 46, 57, 63, 70–1, 77–8, 84, 86, 88, 89, 117, 120, 135, 140–1, 154, 165, 172, 177, 189, 224, 226, 270, 294, 316, 322, 334
Dandavate, Madhu 70, 85, 110, 115, 117, 260
Dar, Maqbool 187, 203, 258
Davos *see* World Economic Forum
debt, deficit 64, 100, 106, 116, 120–1, 136, 237, 243–4, 247, 269
decentralization 1, 50, 64, 71, 80, 81, 92, 115, 116, 137, 182, 198, 302, 306, 322, 338
Defence of India Rules 56
Delhi 149, 153
Delhi High Court 210–11, 236, 260
demobilization 300, 336
democratic centralism 54
Democratic Socialist Party 89, 95
Deng Xiaoping 246
Department of Animal Husbandry and Dairying 187
Dependency and Development in Latin America (Cardoso and Faletto) v

INDEX

Desai, Kanti 78
Desai, Morarji 61, 73, 77, 80, 81, 82, 83, 84, 85, 119, 121, 274, 320, 335
Deva, Narendra 49
Deve Gowda, Haradanahalli Doddegowda 11, 12, 166–8, 170, 172, 174, 175, 179, 181, 184–8, 191, 196, 198, 200, 202–10, 213, 224–8, 232, 236, 242, 245, 248, 250–3, 260, 292, 301, 312, 335, 338
Development of Women and Children in the Rural Areas Scheme 130
developmentalism 46
Devi, Rabri 261, 269
devolution 1, 14, 64, 71, 80, 91–4, 115, 144, 161, 177, 198, 200, 202, 322, 330
Dhabi, Chandubhai 221
Dharia, Mohan 66
dharma 13, 289
Directive Principles of State Policy 47
Disinvestment Commission 133, 178, 197, 265
Dixit, Sheila 229
Downs, Anthony 19
Dravida Munnetra Kazhagam (DMK) 57–8, 61, 65, 70–1, 99, 105, 107, 110, 132, 143, 144, 158, 161, 162, 168, 171, 204–6, 214, 226, 232, 245, 252, 254, 261, 273–5, 277, 280, 282, 284, 290, 294, 295, 297, 300, 301, 320
Dravidianism 57, 277
Duverger's law 25

e-governance 295
Eastern Europe 31
Economic and Political Weekly 314
economic liberalization 4, 10, 11, 12, 13, 28, 84, 95, 96, 98, 108, 115–15, 126–45, 161, 165, 174, 177, 193–8, 208, 237–43, 292, 295–6, 303, 308, 322, 339
economic reform 242–7, 264–9
Economic Survey 243
education 46, 51, 53–4, 83, 91, 96, 106, 117, 120, 133, 137, 142, 154, 157, 197, 238, 298, 305–6, 322, 332
Eighteenth Brumaire of Louis Bonaparte, The (Marx) v
Election Commission (EC) 48, 68, 69, 70, 104, 128, 129, 131, 257, 271, 283–4, 293–4, 299–300, 314
elections
  1937 provincial elections 49, 52
  1951 Tamil Nadu state assembly elections 57
  1951–1952 general election 45
  1957 Kerala state assembly election 51, 53
  1962 general election 54, 55
  1967 general election 45, 56, 57
  1977 general election 66, 67, 79, 80
  1978 Karnataka by-elections 83
  1980 general election 87
  1983 Andhra Pradesh state assembly elections 92
  1983 Jammu & Kashmir state assembly elections 92
  1984 general election 10, 29, 94, 98
  1987 Jammu & Kashmir state assembly elections 98
  1988 Uttar Pradesh local elections 100
  1989 general election 6, 7, 16, 20, 29, 30, 101, 103, 109, 183
  1989 Karnataka state assembly elections 183
  1989 Tamil Nadu state assembly elections 105
  1990 Meham by-election 119

457

INDEX

1991 general election 121–2, 141, 209
1993 Uttar Pradesh state assembly election 140
1994 Andhra Pradesh state assembly elections 158
1993–1995 assembly elections 126–7
1996 general election 2, 129, 138, 147–58, 191, 193, 208
1996 Jammu & Kashmir state assembly elections 200–4, 222–3
1996 Uttar Pradesh state assembly election 188, 223–6
1997 Punjab state assembly elections 237
1998 general election 13, 275–85
1999 general election 3, 291–4
2002 Jammu & Kashmir state assembly elections 223, 295
2004 general election 3, 298
2007 Uttar Pradesh state assembly elections 306
2009 general election 3, 310–17
2011 Kerala and West Bengal state assembly elections 3
2013 Delhi assembly elections 342
2014 general election 3, 27, 342
electricity 206–8, 262, 263
elite politics 126
Emergency (1975–1977) 1, 9, 10, 64–6, 70, 71, 74, 78, 79, 100, 107, 115, 175, 320, 323, 326, 337
Employment Guarantee Bill 120
Engels, Friedrich 241
Eurocommunism 14, 336
Europe 14, 17, 23, 26, 31, 37, 327, 329, 331, 335–6
executive power 6, 7, 13, 24, 66, 74, 77, 171, 175, 182, 259, 271, 288, 321, 323, 328, 337–8

Fabian socialism 49

Faletto, Enzo v
Farraka dam 82, 231
fascism 65, 331, 337
Federal Front 171
federal nationalism 5, 325
federal system 1–2, 5–6, 10–11, 13–14, 25–9, 46, 49, 62–3, 71, 79, 92, 95, 106–7, 114, 127, 141–5, 177, 226, 316, 325, 326, 334, 339, 341
Federation of Indian Chambers of Commerce and Industry (FICCI) 214
Fernandes, George 64, 81, 107, 113, 142, 288
Fifth Central Pay Commission Report 137, 237, 266–7, 268, 338
Finance Commission 93, 97, 243, 269, 295
Finance Ministry 272
first-past-the-post (FPTP) 5, 6, 23, 25–6, 27, 29, 46, 67, 138, 149, 277, 325, 326, 343
Fissile Material Control Regime (FMCR) 133
Five-Year Plans 46, 81, 322
  Sixth (1978–1983) 81
  Fifth (1974–1979) 81
  Eighth (1992–1997) 322
fodder scam 187, 218, 228, 233, 259
food aid 57
food subsidies 174
forced sterilization 70
foreign direct investment (FDI) 135
Foreign Exchange Management Act (FEMA) 266
Foreign Exchange Regulatory Act (FERA) 244
Foreign Investment Promotion Board 233
Foreign Investment Promotion Council 214
foreign policy 5, 11, 20, 21, 22,

113–15, 181, 191–3, 218–19, 230, 249, 262–4, 307–10, 321, 325
Fotedar, Makhan Lal 229
Fourth Front 311
France 241, 308, 310, 336
Freedom of Religion Bill (1978) 85

G-15 summit
  1996 Zimbabwe 232
  1997 Malaysia 272
Gadgil, V.N. 236
game theory 37
Gandhi, Indira 1, 55, 57, 61–6, 71, 78, 79–80, 81, 82, 83, 86, 87, 90, 91, 93–4, 107, 119, 129, 175, 273, 326, 343
Gandhi, Maneka 184
Gandhi, Mohandas Karamchand 49, 52, 53, 64, 219, 332
Gandhi, Rajiv 94, 95–6, 97, 98, 101, 103, 115, 121, 122, 130, 132, 230, 236, 257, 273–4
Gandhi, Sanjay 66, 86, 115
Gandhi, Sonia 130, 131, 211, 229, 236, 236, 257–8, 274, 277, 278, 282, 291, 292, 298, 301, 304, 310, 341
Ganga river 82, 88, 107
Ganga Waters Accord (1996) 230, 231, 324
garibi hatao (abolish poverty) 62
Geelani, Syed Shah 223
General Agreement on Tariffs and Trade (GATT) 174
Geneva, Switzerland 190
Germany 26, 52, 53, 98, 327
gheraos 59
Giri, Varahagiri Venkata 61
globalization 133, 214, 219, 298
Goa 149, 161, 209, 342
Godhra, Gujarat 297–8
Golden Temple, Amritsar 94
"*goonda raj*" (rule of thugs) 306, 335

Gopalaswamy, Vaiko 144, 278
Gramsci, Antonio 139, 333
Green Party (UK) 330
Green Revolution 57, 322, 331
gross fixed investment 296
Gujarat 3, 12, 13, 63, 65, 104, 105, 141, 149, 153, 200, 214, 219, 221, 279, 290, 295–7, 300, 305
Gujarat Janata Parishad 221
Gujral, Inder Kumar 12, 115, 174, 176, 231, 232, 249, 250–4, 258–60, 262–4, 266, 268, 270–6, 289, 325, 335, 338
Gujral doctrine 12, 218–19, 263, 287, 289, 325
Gulf War (1990–1991) 121
Gupta, Indrajit 184, 187, 203, 204, 225, 248, 251, 260, 202, 208–9, 267, 268, 270, 338

harijan 120
Haryana 58, 77, 84, 87, 96, 104, 118, 134, 153, 157, 280, 305
Haryana Janhit Congress (HJC) 311
Haryana Lok Dal (Rashtriya) 141
Haryana Vikas Party (HVP) 134, 153, 157, 280, 281, 285
Hasina, Sheik 230
hawala scandal 130–2, 167, 172, 209, 258
HDW (Howaldtswerke-Deutsche Werft) submarines 98
healthcare 91, 117, 137, 154, 197, 244, 261, 303, 305, 322, 332
Hegde, Ramakrishna 92, 107, 120, 174, 182, 184, 261, 276, 280
Henry J. Hyde United States-India Peaceful Atomic Energy Cooperation Act (2006) 307, 308
Hi-Tec City 295
Himachal Pradesh 77
Hindi 1, 57, 69, 73, 88, 103, 105, 153, 253, 279, 310, 317, 321
Hinduism 83, 90, 94, 97, 219

459

chauvinism 58, 60, 73, 85, 285
extremism 99, 169, 311
nationalism 1–3, 11, 13, 18, 21, 34, 45–6, 49, 58, 63, 100, 126, 136, 159–60, 177, 277, 280, 288–9, 296–7, 311, 341, 343
rashtra 109
revivalism 280
right 2, 10, 13, 58, 59, 60, 71, 73, 77, 80, 85, 89, 97, 99, 100, 101, 104, 107, 121, 125–6, 320, 321, 333
Hinduja brothers 236
Hindutva 2, 11, 120, 125, 133, 134, 138, 153, 154, 282, 285, 287, 289, 290, 298, 320, 321, 337
Hitler, Adolf 52
Home Guards 66
Howitzer field guns 98
Hungary 55
Hurriyat *see* All-Party Hurriyat Conference (APHC)
Hyde Act 307, 308
Hyderabad, Andhra Pradesh 50, 207
hydroelectric power 206–8

Ibrahim, Chand Mahal 183, 224, 225
identity cards 133
illegal immigrants 90, 133
Illegal Migrants (Determination of Tribunal) Act (1983) 98, 227
immigration 90, 133
imperialism 31, 50, 52–3, 108, 138, 190, 289, 301, 304, 308–9, 315, 340
import-substitution-industrialization (ISI) 46
India
1937
provincial elections 49, 52
1942
launch of Quit India movement 53

1946
Cabinet Mission Plan 49
1947
Partition 48, 53
1948
assassination of Mohandas Gandhi 53
1949
Communist Party of India (CPI) calls for rail strike and insurrection 50
1950
Constitution 46
agrarian revolt in Telengana 50
Preventative Detention Act 50
Indo-Nepal Treaty of Peace and Friendship 263
1951
Tamil Nadu state assembly election 57
CPI party congress 51
1951–1952
general election 45
1952
Delhi Agreement 200
1955
Awadi declaration 54
1957
CPI elected in Kerala state assembly election 51, 53
1959
Nagpur resolution 55
dismissal of CPI government in Kerala 55
Sino-Indian border clash 55
1962
general election 54, 55
Sino-Indian war 55, 57
1964
CPI split; foundation of Communist Party of India–Marxist (CPI-M) 56, 175, 336
death of Jawaharlal Nehru 57
1965–1966
droughts 57

INDEX

1967
general election 45, 56, 57
foundation of Bharatiya Kranti Dal 59
Land Reforms Amendment Act in Kerala 59, 61
1969
Indira Gandhi supports V.V. Giri's candidacy for president 61
1971
Maintenance of Internal Security Act (MISA) 62, 65
Indo-Soviet Treaty of Peace, Friendship and Cooperation 81
Indo-Pakistan war 79, 82
1973
global oil price shock 63
Muzaffarpur session 72
1974
resignation of Chimanbhai Patel ministry 63
Jayaprakash Narayan holds rally in Patna 64
national rail strike 64
construction of Farraka dam 82
Henry Kissinger makes visit to India 82
1975
Allahabad High Court charges Indira Gandhi with electoral malpractice 64
Delhi Accord 80
1975–1977
Emergency 1, 9, 10, 64–6, 70, 71, 74, 78, 79, 100, 107, 115, 175, 320, 323, 326, 337
1976
resumption of diplomatic links with China and Pakistan 82
Bachawat Tribunal Award 206, 207
1977
general election 66, 67, 79, 80
Industrial Policy Resolution 81
massacre of Dalits in Belchi 78
five-year agreement with Bangladesh over Ganga water conflict 82
1978
Jimmy Carter makes state visit 82
Atal Bihari Vajpayee makes visit to Pakistan 82
launch of Operation Barga in West Bengal 90
CPI-M party congress 72
Ashok Mehta Committee submit report 79
Atal Bihari Vajpayee makes visit to Peking 82
Indira Gandhi wins by-election in Karnataka 83
Bindheshwari Prasad Mandal appointed to head Second Backward Classes Commission 83
Charan Singh mobilizes one million farmers in New Delhi 84
Freedom of Religion Bill 85
1979
Charan Singh appointed finance minister and deputy prime minister 84
global oil price shock 84
Madhu Limaye organizes conference on "third force" 85
foundation of Janata Party-Secular (JP-S) 85
Y.B. Chavan tables no-confidence motion in the Lok Sabha 85
resignation of Morarji Desai; Charan Singh becomes prime minister 86
death of Jayaprakash Narayan 87
1979–1980
Union budget 84
1980
general election 87
revival of National Integration Council 93
V.P. Singh appointed chief minister of Uttar Pradesh 107

461

INDEX

1982
foundation of Telugu Desam Party (TDP) 91
1983
Nellie massacre 90
Illegal Migrants (Determination of Tribunal) Act 98, 227
1984
Operation Blue Star 94
assassination of Indira Gandhi 94, 129
anti-Sikh pogrom 94, 96, 114, 129
general election 10, 29, 94, 98
Bajrang Dal resume *Ram mandir* campaign 97
1985
Anti-Defection Law 17, 95, 97, 334
Rajiv Gandhi-Harcharan Singh Longowal Accord 96
Assam accord 96
Shah Bano case 96, 97
1985–1986
Union budget 95
1986
Muslim Women (Protection of Rights on Divorce) Act 97
1987
Jammu & Kashmir assembly polls 98, 203
Bofors scandal breaks; resignation of V.P. Singh from Congress 98
launch Jan Morcha (People's Front) 98
1988
Lakhubhai Pathak files fraud complaint against Chandraswami 131
foundation of Janata Dal (JD) 99
Vishwa Hindu Parishad (VHP) conduct *Ram shila pujans* 97, 99
CPI-M party congress 99

Uttar Pradesh local elections 100
1989
general election 6, 7, 16, 20, 29, 30, 101, 103, 109, 183
Jammu and Kashmir Liberation Front (JKLF) kidnaps Rubaiya Saeed 113
St. Kitts forgery affair 131, 169, 210
1990
talks with Nepal 115
Meham by-election 119
withdrawal of Indian Peace Keeping Force (IPKF) from Sri Lanka 115, 324
Bhim Rao Ambedkar awarded Bharat Ratna 117
Agricultural and Rural Debt Scheme 120
Prasar Bharati (Broadcasting Corporation of India) Act 120
Jagmohan recalled from Jammu & Kashmir 113
National Commission for Women Act 117
launch of Cauvery Waters Dispute Tribunal 114, 204
Constitution (65th Amendment) Act 117
Armed Forces Special Powers Act (AFSPA) 113
L.K. Advani launches *rath yatra* in support of *Ram mandir* campaign 120
Chandra Shekhar becomes prime minister 121
financial crisis 121
1990–1991
Union budget 115, 116
1991
resignation of Chandra Shekhar 121, 209
assassination of Rajiv Gandhi 122, 130
constitution of Jain Commission 273

INDEX

general election 121–2, 141, 209
initiation of bilateral negotiations with Pakistan 191, 325
balance of payments crisis 126, 133
hawala scandal breaks 130–2, 167, 172, 209
V.P. Singh diagnosed with cancer 143
1992
revival of Lok Dal 141
demolition of *Babri Masjid* 2, 125–6, 129, 170, 177
1992–1993
Bombay riots 170
1993
Jharkhand Mukti Morcha (JMM) supports Congress in motion of confidence 130, 188, 211
Uttar Pradesh state assembly election 140
Samajwadi Janata Party (SJP) rechristened Haryana Lok Dal (Rashtriya) 141
Peace and Tranquility Agreement with China 230
1993–1995
assembly elections 126–7
1994
CPI-M industrial policy statement 137–8
Bommai judgment 177
1995
Bahujan Samaj Party (BSP) withdraws support for coalition with Samajwadi Party (SP) 189
Ramesh Chandra committee report 189
Shankersingh Vaghela threatens to split Bharatiya Janata Party (BJP) 219
1996
death of N.T. Rama Rao 143, 144
Congress rejects new pact with All Indian Anna Dravida Munetra Kazhagam (AIADMK) 132
JMM pay-off scandal breaks 131
Lalu Yadav appointed head of Janata Dal (JD) Political Affairs Committee 183
foundation of Madhya Pradesh Vikas Congress (MPVC) 132
fodder scam breaks 187, 218
resumption of bilateral negotiations with Pakistan 191
general election 2, 129, 138, 147–58, 191, 193, 208
Prabhakar Rao linked to urea scam 179, 188
Narasimha Rao linked to JMM pay-off scandal 131
Congress (I) enters pact with BSP 188
United States presses India to sign Comprehensive Test Ban Treaty (CTBT) 190
Narasimha Rao questioned over Lakhubhai Pathak cheating case 131, 208, 210
conflict between Karnataka and Tamil Nadu over Cauvery waters 204–6
conflict between Karnataka and Andhra Pradesh over Rapid Irrigation Benefit Programme 206
Shankersingh Vaghela expelled from BJP; party splits 219, 221
I.K. Gujral makes visit to Bangladesh 231
Narasimha Rao questioned over JMM pay-off scandal 211
Narasimha Rao charged in Lakhubhai Pathak cheating case 211
Narasimha Rao resigns from presidency of All-India Congress Committee (AICC) 211, 213

463

INDEX

Union finance bill 214
Constitution (81ˢᵗ Amendment) Bill 214
proposals for Women's Bill 261
Lok Pal Bill 218
Jammu & Kashmir state assembly elections 200–4, 222–3
Hurriyat leaders arrested 223
Uttar Pradesh state assembly election 223–6
Deve Gowda tours northeastern states 226
revival of Foreign Investment Promotion Board 233
India attends G-15 summit in Zimbabwe 232
Jiang Zemin makes state visit 230
cyclone hits Andhra Pradesh 232
Ganga Waters Accord 230, 231, 324
Treaty of Trade with Nepal 263
Patna High Court reprimands CBI over fodder scam inquiry 233
Narasimha Rao resigns from AICC 233

1996–1997
Union budget 197

1997
Sitaram Kesri becomes president of AICC 233
Sitaram Kesri grilled by CBI over assets 236
resumption of Bofors scandal inquiry 236
Palaniyappan Chidambaram announces economic reforms 242
Punjab assembly elections 237
resumption of negotiations with Pakistan 223, 249, 263–4
CBI declares sufficient evidence to charge JMM pay-off suspects 248
Mayawati becomes chief minister of Uttar Pradesh 249

Congress (I) withdraws support for Government 250
Sonia Gandhi joins AICC 257
Rajiv Gandhi charged in Bofors scam 257
CBI recommends indictment of fodder scam suspects 259
calls for Lalu Yadav to quit office 259
I.K. Gujral appoints Bhabani Sengupta as officer on special duty 258
reintroduction of proposals for Women's Bill 261
I.K. Gujral attends South Asian Association for Regional Cooperation (SAARC) 262
JD hold organizational elections 260
trade pact with Nepal 263
ISC standing committee addresses Article 356 of the constitution 269
Lalu Yadav advocates statehood for Jharkhand 261
I.K. Gujral endorses petroleum price rise 265
Kocheril Raman Narayanan becomes president 271
withdrawal of Insurance Regulatory Authority Bill 265
Foreign Exchange Management Act (FEMA) 266
public sector unions threaten strike 267
amendment of Broadcasting Corporation of India (Prasar Bharati) Bill 272
Muthuvel Karunanidhi threatens to intervene in Cauvery waters dispute 272
media leaks excerpts from Jain Commission Report 273
demands for expulsion of Dravida Munnetra Kazhagam

INDEX

(DMK) from United Front 273–4
Supreme Court rules to enhance autonomy of CBI 260
Naveen Patnaik breaks up JD unit in Orissa 276
1997–1998
Union budget 12, 237, 243–7, 253, 264
1998
general election 13, 275–85
Pokhran II nuclear bomb tests 289, 307
Congress (I) make Pachmarhi declaration 291
foundation of Rashtriya Loktantrik Morcha 292
1999
India pressured to sign Comprehensive Test Ban Treaty (CTBT) 288
Lahore Declaration 289
BJP calls for greater coalition *dharma* 289
Kargil war 292
foundation of Nationalist Congress Party (NCP) 291
general election 3, 291–4
free trade agreement with Sri Lanka 289
2000
Bangaru Laxman becomes president of BJP 294
Chhattisgarh, Jharkhand and Uttarakhand receive statehood 294
2001
A.B. Vajpayee meets with Pervez Musharraf in Agra 295
2002
state finance ministers agree to adopt uniform central value added tax 296
anti-Muslim pogrom in Gujarat 3, 13, 296–7, 342

Prevention of Terrorism Act 297, 301
death of G.M.C. Balayogi 297
Chandrababu Naidu demands resignation of Narendra Modi 297
Jammu & Kashmir state assembly elections 223, 295
2003
Chandrababu Naidu supports NDA during no-confidence motion 297
Congress (I) Shimla party session; reversal of Pachmarhi declaration 301
2004
general election 3, 298
2005
Right to Information Act 303
National Rural Employment Guarantee Act 304
CPI-M party congress 304
Manmohan Singh and George W. Bush announce intention to forge "global partnership" 307
India backs IAEA resolution on Iran 307
2006
Scheduled Tribes and Other Traditional Forest Dwellers (Recognition of Forest Rights) Act 304
India backs IAEA resolution on Iran 307
formalization of New Framework for the US-India Defense Relationship 307
revival of Jan Morcha in Uttar Pradesh 306
West Bengal assembly polls 306
Henry J. Hyde United States-India Peaceful Atomic Energy Cooperation Act 307
2007
Nandigram violence 305

Uttar Pradesh state assembly elections 306
foundation of United National Progressive Alliance (UNPA) 306
joint naval exercises with Australia, Japan and Singapore 308
2008
Indo-US Civil Nuclear Agreement 3, 13, 308, 309–10, 339
anti-Christian violence in Kandhamal region 311
demonstrations in Singur 305
2009
general election 3, 310–17
2010
Civil Liability for Nuclear Damages Bill 310
2011
communists lose office in Kerala and West Bengal 316
2012
CPI-M party congress 316
2013
Delhi assembly elections 342
2014
general election 3, 27, 342
CPI-M announces eleven-party bloc in Lok Sabha 342
India Against Corruption 341
India Infrastructure Report, 244
"India Shining" 298
Indian Administrative Service (IAS) 47
Indian Civil Service (ICS) 47
Indian left *see* Left
Indian National Congress (INC) 1, 2, 3, 4, 9–10, 12, 13, 17, 18, 28, 29–30, 32, 45–66, 67–70, 86, 99, 160, 335
   All-India Congress (I) Committee (AICC) 130, 233, 257
   Congress (I) Parliamentary Party (CPP) 130, 226, 233, 251
   Congress for Democracy (CFD) 71, 73, 74
   Congress Parliamentary Party 160
   Congress Socialist Party (CSP) 49, 51, 52, 333
   Congress Working Committee (CWC) 160, 188, 208, 211, 225, 226, 229, 233, 237, 251, 252, 257, 274
   Indira (I) *see under* Congress (I)
   Nationalist Congress Party (NCP) 291, 300, 301, 311
   Organization (O) 61, 67, 73, 74, 77, 85, 183
   Requisitionists (R) 61, 62, 65, 66, 69–70, 71, 73, 77, 79, 80, 82, 85
   Socialist (S) 89, 95, 99, 110
   Syndicate 57, 61, 93
   Tiwari (T) 131, 161, 187, 188, 224, 229, 307
   Urs (U) 83, 85, 86, 87
Indian National Lok Dal (INLD) 307
Indian National Trade Union Congress (INTUC) 47
Indian Ocean 308
Indian Peace Keeping Force (IPKF) 115, 273, 324
Indian Police Service (IPS) 209
indigenous peoples 46, 227, 327; *see also* Adivasis
Indira hatao (get rid of Indira Gandhi) 62
Indo-Nepal Treaty of Peace and Friendship (1950) 263
Indo-Nepal Treaty of Trade (1996) 263
Indo-Pakistan war (1971) 79, 82
Indo-Soviet Treaty of Peace, Friendship and Cooperation (1971) 81
Indo-US Civil Nuclear Agreement (2008) 3, 13, 308, 309–10, 339
Indra Sawney v. India 142
Industrial Policy Resolution (1977) 81

INDEX

industry, industrialization 4, 21, 46, 51, 58–9, 63, 65–6, 72, 78, 81, 84, 88, 91, 98, 101, 106, 108, 116, 118, 133, 136–7, 178, 214, 233, 237–8, 242–6, 264, 276, 295, 303, 305, 306, 315, 331, 341
inflation 57, 63, 95, 96, 105, 110, 135, 137, 197, 298, 341
infrastructure 84, 134, 136–7, 178, 197, 202, 214, 226, 238, 243–4, 246–7, 264, 269, 343
Infrastructure Development Finance Company (IDFC) 197
Insurance Regulatory Authority Bill 265
Integrated Rural Development Programme 130
Intelligence Bureau (IB) 210, 236
Inter-State Council (ISC) 93, 106, 114, 177, 226, 269, 295, 323, 324
International Atomic Energy Agency (IAEA) 307, 308, 309
International Monetary Fund (IMF) 136, 163, 247
investment 4, 28, 68, 84, 106, 108, 116–17, 120, 126, 134–5, 137–8, 174, 178, 193, 196–7, 202, 214, 233, 237–47, 263–7, 269, 276, 295–6
Iran 307, 308, 310
Iraq 121
irrigation 61, 84, 137, 197, 205–8
Islam 3, 70, 84, 88, 89, 90, 96, 97, 104, 141, 147, 169, 170, 219, 280, 291, 292, 296–7, 312, 316
Islamabad, Pakistan 82, 264
Islamism 230, 295, 315
Israel 297
Italy 26, 98, 236, 336, 337–8

Jagmohan (Jagmohan Malhotra) 113
Jain Commission 273–4, 278, 279, 301
Jain, Milind Chand 273
Jain, S.K. 130
Jaish-e-Mohammed 295
Jalandhar, Punjab 276
Jamaat-e-Islami 65
Jammu & Kashmir 10–12, 48, 79–80, 92–4, 98, 110, 113, 133, 157, 176–7, 187, 191, 200–4, 214, 219, 222–3, 228, 249, 259, 262–4, 269, 280, 284, 292, 295, 298, 300, 321, 323
Jammu and Kashmir Liberation Front (JKLF) 113
Jammu and Kashmir National Conference (JKN) 79, 92, 94, 98, 200, 202–4, 222–3, 284
Jan Dal 306
Jan Morcha (People's Front) 98–9, 278–9, 306
Jan Sangh 18, 58, 59, 62, 63, 65, 67, 70, 73, 74, 77, 80, 81, 82, 83, 85, 87, 89, 121, 133, 134, 169, 191, 321, 333, 337
Janata Dal (JD) 1, 2, 32, 323, 334
  1980–1989 99, 101
  1989–1991 103, 107, 110, 113, 119, 122
  1991–1996 134–5, 140–5, 154, 157, 164, 166–7, 172, 174, 178, 182–7, 189, 196, 198, 200, 203, 206, 209–11, 219, 224, 233
  1997–1998 245, 247, 258–62, 270, 272, 276, 277–8, 282, 284, 287
  1998–2012 289, 292, 295, 311
Janata Dal–Secular (JD-S) 292, 311
Janata Dal–United (JD-U) 292, 297, 312, 314, 334, 342
Janata National Party-Secular (JNP-S) 87
Janata *parivar* 11, 18, 88, 141–2, 162, 167, 253, 312, 314, 333, 334, 337
Janata Party (JP) 1, 2, 4, 5, 7, 10,

INDEX

11, 14, 17, 21, 32, 33, 34, 66, 67–88, 89–90, 92, 95, 96, 99, 100, 106, 107, 119, 121, 141, 162–3, 167, 178, 179, 253, 274, 280, 319–25, 328–9, 333, 335, 337, 341
Janata Party–Secular (JP-S) 85–6, 89
Japan 308
jatis 141, 334, 335
Jats 58, 59, 118, 119, 232
Jayalalitha, Jayaram 132, 290, 294, 232
Jayawardene, Junius Richard 115
Jena, Srikant 184
Jethmalani, Ram 169
Jharkhand 261, 294
Jharkhand Area Autonomous Council (JAAC) 261
Jharkhand Mukti Morcha (JMM) 122, 130, 131, 188, 211, 248, 261, 282
Jharkhand Vikas Morcha 307
Jiang Zemin 230
Joshi, Murli Manohar 170, 289
Joshi, Puran Chand 50, 51
Journal of Parliamentary Studies (Kumar) 127, 128

Kalam, Avul Pakir Jainulabdeen Abdul 309
Kalapani border dispute 263
Kalelkar, Kakasaheb 83
Kammas 59
Kandhamal, Odisha 311
kar sevaks 121
Karachi, Pakistan 191
karans 276
Karat, Prakash 304, 311, 316–17
Kargil war (1999) 292
Karnataka 59, 63, 69–70, 79, 83, 89, 92–3, 95–6, 98, 104, 110, 114, 135, 141, 144, 148–9, 153, 157, 166–7, 172, 174, 181–4, 187, 198, 202, 204–8, 209, 251, 261, 272, 274, 276, 278, 280, 292, 295, 306, 311–12, 323, 334
Karnataka Congress Party (KCP) 161
Karnataka High Court 183
Karunakaran, Kannoth 160, 229, 233, 252, 273
Karunanidhi, Muthuvel 162, 172, 204–6, 254, 261, 272, 273, 294
Kashmir 10–12, 48, 79–80, 92–4, 98, 110, 113, 133, 157, 176–7, 187, 191, 200–4, 214, 219, 222–3, 228, 249, 259, 262–4, 269, 280, 284, 292, 295, 298, 300, 321, 323
Kashmiriyat (Kashmiri cultural identity) 113
Kerala 3, 20, 48, 51, 53, 56, 58, 59, 61, 69, 70, 90–1, 96, 99, 108, 114, 137, 139, 144, 149, 157–8, 165, 166, 172, 176, 184, 187, 233, 277, 279, 300–1, 305–6, 311, 312, 315, 316, 339, 341
Kerala Congress (KEC) 147, 307
Kesri, Sitaram 211, 213–14, 225, 229, 233, 235, 236, 250–3, 257–8, 273–4, 278, 279
Keynesianism 339
KHAM (Kshatriyas, Harijans, Adivasis, Muslims) 63
Khan, Arif Mohammad 98
Khandayats 276
Khrushchev, Nikita 53, 55
Kidwai, Akhlaqur Rahman 259–60
Kisan Mazdoor Praja Party (KMPP) 49
kisan politics 109, 320
Kissinger, Henry 82
Koeris 58, 142
Kozhikode, Kerala 316, 341
Kripalani, Jivatram Bhagwandas 49, 77, 78
Krishak Samiti (Peasant's Organization) 60
Krishna River 206

Kshatriyas 63, 142
kulaks (rich peasants) 85, 164
Kumar, Nitish 142, 312
Kumar, Sanjay 127, 128
Kumaramangalam, Mohan 62
Kurmis 58, 142
Kuruvai paddy crop 204
Kuwait 121

labor, labor movement 47, 52, 58–9, 64, 72–3, 77–8, 84, 88, 90–1, 101, 108, 127, 136, 267–8, 322
Labour Party (UK) 272, 330
Ladakh 200, 202
Laffer curve 243
Lakhubhai Pathak cheating case 131, 208, 210, 211
Lal, Bhajan 311
Lal, Devi 107, 109, 118, 119, 320, 335
land acquisitions, seizures 62, 114, 218, 305
Land Ceiling Act 305
land reform 64, 74, 120
Land Reforms Amendment Act (1967) 59, 61
landless laborers 74, 77–8, 84, 90, 137, 141, 305–6, 322
language 6, 16, 27, 326, 327; *see also* Hindi, Tamil, Telugu, Urdu
large-N studies 329
Lashkar-e-Taiba 295
Latin America 315
Laxman, Bangaru 294
Left 1, 3, 5, 7, 9, 11, 13, 14, 21, 36, 45–66, 85–6, 100, 107, 136–9, 144, 157–8, 193, 208–9, 232, 238, 241, 250, 265, 268, 277, 278, 282, 285, 290, 294, 301–2, 304–6, 308–17, 320, 321, 342–3
socialist 2, 4, 6, 8, 10, 13, 14, 21, 37, 45, 50–4, 57, 60–1, 67, 70, 73–4, 81, 85–6, 95, 108, 109, 115, 121, 141, 320, 331–3

communist *see under* communism
Left Democratic Front (LDF) 91, 108, 137, 305, 316
Left Front (LF) 1, 3, 20, 59, 73, 90, 95–6, 109, 110, 122, 134, 137, 158, 162, 187, 193, 208, 287, 305–6, 308, 311, 316, 333
Leghari, Farooq 223
Lenin, Vladimir 53, 316
Leninism 31, 54
Liberal Democrats (UK) 329
Liberal Party (Canada) 330
Liberation Tigers of Tamil Eelam (LTTE) 115, 122, 132, 273
Limaye, Madhu 61, 85
Line of Actual Control 230, 259, 264
linguism 93
literacy 68, 137
Lodhis 142
Lodhs 59
Lohia, Ram Manohar 49, 51, 53, 54, 59, 67, 215, 334
Lok Dal 87, 88, 89, 95, 96, 99, 107, 109, 119, 141
Lok Janshakti Party (LJP) 297, 306, 311, 312, 342
Lok Pal 68, 341
Lok Pal Bill (1996) 106, 218, 235
Lok Sabha 26–7, 67, 69, 71, 85, 86, 87, 113, 127, 147, 158, 159, 208, 214–15, 226, 258, 269, 275, 302, 309, 342, 343
lok shakti (people power) 64
Lok Shakti 280, 281, 282, 292
Lokayukta 68
Loktantrik Congress 280
Lone, Abdul Gani 223
Lucknow Development Authority 271
Ludhiana, Punjab 114

Machiavelli, Niccolò 34, 210
Madhya Pradesh 58, 59, 69, 77, 104, 148, 149, 153, 184, 188,

INDEX

233, 279, 290, 291, 294–5, 301, 296, 298, 300
Madhya Pradesh Vikas Congress (MPVC) 132, 161, 229
Madras, Tamil Nadu 58, 110, 205
Maha Gujarat Janata Parishad (MGJP) 221
Mahadalits 312
Maham, Haryana 118
Mahanta, Prafulla Kumar 172
Maharaj, Satpal 229
Maharashtra 56, 62, 99, 104, 134, 144, 148–9, 153, 158, 160, 170, 184, 191, 233, 235, 251, 274, 277–80, 285, 295, 300, 301, 305, 315
Maharashtra Navnirman Sena 315
Maharashtravadi Gomantak (MAG) 161
Maharashtrawadi Gomantak Party (MGP) 278, 342
Mahato, Shailendra 211
Mahato, Sudhir 248
Mahato, Surendra 188
Maintenance of Internal Security Act (MISA) 62, 65, 67
majority governments 3, 7, 20, 25–6, 27
Malaysia 272
Maldives 218, 262
Malik, Yasin 223
Malkani, Kewalram Ratanmal 120
Mandal Commission Report 109, 117–20, 121, 126, 138, 140, 141, 142, 162, 167, 215, 262, 276, 321, 326
Mandal, Bindheshwari Prasad 83, 117, 141
Mandal, Suraj 211
Mao Tse-Tung 31
Maoism 305, 315
Maran, Murasoli 214, 270, 272
Marandi, Simon 211
Marathas 59
Marumalarchi Dravida Munnetra Kazhagam (MDMK) 144, 278, 280, 294, 307
Marx, Karl v
Marxism 19, 49, 85, 164, 305, 317, 332, 339; *see also* communism
mass politics 126
Mayawati Kumari 189, 224, 226, 249, 250, 270, 271, 272, 290, 292
McKinsey 305
Mehta, Ashok 54
Mehta, Suresh 221
methodology 9, 15–18, 36, 39–41
minimum alternative tax (MAT) 197
minimum connected winning coalitions 19, 24, 25
minimum wage 116, 137, 238
minimum winning coalitions 24, 26, 27
Minorities Commission 133, 135, 172, 294
minority governments 1, 2, 6, 7, 10, 13, 16, 19, 24, 25, 26, 27, 29
Mishra, Chaturanan 184, 187, 272
Mishra, Janeshwar 184
Missile Technology Control Regime (MTCR) 133
mobilization 6, 8, 11, 18, 21, 27, 34, 47, 51–3, 57–8, 60, 62, 65, 70–2, 84, 93, 96–7, 99, 101, 106, 110, 118–19, 139, 141, 149, 154, 164, 167, 178, 225, 288, 301, 311–12, 326, 331, 332
Modi, Narendra 3, 297, 342
Mohanta, Prafulla Kumar 166
Moily, Veerappa 206
monsoons 84, 96, 204, 214, 298
Mookerjee, Chittatosh 204
Moopanar, Govindaswamy Karuppiah 132, 166, 229, 236, 252, 254, 268, 274, 278
Moscow, Russia 254
mosque of Babur *see Babri masjid*
Movement for Regeneration 63

INDEX

Mujib-ur-Rahman, Sheik 230
Mukherjee, Ajoy 58
Mukherjee, Pranab 273, 291
multiple bipolarities 6, 27, 128, 326
Mumbai riots (1992–1993) 170
Mumbai, Maharashtra 191
Musharraf, Pervez 295, 298
Muslim League 61, 86
Muslim Women (Protection of Rights on Divorce) Act (1986) 97
Muzaffarpur session 72

Nagaland 48, 295
Naidu, Chandrababu 145, 162, 166, 172, 174, 206, 207, 208, 232, 238, 259, 275, 284, 285, 295, 297
Namboodiripad, Elamkulam Manakkal Sankaran 53, 56, 59, 99, 122, 163
Nandigram, West Bengal 305, 314
Narain, Raj 73, 85, 89
Narayan, Jayaprakash 1, 49, 51, 52, 53, 54, 63–4, 65, 66, 68, 70–4, 78, 80, 84, 87, 99, 163, 335
Narayanan, Kocheril Raman 271, 275, 284
narrative 8–9, 18, 36, 39–40, 321
Natarajan, Jayanthi 261
National Advisory Council (NAC) 301, 303
National Bank for Agricultural and Rural Development (NABARD) 197, 243, 247
National Commission for Scheduled Castes and Scheduled Tribes 117
National Commission for Women Act (1990) 117
National Common Minimum Programme (NCMP) 303
National Democratic Alliance (NDA) 3, 7, 13, 32, 33, 95, 287–98, 301, 307, 311, 314, 322, 324–5, 328–9, 338, 342, 343

National Development Council (NDC) 81, 93, 97, 106, 177, 196, 245, 323
National Front (NF) 1, 2, 4, 5, 7, 10, 14, 17, 21, 32, 34, 98–101, 103–122, 134–5, 136, 140, 142, 143, 154, 157, 161–3, 167, 171, 176, 179, 184, 191, 193, 203, 215, 272, 273, 319–26, 328–9, 333, 335, 337
National Integration Council 93
National Judicial Commission 120
National Minorities Commission 133, 135, 172, 294
national power 2–3, 11–14, 32, 51, 80, 87, 101, 114, 119, 154, 157, 159, 170, 175, 189, 213–33, 279–81, 287, 290–1, 321, 324, 342
National Rural Employment Guarantee Act (2005) 304
nationalism 1–3, 11, 13, 18, 21, 34, 45–6, 49, 58, 63, 100, 126, 136, 159–60, 177, 277, 280, 288–9, 296–7, 311, 341, 343
Nationalist Congress Party (NCP) 291, 300, 301, 311
navaratnas (nine jewels) 244
Naxalbari, West Bengal 60
Naxalites 60, 64, 305
Nayanar Erambala Krishnan 172
Nazi Germany 52, 53
Nehru, Arun 98, 110
Nehru, Jawaharlal 46, 48, 49, 50, 55, 57, 59, 64, 80, 83, 93, 219
Nellie massacre (1983) 90
neoliberalism 3, 193, 247, 301–2, 306, 315–16, 343
neopatrimonialism 210
Nepal 82, 114–15, 218, 230, 263, 324
New Age 53
new democracy 31
New Democratic Party (Canada) 330

INDEX

New Framework for the US-India Defense Relationship 307
Nirman, Nav 63
non-alignment 21, 55, 82, 190
non-Brahmanism 99
non-Plan loans 244
Non-Proliferation Treaty (NPT) 176, 190, 288, 307, 310
Non-Resident Indians (NRIs) 244
noncooperative game theory 37
Northeast India 12, 48, 90, 133, 214, 226–8, 231, 269, 279, 295
nuclear power/weapons 3, 13, 21, 133, 176, 190–3, 288, 307–10
Nuclear Suppliers Group (NSG) 308

Obama, Barack 310, 317
Odisha 311
oil pool deficit 196–7, 265, 266
oil price shocks
 1973 63
 1979 84
Ola, Sis Ram 229
Operation Barga 90
Operation Blue Star (1984) 94
Opposition 29, 46, 49, 51, 56, 61, 62, 63, 64–6, 71, 79, 94, 98, 103, 248, 288, 297
Orissa 70, 77, 104, 113, 141, 144, 157, 158, 172, 276, 280, 311, 334, 342
Other Backward Classes (OBCs) 2, 10, 61, 63, 73, 83, 88, 106, 109, 117–20, 140–2, 154, 165, 167, 177, 189, 215, 224, 228, 232, 254, 259, 261, 276, 291, 292, 312, 334

Pachmarhi, Madhya Pradesh 291, 301
Pakistan 12, 81, 83, 113, 176, 190, 191–3, 204, 219, 249, 262–4, 288, 292, 295, 298, 307, 323, 324, 325
 1971 Indo-Pakistan war 79
 1976 resumption of diplomatic links with India 82
 1977 military coup 82
 1978 Atal Bihari Vajpayee makes visit to Islamabad 82
 1991 initiation of bilateral negotiations with India 191
 1995 receipt of nuclear ring magnets from China 190
 1996 resumption of bilateral negotiations with India 191
 1996 Benazir Bhutto dismissed from office on graft charges 223
 1997 general election; resumption of negotiations with India 223, 249, 263–4
 1999 Lahore Declaration 289
 1999 Kargil war 292
 2001 Pervez Musharraf meets with A.B. Vajpayee in Agra 295
Palkhivala, Nani 81
Palshikar, Suhas 314
panchayats 79, 90, 106, 108, 214–15, 306, 312
Parti Communiste Français (PCF) 336
Partido Comunista de España (PCE) 336
Partido dos Trabalhadores (Brazil) 340
Partito Comunista Italiano (PCI) 336, 337–8
parliamentary government 1–7, 10–11, 13–14, 17, 19, 24–30, 32–7
party leaders 2–3, 5–10, 12, 17–18, 20–3, 27–34, 36, 38, 40
Parvati, Lakshmi 145
Paswan, Ram Vilas 172, 224, 259, 292, 297
Patel, Chimanbhai 63, 141
Patel, H.L. 221

INDEX

Patel, Jayadevappa Halappa 172, 183, 205, 207, 272, 292
Patel, Keshubhai 219
Pathak, Lakhubhai 131, 208, 210
Patna High Court 228, 233, 260
Patna, Bihar 64
Patnaik, Biju 70, 107, 120, 172, 174, 196, 198, 276
Patnaik, Naveen 276
patrimonialism 210
Pattali Makkal Katchi (PMK) 277, 280, 294
Pawar, Sharad 160, 209, 229, 233, 235, 236, 251, 273, 291
Pay Commission *see* Central Pay Commission
Peace and Tranquility Agreement (1993) 230
Peasant's Organization 60
Peasants & Workers Party (PWP) 86, 144
Peking, China 78
people's democracy 31
People's Democratic Front 226
People's Democratic Party (PDP) 223, 300, 314
People's Republic of China *see* China
People's United Left Front (PULF) 58
perestroika 105
petroleum sector 244, 265
Planning Commission 93, 207, 303
pluralism 19, 136
plurality-rule elections 6, 26–7, 56, 95, 138, 153, 281, 326, 328
Pokhran II nuclear bomb 289
policy-realization theories 19, 21, 22
political judgment 7–9, 33–9, 41, 168, 175, 208, 223, 242, 252, 281, 298, 301, 309, 331, 335–40
political parties 1–4, 6–14, 15–41
   communist *see under* communism
   national 11

regional parties 1, 2, 4, 6, 11, 12, 14, 22, 27, 63, 71, 89–101, 164
   secular 13
   state-based 2, 6, 10, 26, 29
   socialist *see under* socialism
Pondicherry 114
poverty v, 1, 2, 4, 64, 81, 90, 130, 132, 137, 177, 187, 241, 242, 243, 305–6, 322, 331
power 7–8, 17–23, 29, 38, 50, 52, 60–1, 77, 79, 85, 87, 92, 106, 109, 118, 131–2, 135, 140, 142, 159, 166, 182, 184, 241, 252, 270, 290–1, 301–2, 309, 312, 319, 323, 325–44
   balance of 6, 60, 74, 143, 181, 224, 323, 327, 338
   executive 6, 7, 13, 24, 66, 74, 77, 171, 175, 182, 259, 271, 288, 321, 323, 328, 337–8
   national 2–3, 11–14, 32, 51, 80, 87, 101, 114, 119, 154, 157, 159, 170, 175, 189, 213–33, 279–81, 287, 290–1, 321, 324, 342
   separation of powers 67, 259
   sharing 6, 9, 12, 13, 26, 29–33, 49, 52, 59, 77, 101, 109, 157, 169, 175, 248–9, 270, 287, 289, 291, 292, 321, 327–8
   struggles 4, 11, 16, 18–20, 23, 27, 37, 54, 83, 119, 141, 143, 223, 274, 326, 329, 332–4
   zero-sum conception 7, 60, 166, 198, 232, 310, 333
power maximization theories 18, 19, 20, 22
Pradesh Congress Committees (PCC) 47
Praja Rajyam Party 315
Praja Socialist Party (PSP) 49–50, 54, 59, 61, 63, 332
Prasad, Jitendra 273
Prasar Bharati (Broadcasting

473

Corporation of India) Act (1990) 120, 323
Premadasa, Ranasinghe 115
President of India 30
Preventative Detention Act (1950) 50
Prevention of Terrorism Act (2002) 297, 301
Prithvi medium-range ballistic missiles 262
privatization 127, 136, 178, 238, 241, 244, 265, 322
process tracing 39
Progressive Conservative Party (Canada) 330
Progressive Democratic Front 59
property 47, 61–2, 68, 114
proportional representation (PR) 23, 26, 327, 329
Protection of Places and Worship Act 177
Przeworksi, Adam v
public distribution system (PDS) 233, 242
public fixed investment 84
Public Premises (Eviction) Act 135
Public Sector Enterprises (PSEs) 193–6
public sector utilities (PSUs) 133, 136, 238, 241, 265
Punjab 10, 55, 56, 58, 71, 87, 97, 94, 114, 110, 149, 153, 172, 235, 237, 250, 276, 279, 280, 305, 323
Puthiya Tamilagam 277

Quattrocchi, Ottavio 98, 236
Quebec 327, 330
Quit India movement 49, 53
quota politics 73, 109, 320

Raja, D. 245
Rajagopalachari, Chakravarti 219
Rajasthan 69, 77, 104, 105, 149, 153, 200, 274, 279, 296, 298

Rajiv Gandhi-Harcharan Singh Longowal Accord 95, 97
Rajputs 107
Rajya Sabha 215, 226, 253, 343
Ram mandir 97, 99, 120–1, 125–6, 129, 133, 167, 169, 177, 248, 281, 287, 290, 321, 326, 337
Ram, Jagjivan 71, 77, 78, 86, 87, 89, 290, 320, 335
Ram, Kanshi 120, 224, 292
Ramachandran, Marudhur Gopalan 273
Ramaiah, B.B. 242
Ramjanmabhoomi movement 97, 99, 107, 120–2, 249, 337
Ramoowalia, Balwant Singh 172
Ranadive, Bhalchandra Trimbak 50, 58, 163
Rao, Nandamuri Taraka Rama 91, 105, 143, 144
Rao, P.V. Prabhakar 179, 188
Rao, Pamulaparti Venkata Narasimha 12, 122, 126, 130–1, 144, 160, 161, 167, 168, 169, 170, 174, 184, 188, 189, 190, 197, 206, 208–11, 213–14, 219, 226, 232–3, 236, 251, 273, 289
Rao, Rajeshwar 50
Rapid Irrigation Benefit Programme 206–8
Rashtriya Janata Dal (RJD) 258–62, 276, 279, 282, 285, 292, 300, 302, 306, 311, 312, 334
Rashtriya Janata Party (RJP) 221, 279
Rashtriya Loktantrik Morcha 292
Rashtriya Nav Nirman Vedike 183
Rashtriya Swayamsevak Sangh (RSS) 63, 65, 71, 74, 85, 99, 101, 120, 126, 154, 221, 288, 289, 297, 321, 337
rath yatra (chariot journey) 120
rational choice theories 8, 16, 36–9, 41, 339
Ray, Dilip Kumar 187

INDEX

realists 35, 36, 231, 242
"red-green" alliance 14, 331
Reddy, Kotla Vijaya Bhaskara 206, 273
Reddy, Neelam Sanjiva 86, 119
Reddy, Sudini Jaipal 178–9, 258, 260, 272
Reddys 59
Reform Party (Canada) 330
region, regionalism 1, 2, 3, 4, 5, 6, 10, 11, 12, 13, 14, 16, 20, 21, 27, 63, 93, 108, 127, 147, 326, 330
regional parties 1, 2, 4, 6, 11, 12, 14, 22, 27, 63, 71, 89–101, 164
regional type approaches 24
religion 9, 21, 89, 93, 99, 135, 139
   Christianity 3, 135, 177, 290, 311
   Hinduism *see under* Hiduism
   Islam 3, 70, 84, 88, 89, 90, 96, 97, 104, 141, 147, 169, 170, 219, 280, 291, 292, 296–7, 312, 316
   Sikhism 94, 96, 114, 129, 172, 237, 276
Republican Party of India (RPI) 278, 279
Research and Analysis Wing (RAW) 98, 263
Reserve Bank of India (RBI) 242, 243, 264
Revolutionary Socialist Party (RSP) 136, 158, 188, 290, 291
Right to Information Act (2005) 303
Rolling Plan 81, 84
Roosevelt, Franklin Delano 65
Roy, Manabendra Nath 51
Rural Infrastructure Development Fund (RIDF) 197
Russian Federation 254, 308, 310

Saeed, Mufti Mohammed 110, 113
Saikia, Hiteswar 227
Sainis 232
Salt March 64
samajik samarasata (social integration) 270, 335
Samajwadi Janata Party (SJP) 109, 141, 183
Samajwadi Party (SP) 135, 140–2, 144, 157, 167, 184, 187, 189, 203, 209, 219, 224, 245, 249, 270, 271, 274, 277–80, 282, 285, 306, 307, 309, 312, 332, 342
Samata Party (SAP) 134, 142, 153, 157, 159, 224, 281, 288
Samba paddy crop 204
sampoorna kranti (total revolution) 64–5
Samyukta Socialist Party (SSP) 54, 58, 59, 62, 63, 332
Samyukta Vidhayak Dal (SVD) 58, 60–1, 333
sangh parivar 133–4, 169, 191, 280, 287, 288, 290, 300, 343
Sangma, Purno Agitok 170, 248, 275
sanitation 137
Sanskritic culture 107, 109, 142, 280
saptakranti (seven-fold revolution) 49
Sarkaria Commission 143–4, 177, 198, 221
Sarkaria, Ranjit Singh 93
Sarvodaya (collective welfare through constructive work) 53, 63, 78, 335
satraps 18
satyagraha (non-violent direct action) 52
Savanoor, Ratnamala 261
Sayeed, Mufti Mohammad 203, 223
SC/ST (Prevention of Atrocities) Act 270
Scandinavia 25
Scheduled Castes (SCs) 46, 70, 117–18, 140, 141, 189, 215, 270

475

Scheduled Castes and Scheduled Tribes Commission 106
Scheduled Tribes (STs) 46, 70, 117–18, 140, 141, 215, 270, 304
Scheduled Tribes and Other Traditional Forest Dwellers (Recognition of Forest Rights) Act (2006) 304
Schopenhauer, Arthur 40
Schumpeter, Joseph Alois 19
Scindia, Madhavrao 132, 229, 233, 258
Scottish National Party 330
Second Backward Classes Commission 83
secularism 13, 46, 89, 101, 126, 135, 136, 165, 177, 225, 280, 298, 307, 316, 341
Sengupta, Bhabani 258–9, 262
separation of powers 67, 259
September 11 attacks (2001) 297
Seshan, Tirunellai Narayana 122
Shah Bano case (1985) 96, 97
Shah Commission 107
Shah, Jayantilal Chhotalal 78
Shah, Shabir 223
Sharif, Nawaz 262, 289
Sharma, R.C. 260
Sharma, Shankar Dayal 158–9, 166, 168–9, 170, 171, 209, 250, 261
Shastri, Lal Bahadur 57
Shekhar, Chandra 66, 87, 107, 109, 116, 119, 121, 122, 167, 209, 320, 321, 335
shilanyas (foundation stone laying ceremony) 97, 99
Shiromani Akali Dal (SAD) 71, 74, 96, 159, 237, 276, 281
Shiv Sena 99, 134, 153, 170, 280, 281, 285
shudra castes 57, 61, 83, 166
Siachen glacier 259, 263
Sikhism 94, 96, 114, 129, 172, 237, 276

Sikkim 149
Singapore 308
Singh, Ajit 109, 110, 116, 141, 224
Singh, Arjun 248, 258, 273, 291
Singh, Charan 59, 73–4, 77, 78, 84, 85, 86, 87, 121, 320, 321, 331
Singh, Dijvijay 188
Singh, Jaswant 266, 289
Singh, Joginder 209, 228, 236, 257, 260
Singh, Kalyan 125–6, 224, 249, 270, 271, 291, 292, 312
Singh, Krishna Pal 221
Singh, Kunwar Natwar 229
Singh, Manmohan 238, 245, 301, 303, 307, 308, 309, 341
Singh, Raghuvansh Prasad 187, 260
Singh, Vishwanath Pratap 95, 98–100, 107–21, 143, 162–3, 166, 171, 172, 174, 210, 219, 221, 253, 254, 273, 274, 276, 306, 320, 335, 337
single-member simple-plurality (SMSP) 23, 46
single-party majority governments 3, 7, 20, 25–6, 27
Singur, West Bengal 305, 314
Sinha, Kamala 261
Sinha, Yashwant 246
Sino-Indian war (1962) 55, 57
Sino-Vietnamese war (1979) 82
Sir Creek 263
Sitaram Yechuri 232
slum clearances 66, 70
social democracy 1, 4, 14, 37, 91, 100, 165, 244, 322, 331–2, 339–40, 344
social justice 1, 53–4, 85, 106, 135–6, 138, 140, 142, 144, 169, 177–8, 312, 315
social welfare 64, 68, 105, 116, 183, 303, 315, 322, 341, 343
socialism 2, 4, 6, 8, 10, 13, 14, 21, 37, 45, 50–4, 57, 60–1, 67, 70, 73–4, 81, 85–6, 95, 108, 109, 115, 121, 141, 320, 331–3

INDEX

Socialist Party 54, 67, 70, 73, 74, 85, 86, 321
Soren, Shibu 211
South Asian Association for Regional Cooperation (SAARC) 262
Southeast Asia 226, 266, 272
Soviet Union 50, 51, 52, 53, 55, 81, 105, 324, 336, 337, 340
   1939 Molotov–Ribbentrop Pact 53
   1941 German invasion 53
   1956 Khrushchev denounces Stalin 53
   1956 invasion of Hungary 55
   1948 breaking of ties with Yugoslavia 55
   1971 Indo-Soviet Treaty of Peace, Friendship and Cooperation 81
Spain 336
Special Economic Zones (SEZs) 306
Sri Lanka 82, 115, 218, 232, 273, 289, 324
Srikrishna Commission 170
Srinagar, Jammu & Kashmir 93, 94, 200
St. Kitts forgery affair (1989) 131, 169, 210
Stalin, Joseph 53, 65
Stalinism 336, 340
States Reorganization Committee 49
sterilization 70
strategy and tactics 31, 37
strikes 31, 47, 50, 64, 137, 233, 267–8
Study of Developing Societies (CSDS) 40
subaltern groups 1, 8, 106
submarines 98
Sundarayya, Puchalapalli 52
Supreme Court 66, 79, 97, 131, 135, 142, 177, 204, 206, 211, 221, 228, 260, 271
Surjeet, Harkishan Singh 162, 165, 167, 202, 203, 225, 245, 254, 266, 282, 304
surplus multiparty governments 20, 25
swadeshi (self-reliance) 68, 133–4, 238
Swatantra Party 59, 62, 63, 219
Sweden 98
Switzerland 190, 236, 327
Syndicate 57, 61, 93

Talbott, Strobe 289
Tamil 57
Tamil Maanila Congress (TMC) 132, 158, 161–2, 166, 171, 174, 178, 226, 229, 232, 236, 245, 252, 254, 258, 261, 268, 274, 277, 278, 282
Tamil Nadu 57–9, 61, 65, 69, 70, 71, 95, 105, 114, 132, 143, 144, 148, 149, 158, 168, 172, 204–7, 214, 261, 272, 277–80, 284, 285, 294, 295, 300, 301, 305, 306, 311, 323, 342
Tamilaga Rajiv Congress (TRC) 280
Targeted Public Distribution System (TPDS) 242
Taslimuddin, M. 187
Tata 305
tax evasion 98, 117
taxation 47, 64, 68, 81, 84, 95, 116, 174, 197–8, 243, 244, 246, 296
Telangana 311
telecommunications 244
Telengana Rashtra Samiti (TRS) 300, 311
Telengana, Hyderabad 50
Telugu 49
Telugu Desam Party (TDP) 91–2, 95, 99, 105, 107, 110, 143, 144, 145, 158, 161, 166, 167, 168, 171, 174, 206, 226, 238, 245, 252, 259, 261, 275, 277, 280,

477

INDEX

282, 284, 285, 287, 294, 295, 296, 298, 311, 312, 315, 320, 342
Ten Point Programme 61
terrorism 97, 113, 297
Thackeray, Bal 170
Thakre, Kushabhau 288
third electoral system 1, 13, 105, 296, 301–2, 331, 342
third force 2, 3, 4, 5, 6, 8, 11, 13, 40, 45, 50, 56, 58, 85, 125–45, 160, 171, 175, 179, 272, 277, 287–317, 342
Third Front 2, 3, 4, 6, 11, 14, 29, 134, 136–45, 154–5, 157–8, 160–70, 290, 291, 295, 302, 311–12, 322, 327, 330, 337
Thirunavukkarasu faction (AIADMK) 277
Thucydides 34
Tiananmen Square massacre (1989) 105
Tibet 53, 55, 115
Tikait, Mahendra Singh 224
Tiwari, Narayan Datt 131, 188, 258
Towards Social Transformation 116
trade unions 47, 59, 64, 72, 91, 267–8
Treasury Bills 243
Trinamool Congress 278, 280, 281, 285, 297, 298, 305, 315, 316, 342
Tripura 20, 139, 149, 158, 277, 311, 339
Twenty Point Programme for Economic Progress 64, 65
two-party systems 25, 56

Uniform Civil Code (UCC) 100, 133, 169, 281, 287, 290, 321
uniform sales tax 296
Union budget
1979–1980 84

1985–1986 95
1990–1991 115, 116
1996–1997 197
1997–1998 12, 237, 243–7, 253, 264
Union finance bill (1996) 214
Union Muslim League (IUML) 147
Union of Soviet Socialist Republics (USSR) *see* Soviet Union
Union Territories (UTs) 148
United Communist Party of India (UCPI) 279
United Democratic Front (UDF) 57, 91, 316
United Front (UF) 2–3, 4, 5, 7, 11, 12–13, 14, 17, 21, 33, 34, 95, 319–26, 328–9, 335, 337, 338, 339, 340, 341
1991–1996 147–79, 181–8, 190–3, 196, 198, 202, 205, 206, 208–10, 213–15, 218–19, 221–4, 226, 229, 232, 233
1997–1998 235–6, 238, 247, 250, 252, 253, 254, 257–85, 287
1998–2012 290, 298, 307
united front strategy 31, 52, 71, 175, 304, 311
United Goans Democratic Party (UGDP) 161, 278
United Kingdom 26, 53, 329–30, 310
United Kingdom Independence Party (UKIP) 330
United Left Front (ULF) 11, 58
United Liberation Front of Assam (ULFA) 227
United National Progressive Alliance (UNPA) 306–7, 309
United Nations (UN) 190, 307
United Progressive Alliance (UPA) 3, 7, 13, 29, 32, 33, 298–310, 311, 314–15, 322, 324, 328–9, 338, 339, 340, 341, 343
United States 50, 81, 82–3, 113,

INDEX

138, 176, 219, 289, 304, 307–8, 324
1954 Atomic Energy Act 307
1974 Henry Kissinger makes visit to India 82
1978 Jimmy Carter makes state visit to India 82
1996 United States presses India to sign Comprehensive Test Ban Treaty (CTBT) 190
2001 September 11 attacks 297
2005 Manmohan Singh and George W. Bush announce intention to forge "global partnership" 307
2006 formalization of New Framework for the US-India Defense Relationship 307
2006 Henry J. Hyde United States-India Peaceful Atomic Energy Cooperation Act 307
2008 Indo-US Civil Nuclear Agreement 3, 13, 308, 309–10, 339
untouchable castes *see Dalits*
Urdu 88
urea scam 179, 188
Urs, Devraj 83, 89
Uttar Pradesh 2, 12, 56, 58–9, 63–4, 69, 73, 77, 84, 87, 97, 99, 100, 103, 105, 107, 109–10, 119, 120–2, 125–6, 129, 131, 133–4, 140, 142–4, 148–9, 153–4, 157, 167, 169, 172, 177, 188–9, 200, 209, 214, 223–6, 231, 235, 247–50, 270–2, 274, 277–81, 287, 291, 295–6, 300, 306, 309, 311–12, 321, 334, 342
Uttarakhand 232, 294

Vaghela, Shankar Singh 219, 221, 222, 279
Vaiko (V. Gopalasamy) 144, 278
Vajpayee, Atal Bihari 74, 81, 82, 83, 86, 99, 101, 134, 159, 163, 168, 169, 170, 189, 190, 219, 281, 284, 285, 289, 295, 297, 298, 321
varna classifications 141
Vellalas 59
Venkatraman, Ramaswamy 103, 121, 159, 168
Verma, Beni Prasad 187
Verma, C.P. 259
Vienna, Austria 307
Vietnam 82
Vijayawada, Andhra Pradesh 93, 94
vikaas nahin, sammaan chahiye ("we need dignity, not development") 312
Vishwa Hindu Parishad (VHP) 97, 99, 126
Vokkaliga, Karnataka 207
Vokkaligas 59
Voluntary Income Disclosure Scheme (VIDS) 264
vote-seeking theories 19, 22

Wakf properties 135, 294
water disputes 82, 114, 204–8, 272, 323, 324
Weber, Max 34
West Bengal v, 3, 12, 20, 36, 55–6, 58–60, 71, 73, 90, 95–6, 98–9, 101, 105, 108, 119, 137–9, 144, 149, 153, 158, 162–6, 172, 176, 179, 187, 197, 207, 231, 241, 277–9, 280, 295–6, 300–1, 305–6, 308, 311, 314–16, 332, 337, 339, 342
West Bengal Industrial Development Corporation 91, 137
West Bengal United Front 163
Western Europe 17, 23, 26, 327, 329, 331, 335–6
Westminster-style governments 26, 329, 330
Women's Bill (1996) 214–15, 235, 261–2
World Bank 116, 136, 163, 238
World Economic Forum 246

479

INDEX

World War II (1939–1945) 52–3, 65
Wullar Barrage 263

Yadav, Beni Prasad 189
Yadav, Dharam Pal 258
Yadav, Lalu Prasad 120, 142, 167, 172, 182, 183, 187, 228, 233, 253, 254, 258, 259, 260, 261, 276, 278, 279, 292, 312
Yadav, Mulayam Singh 109, 119, 121, 141, 142, 167, 189, 203, 209, 224–5, 248, 254, 267, 271–2, 274–5, 282, 291–2, 309
Yadav, Sharad 142, 183, 203, 224, 254, 260, 261, 292
Yadav, Yogendra 314
Yadavs 59, 140–2, 334
Yechury, Sitaram 178, 311
Yojana, Indira Awas 130
Young Turks 61, 66, 77
Youth Congress 62
Yugoslavia 55

zamindari 47, 57, 331
zero-tax companies 197
Zimbabwe 232